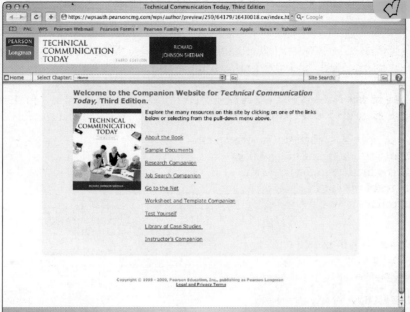

In addition to the **Go to the Net** resources available at **www.pearsonhighered.com/johnsonweb3**, you will also find:

Sample Documents—Find an array of new, web-only models that you can use in crafting your own technical writing documents.

Worksheet and Template Companion—Analyze readers, invent ideas, design documents, and revise your work by downloading worksheets and templates.

Test Yourself—Improve your retention of key concepts with flashcards and practice important skills by using the ready-made tutorials, quizzes, and essay questions tailored for each chapter.

Library of Case Studies—Discover a wealth of case studies that reflect on and discuss real workplace situations that present ethical challenges. Talk about them with your instructor and devise your own solution, using the tools you have learned in your reading and your own experiences as springboards.

Research Companion—Start your research through this invaluable asset, which links you to other powerful search engines, online libraries, reference tools, and strategies for developing effective research methodologies.

Job Search Companion—Begin your search for employment opportunities now. Go to this one-stop spot for finding and landing a job. Here, you will also find links to sites that offer a variety of tools and templates for developing winning résumés and letters of application; excellent advice on effective interviewing; online job search engines; and additional advice about finding jobs in technical disciplines.

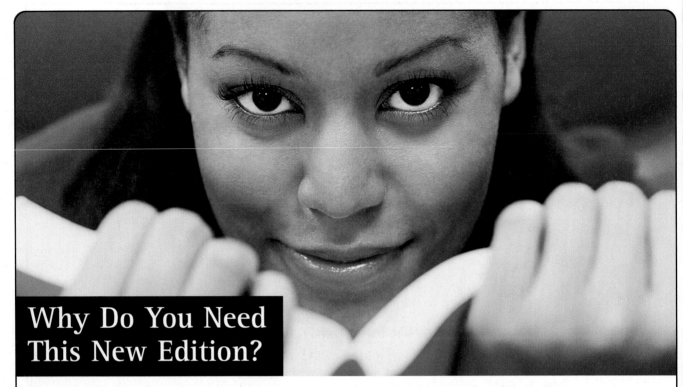

Why Do You Need This New Edition?

The world of technical communication is constantly evolving. If you're wondering why you should buy this new edition of *Technical Communication Today*, here are a few great reasons:

❶ Highly current and innovative discussions provide you with the tools and strategies to succeed in today's wired workplace, such as utilizing instant messaging at work (Ch.13); social networking for personal and professional use (Ch.15); and using iPods, Smartphones, and PDAs to deliver presentations (Ch.16).

❷ New sample workplace documents provide you with even more examples of the types of communications you will be generating and reading in the workplace, including a memo (Ch. 3), a planning calendar (Ch. 4), a performance review form (Ch. 4), and a chronological résumé (Ch.15).

❸ Expanded coverage of cross-cultural communication and internationalized work environments helps you prepare to enter today's diverse and global companies, with discussions on understanding and communicating with different cultures (Ch. 6), style considerations for international e-mail (Ch.13), online translation software (Ch.14), and collaboration tools for employees and clients working at different sites (Ch. 23).

❹ Expanded coverage of using visuals and media effectively includes adding podcasts to electronic documents and new examples of incorporating screen shots, tables, graphs, bar charts, and pie charts in technical documents (Ch.11).

❺ Six new case studies offer you the chance to reflect on and discuss in class real workplace situations that present ethical challenges (Chs. 2, 5, 7, 11, 14, 19).

❻ New and pragmatic discussions about the rhetorical context of a document and the conventions of its genre help you shape ideas and structure information in ways that fit readers' expectations and communication situations (Ch. 2).

❼ Updated screen shots throughout ensure that you have the most current, up-to-date information available.

❽ The Companion Website gives you the support you need to write for a variety of workplace situations and audiences and includes annotated sample documents, case studies, chapter quizzes, and flashcards.

❾ Access the e-book version of *Technical Communication Today* 3e through MyTechCommLab **mytechcommlab**. Completely reorganized with a wealth of new content specific to technical communication, MyTechCommLab now includes a new section on document design and graphics, a tutorial on writing formal reports, and new model documents and activities. Visit mytechcommlab.com to save time and improve results with the best available multimedia resources for technical writing in one easy-to-use place.

PEARSON

Technical Communication Today

THIRD EDITION

Richard Johnson-Sheehan
Purdue University

Longman

New York   San Francisco   Boston
London   Toronto   Sydney   Tokyo   Singapore   Madrid
Mexico City   Munich   Paris   Cape Town   Hong Kong   Montreal

To Tracey, Emily, and Collin

Publisher: Joseph Opiela
Acquisitions Editor: Lauren Finn
Senior Development Editor: Anne Brunell Ehrenworth
Marketing Coordinator: Bonnie Gill
Senior Supplements Editor: Donna Campion
Senior Media Producer: Stefanie Liebman
Production Manager: Jacqueline A. Martin
Project Coordination and Electronic Page Makeup: Elm Street Publishing Services
Text Design Adaptation of Second Edition Design: Elm Street Publishing Services

First Edition Text Design: Dorling Kindersley, UK; DK Design Director Stuart Jackman; Project Manager, DK Designs: Nigel Duffield; Senior Project Art Editor: Anthony Limerick; Senior Designer: Lee Redmond; Designer: Yvonne Cornes
Cover Design Manager: John Callahan
Cover Designer: Maria Iliad
Cover Photos: iStockphoto and Getty Images
Photo Researcher: Julie Tesser
Senior Manufacturing Buyer: Alfred C. Dorsey
Printer and Binder: Courier/Kendallville
Cover Printer: Coral Graphics Services, Inc.

Library of Congress Cataloging-in-Publication Data
Johnson-Sheehan, Richard.
    Technical communication today/Richard Johnson-Sheehan.—3rd ed.
        p.   cm.
    Includes bibliographical references and index.
    ISBN 0-205-63244-0
    1. Communication of technical information—Data processing.   I. Title.
T10.5.J65 2010
601'.4—dc22
                                                        2008036896

Longman
is an imprint of

www.pearsonhighered.com

1 2 3 4 5 6 7 8 9 10—CRK—12 11 10 09

ISBN-13 978-0-205-63244-2
ISBN-10 0-205-63244-0

Contents

CHAPTER

2

The Technical Writing Process Today   *16*

CHAPTER

3

Readers and Contexts of Use   *38*

CHAPTER

4 | Working in Teams   65

CHAPTER

5 | Ethics in the Technical Workplace   91

Part 2: Communicating in the Technical Workplace   *117*

CHAPTER 9

Using Plain and Persuasive Style   *205*

Part 3: Working in the Wired Workplace   *329*

CHAPTER 13

Using E-Mail and Instant Messaging   *330*

CHAPTER 14

Designing Websites   *348*

CHAPTER

15 | Starting Your Career   *374*

CHAPTER

16

Preparing and Giving Presentations   *409*

Part 4: Genres of Technical Communication   449

CHAPTER

19

Technical Descriptions   *515*

CHAPTER
22 Activity Reports · **634**

CHAPTER
23 Analytical Reports   **659**

Appendixes   *A-1*

Appendix A: Grammar and Punctuation Guide   *A-2*

Appendix B: English as a Second Language Guide   *A-19*

Appendix C: Documentation Guide   *A-23*

Preface

The controlling idea in this book is that people use their computers to help them think, research, compose, design, and edit. Computers do more than replace the typewriter in the writing process: They are essential *thinking tools* that powerfully influence how we develop, produce, design, and deliver technical documents and presentations. By centralizing computers in the writing process and exploring how we use them to join the ongoing electronic conversation around us, this book shows students and professionals how to take full advantage of these important workplace tools.

The third edition of *Technical Communication Today* continues to keep up with the evolution of technology in the technical workplace. New media and communication technology are dramatically altering technical fields at an astounding rate. We are working more efficiently, more globally, and more visually. These changes are exciting, and they will continue to accelerate in the technical workplace, making books like this one increasingly critical to success in technical fields. With this new edition, my goal is to share the excitement of technology and technical communication with people who want to work in technical fields.

Traditionally, books like this one have been addressed to engineers and engineering students. Today, as the technical workplace has expanded, almost all professionals find themselves needing to communicate technical information. To meet this need, this book addresses a broader range of people, including those who need to communicate in the natural sciences, the social sciences, computer science, medicine, public relations, law, business — and engineering.

What's New in the Third Edition?

The third edition of *Technical Communication Today* includes new material in virtually every chapter. Highlights of this text include the following new features.

Major Changes

- **Highly current and brand new discussions of how we use technological tools** to think, invent, plan, collaborate, draft, revise, and deliver technical communication in the workplace have been added throughout the text.

  - **New Help boxes** on giving presentations using an iPod, MP3, PDA, or Smartphone (Ch. 16); using a spreadsheet to rough out a proposal budget (Ch. 21); getting up to speed quickly on new software applications (Ch. 1); stopping online harassment (Ch. 5); and providing students with the strategies they need to succeed in today's wired workplace.
  - **New examples of using software applications** as invention tools help students create and draft technical documents (Ch. 2).
  - **A new section on using instant messaging and text messaging at work** provides best practices for both (Ch. 13).
  - **New coverage of social networking websites** addresses both how job seekers can use sites for professional networking and the importance of keeping one's personal site "clean" for hiring managers (Ch. 15).

- **Expanded coverage of copyright law and information sharing** highlights ethical and legal considerations students will encounter (Ch. 5).
- **Advice on how to compress image files for faster loading** helps students design more reader-friendly websites (Ch. 14).

- **New rhetorical discussion of genre asks students to consider genres as tools** they can use to shape ideas and structure information in ways that fit reader expectations and communication situations (Ch. 2).
- **New sample documents** include a planning calendar that outlines stages of a project and intermittent due dates (Ch. 4); a performance review form for evaluating team members' contributions (Ch. 4); a chronological résumé of a student returning to college (Ch. 15); a white paper (Ch. 22); a memo from the U.S. Department of Homeland Security regarding an internal management survey (Ch. 3); and an extended definition that demonstrates the importance of technical definitions to effective communication (Ch. 18).
- **Expanded coverage of cross-cultural communication and internationalized work environments** helps students prepare to enter today's diverse and global companies.

  - **New discussions of translation tools such as Yahoo! Babel Fish (Ch. 9) and Google Translate (Ch. 14)** show students how they can use simple tools to create text that can be easily translated for international users.
  - **Collaboration tools such as Google Docs (Ch. 23) and style considerations for international email use (Ch. 13)** appear in new, easy-to-reference Help boxes that facilitate working with international colleagues and customers.
  - **New guidance on understanding audience needs and using persuasive strategies** helps students communicate within cultures that often emphasize needs of the community over needs of the individual (Ch. 6).
  - **New section on writing documentation for readers of different cultures** contains guidance on how to adjust a document's language for international readers, specifically in terms of verbal and design considerations (Ch. 20).

- **Expanded coverage of using visuals and media effectively** includes updated guidance on adding audio to electronic documents; a new discussion of design principles to consider when laying out visual elements; and new examples of screen shots, tables, graphs, bar charts, and pie charts—graphics integral to most technical documents (Ch. 11).
- **Six new case studies** present opportunities for reflection and rich class discussion, preparing students for real workplace situations by presenting ethical challenges that include:
  - Going over an immediate supervisor to submit a proposal laden with errors (Ch. 2);
  - Deciding whether to work with a prestigious company that conducts controversial research (Ch. 7);
  - Presenting distorted information to a company's upper management (Ch. 11);
  - Determining whether to be a whistleblower when one's company purposely "leaks" information to head off the competition (Ch. 19);
  - Informing a manager of the impact a new website is having on clients who cannot access it due to their outdated equipment (Ch. 14);

- Implanting computer chips into employees' hands in an effort to ensure security (Ch. 5).
- **New, updated screen shots** throughout the text reflect the most up-to-date information and images students will see on the web.

Specific Chronological Changes

- **In Chapter 1,** a new Help box, "How to Learn New Software Quickly," will help students get up to speed on new software applications they will need to create technical documents.
- **In Chapter 2,** two new sections explore using a genre-based approach to the writing process. These new discussions help students understand the importance of organizing ideas in ways that meet readers' expectations and the rhetorical situations in which their documents will be used. Additional changes include a new example of logical mapping (an invention strategy) and a new case study.
- **In Chapter 3,** a new memo from the U.S. Department of Homeland Security provides an example of a document that fails to consider tertiary readers.
- **In Chapter 4,** there is detailed guidance on the initial steps to creating a successful team; a new project planning calendar that will help students track every phase of their project; new guidance on software for people working in virtual teams and a new example of groupware students can utilize when working on collaborative projects; and new guidance on assessing and evaluating the performance of team members that includes a new Team Performance Review form.
- **In Chapter 5,** a new Help box, "Stopping Cyberbullying and Computer Harassment," shows how students can respond effectively to online threats. In addition, laws governing the use of information and text, such as copyright law and information sharing, receive more prominent treatment, and the chapter concludes with a new case study.
- **Chapter 6,** now titled "Planning and Persuasion," has been revised and shows students the importance of strategic planning in producing high-quality documents and presentations that will persuade readers. A new example of reframing using common values shows students the importance of knowing your reader or audience profile in order to effectively persuade. New guidance explores how to persuade readers from high-context cultures such as Asia and the Middle East, where emphasis can often be more on community than on individuals.
- **In Chapter 7,** new tips help students make the most of listservs and online databases when conducting research. A new example of an interview provides a model of well-crafted questions, and the chapter ends with a new case study.
- **In Chapter 9,** a new Help box, "Intercultural Style and Translation Software," discusses how students can use free, online translators to create documents that can be read and understood by international users. In addition, there are new examples of using metaphors—techniques to persuade readers—in technical documents, and a new example of a persuasive argument that shows students how to incorporate style techniques into a document.
- **In Chapter 10,** a new, detailed example of effective headers and footers should help students increase the usability of their workplace documents, often read by busy colleagues and clients looking to immediately orient themselves within a document and find critical information.

- **In Chapter 11,** new guidance on properly labeling graphs is accompanied by new examples of graphics that students will use in their technical documents: a table containing numerical data, a graph that tells a simple story, a bar chart, and a pie chart. Further new guidance on design and layout includes a discussion on adhering to the "Rule of Thirds" when incorporating photographs and a new discussion on incorporating screen shots into technical documents. The chapter also includes updated instruction on adding audio—such as podcasts—to electronic documents and includes a new case study.
- **In Chapter 13,** there is a new example and abbreviated guidance on properly formatting a business e-mail correspondence. A new Help box, "Using E-mail Internationally," provides tips on the types of conventions students should use when corresponding with international clients and co-workers. Finally, two new sections titled "Using Instant Messaging at Work" and "Using Text Messaging at Work" distinguish between the two methods of communication and best practices for both in the technical workplace.
- **In Chapter 14,** revisions include new tips for using Google Translate when creating a website that will be viewed by international readers and includes new guidance on designing and uploading reader-friendly websites (including advice on how to compress image files for quicker loading) and a new case study.
- **In Chapter 15,** social networking sites are discussed as job-search tools, and the discussion includes the importance of maintaining personal information that is stored online in a professional manner, as it can be viewed by potential employers. A new example of a chronological résumé shows how one job seeker, who returned to college to earn an additional degree, formatted her résumé to showcase both work and educational experience.
- **In Chapter 16,** a new Help box, "Giving Presentations with Your iPod, MP3, PDA, or Smartphone," discusses the ease of using these portable devices to deliver a presentation.
- **In Chapter 17,** the new guidance on creating an informative body for a letter or memo encourages writers to consider audience needs and create effective bodies that allow readers to take action or make a decision.
- **In Chapter 18,** the concept that defining terms is crucial to effective technical communication is highlighted through new discussions on parenthetical definitions and an example of an extended definition to help writers craft clear, easy-to-read technical documents.
- **In Chapter 19,** a new discussion of patents frames the importance of detail in technical description; a new case study closes the chapter.
- **Chapter 20** has been expanded and is now titled "Instructions and Documentation." This chapter discusses the three most commonly written and used kinds of documentation in the technical workplace: instructions, specifications, and procedures and protocols. It includes new organizational charts that clearly illustrate the distinguishing components of these documentations and a new section on planning documentation for cross-cultural readers, with guidance on verbal and design considerations.
- **In Chapter 21,** "Roughing Out a Budget with a Spreadsheet," a new Help box walks students through one of the most important but least understood components of a proposal.

- **In Chapter 22,** a new white paper provides a model of a type of activity report that is both concise and unbiased.
- **In Chapter 23,** "Using Google Docs to Collaborate with International Teams," a new Help box, highlights free, web-based software that contains a common space for composing and editing technical documents with co-workers both in the same office and abroad.

Guiding Themes

In times of accelerated change, we must quickly adapt to new communication tools and strategies, while retaining proven approaches to writing and speaking. In this book, I have incorporated the newest technology in workplace communication. But the basics have not been forgotten. You will also find that the book is grounded on a solid core of rhetorical principles that have been around for at least two and a half millennia. In fact, these core principles hold up surprisingly well in this Information Age and are perhaps even more relevant as we return to a more visual and oral culture.

My intent was to develop a book that teaches students the core principles of rhetoric, while showing them how to use computers in a rapidly evolving information-based society.

Computers as Thinking Tools

The foremost theme of this book is that computers are integral and indispensable in technical communication. This premise may seem obvious to many readers; yet the majority of technical communication textbooks still do not successfully integrate computers into their discussions of workplace communication. These textbooks often limit computers to their word-processing abilities. They do not adequately show students how to fully use their computers to succeed in a networked technical workplace.

This book reconceptualizes the computer as a thinking tool in the technical workplace and in student learning. We need to recognize that students use their computers as thinking tools from beginning to end, inventing their ideas and composing text at the same time. In this book, the writing process has been redefined with the computer as a communication medium. As a result, the writing process described here is far more in line with the kinds of computer-centered activities in the technical workplace.

Genres as Pathways for Interpretation and Expression

This book follows a genre-based approach to writing and speaking in technical workplaces. Genres are relatively stable patterns that help people accomplish their goals in a variety of common rhetorical situations. Genres are not formulas or recipes to be followed mechanically. Instead, they offer flexible approaches that allow people to bring order to the evolving reality around them.

Genres can be used to interpret rhetorical situations, helping people in technical workplaces make decisions about what kinds of information they need to generate or collect. They can help individuals and collaborative teams plan out projects and develop rhetorical strategies for responding appropriately to complex situations. They can then be used to guide invention, organization, style, and design.

A genre-based approach to technical communication provides students with a "genre set" that is applicable to a variety of technical communication situations. While practicing these genres, students can also learn how to adapt genres and cross genres in ways that help them respond appropriately to situations that are unique or new to them.

Visual-Spatial Reading, Thinking, and Composing

This book also reflects an ongoing evolution in technical communication from *literal-linear* texts toward *visual-spatial* documents and presentations. We now see documents as "spaces" where information is stored and flows. Visual-spatial reading, thinking, and composing involve interacting with text in three dimensions.

This book addresses this evolution toward visual-spatial thinking in four ways:

- First, this book shows writers and speakers how to use visual-spatial techniques to research, invent, draft, design, and edit their work.
- Second, it teaches students how to write and speak visually, while designing highly navigable documents and presentations.
- Third, the book shows how to compose visual-spatial documents like hypertexts, websites, and multimedia presentations. Writing in these environments is becoming increasingly important as companies move their communications and documentation online.
- Finally, it practices what it preaches by presenting information in a visual-spatial way that will be more accessible to today's students. Clearly, students learn differently now than they did even a couple of decades ago. This book reflects their ability to think visually and spatially.

This visual-spatial turn is an important intellectual shift in our culture—one that we do not fully understand at the moment. We do know, however, that communicating visually and spatially involves more than adding headings and charts to documents or using PowerPoint to enhance oral presentations. Instead, we must recognize that the advent of the computer, which is a visual-spatial medium, is revolutionizing how we conceptualize the world and how we communicate. Increasingly, people are thinking visually and spatially in addition to literally and linearly. This book incorporates this important change.

The International, Cross-Cultural Workplace

This edition of *Technical Communication Today* includes expanded coverage of international and cross-cultural issues. I have met with hundreds of technical communication instructors and have learned that they want even more coverage of the globalized, cross-cultural workplace.

International and cross-cultural issues are integrated into the main discussion rather than shunted off into special sidebars, because issues of globalization are no longer separable from technical communication. Today, we always need to think globally, because computers greatly expand our reach into the world.

The Activity of Technical Communication

In this computer-centered age, people learn by doing, not by passively listening or reading. This book continues to stress the *activity* of technical communication—producing effective documents and presentations. Each chapter follows a process approach that mirrors how professionals communicate in the technical workplace. Meanwhile, the book shows students how to pay close attention to the evolving workplace contexts in which communication happens.

Perhaps this theme comes about because of my experiences with students and my observations of people using books like this one. As someone who has consulted and taught technical communication for nearly two decades, I realize that today's students rarely read their textbooks. Instead, they *raid* their textbooks for the specific information they need to complete a task. They use their textbooks like they use websites. They ask questions of the text and then look for the answers.

Features of Chapters

Each chapter includes regular features that students and instructors will find helpful and engaging.

Chapter Objectives and Chapter Review

At the beginning of each chapter, the *Chapter Objectives* are listed so that students know the exact learning objectives for the chapter. Then, at the end of each chapter, a *Chapter Review* summarizes the main points to help students review what they have learned. These bookend features should help instructors and students identify the most important concepts and issues covered in each chapter. They will also help students identify the key points that they should pay attention to as they use the chapter.

Help Boxes

Each chapter includes a *Help* box that describes how a particular computer application or computer-related strategy can be used to improve the production of documents and presentations or communications in the workplace. The *Help* boxes describe applications that are common to most computer software packages. In many cases, they offer tips and strategies to help students enhance and streamline their researching, drafting, editing, and communicating using computers. See the Table of Contents for each chapter's *Help* box.

Go to the Net

At the bottom of pages, students will find helpful *Go to the Net* notes. These notes refer students to the book's Companion Website. At this website, students will find a wealth of additional materials, including worksheets, sample documents, and links to helpful websites.

At Work Boxes

In the *At Work* boxes in each chapter, professionals in technical workplaces answer questions about technical communication. These boxes offer helpful tips and strategies for improving technical communication, but they also offer students a glimpse into the challenges faced in real technical workplaces. These boxes feature interviews with working professionals discussing communication challenges.

Do writers in the technical workplace *really* follow a writing process as they write?   33

What are some strategies for communicating with people from another culture?   56

What is a good way for virtual teams to develop and organize a project?   85

Why should technical professionals learn about ethics?   108

At a Glance Boxes

The *At a Glance* boxes in each chapter summarize key content for quick review of major concepts. They are designed to make basic information highly accessible to the scanning reader. Moreover, they provide access points at which readers can enter the text and read further on important subjects.

These boxes, which are interspersed throughout each chapter, serve as study aides and self-check reviews of major topic discussions:

Links

Links are marginal items that alert students to additional information on a related topic that is available elsewhere in the book. This textbook is designed to be used like hypertext. These *Links* allow students to move spatially through the text to find the information they need on demand.

Annotated Sample Documents

In each chapter, sample documents and texts with annotations are provided. The samples show students good examples of documents that have worked in the technical workplace. The annotations point out the documents' major features. At the bottom of the page, *Go to the Net* notes will help students find more examples on the book's Companion Website.

Exercises and Projects

Every chapter ends with several *Exercises and Projects.* The "Individual or Team Projects" can be completed either individually or in small teams. The "Collaborative Project" is a larger project designed for larger teams. The *Exercises and Projects* are designed to challenge students with realistic situations, allowing them to put the information in the chapter into practice.

Revision Challenges

The chapters that discuss specific genres of technical communication (Chapters 17–23) each include a Revision Challenge text. These texts are one or two revisions away from being finished. They can be used as discussion subjects or students can practice revision—a crucial skill—by reworking, rewriting, and redesigning the sample texts.

Case Studies

The *Case Studies* at the end of each chapter present situations typical of those that students are likely to encounter in their jobs, usually involving an ethical dilemma. Students are encouraged to consider the personal, social, political, ethical, economic, and environmental issues that shape decisions in the technical workplace. These *Case Studies* are especially helpful as topics of class discussions, whether held online or in the classroom.

Supplements to the Book

Accompanying this book are important tools that instructors and students will find especially helpful.

For Instructors

Instructor's Manual

The *Instructor's Manual* offers teaching strategies for each chapter while also providing prompts for class discussion and strategies for improving student writing and presentations. The *Instructor's Manual* is also available online and offers additional materials for downloading, such as slides and PowerPoint lectures. The online version offers additional ideas for assignments and projects.

Companion Website

The Companion Website (http://www.pearsonhighered.com/johnsonweb3) includes a wealth of sample documents, materials, and links that greatly expand on the content of the book. The website is designed to be used side by side with the book. The special "For Instructors Only" area has materials and teaching strategies to maximize the use of *Technical Communication Today*.

For Students

Companion Website

The Companion Website (http://www.pearsonhighered.com/johnsonweb3) includes a wealth of sample documents, materials, and links that greatly expand on the content of the book. The website is designed to be used side by side with the book. Students are able to look at more samples, download worksheets, and learn from the excellent communication-related websites available on the Internet.

Longman Grammar and Documentation Study Card
(ISBN 0-321-29203-0)
This colorful study card is an eight-page guide to key grammar, punctuation, and documentation skills and provides students with a useful tool for future reference.

MyTechCommLab mytechcommlab
Save time and improve results! When you use the e-book version of *Technical Communication Today* 3e with this dynamic, comprehensive site, you get the best multimedia resources for technical writing in one, easy-to-use place. Visit mytechcommlab.com for engaging and interactive content that will help you

improve the technical communication skills you will need most—writing, research, and document design.

Resources for Technical Communication, Second Edition

(ISBN 0-321-45081-7)
A complement to Longman's introduction to technical communication textbooks, *Resources for Technical Communication,* Second Edition, offers a wide range of sample documents for use in technical communication courses, including letters, proposals, reports, memos, e-mails, résumés, abstracts, instructions, descriptions, and slide presentations.

Acknowledgments

The third edition of *Technical Communication Today* has given me the opportunity to work with many people at Pearson and at universities around the country. My name is on the cover, but many, many people have collaborated with me on the book. I would like to thank them here:

Teresa Aggen, Pikes Peak Community College; Sherrie L. Amido, California Polytechnic State University–San Luis Obispo; Martha Bachman, Camden County College; Russell Barrett, Blinn College; Eric Bateman, San Juan College; Norman Douglas Bradley, University of California–Santa Barbara; Lee Brasseur, Illinois State University; Stuart Brown, New Mexico State University; Ellie Bunting, Edison College; Roger Friedman, Kansas State University; Jeffrey Jablonski, University of Nevada–Las Vegas; Rebecca Jackson, Texas State University; Leslie Janac, Blinn College–Bryan Campus; Miles A. Kimball, Texas Tech University; Barry Lawler, Oregon State University; Barbara L'Eplattenier, University of Arkansas–Little Rock; Anna Maheshwari, Schoolcraft College; Barry Maid, Arizona State University; Jodie Marion, Mt. Hood Community College; Mary S. McCauley, Wake Technical Community College; Jacqueline S. Palmer, Texas A&M University; Andrea M. Penner, San Juan College; Cindy Raisor, Texas A&M University; Sherry Rankins-Robertson, Arizona State University; Jennifer Sheppard, New Mexico State University; Nancy Small, Texas A&M University; Krista Soria, University of Alaska Anchorage; Karina Stokes, University of Houston–Downtown; Valerie Thomas, University of New Mexico; Christopher Toth, Iowa State University; Jack Trotter, Trident Technical College; Alan Zemel, Drexel University.

Editors Joseph Opiela and Anne Brunell Ehrenworth were essential in the revision and expansion of the book. Joe offered excellent ideas for strengthening the book, and Anne did a wonderful job of helping me implement those ideas. I would like to thank my colleagues, Professors Scott Sanders, Charles Paine, and David Blakesley. Our collaborations and discussions led to many of the concepts and strategies discussed in this book. Also, I would like to thank Christina Saidy and Mark Pepper, who helped me research new topics and find new sample documents.

Most important, I would like to thank my wife, Tracey, and my children, Emily and Collin, for being patient while I wrote and developed a new edition of this book. This book is for them.

RICHARD JOHNSON-SHEEHAN
PURDUE UNIVERSITY

PART

1

Elements of Technical Communication

CHAPTER

1

Communicating in the Workplace

In this chapter, you will learn:

- The importance of communication in today's technical workplace.

- To define technical communication as a process of managing information in ways that allow people to take action.

- To recognize technical communication as a blend of action, words, and images.

- To distinguish technical communication from other kinds of communication, especially expository writing.

- That technical communication is interactive and adaptable, reader centered, team centered, and highly visual.

- The importance of ethical, legal, political, international, and cross-cultural factors in technical communication.

- The importance of effective written and spoken communication to your career.

When new college graduates begin their careers in technical and scientific fields, they are often surprised by the amount of writing and speaking required in their new jobs. Of course, they knew technical communication would be important, but they never realized it would be so crucial to their success.

Effective communication is the cornerstone of the technical workplace, whether you are an engineer, scientist, doctor, nurse, psychologist, social worker, anthropologist, architect, technical writer, or any other professional in a technical field. People who are able to write and speak effectively tend to succeed. People who cannot communicate well often find themselves wondering why they did not get the job or why they were passed over for promotions.

Today, effective technical communication is more important than ever. We live in an age in which entire industries are built around the development, retention, and application of information. The ability to communicate effectively is crucial if you plan to survive and succeed.

Computers are the central nervous system of the information-based workplace. Throughout your career, you will find yourself using computers to communicate with others:

- Your desktop and laptop computers will be networked to local area networks (LANs), wireless networks, and the Internet.
- Your personal digital assistant (PDA) will keep track of your schedule for you, receive and send e-mail and instant messages, and allow you to browse the Internet.
- You will use e-mail, websites, instant messaging, and blogs to communicate with people inside and outside your office.
- You will use your wireless phone to talk to people around the world.
- You will use social networking sites like Facebook and virtual worlds like Second Life to do business.
- Wireless networks will allow you to use your computer, PDA, and other communication tools to keep in touch with your colleagues and co-workers.

In essence, computers are communication tools. And, as they have become more central to our lives, the ability to communicate with computers has become an essential part of our careers. Today, the ability to communicate effectively and efficiently with computers might be the most important skill you need for a successful career in technology.

How Important Is Technical Communication?

Surveys regularly show that oral and written communication skills are among the most important in the technical workplace. For example, in a 2003 survey, members of the American Institute of Aeronautics and Astronautics (AIAA) were asked to evaluate their educational preparation for their jobs (AIAA, 2003). The AIAA report that discussed the survey recommended that the following areas be improved in undergraduate education for design engineers:

Areas of Needed Improvement in Education for Engineers

1. Oral communication

2. Visualization in three dimensions

To see the AIAA survey, go to
www.pearsonhighered.com/johnsonweb3/1.1

3. Technical writing

4. Understanding the processes of fabrication and assembly

5. Using CAD, CAM, and solid modeling

6. Estimating solutions to complex problems without using computer models

7. Sketching and drawing

The membership of the AIAA is made up mostly of engineers, so it is interesting that two out of three of the top skills they listed stress the importance of technical communication. In fact, 34 percent of the surveyed AIAA members ranked oral communication "top most important," while 54 percent ranked it "very important." Meanwhile, writing was ranked "top most important" by 14 percent of respondents, while 61 percent ranked it "very important" (Figure 1.1).

How AIAA Members Rated Skills in the Workplace (by percentage of respondents)

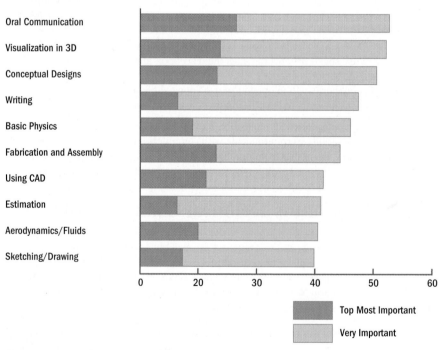

Figure 1.1: Professionals in technical fields regularly put communication skills at the top of their lists of important workplace skills.

Source: AIAA, 2003.

To see other surveys concerning workplace writing, go to
www.pearsonhighered.com/johnsonweb3/1.2

This survey's results are very much in line with survey results in other technical fields. An article in the engineering journal *Professional Issues in Engineering Education and Practice* reported that:

> Within 2–3 years of graduation, engineers spend about 30 percent of their time on the job writing; engineers in middle management spend 50–70 percent; and engineers in senior management spend 70 percent or more—up to 95 percent. (Silyn-Roberts, 1998)

These kinds of results highlight an important paradox in the technical workplace. Obviously, technical communication is very important; yet managers regularly report that college graduates do not have adequate communication skills. Corporations spend billions each year to improve the writing skills of their employees, according to the 2004 report "Writing: A Ticket to Work," from the National Commission on Writing. Poor writing skills are the "kiss of death," according to the report, because 51 percent of companies say they "frequently or almost always take writing into consideration when hiring salaried employees." (p. 9)

Fortunately, you can learn how to write and speak effectively in the technical workplace. The ability to communicate effectively is not something people are born with, and with guidance and practice, anyone can learn to write and speak well.

The purpose of this book is to help you acquire the communication skills you need for success in the technical workplace. You will find throughout your career that people who write and speak clearly and persuasively are much more likely to succeed than those who don't. Right now, you have a golden opportunity to develop these important technical communication skills. They will help you land the job you want, and they will help you succeed.

What Is Technical Communication?

Not long ago, a reasonable definition of technical communication might have said something about *translating* complex ideas into written text. This concept of translation is still important to effective communication in the workplace. But the centralization of computers in the workplace has added new dimensions and responsibilities.

Today, with computers, you need to be an *information manager* in the workplace. Experts are no longer just the people who know a great amount about a subject. Rather, experts are the people who can effectively manage and utilize the large amounts of information that are generated and stored on computers.

To reflect the importance of this information management, this book will use the following definition of technical communication: Technical communication is a process of managing technical information in ways that allow people to take action. The key words in this definition are *process, manage,* and *action.* In this book, you will learn the *process* of technical communication so you can *manage* large amounts of information in ways that allow you to take *action.*

In this Information Age, managing information means we are primarily concerned with how information *flows.* In other words, technical communication is about guiding the flow of information into useful areas so people, including ourselves, can use that information.

 GO TO THE NET
For other definitions of technical communication, go to
www.pearsonhighered.com/johnsonweb3/1.3

What Is Technical Communication?

5

Technical Communication: Actions, Words, Images

If you are a college student, much of your writing experience until now has probably been *expository* in nature. The root word for expository is "exposition," meaning you have been taught to *display* or *exhibit* information. Primarily, you have been taught how to demonstrate that you have acquired and retained knowledge on specific subjects. Your readers have mostly been teachers and professors.

Technical communication is different from expository writing, because it puts more focus on taking action with words and images. It focuses more on achieving a specific *purpose* with language. It also puts a much heavier emphasis on anticipating the needs of the readers and communicating information clearly and persuasively. Moreover, issues of ethics, legality, politics, and culture in technical communication are often much more tangible, because technology has such an immediate impact on people. Figure 1.2 shows some of the characteristics of technical communication that can set it apart from expository writing.

Fortunately, the skills you learned for expository writing are all adaptable to technical communication. The main difference is that technical communication puts a much greater emphasis on achieving a specific purpose with words and images.

Technical Communication Is Interactive and Adaptable

One of the most significant changes brought about by computers is the amount of *interactivity* among people in the technical workplace. In the computer-networked workplace, people are constantly communicating with each other and sharing their ideas.

The Qualities of Technical Communication

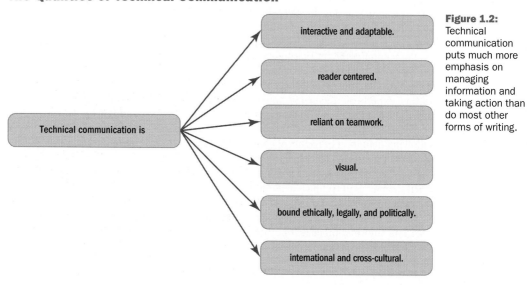

Technical communication is
- interactive and adaptable.
- reader centered.
- reliant on teamwork.
- visual.
- bound ethically, legally, and politically.
- international and cross-cultural.

Figure 1.2: Technical communication puts much more emphasis on managing information and taking action than do most other forms of writing.

Sample Webpage

Figure 1.3: Websites are highly interactive, allowing readers to follow a variety of different paths to find information.

Ethics and politics are an important concern in all technical documents.

Document is highly scannable and adaptive to readers' needs by using small columns.

Links make the text highly interactive.

The webpage is highly visual, including color.

Links take readers to more information about a subject.

Source: National Human Genome Research Institute, http://www.genome.gov.

As a result of this interactivity, it is possible for you to quickly adapt documents and presentations to fit the specific needs of many different kinds of readers and situations. Websites are an especially interactive form of technical communication (Figure 1.3). Using a website, people can find the information that is most helpful to them. And, if they cannot find the information they need on the website, they can send an e-mail to get the answers they need.

Similarly, paper-based documents can be adapted to the changing needs of readers. Before computers, it was difficult to adjust and revise paper-based documents. Once they were printed, documents were hard to change. Today, with computers, you can easily update documents to reflect changes in your company's products and services. Or, you can quickly revise documents to address unexpected changes in the workplace.

As illustrated in Figure 1.4, the relationship between the producers and users of documents has become highly interactive. The document itself, whether it is a website, a manual, a report, a proposal, or an oral presentation, is simply the meeting space between its producers and users. That space can be changed to suit the changing situations in which the document is needed.

A Document as a Meeting Space

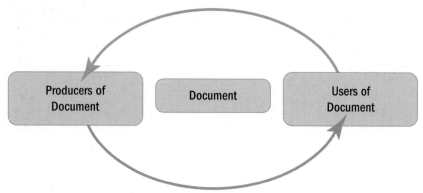

Figure 1.4: Today, producers and users of documents interact when needed, allowing the writer to adjust the text to suit the needs of specific users and contexts.

The highly interactive nature of communication means you will be continually adapting your documents to the changing needs of the readers and the situations in which they will use your materials.

Technical Communication Is Reader Centered

In technical communication, readers play a much more significant role than they do in other kinds of writing. When writing a typical college essay, you are trying to express *your* ideas and opinions. Technical communication turns this situation around. It concentrates on what the readers "need to know" to take action, not only what you, as the writer, want to tell them.

Because it is reader centered, effective technical communication tends to be highly pragmatic. Effective technical communication is:

- **Efficient**—providing only the information that readers need to know.
- **Easy to understand**—employing an organization and style that are easy for readers to follow and comprehend.
- **Accessible**—providing visual cues like headings, lists, and graphics to help readers quickly access important information.
- **Action oriented**—highlighting how the information should be used.
- **Adaptable**—allowing documents to be freely changed to suit the specific needs of different readers.

Link

To learn about adapting texts to readers and contexts, go to Chapter 3, page 46.

Of course, your needs as the writer are important too. Your *need to tell* and the readers' *need to know* should be aligned. When the needs of both writer and readers are met, you will have produced an effective document.

Technical Communication Relies on Teamwork

You would be hard-pressed to find a technical career in which you would work alone. Technical workplaces are highly collaborative, meaning you will likely work with a team of specialists on almost every project. Writing and presenting with a team are crucial skills in any technical workplace.

Computers have only heightened the team orientation of the technical workplace. Today, because documents can be shared through e-mail or the Internet, it is common for many people to be working on a document at the same time. In some cases, your team might be adjusting and updating documents on an ongoing basis.

While in school, you might have been given the false impression that you can successfully work alone. After all, writing essays and taking tests are usually individual activities. However, in the workplace, you will almost always work in a team. Knowing how to communicate and negotiate with others is a strength that will be an important part of your professional career.

Link

For more information on working in teams, see Chapter 4, page 66.

Working with a Team

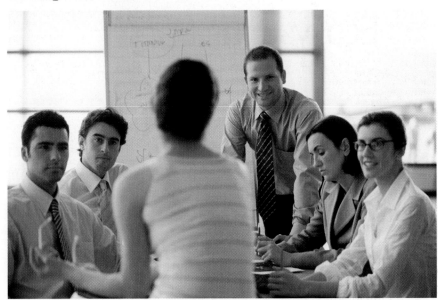

Working with a team can be fun and rewarding. Teams take advantage of the strengths and knowledge of different people to succeed.

Technical Communication Is Visual

By making texts highly visual, you can help readers quickly locate the information they need. Visual cues, like headings, lists, diagrams, and margin comments, are common in technical documents (Figure 1.5). Graphics also play an important role in technical communication. By using charts, graphs, drawings, and pictures, you can clarify and strengthen your arguments in any technical document. Today's readers grow quickly impatient with large blocks of text. They prefer graphics that reinforce the text and help them quickly gain access to important information.

One important change in our computer-centered society is that people think more visually. Your readers grew up with television and computers, so they are used to receiving information through visual media. Therefore, technical documents, whether they are manuals, reports, or websites, need to present information visually.

Link

For more information on visual design, see Chapter 10, page 232.

Link

To learn about using graphics in documents, turn to Chapter 11, page 275.

The Importance of Visual Design

Visuals add color and emotion.

An easy-to-read title identifies the document's subject.

Headings make the text highly scannable.

The two-column format makes the text easy to scan.

Figure 1.5: Visual design is an essential part of technical communication.

Ecosystems

NOS Releases Two National Progress Reports on Reef Conservation

This year, NOS released two major progress reports on coral reef research, monitoring and management. *The State of Coral Reef Ecosystems of the United States and Pacific Freely Associated States: 2005* established the first quantitative baseline of the conditions of shallow water coral reef ecosystems in the U.S., the Republic of Palau, the Republic of the Marshall Islands, and the Federated States of Micronesia. More than 160 scientists and resource managers contributed to the report, which documents the geographic extent of reef ecosystems and the status of water quality, benthic habitats, associated biological communities and key threats to coral ecosystem health. The second report, *Implementation of the National Coral Reef Action Strategy: Report on U.S. Coral Reef Agency Activities from 2002 to 2003,* highlights the activities of NOAA and the U.S. Coral Reef Task Force under each of the 13 national conservation goals defined by the 2002 U.S. National Coral Reef Action Strategy. The report indicates that collective research and management actions are moving in the right direction, citing examples like the creation of 14 new coral reef protected areas and the creation of Local Action Strategies for conservation.

Tortugas Ecological Reserve Show Signs of Species Abundance

Four years after the establishment of the Tortugas Ecological Reserve, NOS scientists are studying how the ecosystem is changing as a result of reserve status. This year, scientists conducted 253 dives to collect data and fish samples, and found that certain fish species are increasingly abundant.

New Tide and Water Quality Monitoring Station Includes Multiple Features

In August, NOS installed a tide and water quality monitoring station at the Wells National Estuarine Research Reserve (NERR), in Wells, Maine. The station combines the capabilities of the National Water Level Observation Network (NWLON) and the System-wide Monitoring Network. The station, which is the first of its kind installed at a NERR, includes primary and backup water level sensors, a suite of meteorological sensors, and a water quality sensor that measures several parameters. The NWLON technology allows Wells NERR staff to access water level, weather, and water quality data all from the same platform at the same time. Products generated from these data will benefit both short-term (such as habitat restoration) and long-term (such as sea level trends) applications, as well as research and education objectives.

Restoration Efforts at Blackwater National Wildlife Refuge

NOS and NOAA Fisheries are working with the U.S. Geological Survey (USGS), U.S. Army Corps of Engineers, U.S. Fish and Wildlife Service, National Aquarium in Baltimore, and others to restore 8,000 acres of wetlands at the Blackwater National Wildlife Refuge in eastern Maryland. Under common observing and data management principles of the Integrated Ocean Observing System, the partners are collecting water level data so that NOAA can process and conduct analyses of the data to apply to the restoration project. The Refuge also hosted a workshop on the importance of geodetic control for tidal analysis and applications. After the workshop, a global positioning system survey was conducted to connect NOAA's and USGS's water level stations, and USGS's surface elevation tables to the same geodetic network.

NOAA'S NATIONAL OCEAN SERVICE: ACCOMPLISHMENTS 2005

Source: National Ocean Service, 2005.

GO TO THE NET

For more on the importance of visual design in technical communication, go to **www.pearsonhighered.com/johnsonweb3/1.6**

Technical Communication Has Ethical, Legal, and Political Dimensions

In the increasingly complex technical workplace, issues involving ethics, laws, and politics are always present. With the increased freedom allowed by computers, we now have an increased ability to intentionally and unintentionally violate ethical and legal standards. Moreover, computers have created new micro- and macropolitical challenges that need to be negotiated in the workplace.

Employees at all levels need to know how to negotiate ethical, legal, and political issues. As management structures become flatter—meaning there are fewer layers of management—employees are being asked to take on more decision-making responsibilities than ever. In most corporations, fewer checks and balances exist, meaning all employees need to be able to sort out the ethical, legal, and political aspects of a decision for themselves.

- All technical documents involve ethical issues of rights, justice, and fairness. When you are communicating with others, you need to know how to analyze the ethical issues at stake.
- Legal issues are especially important in technical communication. Issues of product liability, copyright, and proprietary ownership play important roles in how documents are written and how information is conveyed.
- Political issues involve negotiating among the interests and needs of many stakeholders. The politics of the technical workplace can be complex, influencing how you communicate and how your messages are received.

To communicate effectively in the technical workplace, you need to be aware of the ethical, legal, and political issues that shape your writing and speaking.

Link

Ethical, legal, and political issues are discussed in Chapter 5, starting on page 92.

Technical Communication Is International and Cross-Cultural

Computers have also increased the international nature of the technical workplace. Today, it is common for professionals to regularly communicate with people around the world. Almost all technical companies and institutions compete in a global marketplace. Many have offices, communication hubs, and manufacturing sites in Europe, Asia, Africa, Australia, and South America.

The growth of international trade means you will find yourself working with people who speak other languages and have other customs. They will also hold different expectations about how technical documents and presentations should work.

Always keep in mind that communication practices that North Americans might consider "normal" or "common sense" can be strange or even offensive to people from other cultures. You need to understand the expectations of your readers in other countries.

Cross-cultural communication within North America is also a challenge. Our society is becoming more diverse than ever. As a result, readers will interpret your documents from a variety of perspectives. Your readers may all be North Americans, but they will likely be reading your document from cultural perspectives that are different from yours. You need to be aware of these perspectives so you can adjust your message to suit their needs.

AT A GLANCE

Qualities of Technical Communication

Technical communication is:

- interactive and adaptable.
- reader centered.
- produced in teams.
- visual.
- influenced by ethics, laws, and politics.
- international and cross-cultural.

Link

To learn about communicating internationally and cross-culturally, go to Chapter 3, page 53.

In the global market, the ability to communicate is the key to success.

Your Career and Technical Communication

It is typical for students in technical disciplines to wonder why they need to learn how to write and speak effectively. But, if you ask professionals who work in technical fields, you will quickly find that communication is one of their most valued skills (Silyn-Roberts, 1998). After all, in technical workplaces, almost every activity involves writing or speaking. So, writing and speaking clearly and persuasively often make the difference between success and failure.

Take a look at the help wanted advertisements on Internet job sites or in your local paper. You will notice that many technical jobs stress communication skills. Managers in these disciplines consistently rate effective communication as a "very important" skill for employees (Fagen & Coish, 1999). Employers have little patience with employees who cannot write and speak effectively.

Effective technical communication will be essential to your career in a variety of ways:

- Your managers will immediately notice whether you can communicate clearly and efficiently, because your writing and speaking skills are the first outward signs of your abilities.
- Your performance evaluations will often be based on the documents (memos, reports, proposals, websites, presentations, etc.) that you wrote or helped produce.

Need help fighting cyberphobia? Go to
www.pearsonhighered.com/johnsonweb3/1.8

How to Learn New Software Quickly

Starting a course in technical communication is similar to starting a new job in the technical workplace. Your professor and future supervisors will expect you to learn new software programs to complete your assignments and do your job. And the odds are good that they won't have time to teach you how to use this new software.

So, how can you learn new software quickly on your own? Here are some tips:

Ask experts—New students and employees are often reluctant to ask others for help because they don't want to look unprepared or unqualified. Your university or employer, however, should have "information technology" (IT) specialists on staff who provide training and can work with you one-on-one. Locate these people and schedule a training session or one-on-one tutorial.

Learn by exploring and playing—Set aside an hour or two to mess around with the software. Explore the menu bar. Try out some of the features. Click on buttons in the palettes and windows.

Use Help—Software programs usually have a preloaded Help function that provides tutorials, answers questions, and offers advice. You should spend some time taking any tutorials and searching through the Help contents. When you have questions, click on Help right away to see what advice is available.

Start with simple tasks—You are not going to be able to learn everything at once, so start out using the software to do simple tasks. If you have a large project to complete, identify the simpler tasks and do them first. Then, work your way up to more complex tasks as you grow more familiar with the software.

Go to the software company's website—Software publishers often include a variety of product demonstrations, tutorials, and training tools that were created after the software was released. These online materials tend to be written for people who are not familiar with the software, so you should find the information especially helpful.

Run an Internet search—Hey, you're not the first person who needed to learn how to use this software. So, run Internet searches through Yahoo, Ask.com, or Google to see if anyone else has put some helpful advice on the web.

Do it over and over—Each time you learn something new about the software, do it a few times to master that skill. Try to do it better or faster each time. Repetition will help you cement that skill into your mind.

Learning new software can be frustrating, especially when you are trying to make a good first impression with your professor or your new boss. The key is to relax. Showing frustration will make a far worse first impression than simply asking for help or spending some time playing around with the software. A positive attitude—even when you're not sure how to do something—will go a long way toward building a positive relationship with your professor or boss.

- Your colleagues will judge your abilities by the effectiveness of your documents or presentations.
- Your company's clients will judge the quality of your entire company by the quality of your documents and the presentations you give them.

Here is your golden opportunity to master these important writing and speaking skills. If you are reading this book, we can safely assume that you are in a class on technical communication or are looking to improve your skills in the technical workplace. This book will give you the tools you need for success.

- Computers, the Internet, and instant forms of communication have had an enormous impact on communication in the technical workplace.

- Technical communication is defined as a process of managing technical information in ways that allow people to take action.

- Technical communication is a blend of actions, words, and images. Readers expect technical documents to use writing, visuals, and design to communicate effectively.

- Technical communication is interactive, adaptable, reader centered, and often produced in teams.

- Technical communication has ethical, political, international, and cross-cultural dimensions that must be considered.

- Effective written and spoken communication will be vital to your career.

EXERCISES AND PROJECTS

Individual or Team Projects

1. Locate a document that is used in a technical workplace through a search engine like Google.com, Altavista.com, or Yahoo.com. To find documents, type in keywords like "report," "proposal," "instructions," and "presentation." Links to sample documents are also available at www.pearsonhighered.com/johnsonweb3/1.10.

   What characteristics make the document you found a form of technical communication? Develop a two-minute presentation for your class in which you highlight these characteristics of the document. Compare and contrast the document with academic essays you have written for your other classes.

2. Using a search engine on the Internet, locate a professional who works in your chosen field. Using e-mail, ask that person what kinds of documents or presentations he or she needs to produce. Ask how much time he or she devotes to communication on the job. Ask whether he or she has some advice about how to gain and improve the communication skills that you will need in your career. Write a memo to your instructor in which you summarize your findings.

3. Make a list of your goals for this class. What are some skills you would like to learn? What are some experiences you would like to have? In a memo to your instructor, briefly summarize these goals and discuss how you are going to reach them.

For sites that show examples of mission statements, go to
www.pearsonhighered.com/johnsonweb3/1.10

4. In a memo or e-mail, write a short profile of someone in your class. Tell your readers what you found interesting about this person, concentrating on information others in your class would likely want to know. Then, turn your memo into a one-minute presentation in which you introduce your classmate to the rest of your class.

5. Using the information in this chapter, write a memo to your instructor in which you compare and contrast the kinds of writing you have done for classes in the past (e.g., essays, short answer, short stories) with the kinds of writing you expect to do in your career. Then, tell your instructor how this class would best help you prepare for your career in a technical workplace.

Collaborative Project:
Writing a Course Mission Statement

As you begin this semester, it is a good idea for your class to develop a common understanding of the course objectives and outcomes. Companies develop mission statements to help focus their efforts and keep their employees striving toward common ends. Corporate mission statements are typically general and nonspecific, but they set an agenda or tone for how the company will do business internally and with its clients.

Your task in this assignment is to work with a group to develop a "Course Mission Statement" in which you lay out your expectations for the course, your instructor, and yourselves. To write the mission statement, follow these steps:

1. Go to www.pearsonhighered.com/johnsonweb3/1.10 to find links to sample mission statements. Or, you can use an Internet search engine to find your own examples of mission statements. Just type "mission statement" in Lycos.com, Webcrawler.com, or Google.com.

2. In class, with your group, identify the common characteristics of these mission statements. Pay special attention to their content, organization, and style. Make note of their common features.

3. With your group, write your own course mission statement. Be sure to include goals you would like the course to meet. You might also want to develop an "ethics statement" that talks about your approach to ethical issues associated with assignments, course readings, and attendance.

4. Compare your group's course mission statement with other groups' mission statements. Note places where your statement is similar to and different from their statements.

When your course mission statement is complete, it should provide a one-paragraph description of what you are trying to achieve in your class.

> For additional technical writing resources, including interactive sample documents, document design tutorials and guidelines, and more, go to **www.mytechcommlab.com.**

CHAPTER

2

The Technical Writing Process Today

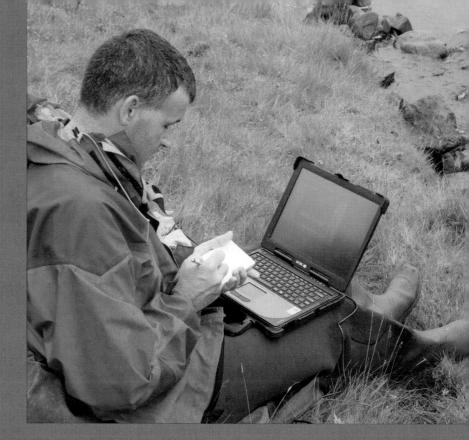

In this chapter, you will learn:

- How to use a writing process that takes advantage of your computer's capabilities.

- Strategies for defining a document's *rhetorical situation*.

- The importance of defining your purpose.

- Techniques for organizing and drafting your ideas with common genres.

- How to use your computer to overcome writer's block.

- The importance of style and design in informative and persuasive documents.

- The basics of revising and editing your work.

W ork smarter, not harder. That's the basic message of this chapter. Computers have made technical communication easier in countless ways. Today, many of us cannot even imagine working without tools like e-mail, instant messaging, presentation software, word processors, and the Internet.

But one problem with writing on computers is that many people, especially novice writers, have not developed a dependable "writing process." Instead, they sit down at their computer and hammer out a document. The printed words on the screen give the false impression that the document is "finished" when it is really just a rough, rough draft. As a result, documents written on computers sometimes lack the proper quality and depth required in the technical workplace.

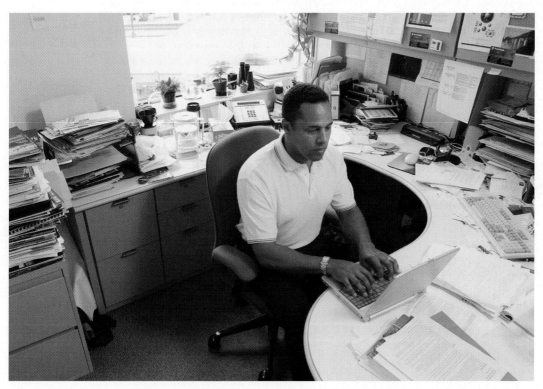

Developing a consistent writing process is a key to success in the technical workplace.

Before computers, people were forced to go through a writing process using pen, paper, and typewriter. Writers would use their pens to take notes and freewrite. They would compose "throwaway drafts" and handwrite more drafts on paper. Then, they would revise and proofread their final drafts while typewriting. Each stage of their writing process would invite them to reflect more deeply about their documents' purpose and readers (Elbow, 1981; Perl, 1994).

With computers, a writing process is still essential for producing quality technical communication, because most technical documents are too complex to be handled in one drafting stage. The purpose of this chapter is to show you how to develop your own writing process—one that suits your individual work habits. A good writing process will help you work smarter when you are writing technical documents because it will allow you to take full advantage of your computer's capabilities.

Genres in Technical Communication

Of course, we all have our own ways of doing things. Writing is no different. Each of us develops a writing process that fits our own needs and work habits.

In this book, you will be learning a *genre-based approach* to the writing process. In technical communication, genres are relatively stable patterns that reflect the activities and practices of the workplace. A genre shapes a document's content, organization, style, design, and the medium in which it is delivered. It also helps you anticipate the needs of your readers and the situations in which they will use your text.

For example, an "analytical report" is a different genre than a "specification" (Figure 2.1). Analytical reports and specifications are written for different kinds of readers and situations. They include different information and follow their own organizational patterns. The style and design of these two genres are distinctly different. Yet, someone working in a technical workplace would need to know how to use both of these genres.

Two Different Genres

Figure 2.1: Each genre has its own content, organization, style, and design. Here are the outlines of two distinctly different genres set side by side.

For more discussions of the writing process, go to
www.pearsonhighered.com/johnsonweb3/2.1

Genres do much more than help you organize your ideas. They help you interpret and make sense of what is happening in the technical workplace. For example, if you know you need to write an analytical report, the genre will help you figure out what kind of information you need to collect, how that information should be arranged, and how it should be presented. Your readers, meanwhile, will interpret your ideas through the genre. If you call something a "report," they will have specific expectations about the content, organization, style, design, and medium of the document.

Genres are not formulas or recipes to be followed mechanically. They are not containers into which we pour some sentences and paragraphs. Instead, genres in technical communication reflect the activities and practices of scientific and technical workplaces. Each genre should be adapted to fit the readers and the situations in which the document will be used.

Genres and the Technical Writing Process

A consistent *writing process* is important to using technical communication genres effectively. When learning to write for the workplace, it helps to view writing as a process that follows a series of stages (Figure 2.2). These stages include the following:

The Technical Writing Process

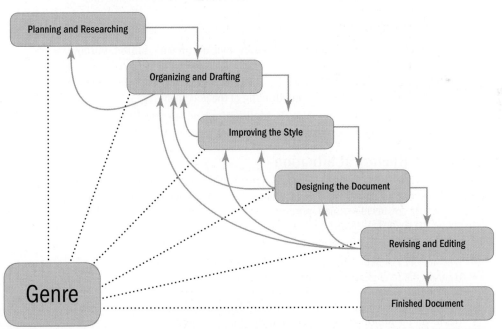

Figure 2.2: The technical writing process involves moving back and forth among several stages. Each stage is shaped by the genre of the document.

- **Planning and researching**—planning the project, collecting information, and developing your own ideas.
- **Organizing and drafting**—deciding how to arrange your information and then turning those ideas into sentences, paragraphs, and sections.
- **Improving the style**—writing clearly and persuasively for your readers.
- **Designing the document**—developing an appropriate page design that improves the usability and attractiveness of your document.
- **Revising and editing**—improving the quality of your work by re-visioning, rewriting, and proofreading your writing.

The boundaries among these stages are blurred, and you will find yourself working back and forth among them, as shown in Figure 2.2. While drafting, for example, you may discover that you need to do more research on your topic. While editing, you may decide that you need to draft another section. Overall, though, these stages will lead you from the beginning of a project to the end.

Meanwhile, the genre you are using will guide and influence each stage. The document's genre helps you make decisions about what the document should contain, how it should be organized, what style would be appropriate for the readers, and what design and medium would be best for presenting your ideas.

Stage 1: Planning and Researching

Before you begin drafting the document on your computer, you should give yourself time to plan and do research. Even if time is short, you should spend some time planning and researching before you start drafting. After all, half-baked ideas only confuse and frustrate the readers.

When planning and researching, you should spend some time doing three activities:

Defining the rhetorical situation—Identify your document's subject, purpose, readers, and context of use.

Defining your purpose—Sharpen your purpose into a one-sentence statement that will guide your research and drafting of the document.

Researching your subject—Use electronic, print, and empirical sources to collect information on your subject.

Defining the Rhetorical Situation

A good first step is to define the *rhetorical situation* that will shape the content, organization, style, and design of your document. Understanding the rhetorical situation means gaining a firm grasp of your document's subject, purpose, readers, and context of use (Figure 2.3).

To define the rhetorical situation, you should start out by asking the *Five-W and How Questions:* who, what, why, where, when, and how.

- *Who* are our readers, and who else is involved with the project?
- *What* do the readers want and need, and what do we want and need?
- *Why* do the readers need the information in this document?
- *Where* do they need the information, and where will they use it?
- *When* will the information be used, and when is it needed?
- *How* should we achieve our purpose and goals?

Want to know more about rhetoric? Go to
www.pearsonhighered.com/johnsonweb3/2.3

Defining the Rhetorical Situation

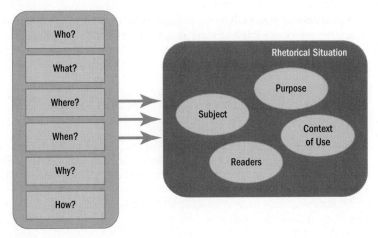

Figure 2.3:
The Five-W and How Questions can help you determine the rhetorical situation for your technical document or presentation.

The Five-W and How Questions

AT A GLANCE

- Who are the readers of this document?
- What am I writing about?
- Why am I writing this document or presenting this material?
- Where will the document be used?
- When will the document be used?
- How will the document be used?

The Five-W and How Questions should give you an overall sense of your document's rhetorical situation. Now you can use that information to more firmly define your document's subject, purpose, readers, and context of use.

Spend some time taking notes on the following four issues:

Subject—What is the document about? What is it *not* about? What kinds of information will my readers need to make a decision or complete a task? What is the scope of the project?

Purpose—Why is this document needed? What does it need to achieve or prove?

Readers—Who are the readers of this document? What are their needs and interests?

Context of use—Where and when will this document be used? What physical, economic, political, and ethical constraints will shape this text?

Link

For more information about defining your readers and context of use, see Chapter 3, page 46.

As you begin the writing process, you should consider these four categories separately. In some cases, especially with larger documents, it is helpful to write down your answers so you can keep yourself focused and on track.

Defining the rhetorical situation may seem like an added step that will keep you from writing. Actually, knowing your document's rhetorical situation will save you time and effort, because you will avoid dead ends, unnecessary revision, and writer's block.

Defining Your Purpose

Among the four elements of the rhetorical situation, your document's purpose is probably the most important. It is what you want to do—and what you want the document to achieve.

Your purpose statement is like a compass for the document. Once you have clearly defined it for yourself and your readers, you can use that purpose statement to guide your decisions about the content, organization, style, and design of your document.

When defining your purpose, try to express exactly what you want your document to achieve. Sometimes it helps to find an appropriate action verb and then build your purpose statement around it. Here are some useful action verbs that you might use:

Informative Documents	Persuasive Documents
to inform	to persuade
to describe	to convince
to define	to influence
to review	to recommend
to notify	to change
to instruct	to advocate
to advise	to urge
to announce	to defend
to explain	to justify
to demonstrate	to support

Two Types of Purpose in Technical Documents

AT A GLANCE

- To inform—to give readers the information they need to understand a subject or make a decision.
- To persuade—to attempt to convince readers that they should hold a particular opinion or take a specific action.

Once you have chosen an action verb, try to state your purpose in one sentence. It might help to finish the phrase "The purpose of my document is to...."

The purpose of my report is to review the successes and failures of wolf reintroduction programs in the western United States.

The purpose of my proposal is to recommend significant changes to flood control strategies in the Ohio River Valley.

Hammering your purpose statement down into one sentence is hard work but worth the effort. Your one-sentence purpose statement will focus your writing, saving you time.

Researching Your Subject

Solid research is your next step. While researching your subject, you need to gather information from a variety of sources, including the Internet, print documents, and empirical methods (e.g., experiments, surveys, observations, interviews).

Link

For more information on research methods, go to Chapter 7, starting on page 141.

Computers have significantly changed the way we do research in technical workplaces. Before computers, finding sufficient information was usually a writer's main challenge. Today, there is almost too much information available on any given subject. So, it is important that you learn how to *manage* information, sorting through all the texts, scraps, junk, and distortions to uncover what you need. Your documents should give your readers only the information they need to make a decision or take action. Leave out anything else.

For helpful tools for doing research online, go to
www.pearsonhighered.com/johnsonweb3/2.5

Stage 2: Organizing and Drafting

Organizing and drafting is usually the hardest part of the writing process. While organizing and drafting, you are essentially doing two things at the same time:

Organizing the content—Using common genres to shape your ideas into documents that will be familiar to the readers.

Drafting the content—Generating the content of your document by including facts, data, reasoning, and examples.

As you write your document, you will move back and forth between organizing and drafting.

Organizing the Content

Before drafting, you need to decide how to organize the information you collected in a way that achieves your purpose.

INTRODUCTIONS, BODIES, AND CONCLUSIONS   All technical documents tend to have a beginning (introduction), a middle (body), and an end (conclusion).

The **introduction** sets a context for the document by telling readers your subject, purpose, and main point. It might also offer background information on the subject, while stressing the importance of the subject and forecasting the structure of the rest of the document.

The **body** presents the content of the document. The body includes facts, data, reasoning, and examples that help the document achieve its purpose and prove its main point.

The **conclusion** reestablishes the context for the document by restating the main point, restressing the importance of the subject to the readers, and looking to the future.

A well-written technical document begins by offering contextual information up front in the introduction (Figure 2.4). Basically, the introduction tells readers the who, what, where, when, why, and how. Then, the body provides the content (the need-to-know information) that readers want. The conclusion reframes the discussion by circling back to the contextual information stated in the introduction.

GENRES   Technical documents also follow patterns of organization called *genres*. A genre is a familiar pattern that readers will expect the document to follow.

Figure 2.5, for example, is immediately recognizable as a set of instructions because it follows a common genre. Readers immediately know how to use this document.

Link

For more information on writing introductions and conclusions, see Chapter 8, page 182.

AT A GLANCE

Genres in Technical Communication

Here are the genres covered later in this book:

- Letters and Memos— Chapter 17
- Technical Definitions— Chapter 18
- Technical Descriptions— Chapter 19
- Instructions and Documentation— Chapter 20
- Proposals—Chapter 21
- Activity Reports— Chapter 22
- Analytical Reports— Chapter 23

Basic Organization of a Document

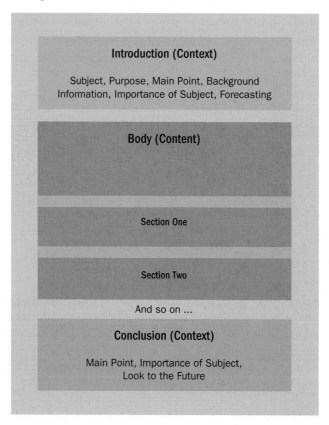

Figure 2.4: Technical documents should have a beginning, middle, and end. The introduction sets a context. The body includes the content. The conclusion resets the context.

Chapters 17 through 23 describe how to use the most common genres in technical workplaces. In most situations, you will know which genre you need because your supervisor or instructor will ask you to write a "report," a "proposal," or a "letter." But, if you are uncertain which genre suits your needs, pay attention to your document's purpose. Then, find the genre that best suits the purpose you are trying to achieve.

Drafting the Content

When you start drafting your document, remember that the purpose of drafting is *not* to create the final version of your document. The purpose of drafting is to put your ideas on the screen so you can work with them. Don't worry about "getting it right" or writing in a polished way.

To find websites that discuss various genres, go to
www.pearsonhighered.com/johnsonweb3/2.7

Sample of Genre: Instructions

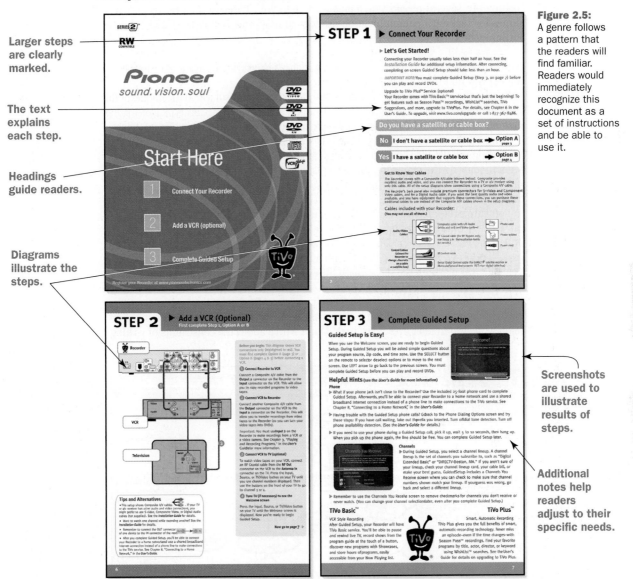

Larger steps are clearly marked.

The text explains each step.

Headings guide readers.

Diagrams illustrate the steps.

Figure 2.5: A genre follows a pattern that the readers will find familiar. Readers would immediately recognize this document as a set of instructions and be able to use it.

Screenshots are used to illustrate results of steps.

Additional notes help readers adjust to their specific needs.

Source: TiVo.

Sometimes the hardest part of drafting is just putting words on the screen. Fortunately, several *invention techniques* can help you. Three of the best techniques for technical communication are logical mapping, freewriting, and outlining/boxing.

LOGICAL MAPPING   Logical mapping is a highly visual way to invent your ideas, helping you find their logical relationships.

To map the content of your document, start by putting your subject in the middle of the screen or a piece of blank paper. Put a circle or box around it. Then, start typing or writing your other ideas around the subject, and put circles or boxes around them (Figure 2.6).

Now, fill the screen or page with words and phrases related to the subject. Start connecting related ideas by drawing lines among them. As you draw lines, you will begin to identify the major topics, concepts, or themes that will be important parts of the document you are writing. These major issues can be found in the clusters of your map.

Software programs such as Inspiration (shown in Figure 2.6), Visio, MindManager, and IHMC Concept Mapping Software can help you do logical mapping on screen. Otherwise, you can use the Draw function of your word processor to create "text boxes" and draw lines among them. With a little practice, you will find that you can create logical maps on the screen with little effort.

FREEWRITING   Freewriting is easy. Simply put your fingers on the keyboard and start typing into a document file in your word processor. Type for five to ten minutes before you stop to look over your work. Don't worry about the constraints of writing

Figure 2.6: Logical mapping helps you break out of the usual restrictions without using sentences. You can begin connecting together your ideas as you put them on the screen. This map was made with a software program called Inspiration.

Source: Diagram created in Inspiration® by Inspiration Software®, Inc.

Freewriting

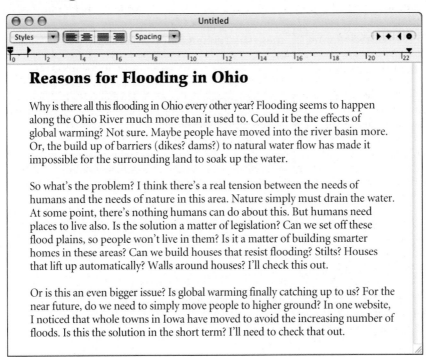

Reasons for Flooding in Ohio

Why is there all this flooding in Ohio every other year? Flooding seems to happen along the Ohio River much more than it used to. Could it be the effects of global warming? Not sure. Maybe people have moved into the river basin more. Or, the build up of barriers (dikes? dams?) to natural water flow has made it impossible for the surrounding land to soak up the water.

So what's the problem? I think there's a real tension between the needs of humans and the needs of nature in this area. Nature simply must drain the water. At some point, there's nothing humans can do about this. But humans need places to live also. Is the solution a matter of legislation? Can we set off these flood plains, so people won't live in them? Is it a matter of building smarter homes in these areas? Can we build houses that resist flooding? Stilts? Houses that lift up automatically? Walls around houses? I'll check this out.

Or is this an even bigger issue? Is global warming finally catching up to us? For the near future, do we need to simply move people to higher ground? In one website, I noticed that whole towns in Iowa have moved to avoid the increasing number of floods. Is this the solution in the short term? I'll need to check that out.

Figure 2.7: While freewriting, just get your ideas on the screen. Simply writing ideas down will help you locate important ideas and directions for research.

such as sentences, paragraphs, grammatical correctness, or citations. Just keep typing. Eventually, you will find that you have filled one or more screens with words, sentences, and fragments of sentences (Figure 2.7).

You may or may not end up using many of the words and sentences in your freewriting draft, but the purpose of freewriting is to put your ideas on the screen. It helps you fight through writer's block.

Once you have done some freewriting, identify the two to five major items in your text that seem important. Then, spend five to ten minutes freewriting about each of these items separately. Like magic, within half an hour to an hour, your freewriting will probably give you more material than you need to write your text.

OUTLINE OR BOXING   Outlines can be used throughout the drafting process. Most word-processing programs will allow you to draft in *Outline* mode or *Document Map* mode (Figure 2.8). Sometimes it helps to sketch an outline before you start drafting. That way, you can see how the document will be structured.

When revising, you might find it helpful to re-outline your document or print out the computer-generated outline in Outline mode. This outline will help you find places where need-to-know information should be added or unnecessary information taken out.

Outlining or Document Mapping

Outline of document →

Text of document →

Figure 2.8: In Document Map mode, the computer automatically outlines your document on the left. This feature allows you to keep the whole structure of the document in mind as you work on individual parts.

Boxing is less formal. As you plan your document, draw boxes on the screen or a piece of paper that show the ideas or topics in your document (Figure 2.9). Then, type your ideas into the boxes. If you want to make multiple levels in your text, simply create boxes within boxes. You can use the Table function in your word-processing software to make boxes. When using the Table function, start out with a few boxes. Then add cells to the table as you need more boxes.

Boxing

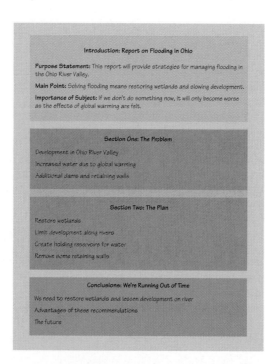

Figure 2.9: Boxing is like outlining. Each of the cells can be filled with your ideas. Then, you organize these ideas into a more structured document.

These invention tools don't work for you? To learn about other invention techniques, go to
www.pearsonhighered.com/johnsonweb3/2.9

Using Your Computer to Overcome Writer's Block

Have you ever stared at a blank computer screen and it just stared back? Many people have trouble starting the writing process. Here are ten quick tips for overcoming writer's block:

Just write something—Don't wait to be inspired. Those moments are rare. Instead, just go ahead and write, even if you are putting garbage on the screen. The act of writing will get you going.

Write something every day—Writer's block is directly linked to anxiety, which is often linked to an impending deadline or due date. If you write for just half an hour to an hour every day (that's all), you will usually finish all your documents with time to spare.

In a notebook or a PDA, keep records of your ideas. Good ideas can come to you at any time. Make sure you keep track of them.

(continued)

Lower your standards while drafting—Stop trying to get it right on the first try. Instead, put your ideas on the screen without worrying about whether they are intelligent or grammatically correct. Then, spend more time during the revision phase turning those ideas into a document that embodies the quality your readers expect.

Talk it out—Professional writers can often be found talking to themselves or friends about their work. Sometimes it helps to say out loud what you want to write. Then, after you have rehearsed what you want to say a few times, write it down.

Use a notebook or your PDA—Jot down notes in a notebook or your personal digital assistant (PDA) as they come to mind. Then, when you are ready to write, you will find a cache of ideas to draw from.

Use both sides of your brain—The right side of your brain is more visual than your left. Use logical mapping, freewriting, and boxing to tap into that visual creativity. These invention techniques will help you put your ideas on the screen. Then, the left side of your brain can organize them into sentences and paragraphs.

Surf and dump—Surf the Internet, looking for items related to the subject of your document. Cut and paste, or "dump," the information you find into your document. Then start writing. Remember to cite your sources, though, and don't plagiarize the works of others.

"What I really mean is…"—Whenever you are blocked, finish the sentence "What I really mean is…" You will find that simply finishing this sentence will help you get past the temporary block.

Write an e-mail—Start writing your document as an e-mail to a friend or yourself. E-mail is often a more familiar writing environment, allowing you to relax as you write.

Stop procrastinating—Procrastination is the usual culprit behind writer's block. The pressure of a deadline can cause your brain to freeze. So, start each project early, and write a little every day. Your writer's block will evaporate.

Having trouble with writer's block?
For more help, go to
www.pearsonhighered.com/johnsonweb3/2.10

Stage 3: Improving the Style

All documents have a style, whether a style is chosen or not. If properly chosen, good style can make your documents easier to read and more persuasive. But, if you don't pay attention to style, your readers might find your documents chaotic and difficult to follow. Also, today's readers expect technical documents to do more than present content. They expect technical documents to grab and keep their attention.

Have you ever read a text that just didn't feel right? More than likely, you were reacting to its erratic style. Good style is a choice you can and should make. In Chapter 9 of this book, you will learn about two kinds of style that are widely used in technical documents: plain style and persuasive style.

> **Plain style**—This style stresses clarity and accuracy. By paying attention to where words appear in sentences and paragraphs, you can make your ideas clearer and easier to understand.
>
> **Persuasive style**—Using persuasive style, you can motivate readers by appealing to their values and emotions. You can use similes and analogies to add a visual quality to your work. You can use metaphors to change your readers' perspective on issues. Meanwhile, you can use tone and pace to add energy and color to your work.

Most technical documents are written in the plain style, but many documents need the extra energy and vision provided by the persuasive style. Your goal should always be to make information as clear and concrete as possible. When persuasion is needed, you will want to energize your words with persuasive style techniques.

Link

For more about improving style, go to Chapter 9, page 206.

Stage 4: Designing

With computers, you can lay out a page in minutes and create graphics with a few clicks of a button. So, design is not only possible—your readers will *expect* your technical documents to be well designed. You should make document design a regular part of your writing process.

Today, few people read documents word for word. In fact, right now you are probably scanning this chapter, looking for the highlights. You're probably not reading the chapter word for word, sentence by sentence. If so, you are like most readers of technical documents. You are a "raider" for information as well as a reader.

So, as you think about the design of your document, keep this saying in mind: *Readers of technical documents are "raiders" for information.* Your readers, like you, want the important parts highlighted for them. They prefer documents that use graphics and document layout to make the information more accessible, interesting, and attractive (Figure 2.10).

Chapter 10 discusses document design techniques and strategies. Chapter 11 discusses the creation and use of graphics in documents. As you draft and revise your document, keep looking for places where you can use design to help your readers find the information they need. Look for places where graphics might support or reinforce the written text. And think about how the layout of the document can improve its readability and your readers' interest in what you are telling them.

Link

For more ideas about designing documents and screens, go to Chapter 10, page 232.

Link

To learn about creating and using graphics, turn to Chapter 11, page 275.

 Want to improve your writing style? Go to
www.pearsonhighered.com/johnsonweb3/2.11

**Stage 4:
Designing** 31

Document Design Is Very Important

The organization's name is easy to locate.

Topics are listed clearly.

Text is highly scannable because it is in groups.

Pictures add a human quality.

Color adds energy.

Figure 2.10: Because readers are raiders of information, you want the design of your document to be visually accessible.

Source: American Red Cross, http://www.redcross.org.

Stage 5: Revising and Editing

When you have finished drafting and designing the document, you are about two-thirds finished. In technical communication, it is crucial to leave plenty of time for revising, editing, and proofreading. Clarity and accuracy are essential if your readers are going to understand what you are trying to tell them.

In Chapter 12, you will learn about four levels of revising and editing:

Revising and Editing

AT A GLANCE

• Level 1: Revising
• Level 2: Substantive Editing
• Level 3: Copyediting
• Level 4: Proofreading

Level 1: Revising—Once you have drafted the document, it is time to reconsider your subject and purpose while thinking again about the information your readers need to know and the places where they will use your document.

Level 2: Substantive editing—Look at the content, organization, and design of the document. The content of your document should be organized with a genre that is familiar to your readers. Make sure your document has an effective beginning (introduction), middle (body), and end (conclusion).

GO TO THE NET

If you want more information on document design, go to
www.pearsonhighered.com/johnsonweb3/2.12

Tim Raine

RESEARCHER/WRITER, RAINE TITLE SEARCH, LLC.

Raine Title Search, in Anamosa, Iowa, researches properties for sale and writes technical reports for its clients.

Do writers in the technical workplace *really* follow a writing process as they write?

Yes. Experienced writers may not consciously consider each step required to complete daily writing tasks. Like any skill, when you learn the fundamentals of technical writing and then practice what you learn on a daily basis, the writing process becomes a habit that you don't think about. The more you write and the more you pay attention to the process you follow as you write, the easier it becomes to recognize and produce high-quality writing.

Learning technical writing is like learning any workplace skill. When you're acquiring a new skill, you pay great attention to the process. As you grow more and more familiar with the skill, you tend to stop thinking so much about the process and pay more attention to achieving your goals. You don't forget the process—it just becomes a habit that you can rely on.

However, good writers also know their own writing process and routinely evaluate it. Thinking critically about how you write allows you to streamline your process to handle the task at hand as efficiently as possible.

Knowing your writing process also gives you a frame of reference for estimating the time and effort required to complete specific writing tasks. The ability to make the most of your time and to estimate what it takes to produce high-quality writing is extremely useful in any work environment.

AT WORK

Level 3: Copyediting—Effective writers often spend a significant amount of time copyediting the sentences, paragraphs, and graphics in their documents. Copyediting helps make the document easier to read and more persuasive. It also ensures that graphics are accurate and support the written text.

Level 4: Proofreading—Finally, don't forget to carefully proofread your document. In the technical workplace, quality is taken very seriously. When readers encounter a document that has grammar problems, typos, spelling errors, and usage problems, they will question the quality of the information. At a minimum, you can use the spelling and grammar checker on your computer to catch errors as you proofread (Figure 2.11). But, you should also carefully read the document to catch mistakes that slip through.

Link

To learn more about revising and editing, go to Chapter 12, page 306.

At the end of a difficult project, it is sometimes tempting to simply throw up your hands and send in the document. You can even trick yourself into believing that readers won't really care about the quality of your document. They do. It is important to spend time editing your work to meet the readers' high expectations.

Want to know more about editing? Go to **www.pearsonhighered.com/johnsonweb3/2.13**

Stage 5: Revising and Editing

33

Spelling and Grammar Checker

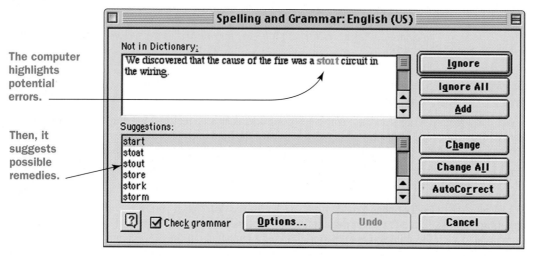

The computer highlights potential errors.

Then, it suggests possible remedies.

Figure 2.11:
The spelling and grammar checker in your word processor software can help locate typos and misspellings. It is less helpful for locating grammatical errors.

Developing Your Own Writing Process

This chapter showed you a writing process that works very effectively in the technical workplace, but there are many other effective writing processes available. The secret is to develop a writing process that suits your own working style. Spend some time thinking about how you prefer to plan, research, organize, draft, and edit your work. Think about your sense of style and design. Then, adapt the writing process shown in this chapter to your own needs.

When you have developed your own writing process, you will find writing technical documents to be easier, more effective, and much more satisfying.

- By consciously developing a writing process, you will learn how to write more efficiently. In other words, you will "work smarter, not harder."

- A useful computer-based writing process includes the following stages: planning and researching, organizing and drafting, improving the style, designing, and revising and editing.

- The key to effective technical writing is to first define your rhetorical situation, putting special emphasis on the purpose of your document.

- Technical writing genres are helpful for organizing information into patterns that your readers will expect.

- When drafting your text, you should use logical mapping, freewriting, or boxing/outlining to put your ideas on the screen.

CHAPTER REVIEW

Some good advice about developing a writing process is available at
www.pearsonhighered.com/johnsonweb3/2.14

- You can use the plain or persuasive style to clarify and energize your message.

- Document design should be used to enhance readability, support the written text, and make the document more attractive.

- Revising and editing a document is important to improve clarity and accuracy while ensuring that the content provides only need-to-know information.

Individual or Team Projects

1. Interview a professional in your field or a professor in your major about his or her writing process. Ask about smaller documents like e-mail, memos, letters, and instructions. Then, ask about larger documents. How does this person's process follow any or all of the stages described in this chapter? Here are some questions you might ask:

   - How much planning and research do you do before drafting?
   - Do you follow any regular organizational patterns in your documents?
   - How much revising do you do?
   - How do you use style to clarify or amplify your message?
   - When do you start thinking about the visuals and design of the document?

   Write a memo to your instructor in which you discuss the results of your interview. Compare and contrast the results of your interview with the writing process described in this chapter.

2. Write an e-mail to your instructor in which you describe your own writing process. Each person has a unique approach to writing. What is your unique approach? How is your writing process similar to the one described in this chapter? How is your writing process different? How can you adapt your writing process to fit the technical workplace?

   Compare your writing process with that of others in your class. How does your writing process differ from theirs? Why do you think each of you follows a different writing process?

3. More than likely, you are working on a document right now. Describe the rhetorical situation in which this document will be used. In other words, (1) define the subject, (2) state your purpose, (3) describe the readers, and (4) describe the document's context of use. In a memo to your instructor, write a description of the rhetorical situation. Also, tell your instructor how this document's rhetorical situation will require you to adapt the information you collect and create to the document's subject, purpose, readers, or context of use.

Collaborative Project

Imagine you and your group are writing a report on sleep deprivation and its effects at your college or workplace. Using the Internet, run searches on sleep deprivation to collect information. You can use search engines like Google.com, Altavista.com, Lycos.com, or Yahoo.com. Once you have gathered some information, have each member of your group do one of the following forms of drafting:

Logical mapping—Put "sleep deprivation" in the middle of your screen or a sheet of paper. Then, use logical mapping to write down all the ideas, topics, and facts that you might include in the report. Connect related ideas.

Freewriting—Write a screen or two of text in which you describe everything you know or have found about the effects of sleep deprivation.

Boxing or outlining—Use boxing or outlining to start developing an organization for your report. When boxing, draw large boxes that define topics. Then, fill those boxes with the ideas, facts, and arguments that will be covered under each topic in the report.

Present the results of your research and drafting to your class. Show how your logical mapping, freewriting, and boxing/outlining might shape your writing process.

For additional technical writing resources, including interactive sample documents, document design tutorials and guidelines, and more, go to **www.mytechcommlab.com.**

36   Chapter 2
     **The Technical
     Writing Process
     Today**

 To learn more about working in teams, go to
www.pearsonhighered.com/johnsonweb3/2.16

CASE STUDY

Changing Bad Habits

Vicki Clark, a design engineer at Cleveland Solutions, was excited to have a summer intern assigned to her. Her colleagues groaned when their supervisor, Brad Tellers, announced in May that a new engineering intern would be joining them later that month. They muttered about "extra work." But Vicki remembered fondly her experiences as an intern between her junior and senior year of college. That internship was her first chance to do a "real job," something more than flipping burgers or babysitting.

The intern, Steve Zimms, made a positive first impression on her. He was eager to learn, and he seemed excited about an opportunity to work in a design engineering firm. He seemed to be a whiz at the computer-aided drafting (CAD) system, and he was a fast learner. She put him right to work.

Steve was able to handle all of the assignments Vicki gave him. He also had a good idea about how they could integrate CAD with fatigue analysis software to better anticipate where and when cracks would appear in the aluminium suspension components that Vicki was currently designing.

Vicki liked Steve's idea. On Tuesday, she and Steve pitched it to her boss, Brad. With a smile, Brad said, "Great idea. Write it up as a proposal and have it on my desk by Monday. If your idea works, Steve, you might have a job waiting for you here when you graduate."

Vicki was very busy, so she let Steve write up the proposal himself. Monday morning, Steve sent the proposal directly to Brad without running it past Vicki first for comments.

Two days later, Brad was down in Vicki's office with the proposal in his hand. "I still like the idea, Vicki, but this proposal is awful. It looks like he wrote it overnight."

After Brad left, Vicki called Steve into her office and asked him about the proposal. Steve admitted that he rushed it. He had waited until Sunday night to put his ideas on paper. "That's how I always write, and I've always gotten A and B grades on my papers. That's just the way I write."

Vicki remembered her college days. She, too, could blast out papers the night before and still get A's. The workplace was different, though. Good ideas needed to be thought out more. Drafting took longer. Large documents, like proposals, needed more revision than the typical college paper. They also needed to be cycled past supervisors for comments and suggestions for improvement.

If you were Vicki, how might you train Steve to develop a more reliable writing process? How could you use the steps described in this chapter to help him slow down and think through the stages of his writing?

To learn about ways to prepare yourself for an internship, go to
www.pearsonhighered.com/johnsonweb3/2.17

CHAPTER

3

Readers and Contexts of Use

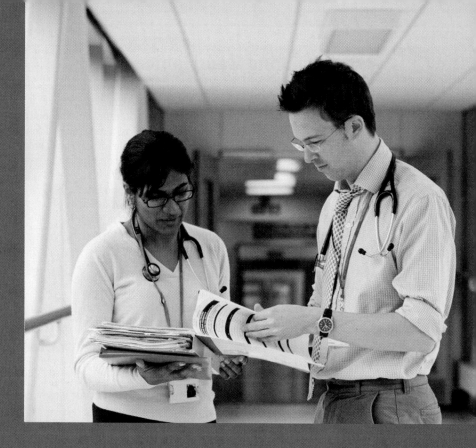

In this chapter, you will learn:

- How to use the computer as a reader analysis tool.

- How to develop a comprehensive profile of a document's readers.

- How to sort your readers into primary, secondary, tertiary, or
  gatekeeper audiences.

- Techniques for identifying readers' needs, values, and attitudes about
  you and your document.

- How to analyze the physical, economic, political, and ethical contexts
  of use that influence how readers will interpret your text.

- How to anticipate the needs of international and cross-cultural readers.

Knowing your readers is essential to effective technical communication. More than ever, readers don't have time to slog through information they don't need. So, you should find out exactly what your readers need to know and how they want that information presented.

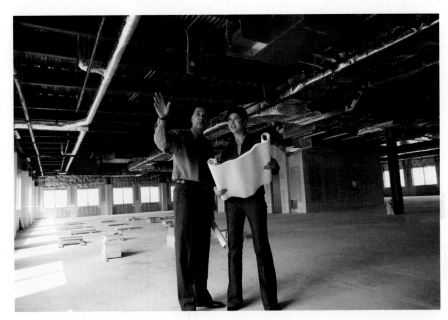

Your readers only want the information they need to make a decision or take action. As the writer, you need to find out what they need and how they want the information presented.

Another concern is the ever-increasing importance of international communication through electronic networks. In technical fields, you *will* find yourself regularly communicating with people who speak other languages, have different customs, and hold different expectations. Computers have broken down many of the geographical barriers that once separated people and cultures. It is now common to communicate with people around the world on a daily basis.

In this chapter, you will learn how to develop *profiles* of your readers, so you can anticipate what they need to know and the contexts in which your documents will be used.

Profiling Your Readers

In technical communication, documents are designed to suit the needs of specific types of readers. For this reason, early in the writing process, you should profile the types of people who might be interested in your document (Figure 3.1).

Developing a Reader Profile

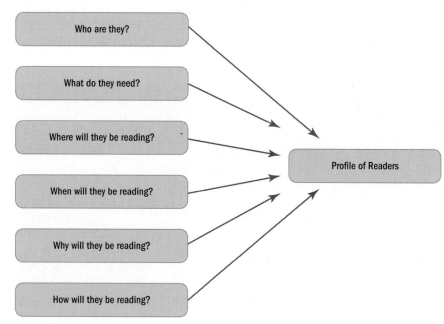

Figure 3.1:
To develop a profile of your readers, use the Five-W and How Questions to look at them from a variety of perspectives.

Reader profiles are sketches of your readers' tendencies, abilities, experiences, needs, values, and attitudes. To build a profile, begin asking yourself the Five-W and How Questions about your readers.

Who might read this document?

What information do they need?

Where will they read the document?

When will they read the document?

Why will they be reading it?

How will they be reading it?

As you answer these questions, keep in mind the following guidelines about your readers and how they prefer to read.

Guideline One: Readers are "raiders" for information—People don't read technical documents for pleasure. Instead, most readers are *raiding* your document for the information they need to make decisions or take action.

Guideline Two: Readers are wholly responsible for interpreting your text—You won't be available to explain what your document means, so your readers need to be able to easily figure out what you are telling them.

For websites that offer other ways to use the Five-W and How Questions, go to
www.pearsonhighered.com/johnsonweb3/3.1
What else do readers want? For answers, go to
www.pearsonhighered.com/johnsonweb3/3.2

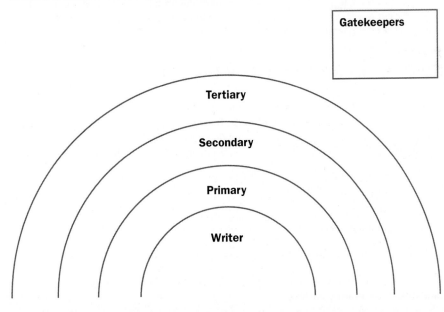

About Your Readers

- Readers are "raiders" for information.
- Readers are wholly responsible for interpreting your text.
- Readers want only "need-to-know" information.
- Readers prefer concise texts.
- Readers prefer documents with graphics and effective page design.

AT A GLANCE

Guideline Three: Readers want only "need-to-know" information—Readers want you to give them only the information they need. Any additional material only makes the information they want harder to find.

Guideline Four: Readers prefer concise texts—The shorter, the better. Usually, the longer the document is, the less likely that people are going to read it. Your readers prefer documents that get to the point and highlight the important information.

Guideline Five: Readers prefer documents with graphics and effective page design—We live in a visual culture. Large blocks of text intimidate most readers. So, include graphics and use page design to make your document more readable.

Think about how you are reading this book. More than likely, you are raiding for need-to-know information. You want the book to be concise and visually interesting. Your readers want these things too.

Identifying Your Readers

You should always begin by identifying the readers of your document. Figure 3.2 shows a Writer-Centered Analysis Chart that will help you locate the various people who might look over your text (Mathes & Stevenson, 1976). You, as the writer, are in the center ring. Each ring in the chart identifies your readers from most important (primary readers) to least important (tertiary readers).

To use the Writer-Centered Chart, begin filling in the names and titles of the primary, secondary, tertiary, and gatekeeper readers who will or might look over your work.

Writer-Centered Analysis Chart

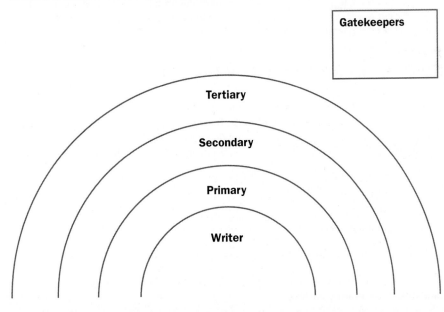

Gatekeepers

Tertiary

Secondary

Primary

Writer

Figure 3.2: A Writer-Centered Analysis Chart starts with you in the center and identifies the various people who may be interested in your document.

PRIMARY READERS (ACTION TAKERS)   The primary readers are the people to whom your document is addressed. They are usually *action takers* because the information you are providing them will allow them to do something or make a decision. Usually your document will have only one or two primary readers or types of primary readers.

SECONDARY READERS (ADVISORS)   The secondary readers are people who *advise* the primary readers. Usually, they are experts in the field, or they have special knowledge that the primary readers require to make a decision. They might be engineers, technicians, lawyers, scientists, doctors, accountants, and others to whom the primary readers will turn for advice.

TERTIARY READERS (EVALUATORS)   The tertiary readers include others who may have an interest in your document's information. They are often *evaluators* of you, your team, or your company. These readers might be local news reporters, lawyers, auditors, historians, politicians, community activists, environmentalists, or perhaps your company's competitors. Even if you don't expect your document to ever fall into these readers' hands, you should keep them in mind to avoid saying anything that could put you or your company at risk. Figure 3.3, for example, shows a memo in which the tertiary readers were not kept in mind.

Types of Readers

- Primary readers: Action takers
- Secondary readers: Advisors
- Tertiary readers: Evaluators
- Gatekeepers: Supervisors

AT A GLANCE

GATEKEEPERS (SUPERVISORS)   The gatekeepers are people who will need to look over your document before it is sent to the primary readers. Your most common gatekeeper is your immediate supervisor. In some cases, though, your company's lawyers, accountants, and others may need to sign off on the document before it is sent out.

Each of these four types of readers will look for different kinds of information. The primary readers are the most important, so their needs come first. Nevertheless, a well-written document also anticipates the needs of the secondary, tertiary, and gatekeeper readers.

Profiling Your Readers' Needs, Values, and Attitudes

Now that you have identified the readers of your document, you should develop profiles that describe their needs, values, and attitudes. Don't assume that your readers have the same needs, values, and attitudes as you do. Readers often have very different characteristics and expectations than the writers of a document.

As you begin considering your readers, think about some of the following issues:

- Readers' familiarity with the subject
- Readers' professional experience
- Readers' educational level
- Readers' reading and comprehension level
- Readers' skill level

With these reader characteristics in mind, you can begin viewing your document from their perspective. To help you deepen this perspective, a Reader Analysis Chart, like the one shown in Figure 3.4, can help you identify your readers' needs, values, and attitudes toward your text.

A Memo That Does Not Consider Tertiary Readers

Figure 3.3:
This memo was leaked to the press by someone at the U.S. Department of Homeland Security (DHS). In it, the DHS Deputy Director discusses the results of a survey of government employees. The survey's results reveal the incredibly low opinion that DHS's employees have of their department and its management. The author tries to spin the results while conceding the obvious. This memo caused some public embarrassment when it slipped out.

From: [DHS] Broadcast
Sent: Tuesday, January 30, 2007
Subject: MESSAGE FROM [DHS] DEPUTY SECRETARY (Michael) JACKSON: DHS FHC SURVEY RESULTS

Importance: High

January 30, 2007

Primary readers are identified here.

MEMORANDUM FOR ALL DHS EMPLOYEES

FROM: MICHAEL P. JACKSON

SUBJECT: Federal Human Capital Survey Results

The purpose and main point are stated up front.

The Office of Personnel Management (OPM) surveyed federal employees last summer about various measures of job satisfaction and agency performance, and the results will be released today. Over 10,400 DHS employees responded and, candidly, what you said shows that DHS is not where any of us wants to be.

The survey results will be posted on the OPM website (www.opm.gov) and our own DHS intranet, and we encourage you to review them in detail. In brief, of 36 peer federal agencies surveyed, DHS ranks as follows:

• 36th on the job satisfaction index
• 35th on the leadership and knowledge management index
• 36th on the results-oriented performance culture index
• 33rd on the talent management index

These results deliver a clear and jolting message from managers and line employees alike. On whole, it is not significantly changed since OPM's 2004 employee survey. Secretary Chertoff and I discussed these results with concern.

Initial details indicate that we get low marks in basic supervision, management and leadership. Some examples are:

This list summarizes the current situation.

• Promotion and pay increase based on merit
• Dealing with poor performance
• Rewarding creativity and innovation
• Leadership generating high levels of motivation in the workforce
• Recognition for doing a good job
• Lack of satisfaction with various component policies and procedures
• Lack of information about what is going on with the organization

I am writing to assure you that, starting at the top, the leadership team across DHS is committed to address the underlying reasons for DHS employee dissatisfaction and suggestions for improvement.

Standing up this new and vital Department is clearly not a walk in the park, but our employees bring a passion for this mission, great professionalism and outstanding performance every single day. DHS employees have shouldered the weight of long hours, complex integration assignments, multiple reorganizations, and no small amount of criticism. In some cases you've had to wait too long for tools you need to suceed.

These are not excuses to rationalize where we stand, rather an acknowledgement on my part of how much our team is doing. And there are good news items in the survey for DHS. As chief operating officer of DHS, I commit to improve results. We will need your help.

Several months ago, the Secretary asked the Homeland Security Advisory Council to study and suggest a strategy for creating a stronger common culture. This month, drawing on the experience of top executives in the private sector, the Council has delivered a set of recommendations for promoting a culture of excellence in DHS.

The author looks to the future.

In the days ahead, our Under Secretary for Management, Paul Schneider, will join the Secretary and me in evaluating carefully the details of the OPM survey and the HSAC report. Our first steps will be to analyze thoroughly the survey data, including specific attention to those government organizations that are recognized for their high performance in these areas, and determine the specific steps to improvement. This process wil include the leadership team in each operating component and every headquarters unit to discuss details of the survey with our workforce. We will do so with a sense of urgency and seriousness.

The memo returns to its main point.

Strengthening core management is one of the Secretary's highest priorities and the key elements are effective communications and proper recognition of our workforce. You deserve nothing less. We will build on some good work that has already been done to chart a path forward on these issues. We will then go where you point us, to improve job satisfaction for the DHS team.

Along the way, I will continue to ask for your help and guidance. Thanks in advance for that assistance, and thanks for what you are doing each day for DHS.

Source: Slate Magazine, http://www.slate.com/id2158997.

Want to see other controversial documents? Go to
www.pearsonhighered.com/johnsonweb3/3.4

Profiling Your Readers 43

Reader Analysis Chart

Readers	Needs	Values	Attitudes
Primary			
Secondary			
Tertiary			
Gatekeepers			

Figure 3.4: To better understand your readers, fill this Reader Analysis Chart with notes about their characteristics.

To use the Reader Analysis Chart, fill in what you know about your readers' needs, values, and attitudes.

NEEDS   What information do your primary readers require to make a decision or take action? What do the secondary readers need if they are going to make positive recommendations to the primary readers? What are the tertiary and gatekeeper readers looking for in your document?

VALUES   What do your readers value most? Do they value efficiency and consistency? Do they value accuracy? Is profit a key concern? How much do they value environmental or social concerns?

ATTITUDES   What are your readers' attitudes toward you, your company, and the subject of your document? Will your readers be excited, upset, wary, positive, hopeful, careful, concerned, skeptical, or gladdened by what you are telling them?

As you fill in the Reader Analysis Chart, you will be making strategic guesses about your readers. Put a question mark (?) in spaces where you aren't sure about your readers' needs, values, or attitudes. These question marks highlight where you need to do more research on your readers.

To find out more, you might interview people who are Subject Matter Experts (SMEs) at your company or who hire themselves out as consultants. These experts may be able to give you insights into your readers' likely characteristics.

Above all, your goal is to view the situation from your readers' perspective. Your profile will help you anticipate how your readers act, react, and make decisions.

GO TO THE NET

For a downloadable version of the Reader Analysis Chart, go to
www.pearsonhighered.com/johnsonweb3/3.5
What is the commonsense trap?
www.pearsonhighered.com/johnsonweb3/3.6

Profiling Readers with Search Engines

Search engines on the Internet are great tools for collecting information on just about any topic, including your readers and the contexts in which they will use your document.

The most popular search engines are Google, Ask, Yahoo, Lycos, HotBot, Excite, and AllTheWeb. Each of these search engines will allow you to type in keywords, a phrase, or even a whole question (Figure A). The search engine will find any websites with those words in them.

You can learn a great amount about your readers by typing in a few keywords. For example, let's say you are writing a proposal to the Salmon Recovery Program in Washington State. You can type in the phrase:

Salmon Recovery Program in Washington

The search engine will pull up thousands of pages that refer to this subject (Google found over 250,000 pages). The engine will usually rank them for you, trying to present you with the most relevant pages first.

Of course, you couldn't read that many pages in your lifetime. You need to narrow your search with some helpful symbols and strategies. For example, perhaps you

Internet Search Engine

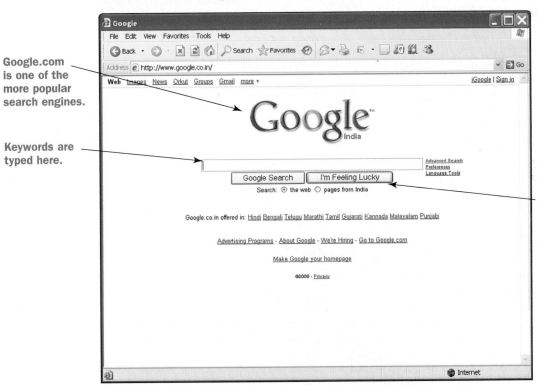

Google.com is one of the more popular search engines.

Keywords are typed here.

Figure A: Type in some keywords, and the search engine will look for information on that subject.

The "I'm Feeling Lucky" button takes you immediately to the most relevant website located by the search engine.

Source: Google, http://www.google.com. GOOGLE is a trademark of Google, Inc.

(continued)

To find websites that help you search for information about people, go to **www.pearsonhighered.com/johnsonweb3/3.7**

notice that Alice Guthrie is the director of the Salmon Recovery Program. So, you refine your search with plus and minus signs.

The + sign—Putting a + in front of words tells the search engine to find only pages that have those words in them.

> Salmon Recovery Program +Washington +Alice +Guthrie

Link

For more information on using search engines, see Chapter 7, page 149.

Here, the search engine will pull up only pages with "Alice" and "Guthrie" in them.

The – sign—Putting a – in front of words will eliminate pages with information you don't want.

> Salmon Recovery Program +Washington +Alice +Guthrie –restaurant

With this minus sign, you can eliminate any pages that refer to Arlo Guthrie's song "Alice's Restaurant."

Quotation marks—If you put a phrase in quotes, the search engine will look for that exact phrase:

> "Salmon Recovery Program" +Washington +Alice +Guthrie –restaurant

Wildcard symbols—Some search engines also have symbols for "wildcards." These symbols are helpful when you know most of a phrase, but not all of it.

> "Salmon ? Program" +Washington +Alice +Guthrie –restaurant

Different search engines use different wildcard symbols. Commonly used symbols include ?,*, and %.

It is amazing how much you can learn about your readers (and other topics) with search engines. These helpful tools will allow you to tailor your documents to the specific needs of your readers.

Profiling Contexts of Use

The places where people will read your document can strongly influence how they interpret what you say. So, you should also build a profile of the *contexts of use* in which they will read or use your document.

Identifying the Context of Use

Perhaps the most obvious concern is the *physical context* in which the document will be used. Will your readers be in their office or at a meeting? Will they be on the factory floor, trying to repair a robotic arm? Or, are they in the emergency room, trying to save someone's life? Each of these physical contexts will alter the way your readers interpret your document.

But context of use goes beyond your readers' physical context. Your readers may also be influenced by the economic, ethical, and political issues that shape how they see the world. To help you sort out these various contexts, you can use a Context Analysis Chart like the one shown in Figure 3.5.

Search engines are available at
www.pearsonhighered.com/johnsonweb3/3.8
Want more search engine tips? Go to
www.pearsonhighered.com/johnsonweb3/3.9

Context Analysis Chart

	Physical Context	Economic Context	Political Context	Ethical Context
Primary Readers				
Readers' Company				
Readers' Industry				

Figure 3.5: Each reader is influenced by physical, economic, political, and ethical concerns. A Context Analysis Chart anticipates these concerns for the primary readers, their company, and their industry.

AT A GLANCE

Context of Use

- *Physical context*—the places where the readers will use your document
- *Economic context*—the money-related issues that will restrain the kinds of actions possible
- *Political context*—the micropolitical and macropolitical trends that will guide your readers
- *Ethical context*—the personal, social, and environmental issues that shape the readers' responses

Here is how to use the Context Analysis Chart: Fill in what you know about the physical, economic, political, and ethical issues that might influence the primary readers, their company, and their industry.

PHYSICAL CONTEXT   Where will your readers use your document? How do these various places affect how they will read your document? How should you write and design the document to fit these places?

ECONOMIC CONTEXT   What are the economic issues that will influence your readers' decisions? What are the costs and benefits of your ideas? How would accepting your ideas change the financial situation of your readers, their company, or their industry?

POLITICAL CONTEXT   What are the political forces influencing you and your readers? On a micropolitical level, how will your ideas affect your readers' relationships with you, their supervisors, or their colleagues? On a macropolitical level, how will political trends at the local, state, federal, and international levels shape how your readers interpret your ideas?

Link

For more help
on identifying
ethical issues,
see Chapter 5,
page 92.

ETHICAL CONTEXT   How will your ideas affect the rights, values, and well-being of others? Does your document involve any social or environmental issues that might be of concern to your readers? Will any laws or rules be bent or broken if your readers do what you want?

Put a question mark (?) in spaces where you don't have specific information about your readers' physical, economic, political, and ethical contexts. You can then turn to the Internet for answers, or you can interview Subject Matter Experts who may have the answers you need.

Using Profiles to Your Own Advantage

You are now ready to use your Reader Analysis and Context Analysis charts to strengthen your writing and make it more informed and persuasive. In your charts, circle or highlight the most important terms, concepts, and phrases. The items you circle are the *tensions* that you will need to address as you collect information and draft the document.

As you draft your document, your analysis of readers and contexts of use will help you

- make strategic decisions about what information to include in your document. Readers don't need all the information you have available. They need only enough to make a decision or take action.
- organize your document to highlight the information that is most important to your readers. Information that your readers consider crucial should be moved up front where they can find it.
- develop a persuasive style that will appeal to your readers. Your notes about their values and attitudes can be used to shape the document to their biases and beliefs.
- design the document for the places it will be used. Your understanding of the context in which the document will be used will help you create the maximum visual impact.

For important documents, this deeper understanding will give your writing extra potency and legitimacy. Readers will appreciate your efforts to see the situation from their point of view.

Figures 3.6 and 3.7 show documents from the same website about the same topic, West Nile virus, that are written to two different types of readers. The first document, Figure 3.6, is written to the general public. Notice how it uses content, organization, style, and design to appeal to this audience.

The second document, Figure 3.7, is written to medical personnel. Notice how the content is far more complex, and the style is less personal in the document for medical personnel. Effective reader analysis and context analysis allowed the author of these documents (probably the same person) to effectively present the same information to two very different kinds of readers.

48   Chapter 3
**Readers and
Contexts of Use**

Want more help with reader analysis? Go to
www.pearsonhighered.com/johnsonweb3/3.11

Document Written to the General Public

Style is action oriented.

Images reinforce written text.

Headings help readers scan document.

Common examples are used to illustrate points.

Fight the Bite!
Avoid Mosquito Bites to Avoid Infection

CDC
CENTERS FOR DISEASE
CONTROL AND PREVENTION

Human illness from West Nile virus is rare, even in areas where the virus has been reported. The chance that any one person is going to become ill from a mosquito bite is low.

Three Ways to Reduce Your West Nile Virus Risk

1. Avoid Mosquito Bites
2. Mosquito-Proof Your Home
3. Help Your Community

Avoid Mosquito Bites!

Apply Insect Repellent Containing DEET (Look for: *N,N-diethyl-meta-toluamide*) to exposed skin when you go outdoors. Even a short time sitting outdoors can be long enough to get a mosquito bite. For details on when and how to apply repellent, see CDC's Insect Repellent Use and Safety page: http://www.cdc.gov/ncidod/dvbid/westnile/qa/insect_repellent.htm

Clothing Can Help Reduce Mosquito Bites. When possible, wear long-sleeves, long pants and socks when outdoors. Mosquitoes may bite through thin clothing, so spraying clothes with repellent containing permethrin or DEET will give extra protection. Don't apply repellents containing permethrin directly to skin. Do not spray repellent containing DEET on the skin under your clothing.

Be Aware of Peak Mosquito Hours. The hours from dusk to dawn are peak mosquito biting times. Consider avoiding outdoor activities during these times -- or take extra care to use repellent and protective clothing during evening and early morning.

Mosquito-Proof Your Home

Drain Standing Water. Mosquitoes lay their eggs in standing water. Limit the number of places around your home for mosquitoes to breed by getting rid of items that hold water. Need examples? Learn more on the Prevention of West Nile Virus question and answer page: http://www.cdc.gov/ncidod/dvbid/westnile/qa/prevention.htm

Install or Repair Screens. Some mosquitoes like to come indoors. Keep them outside by having well-fitting screens on both windows and doors. Offer to help neighbors whose screens might be in bad shape.

Help Your Community

Clean Up. Mosquito breeding sites can be everywhere. Neighborhood clean up days can be organized by civic or youth organizations to pick up containers from vacant lots, parks and to encourage people to keep thier yards free of standing water. Mosquitoes don't care about fences, so breeding sites anywhere in the neighborhood are important.

Source: Centers for Disease Control and Prevention (CDC), http://www.cdc.gov/ncidod/dvbid/westnile/index.htm.

(continued)

Figure 3.6:
This document on West Nile virus was written to the general public. It is action oriented and not very technical. The images also help reinforce the message, and the layout makes it highly scannable.

Report Dead Birds to Local Authorities. Dead birds may be a sign that West Nile virus is circulating between birds and the mosquitoes in an area. Over 110 species of birds are known to have been infected with West Nile virus, though not all infected birds will die.

By reporting dead birds to state and local health departments, the public plays an important role in monitoring West Nile virus. Because state and local agencies have different policies for collecting and testing birds check the Links to State and Local Government Sites page to find information about reporting dead birds in your area: http://www.cdc.gov/ncidod/dvbid/westnile/city_states.htm . This page contains more information about reporting dead birds and dealing with bird carcasses: http://www.cdc.gov/ncidod/dvbid/westnile/qa/wnv_birds.htm

Mosquito Control Programs. Check with local health authorities to see if there is an organized mosquito control program in your area. If no program exists, work with your local government officials to establish a program. The American Mosquito Control Association (www.mosquito.org) can provide advice, and their book Organization for Mosquito Control is a useful reference. More questions about mosquito control? A source for information about pesticides and repellents is the National Pesticide Information Center: http://npic.orst.edu/ , which also operates a toll-free information line: 1-800-858-7378 (check their Web site for hours).

Find out more about local prevention efforts. Find state and local West Nile virus information and contacts on the Links to State and Local Government Sites page.

Links provide
more
information.

Document Written to Experts

Centers for Disease Control and Prevention

West Nile Virus (WNV) Infection
Information for Clinicians

SAFER·HEALTHIER·PEOPLE™

Language is much more technical (and precise).

Clinical Features

Mild Infection

Most WNV infections are mild and often clinically unapparent.
- Approximately 20% of those infected develop a mild illness (West Nile fever).
- The incubation period is thought to range from 3 to 14 days.
- Symptoms generally last 3 to 6 days.

Reports from earlier outbreaks describe the mild form of WNV infection as a <u>febrile illness of sudden onset</u> often accompanied by

- malaise
- anorexia
- nausea
- vomiting
- eye pain

- headache
- myalgia
- rash
- lymphadenopathy

The full clinical spectrum of West Nile fever has not been determined in the United States.

Severe Infection

Approximately 1 in 150 infections will result in severe neurological disease.
- The most significant risk factor for developing severe neurological disease is advanced age.
- Encephalitis is more commonly reported than meningitis.

In recent outbreaks, symptoms occurring among patients hospitalized with severe disease include

- fever
- weakness

- gastrointestinal symptoms
- change in mental status

Document is more focused on providing information than encouraging action.

- A minority of patients with severe disease developed a maculopapular or morbilliform rash involving the neck, trunk, arms, or legs.
- Several patients experienced severe muscle weakness and flaccid paralysis.
- Neurological presentations included

- ataxia and extrapyramidal signs
- cranial nerve abnormalities
- myelitis

- optic neuritis
- polyradiculitis
- seizures

Although not observed in recent outbreaks, myocarditis, pancreatitis, and fulminant hepatitis have been described.

Clinical Suspicion

Diagnosis of WNV infection is based on a high index of clinical suspicion and obtaining specific laboratory tests.

Design is not as scannable.

- WNV, or other arboviral diseases such as St. Louis encephalitis, should be strongly considered in adults ≥50 years who develop unexplained encephalitis or meningitis in summer or early fall.
- The local presence of WNV enzootic activity or other human cases should further raise suspicion.
- Obtaining a recent travel history is also important.

Note: Severe neurological disease due to WNV infection has occurred in patients of all ages. Year-round transmission is possible in some areas. Therefore, WNV should be considered in all persons with unexplained encephalitis and meningitis.

Diagnosis and Reporting

Procedures for submitting diagnostic samples and reporting persons with suspected WNV infection vary among states and jurisdictions. Links to state and local websites are available at
http://www.cdc.gov/ncidod/dvbid/westnile/city_states.htm

Source: Centers for Disease Control and Prevention (CDC), http://www.cdc.gov/ncidod/ dvbid/westnile/index.htm.

(continued)

Figure 3.7: This document was written to clinicians and other medical personnel. It is far more technical in style and design than the document written to the general public. Nevertheless, it contains much of the same information.

Diagnosis and Reporting – *continued*

Diagnostic Testing

WNV testing for patients with encephalitis or meningitis can be obtained through local or state health departments.

- o The most efficient diagnostic method is detection of IgM antibody to WNV in serum or cerebral spinal fluid (CSF) collected within 8 days of illness onset using the IgM antibody capture enzyme-linked immunosorbent assay (MAC-ELISA).
- o Since IgM antibody does not cross the blood-brain barrier, IgM antibody in CSF strongly suggests central nervous system infection.
- o Patients who have been recently vaccinated against or recently infected with related flaviviruses (e.g., yellow fever, Japanese encephalitis, dengue) may have positive WNV MAC-ELISA results.

Reporting Suspected WNV Infection

Refer to local and state health department reporting requirements:
www.cdc.gov/ncidod/dvbid/westnile/city_states.htm

- o WNV encephalitis is on the list of designated nationally notifiable arboviral encephalitides.
- o Aseptic meningitis is reportable in some jurisdictions.

The timely identification of persons with acute WNV or other arboviral infection may have significant public health implications and will likely augment the public health response to reduce the risk of additional human infections.

Laboratory Findings

Among patients in recent outbreaks

- o Total leukocyte counts in peripheral blood were mostly normal or elevated, with lymphocytopenia and anemia also occurring.
- o Hyponatremia was sometimes present, particularly among patients with encephalitis.
- o Examination of the cerebrospinal fluid (CSF) showed pleocytosis, usually with a predominance of lymphocytes.
- o Protein was universally elevated.
- o Glucose was normal.
- o Computed tomographic scans of the brain mostly did not show evidence of acute disease, but in about one-third of patients, magnetic resonance imaging showed enhancement of the leptomeninges, the periventricular areas, or both.

Treatment

Treatment is supportive, often involving hospitalization, intravenous fluids, respiratory support, and prevention of secondary infections for patients with severe disease.

- o Ribavirin in high doses and interferon alpha-2b were found to have some activity against WNV in vitro, but no controlled studies have been completed on the use of these or other medications, including steroids, antiseizure drugs, or osmotic agents, in the management of WNV encephalitis.

For additional clinical information, please refer to Petersen LR and Marfin AA, "West Nile Virus: A Primer for the Clinician[Review]," Annals of Internal Medicine (August 6) 2002: 137:173-9.

For clinical and laboratory case definitions, see "Epidemic/Epizootic West Nile Virus in the United States: Revised Guidelines for Surveillance, Prevention, and Control, 2001," at
www.cdc.gov/ncidod/dvbid/westnile/surv&control.htm

Links provide additional information.

Summaries are used to present important information.

Places where more information can be found are identified.

International and Cross-Cultural Communication

Computers have greatly blurred geographical and political boundaries. Whether you are developing software documentation or describing a heart transplant procedure, your documents will be read and used by people from different cultures. The use of computers, especially the Internet, has only heightened the necessity of working and communicating with people from different cultures (Hoft, 1995; Reynolds & Valentine, 2004). It's all very exciting—and very challenging.

International and cross-cultural issues will affect the content, organization, style, and design of your document.

Differences in Content

Cultures have different expectations about content in technical documentation:

- In China, the content of your documents and presentations should be fact based, and you should focus on long-term benefits for your readers and you, not short-term gains. In business, the Chinese tend to trust relationships above all, so they look for facts in documents and they do not like overt attempts to persuade.
- In Mexico, South America, and many African countries, family and personal backgrounds are of great importance. It is common for family-related issues to be mentioned in public relations, advertising, and documentation. Business relationships and meetings often start with exchanges about families and personal interests.
- In the Middle East, Arabs often put a premium on negotiation and bargaining, especially when it comes to the price of a service or product. As a result, it is crucial that all the details in documents are spelled out exactly before the two sides try to work out a deal. In most cases, though, the first offer you make will rarely be considered the final offer.

Link

To learn more about working in high-context cultures, go to Chapter 6, page 134.

- In Asian countries, the reputation of the writer or company is essential for establishing the credibility of the information (Haneda & Shima, 1983). Interpersonal relationships and prior experiences can sometimes even trump empirical evidence in Asia.
- Also in Asia, contextual cues can be more important than content. In other words, *how* someone says something may be more important than *what* he or she is saying. For example, when Japanese people speak or write in their own language, they rarely use the word *no*. Instead, they rely on contextual cues to signal the refusal. As a result, when Japanese is translated into English, these "high-context" linguistic strategies are often misunderstood (Chaney & Martin, 2004). Similarly in Indonesia, the phrase "Yes, but" actually means "no" when someone is speaking.
- In India, business is often conducted in English because the nation has over a dozen major languages and hundreds of minor languages. So, don't be surprised when your Indian partners are very fluent in English and expect you to show a high level of fluency, especially if you are a native English speaker.
- In several African countries like Tunisia and Morocco, business tends to be conducted in French, even though the official language of the country is Arabic.

Differences in Organization

The organization of a document often needs to be altered to suit an international audience. Organizational structures that Americans perceive to be "logical" or "common sense" can seem confusing and even rude in some cultures.

- In Arab cultures, documents and meetings often start out with statements of appreciation and attempts to build common bonds among people. The American tendency to "get to the point" is often seen as rude.
- Also in Arabic cultures, documents rely on repetition to make their points. To North Americans, this repetition might seem like the document is moving one step back for every two steps forward. To Arabs, American documents often seem incomplete because they lack this repetition.
- Asians often prefer to start out with contextual information about nonbusiness issues. For example, it is common for Japanese writers to start out letters by saying something about the weather. To some Asians, American documents seem abrupt, because Americans tend to bluntly highlight goals and objectives up front.
- In India, the term *thank you* is considered a form of payment. So, if someone has done you a favor, you should not say thank you in the introduction or conclusion of an e-mail or letter. Saying "thank you" suggests you are paying that person for the favor.

Differences in Style

Beyond difficulties with translation, style is usually an important difference among cultures:

- In China, overt attempts at persuasion are often seen as rude and undesirable. Instead, documents and meetings should be used to build relationships and present factual information. Strong relationships lead to good business, not the other way around.

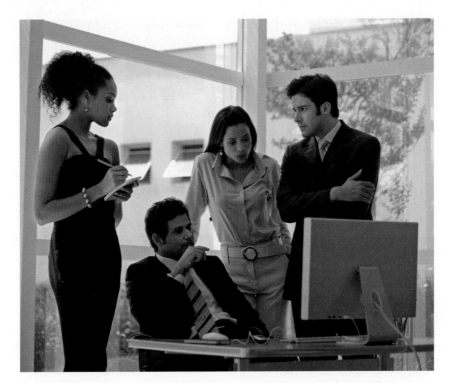

Computers, especially networked computers, have increased opportunities to work across cultures.

- Arabic style may seem overly ornamental to North American tastes, making Arabic documents and presentations seem colorful to non-Arabs. On the other hand, the American reliance on "plain language" can rub against the sensibilities of Arabs, who prefer a more ornate style in formal documents.
- In Mexico and much of South America, an informal style often suggests a lack of respect for the project, the product, or the readers. Mexicans especially value formality in business settings, so the use of first names and contractions in business prose can be offensive.
- In sub-Saharan Africa, readers prefer a document's tone to stress a win-win situation. Your tone, therefore, should imply that both sides will benefit from the arrangement.
- Some Native Americans prefer the sense that everyone had input on the document. Therefore, a direct writing or presentation style will meet resistance because it will seem to represent the opinion of only one person.
- In North America, women are more direct than women in other parts of the world, including Europe. This directness often works to their advantage in other countries, because they are viewed as confident and forward thinking. However, as writers and speakers, women should not be too surprised when people of other cultures resist their directness.

GO TO THE NET Want to learn more about international readers? Go to www.pearsonhighered.com/johnsonweb3/3.12

Carol Leininger, Ph.D.

COMMUNICATIONS MANAGER, F. HOFFMAN-LA ROCHE, SWISS HEADQUARTERS

F. Hoffman-La Roche is a pharmaceutical company that works closely with a partner company, Chugai Pharmaceuticals, in Tokyo, Japan.

What are some strategies for communicating with people from another culture?

Physical distance may be the biggest hurdle in global communication. I believe that distance affects how people work together more than does language or culture. When people are working face to face, they tend to figure out issues by interacting with each other, even if they are not fluent in the same languages. There are awkward moments, but that's to be expected.

Mostly, solid *preparation* is the key to working across distances and languages. Communication needs to be structured and simple. Your readers may be reading your text in their second- or third-best language, so make their job easier by writing as simply and clearly as you can.

All the rules for good technical communication in the United States apply to international communication—only more so. What helps second-language or non-U.S. English speakers?

- State your objectives and purpose clearly.
- Use language consistently (i.e., the same terms for the same things).
- Do not attempt humor until a relationship has been established.
- Rank issues by importance.
- Handle only one message per e-mail or paragraph.
- Use headings and subheadings that convey a specific meaning.
- Minimize use of adjectives and adverbs.
- Minimize prepositional phrases.
- Highlight actions, deadlines, and dates.
- Spell everything correctly (always check the spelling).

Always be as polite as you can by your own cultural standards (e.g., formal language, politeness markers like "please" and "thank you," use of full names in greeting and salutation). Even if your cultural view of what is "polite" is different from that of your audience, your intention to be polite will be recognized by international readers as courtesy and civility.

Differences in Design

Even the design of documents is important when you are working with international and cross-cultural readers:

- Arabic and some Chinese scripts are read from right to left, unlike English, which is read from left to right (Figure 3.8). As a result, Arabic and some Chinese readers tend to scan pages and images differently than do Americans or Europeans.

For information sources on other countries, go to
www.pearsonhighered.com/johnsonweb3/3.13

Different Ways of Scanning a Page

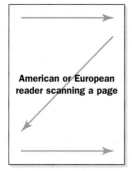

American or European reader scanning a page

Arabic or Chinese reader scanning a page

Figure 3.8: Readers from other cultures may scan the design differently. The design needs to take their preferences into account.

- Some icons that show hand gestures—like the OK sign, a pointing finger, or a peace sign with the back of the hand facing outward—can be highly offensive in some cultures. Imagine a document in which a hand with the middle finger extended is used to point at things. You get the picture.
- In many South American and Asian cultures, the use of the right hand is preferred when handing items (e.g., business cards, documents, products) to people. Therefore, pictures or drawings in documents should show people using their right hands to interact.
- In some Asian cultures a white flower or white dress can symbolize death. As a result, a photograph of white flowers or white dresses can signal a funeral or mourning.
- Europeans find that American texts include too many graphics and use too much white space. Americans, meanwhile, often find that the small margins in European texts make the documents look crowded and cramped.
- Graphs and charts that seem to have obvious meanings to Americans can be baffling and confusing to readers from other cultures. If your international document includes graphs and charts, you should seek out someone from the readers' culture to help you determine whether your visuals will be understood.
- When giving presentations in some Native American cultures, hand gestures should be limited and eye contact should be minimized. Ironically, this advice is exactly the opposite of what most public speaking coaches suggest.

Listen and Learn: The Key to International and Cross-Cultural Communication

With all these differences in content, organization, style, and design, how can you possibly write for international or cross-cultural readers? Here are four helpful strategies:

LISTEN CAREFULLY   A good strategy is to listen carefully to your readers' expectations. Careful listening is a valued quality in all cultures, and you will learn

Want to learn more about Native American cultures? Go to www.pearsonhighered.com/johnsonweb3/3.14

International and Cross-Cultural Communication

57

a great amount by simply paying attention to what your readers expect the document to include and how it should look.

BE POLITE   Politeness in one culture tends to translate well into other cultures. For example, words like *please* and *thank you* are universally seen as polite. Smiles and a friendly tone are almost always welcome. There are subtle differences in how these words and gestures are used in other cultures, but your readers will understand that you are trying to be polite.

RESEARCH THE TARGET CULTURE   Use the Internet to do some research into your readers' cultural expectations for technical documents. On the Internet or at your workplace, you might also find some model texts from the readers' culture. Use them to help guide your decisions about content, organization, style, and design.

TALK TO YOUR COLLEAGUES   You may also seek out coworkers or colleagues who are from the target culture or who have lived there. You can ask them about conventions that might make your document or presentation more effective. They can also help you avoid doing anything awkward or offensive.

Overall, when you are communicating with international readers or people from different cultures, be observant and listen to what they tell you. Do some research into their expectations, and be ready to learn from your mistakes.

CHAPTER REVIEW

- You can use your computer as a reader analysis tool to better tailor your document's content, organization, style, and design to the needs of its readers.

- Early in the writing process, you should begin developing a profile of the types of people who may be interested in your document.

- Your readers will include *primary readers* (action takers), *secondary readers* (advisors), *tertiary readers* (evaluators), and *gatekeepers* (supervisors).

- In your documents and presentations, you should anticipate various readers' needs, values, and attitudes.

- You should anticipate the document's contexts of use, which include the physical, economic, political, and ethical factors that may influence a reader's ideas.

- The emergence of the Internet has heightened the importance of international and cross-cultural communication. You need to adjust the content, organization, style, and design of your text to be sensitive to cross-cultural needs.

GO TO THE NET

For more about using graphics in other cultures, go to
www.pearsonhighered.com/johnsonweb3/3.15

Individual or Team Projects

1. Choose two websites that are designed for very different types of readers. Write a memo to your instructor in which you compare and contrast the websites, showing how they approach their readers differently. How do they use content, organization, style, and design to meet the needs, values, and attitudes of their readers?

   Some pairs of websites you might consider include websites for cars (chevrolet.com versus honda.com), magazines (time.com versus outsidemag.com), or computers (dell.com versus apple.com). Look for websites for products that are similar but pursue different kinds of customers.

2. Consider the advertisement in Figure 3.9 and "reverse-engineer" its reader analysis. Using a Writer-Centered Analysis Chart and a Reader Analysis Chart, identify the primary, secondary, and tertiary readers of the text. Then, make guesses about the needs, values, and attitudes of these readers.

   Write a report to your instructor in which you use your charts to discuss the readers of this document. Then, show how the document anticipates these readers' needs.

Figure 3.9: Who are the intended readers of this advertisement? What are their needs, values, and attitudes?

3. For a document you are writing, conduct a thorough reader analysis. Start out by identifying the primary, secondary, tertiary, and gatekeeper readers. Then, identify these readers' needs, values, and attitudes. And finally, identify the physical, economic, political, and ethical issues that may influence how your readers interpret your document.

   Give a presentation about your readers to your class. Discuss how various readers in various contexts will require you to adjust the content, organization, style, and design of your document.

4. Choose a country or culture that interests you. Then, find two texts written by people from that country or culture. Write a memo to your instructor in which you discuss any similarities or differences between the texts and your own expectations for texts. Pay close attention to differences in content, organization, style, and design of these texts.

Collaborative Project

With a group of people from your class, create a website that explores the needs, values, and attitudes of people from a different country or culture. The website does not need to be complex. On the Internet, identify various websites that offer information on that country or culture. Then, organize those websites by content and create links to them. Specifically, pay attention to the ways in which this country's physical, economic, political, and ethical contexts shape the way its people live their lives.

When you are finished with the website, give a presentation to your class in which your group discusses how this country or culture differs from your own. Answer the following question: If you were going to offer a product or service to the people of this country or culture, what considerations would you need to keep in mind? If you needed to write a proposal or a set of instructions to people from this country or culture, how might you need to adjust it to fit their unique qualities?

For additional technical writing resources, including interactive sample documents, document design tutorials and guidelines, and more, go to **www.mytechcommlab.com**.

Installing a Medical Waste Incinerator

Duane Jackson knew this decision was going to be difficult. As the assistant city engineer for Dover City, he was frequently asked to study construction proposals sent to the city council. He would then write a report with a recommendation. So, when the proposal for constructing a medical waste incinerator crossed his desk, he knew there was going to be trouble.

Overall, the proposal from Valley Medical, Inc., looked solid. The incinerator would be within 3 miles of the two major hospitals and a biotech research facility. And, it would bring about 30 good jobs to the Blue Park neighborhood, an economically depressed part of town.

The problem was that people in Blue Park were going to be skeptical. Duane grew up in a neighborhood like Blue Park, primarily African-American and lower middle class. He knew that hazardous industries often put their operations in these kinds of neighborhoods because the locals did not have the financial resources or political clout to fight them. In the past, companies had taken advantage of these neighborhoods' political weaknesses, leaving the area polluted and unhealthy.

Powerful interests were weighing in on this issue. Dover City's mayor wanted the incinerator badly because she wanted the economic boost the new business would provide. Certainly, the executives at the hospitals and research laboratory were enthusiastic, because a nearby incinerator would help them cut costs. The city councilor who represented Blue Park wanted the jobs, but not at the expense of his constituents' health. Environmental groups, health advocates, and neighborhood associations were cautious about the incinerator.

Analyzing the Readers

After a few weeks of intense study, Duane's research convinced him that the incinerator was not a health hazard to the people of Blue Park. Similar incinerators built by Valley Medical had spotless records. Emissions would be minimal because advanced "scrubbers" would remove almost all the particles left over after incineration. The scrubbers were very advanced, almost completely removing only harmful pollutants such as dioxin and mercury.

Also, the company had a good plan for ensuring that medical waste would not sit around in trucks or containers, waiting to be burned. The waste would be immediately incinerated on arrival.

(continued)

Duane's Writer-Centered Analysis Chart

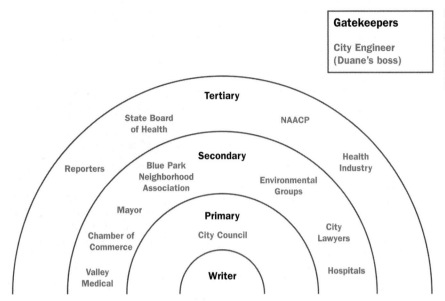

Figure 3.10:
Duane's Writer-
Centered Analysis
Chart showed
how many readers
his report would
have.

Duane decided to write a report to the city council that recommended the incinerator be built. That decision was the easy part. Now he needed to write a report that would convince the skeptics.

After identifying the subject and purpose of the report, Duane decided to do a thorough analysis of his readers and the report's contexts of use. He began with a Writer-Centered Analysis Chart (Figure 3.10). He then used a Reader Analysis Chart to identify the various readers' needs, values, and attitudes (Figure 3.11). Finally, Duane filled out a Context Analysis Chart to identify the physical, economic, political, and ethical issues involved (Figure 3.12).

Duane's Reader Analysis Chart

	Needs	Values	Attitudes
Primary • City Council	Reliable information Environmental impact data Clear recommendation Impartial commentary	Citizen safety Economic development Fairness	Optimistic Cautious
Secondary • Valley Medical • Chamber of Commerce • Mayor • Neighborhood Association • Environmental Groups • City Lawyers • Hospitals	Impartial commentary Specific facts about emissions Cultural and social considerations Valley Medical wants profits	Maintaining character of Blue Park Economic development Environmental safety Safe disposal of waste	Mayor and hospitals are positive and hopeful Neighborhood association and environmental groups are skeptical, perhaps resistant Valley Medical hopeful
Tertiary • Reporters • State Board of Health • NAACP • Health Industry	Impartial decision Reassurance that race or poverty are not factors Basic facts about incinerator	Fairness and honesty Lack of bias Protection of people with little political power	Skeptical Open-minded
Gatekeepers • City Engineer	Reliable information Clear decision based on reliable data	Minimal trouble with mayor and council Low profile; stay out of politics Honesty	Impartial to project Concerned that report may cause tensions

Figure 3.11: Duane filled in a Reader Analysis Chart, noting everything he knew about the various readers of his report. He noticed that most readers wanted impartial and reliable information. Some readers were positive; others were skeptical.

(continued)

For more information about waste incinerators, go to
www.pearsonhighered.com/johnsonweb3/3.17

Case Study   63

Duane's Context Analysis Chart

	Physical Context	Economic Context	Political Context	Ethical Context
Primary Readers • City Council	Initially at their office Further in city council meeting	Looking to improve city economics	Re-election is always an issue Mayor wants it BP city councilor would like it	Exploiting poor neighborhood? Racial issues? Environmental issues
Readers' Company • Dover City	City Hall Engineers' office City website	The jobs are a financial plus. Home values in neighborhood decline? Good for hospitals	Voters are wary Neighborhood and environmental groups draw attention Racial politics?	City liable if mistake? Environmental impact Infringement of people's rights? Legal issues
Readers' Industry • City government	Reports on Internet about project	Job creation Economic growth	Pressure on mayor Shows Dover City is positive about this kind of business	Public relations Don't want to seem exploitative

Figure 3.12: Duane's Context Analysis Chart revealed some interesting tensions between the economics and politics of the decision on the incinerator.

What Should Duane Do Now?

On the facts alone, Duane was convinced that the incinerator would be well placed in the Blue Park neighborhood. But his reader analysis charts showed that facts alone were not going to win over his primary audience, the members of the city council. They would have numerous other economic, political, and ethical issues to consider as well as the facts. In many ways, these factors were more important than the empirical evidence for the incinerator.

One important thing Duane noticed was that the city council members, though the "action takers," were heavily influenced by the secondary readers. These secondary readers, the "advisors," would play a large role in the council's decision.

How can Duane adjust the content, organization, style, and design to better write his report? If you were Duane's readers, what kinds of information would you expect in this kind of report?

Working in Teams

In this chapter, you will learn:

- Why working in teams is essential in technical workplaces.

- The four stages of teaming in the workplace: forming, storming, norming, performing.

- How to use strategic planning to form a team and begin a project.

- Strategies for managing team conflict in the storming stage.

- How to define team roles in the norming stage to improve productivity.

- How to improve performance with Total Quality Management (TQM) strategies.

- How to work as part of a "virtual" team.

Working in teams is an everyday experience in the technical workplace. In fact, when managers are surveyed about the abilities they look for in new employees, they often put "works well with a team" near the top of their list. The ability to collaborate with others is an essential skill if you are going to succeed in today's networked workplace.

Working with a team has several advantages. Teams allow people to:

Concentrate strengths—Teams can divide responsibilities in ways that take advantage of each team member's strengths and abilities.

Foster creativity—Teams can creatively brainstorm and solve problems by drawing from each member's unique perspective and knowledge base.

Share the workload—Teams can divide up the workload, making large projects possible.

Improve morale—People generally enjoy working with others, so the shared responsibilities and rewards boost morale.

Of course, there are also disadvantages to working in teams. Conflicts with team members can be frustrating. Poor planning and time mismanagement can be annoying and demoralizing. Collaboration can lead to uneven-sounding documents. Sometimes other team members don't do their share of the work. These disadvantages can be avoided with good planning and effective communication among team members.

Computers have only increased the ability and necessity to work in teams. Communication tools like e-mail, instant messaging, chat rooms, and websites allow people to work together electronically. Telecommuting, teleworking, and "virtual offices" are becoming more common, and people are increasingly finding themselves working outside the traditional office setting. Now, more than ever, it is essential that you learn how to use computers to help you work effectively with a team of others.

The Stages of Teaming

It would be nice if people worked well together from the start. But, in reality, team members often need time to set goals and adjust to each other's working styles and abilities. In 1965, Bruce Tuckman introduced a model of how teams learn to work together (Figure 4.1). He pointed out that teams go through four stages:

Forming—defining the mission, setting objectives, defining responsibilities, establishing a project schedule with deadlines

Storming—handling disagreements, sensing tension and anxiety, doubting the leadership, addressing conflict, feeling uncertain and frustrated

Norming—forming consensus, refining the team's objectives and outcomes, solidifying team roles, focusing group members around the mission and objectives

Performing—sharing a vision, delegating tasks, feeling a high degree of autonomy, resolving conflicts and differences constructively

During the *forming* and *storming* stages, a team will be reliant on the team leader or manager for guidance and negotiation. As the team reaches the *norming* and *performing* stages, members will become more autonomous, because they will better understand the project's goals and how each person fits into the overall project mission.

To learn more about Tuckman's teaming theories, go to
www.pearsonhighered.com/johnsonweb3/4.1

Tuckman's Four Stages of Teaming

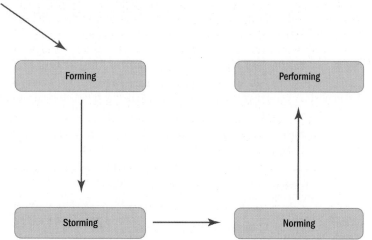

Figure 4.1: A team will typically go through four stages: forming, storming, norming, and performing. Give your team time to properly evolve as a unit.

These stages are not rigid. Instead, a team tends to move back and forth among the stages as the project evolves and moves forward. For example, a team that is performing might need to return to the norming stage if the project's objectives change or someone joins or leaves the team. Occasionally, norming and performing teams return to the storming stage when personalities or goals come into conflict.

You should keep the four stages in mind as you work with your team. Here are a few hints about working well in a team:

- Give your team time to properly form at the beginning of a project by setting objectives, defining responsibilities, and establishing a work plan.
- When storming, don't become too upset or frustrated, because a period of conflict is normal soon after a team is formed.
- When norming, take time to revise team expectations and recognize each member's role in the team, as well as strengths and weaknesses.
- When performing, look for ways to improve quality and refine your team's ability to work at a high level.

By keeping these four stages in mind, you can gain a broader understanding of your team's progress and your role within the team's mission.

Forming: Strategic Planning

Forming is an important part of the team-building process. When a team is first created, the members are usually excited and optimistic about the new project. They are often a little anxious, because each person is uncertain about the others' expectations. So, in the forming stage, members should spend time getting to know each other and assessing each other's strengths and abilities.

When forming a new team, strategic planning is the key to effective teamwork. By working through the following steps at the beginning of a project, you will give your team time to form properly, often saving yourselves time and frustration as the project moves forward.

Step 1: Define the Project Mission and Objectives

Don't just rush in. Take some time to first determine what your team is being asked to accomplish. A good way to start forming is to define the rhetorical situation in which your team is working:

Subject—What are we being asked to develop? What are the boundaries of our project? What are we *not* being asked to do?

Purpose (mission statement)—What is the mission of the project? Why are we being asked to do this? What are the end results (deliverables) that we are being asked to produce?

Readers—Who are our clients? What are their needs, values, and attitudes? Who will be evaluating our work?

Context—What are the physical, economic, political, and ethical factors that will influence this project? How should we adjust to them?

Link

For more help defining the rhetorical situation, turn to Chapter 2, page 20.

Your statement of the purpose is your *mission statement* for the project. You and your team members should first agree on this mission statement before you do anything else. Your mission statement (purpose) is your primary objective. Now, list two to five secondary objectives that you intend to reach along the way.

Our Mission:

The purpose of this Staph Infection Task Force is to determine the level of staph infection vulnerability at St. Thomas Medical Center and to develop strategies for limiting our patients' exposure to staph, especially MRSA.

Secondary Objectives:

• Gain a better understanding of current research on staph and its treatments.

• Raise awareness of staph infections among our medical staff and patients.

• Assess the level of staph risk, especially MRSA, here at the hospital.

• Develop methods for controlling staph on hospital surfaces.

By first agreeing on your team's mission and objectives, you can clarify the goals of the project. You can also avoid misunderstandings about what the team was formed to accomplish.

Step 2: Identify Project Outcomes

Outcomes are the tangible results of your project. Your project outcomes describe the measurable results of the team's efforts. To identify the outcomes of your project, simply convert your objectives into measurable results (Figure 4.2).

For websites that describe other planning strategies, go to
www.pearsonhighered.com/johnsonweb3/4.2

Defining the Project Objectives and Outcomes

Objectives	Outcomes
Gain a better understanding of current research on staph and its treatments	Outcome: Collection of current data and research literature on staph infections. Interview other hospitals about successful control methods. **Deliverable: A report on the findings, due May 25**
Raise awareness of staph infections among our medical staff and patients	Outcome: More awareness of the problem here at the hospital. People taking precautions against staph. **Deliverable: Pamphlets and posters raising awareness and describing staph control procedures**
Assess the level of staph risk, especially MRSA, here at the hospital	Outcome: Experiments that quantify the staph risk at the hospital **Deliverable: Report to the administration about the extent of the problem, due June 2**
Develop methods for controlling staph on hospital surfaces	Outcome: Develop strategies for preventing and containing staph infections. **Deliverable: Create a contingency plan that offers concrete steps for controlling staph. Make training modules to educate staff about the problem.**

Figure 4.2:
By transforming your objectives into outcomes and deliverables, you can show how abstract goals become measurable results.

Then specify the *deliverables* that the project will produce. Deliverables are the products or services that you will deliver to the client during the project and after it is completed.

This process of turning objectives into outcomes and deliverables can take some time. But, in the end, the process will save you time, because everyone in your team will have a clear understanding of the expected results and the products that will be created.

Step 3: Define Team Member Responsibilities

Everyone isn't good at everything. So, once you have defined your objectives, ask each member of the team to talk about his or her abilities and previous experiences. Ask team members to identify ways they could contribute to the project. Discuss any time limitations or potential conflicts with other projects.

If your team's project involves writing a document, you should identify each team member's responsibilities while dividing up the writing task. For example, here are four jobs that you might consider:

> **Coordinator**—The coordinator is responsible for maintaining the project schedule and running the meetings. The coordinator is not the "boss." Rather, he or she is a facilitator who helps keep the project on track.

To see other sample mission statements, go to
www.pearsonhighered.com/johnsonweb3/4.3

Researchers—One or two people in the group should be assigned to collect information. They are responsible for doing Internet searches, digging up materials in the library, and coordinating the team's empirical research.

Editor—The editor is responsible for the organization and style of the document. He or she identifies places where the document is missing content or where information needs to be reorganized to achieve the project's purpose.

Designer—The designer is responsible for laying out the document, collecting images, and making tables, graphs, and charts.

Notice that there is no "writer" among these roles. Instead, everyone in the group is responsible for writing some part of the document.

Step 4: Create a Project Calendar

Project calendars are essential for meeting deadlines. Numerous project management software packages like Microsoft Project, ArrantSoft, and Artemis Project Management help teams lay out calendars for completing projects. These programs are helpful for setting deadlines and specifying when interrelated parts of the project need to be completed (Figure 4.3).

Project-Planning Software

Tasks

Calendar that schedules part of the project

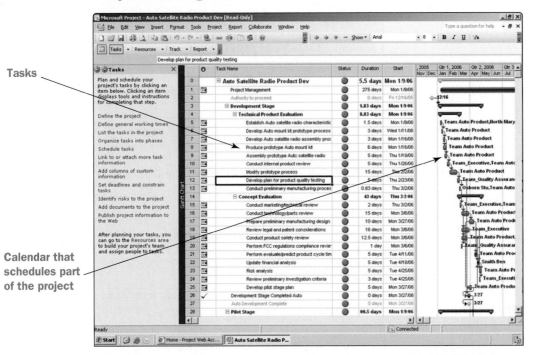

Source: Microsoft Inc., http://www.office.microsoft.com/en-us/project/HA101656381033.aspx.

Figure 4.3: Project-planning software can be helpful when setting a calendar for the team. In this screen, the calendar is represented visually to show how tasks relate to each other over time.

For more on sharing team responsibilities, go to
www.pearsonhighered.com/johnsonweb3/4.4

But you don't need project management software for smaller projects. A reliable time management technique is to use *backward planning* to determine when you need to accomplish specific tasks and meet smaller and final deadlines.

To do backward planning, start out by putting the project's deadline on a calendar. Then, work backward from that deadline, writing down the dates when specific project tasks need to be completed (Figure 4.4).

Backward-Planning Calendar

Project Calendar: Staph Infection Training Modules				
Monday	**Tuesday**	**Wednesday**	**Thursday**	**Friday**
4 Staff Meeting	**5**	**6**	**7**	**8** Complete Collection of Data
11	**12**	**13** Report on Findings Due	**14**	**15**
18	**19** Brochures, Pamphlets Printed	**20**	**21**	**22** Proofread Training Materials
25	**26**	**27** Training Modules Completed	**28**	**29** Deadline: Training Day

Figure 4.4: Backward planning is a process of working backward from the deadline. Starting with the deadline, chart out when each task needs to be completed.

Here is the deadline. Work backward from this date.

The advantage of a project calendar is that it keeps the team on task. The calendar shows the milestones for the project, so everyone knows how the rest of the team is progressing. In other words, everyone on the team knows when his or her part of the project needs to be completed.

Step 5: Write Out a Work Plan

A work plan is a description of how the project will be completed (Figure 4.5). Some work plans are rather simple, perhaps using an outline to describe how the project will go from start to finish. Other work plans, like the one in Figure 4.5, are very detailed and thorough.

Figure 4.5:
A work plan
specifies the
who what,
where, when,
why, and how of
the project.

St. Thomas Medical Center

Date: April 28, 2009
To: Staph Infection Task Force (M. Franks, C. Little, J. Archuleta,
            L. Drew, V. Yi, J. Matthews, J. McManus)
From: Alice Falsworthy, Infections Specialist
Re: Work Plan for Combating Staph Infection

Last week, we met to discuss how we should handle the increase in staph infections here at the St. Thomas Medical Center. We defined our mission, defined major tasks, and developed a project calendar. The purpose of this memo is to summarize those decisions and lay out a work plan that we will follow.

I cannot overstate the importance of this project. Staph infections are becoming an increasing problem at hospitals around the country. Of particular concern are antibiotic-resistant bacteria called methicillin-resistant *Staphylococcus aureus* (MRSA), which can kill patients with otherwise routine injuries. It is essential that we do everything in our power to control MRSA and other forms of staph.

Please post this work plan in your office to keep you on task and on schedule.

Project Mission
Our Mission: The purpose of this Staph Infection Task Force is to determine the level of staph infection vulnerability at St. Thomas Medical Center and to develop strategies for limiting our patients' exposure to staph, especially MRSA.

Secondary Objectives:
• Gain a better understanding of current research on staph and its treatments.
• Raise awareness of staph infections among our medical staff and patients.
• Assess the level of staph risk, especially MRSA, here at the hospital.
• Develop methods for controlling staph on hospital surfaces.

**Purpose of
the work plan**

**Mission
statement**

**Team
objectives**

**Step-by-step
actions**

Deliverables

Project Plan

To achieve our mission and meet the above objectives, we developed the
following five-action plan:

*Action One: Collect Current Research on Staph Infections and Their
Treatments*

Mary Franks and Charles Little will collect and synthesize current research
on staph infections, the available treatments, and control methods. They
will run Internet searches, study the journals, attend workshops, and
interview experts. They will also contact other hospitals to collect
information on successful control methods. They will write a report on
their findings and present it to the group on May 25 at our monthly
meeting.

*Action Two: Create Pamphlets and Other Literature That Help Medical
Personnel Limit Staph Infections*

Juliet Archuleta will begin developing a series of pamphlets and white
papers that stress the importance of staph infections and offer strategies for
combating them. These documents will be aimed at doctors and nurses to
help them understand the importance of using antibiotics responsibly.
Juliet will begin working on these documents immediately, but she will
need the information collected by Mary and Charles to complete her part
of the project. The documents will be completed by July 13, so we can have
them printed to hand out at training modules.

*Action Three: Conduct Experiments to Determine Level of Staph Risk at
St. Thomas Medical Center*

Lisa Drew and Valerie Yi will collect samples around the medical center to
determine the risk of staph infection here at St. Thomas. The Center's
Administration has asked us to develop a measurable knowledge of the
problem at the Center. Lisa and Valerie will write a report to the
Administration in which they discuss their findings. The report will be
completed and delivered to the task force by June 2.

Action Four: Develop Contingency Plans for Handling Staph Instances

John Matthews and Joe McManus will develop contingency plans for
handling instances of staph infections at the Center. These plans will offer

(continued)

concrete steps that the staph task force can take to limit exposure to staph bacteria. These plans will be based on the research collected by Mary and Charles. They will be completed by July 13.

Action Five: Develop Training Modules
When our research is complete, all of the members of this task force will develop training modules to raise awareness of staph infections, offer prevention strategies, and provide information on proper use of antibiotics. Different modules will be developed for doctors, nurses, and custodial staff. Alice Falsworthy and Charles Little will coordinate the development of these training modules. The training module will be ready by August 29.

Project Calendar
Here is a chart that illustrates the project calendar and its deadlines:

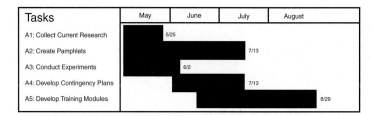

Conclusion
If anyone on the task force would like to change the plan, we will call a meeting to discuss the proposed changes. If you wish to call a meeting, please contact me at ext. 8712, or e-mail me at Alice_Falsworthy@stthomasmc.com.

Deliverables

Timeline for achieving goals

A work plan will do the following:

- Identify the mission and objectives of the project.
- Lay out a step-by-step plan for achieving the mission and objectives.
- Establish a project calendar.
- Estimate a project budget if needed.
- Summarize the results/deliverables of the project.

A work plan is helpful for both small and large projects, because team members need to see the project in writing. Otherwise, they will walk away from meetings with very different ideas about what needs to be accomplished.

By writing up a work plan, your team specifies how the project will be completed and who is responsible for which parts of the project. That way, team members can review the work plan if they are uncertain about (1) what tasks are being completed, (2) when the tasks will be finished, and (3) who is responsible for completing them.

Link

For help writing work plans as proposals, turn to Chapter 21, page 609.

Step 6: Agree on How Conflicts Will Be Resolved

Finally, your team should talk about how it will handle conflicts. Conflict is a natural, even healthy, part of a team project. However, the worst time to figure out how your team will handle conflicts is when you are in the middle of one.

Instead, in advance, talk with your team about how conflicts should be handled.

Will the team take votes on important topics? Will the majority rule?

Will the team rely on the judgment of the team coordinator or the supervisor?

Does the team need to reach full consensus on decisions?

Can any team member call a team meeting to discuss conflicts?

Should agreements be written down for future reference?

Six Steps for Strategic Planning

AT A GLANCE

- Define the project mission and objectives.
- Identify project outcomes.
- Define team member responsibilities.
- Create a project calendar.
- Write out a work plan.
- Agree on how conflicts will be resolved.

You should not shy away from conflict in your team, because conflict is a natural part of the teaming process. Constructive conflict often leads to more creativity and closer bonds among team members. But, destructive conflict can lead to dysfunctional working relationships, frustration, and lower morale. To foster constructive conflict and minimize destructive conflict, you should spend some time during the forming phase talking about the ways conflicts will be resolved.

Storming: Managing Conflict

Not long after the forming stage, a team will typically go through a storming phase. Soon after the project is planned out and the actual work begins, some tension will usually surface among team members. In most cases, though, team members find themselves negotiating, adapting, and compromising to achieve the team's mission.

During the storming stage, team members may:

- resist suggestions for improvement from other members.
- have doubts about the work plan's ability to succeed.
- compete for resources or recognition.
- resent that others are not listening to their ideas.
- want to change the team's objectives.
- raise issues of ethics or politics that need to be addressed.
- believe they are doing more than their share of the work.

Storming is rarely pleasant, but it is a natural part of the teaming process. When storming, teams realize that even the best work plans are never perfect and that people don't always work the same way or have the same expectations. The important thing is to not let small conflicts or disagreements sidetrack the project.

Running Effective Meetings

One way to constructively work through the storming phase is to conduct effective meetings. Nothing is more frustrating to team members than having to waste their time and effort sitting around in an unproductive meeting. By running organized meetings, your team can maintain the structure needed to keep people on track.

CHOOSE A MEETING FACILITATOR   In the workplace, usually a manager or supervisor runs the meeting, so he or she is responsible for setting the time and agenda. An interesting workplace trend, though, is to rotate the facilitator role among team members. That way, everyone has a chance to run the meeting, allowing everyone on the team to take on leadership roles. In classroom situations, your team should rotate the facilitator role to maintain a more democratic approach.

SET AN AGENDA   An agenda is a list of topics to be discussed at the meeting (Figure 4.6). The meeting coordinator should send out the meeting agenda at least a couple days before the meeting. That way, everyone will know what issues will be discussed and decided upon. Begin each meeting by first making sure everyone agrees to the agenda. Then, during the meeting, use the agenda to avoid going off track into nonagenda topics.

START AND END MEETINGS PROMPTLY   If team members are not present, start the meeting anyway. Waiting around for late people can be frustrating, so you should insist that people arrive on time. If people know the meeting will start on time, they will be there on time. Likewise, end meetings on time. Meetings that drag on endlessly can be equally frustrating.

ADDRESS EACH AGENDA ITEM SEPARATELY   Discuss each agenda item before moving on to the next one. Bouncing around among items on the agenda ensures only that the meeting will be inefficient. If someone wants to move ahead to a future agenda item, first make sure the current item of discussion has been addressed.

Want to learn more about resolving conflict in the workplace? Go to
www.pearsonhighered.com/johnsonweb3/4.6

An Agenda

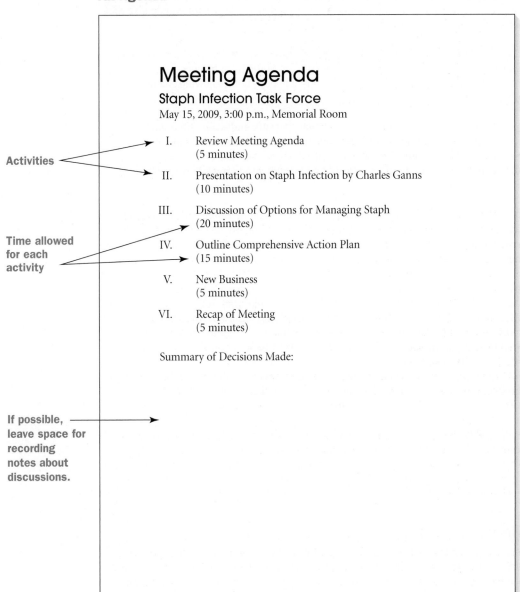

Activities

Time allowed for each activity

Meeting Agenda
Staph Infection Task Force
May 15, 2009, 3:00 p.m., Memorial Room

    I.   Review Meeting Agenda
           (5 minutes)

    II.   Presentation on Staph Infection by Charles Ganns
           (10 minutes)

    III. Discussion of Options for Managing Staph
           (20 minutes)

    IV. Outline Comprehensive Action Plan
           (15 minutes)

    V.   New Business
           (5 minutes)

    VI. Recap of Meeting
           (5 minutes)

Summary of Decisions Made:

If possible, leave space for recording notes about discussions.

Figure 4.6:
A simple agenda is a helpful tool for keeping the meeting on track.

Want to know more about running effective meetings? Go to www.pearsonhighered.com/johnsonweb3/4.7

GO TO THE NET

ENCOURAGE PARTICIPATION   Everyone on the team should say something about each item. If one of the team members has not spoken, the facilitator should give that person an opportunity to speak.

ALLOW DISSENT   At meetings, it is fine to disagree. Active debate about issues will help everyone consider the issues involved. In fact, if the team reaches consensus too quickly on an issue, someone might raise possible objections, allowing a consideration of alternatives.

REACH CONSENSUS AND MOVE ON   People can talk endlessly about each agenda item, even after the team has reached consensus. So, allow any group member to "call the question" when he or she feels consensus has been reached. At that point, you can take an informal or formal vote to determine the team's course of action.

RECORD DECISIONS   During meetings someone should be responsible for keeping the minutes. The minutes record the team's decisions. Minimally, all decisions should be written down. After the meeting, the facilitator should send these notes or the full minutes to the team members, usually via e-mail.

RECAP EACH AGENDA ITEM   At the end of the meeting, leave a few minutes to recap the decisions made by the team and clarify who is doing what. It is not uncommon for teams to have a "great meeting" that still leaves people unsure of what was decided and who is doing what. So, go through each agenda item, summarizing (1) what action will be taken and (2) who is responsible for taking that action.

LOOK AHEAD   Discuss when the team will meet again and the expectations for that meeting. If necessary, clarify what should be accomplished before the next meeting. Also, decide who will be responsible for facilitating the next meeting.

Typically, storming becomes most evident during meetings. People grow frustrated and even angry as the team struggles to accomplish its objectives. That's why running effective meetings is so important. By creating a predictable structure for the meeting, your team will lower the level of frustration, allowing you to get work done.

Mediating Conflicts

Smaller conflicts should be handled using the conflict resolution methods your team discussed when it was forming. Make sure everyone is encouraged to express his or her views openly. Make sure everyone is heard. Then, use the conflict resolution methods (vote, decision of the team leader, appeal to supervisor, reach full consensus) to decide which way to go.

There will be times, however, when personalities or ideas clash in ways that cannot be easily resolved. At these times, you may want to use mediation techniques to help your team move forward (Figure 4.7).

The Steps of Mediation

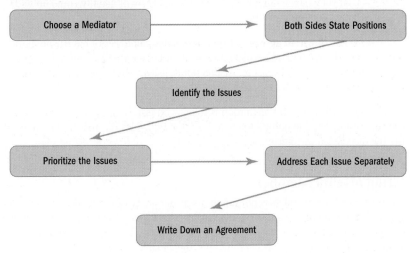

Figure 4.7:
When more formal mediation is needed, you can follow these steps toward a resolution of the problem.

1. **Choose a mediator**—A mediator is like a referee in a game. He or she does not take sides. Instead, it is the mediator's job to keep both sides of an argument talking about the issue. When the dialogue becomes unfriendly or goes off track, the mediator brings both sides back to the issues. In formal mediation, the mediator is someone who is not part of the group and has no connections to either side. In informal mediation, one of the group members who hasn't chosen sides can often serve as the referee.

2. **Ask both sides to state their positions**—Often, conflicts arise simply because each side has not clearly stated its position. Once each side has had a chance to explain its ideas clearly, the conflict should seem more resolvable.

3. **Identify the issues**—Both sides should discuss and identify the "issues" they disagree about. Often, disputes hinge on a small number of issues. Once these issues are identified, it becomes easier to talk about them.

4. **Prioritize the issues from most to least important**—Some issues are more important than others. By prioritizing the issues, both sides can usually find places where they already agree. Or, in some cases, a top priority to one side is not a foremost concern to the other side. By prioritizing the issues, both sides often begin to see room for negotiation.

5. **Address each issue separately, trying to find a middle ground that is acceptable to both sides of the dispute**—Focus on each issue separately and

Link

Ethical issues are often a source of conflict. To learn more about ethics, go to Chapter 5, page 92.

keep looking for middle ground between both sides. When both sides realize that they have many common interests, they will usually come up with solutions to the conflict.

6. **Write down an agreement that both sides can accept**—As the mediation continues, both sides usually find themselves agreeing on ways to resolve some or all of the issues. At this point, it helps to write down any compromises that both sides can accept.

The secret to successful mediation is a focus on issues, not personalities or perceived wrongs. The mediator's job is to keep the two sides discussing the issues, steering the discussion away from other distractions.

Firing a Team Member

Sometimes a team member is not doing his or her share of the work. When this happens, the team might consider removing that person from the project. The best way to handle these situations is to first mediate the problem. The members of the team should meet to talk with this person about their expectations, giving the person a chance to explain the situation.

After hearing this person's side of the story, the team might decide to give him or her a second chance. At that point, a work contract should be written that specifies exactly what this person needs to do for the project.

If the team still wants to let the person go, the supervisor (perhaps your instructor) should be asked about removing the person from the team. The supervisor should be present when the team tells the problematic team member that he or she is being removed.

Norming: Determining Team Roles

The storming period can be frustrating, but soon afterward your team should enter the *norming* stage. In this stage, members of your team will begin to accept their responsibilities and their roles in the project. A sense of team unity will develop as people begin to trust each other. Criticism will become increasingly constructive as team members strive to achieve the project's mission and objectives.

Revising Objectives and Outcomes

The storming stage often reveals the flaws in the work plan. So, when norming, you might find it helpful to revisit and refine the team's original decisions about objectives and outcomes. The team may also want to revise the project schedule and redivide the workload.

You don't need to completely rewrite the work plan from scratch. You should stay with your original work plan in most cases. The plan probably just needs to be revised and refined, not completely redone.

To learn more about mediation, go to
www.pearsonhighered.com/johnsonweb3/4.8
Need to fire a team member? Go to
www.pearsonhighered.com/johnsonweb3/4.9

Virtual Teaming

Increasingly, technical workplaces are turning to virtual teaming to put the right people on any given project. Virtual teaming allows people to work on projects collaboratively through electronic networks, using e-mail, instant messaging, and phones to stay in contact. Electronic networks, called *intranets*, are often used to share information and documents among team members.

Virtual teams, just like teams working together in an office, schedule regular meetings, share ideas and documents, and work toward achieving specific goals. The main differences between on-site teams and virtual teams are how people communicate and where they are located. Software, such as Virtual Office, MS Outlook, Lotus Notes, and CyberMatrix Office, help people stay in touch and collaborate on projects.

Trends in the technical workplace support virtual teaming. Today, more people are "telecommuting" or "teleworking" from home or remote sites. Also, the global economy means people on the same project are sometimes working thousands of miles away from each other. Meanwhile, wireless technologies allow people to work just about anywhere. Chances are you will find yourself working with a virtual team in the near future—if you aren't already.

Interestingly, virtual teaming does not change traditional teaming strategies—it only makes them more necessary. A virtual team will go through the forming, storming, norming, and performing stages, just like an on-site team. If good planning, communication, and conflict resolution are important with on-site teaming, they are even more important in virtual teams. After all, communicating with your virtual team is a little more difficult, because you cannot physically visit each other.

Here are some strategies for managing a successful virtual team:

Develop a work plan and stick to it—Members of virtual teams do not bump into each other in the hallway or the break room. So, they need a clear work plan to keep everyone moving together toward the final goal. Your team's work plan should (1) define the mission, (2) state objectives and measurable outcomes, (3) spell out each stage and task in the project, (4) specify who is responsible for each task, and (5) lay out a project calendar.

Communicate regularly—In virtual teams, the old saying "out of sight, out of mind" now becomes "out of communication, out of mind." Each member of the virtual team should agree to communicate with the others regularly (e.g., two times a day, two times a week). Your team can use e-mail, phones, instant messaging, or chat rooms to contact each other. You and your team members should constantly keep each other up to date on your progress. And, if someone falls out of communication for a day or two, the team leader should track him or her down and urge the team member to begin communicating again.

Hold teleconferences and videoconferences—There are many ways to hold real-time virtual meetings with team members. Your team members can teleconference over the phone, or you can use a chat room or instant messaging to exchange ideas with each other. Increasingly, broadband technology is allowing people to hold videoconferences in which people meet and see each other with

(continued)

HELP

Teleconferencing

A webcam projects images to the other participants in the virtual meeting.

The monitor can be used to show people, documents, or presentations.

Internet cameras (webcams) through computer screens (Figure A). Like on-site meetings, virtual meetings should be preplanned and follow an agenda. The only significant difference between on-site meetings and virtual meetings is that people are not in the same room.

Build trust and respect—One of the shortcomings of virtual teaming is the lack of nonverbal cues (smiles, shrugs, scowls) that help avoid misunderstandings. As a result, people in virtual teams can feel insulted or disrespected much more easily than people in on-site teams. So, it is doubly important that team members learn how to build trust with others and show respect. Trust is built by communicating effectively, meeting deadlines, and doing high-quality work. Respect is fostered by giving compliments and using "please" and "thank you" in messages. When conflicts do arise (and they will), focus on issues and problem solving, not personalities or perceived slights.

Keep regular hours—Time management is always important, even if you are working in a virtual team. During regular office hours, team members should be confident that they can contact each other. You and your team members should be ready to answer the phone, use instant messaging, or answer e-mail as though you were all working together in a typical office.

More than likely, virtual offices, teleworking, and virtual teams will be a common part of the workplace in the near future. Like on-site teaming, virtual teaming requires you to learn how to effectively work with others.

GO TO THE NET

To find websites that discuss virtual teaming, go to
www.pearsonhighered.com/johnsonweb3/4.10

Identifying Team Roles

When planning the project, your team divided up the work, giving each person specific responsibilities. As the team begins norming, though, you will notice that team members tend to take on *team roles* that reflect their personalities, capabilities, and interests.

A management specialist, Meredith Belbin (1981), developed a description of nine team roles that people generally follow. He also carved these nine roles into three categories: people-oriented roles, action-oriented roles, and cerebral roles.

PEOPLE-ORIENTED ROLES   The people in these roles are responsible for managing the activities of the team members:

> The **coordinator** sets the agenda and keeps track of the team's objectives; asks broader questions and occasionally summarizes the team's decisions; keeps an eye on the project calendar and coordinates the work of various team members.
>
> The **resource investigator** goes out to find information, bringing new ideas and strategies into the discussion; looks outside the team for ways to improve the project.
>
> The **team worker** focuses on getting the work done; may not be fully invested in planning the project but will do his or her part of it.

ACTION-ORIENTED ROLES   The people in these roles are responsible for getting things done:

> The **shaper** focuses on team tasks while looking for patterns in team discussions; emphasizes completing the project.
>
> The **implementor** stresses the "how to" nature of the project and is eager to develop methods for turning abstract objectives and plans into real actions.
>
> The **completer/finisher** stresses attention to details and the overall quality of the project; concerned about meeting deadlines and maintaining a schedule.

CEREBRAL ROLES   The people in cerebral roles are responsible for planning, creating, and providing expertise in a project:

> The **monitor/evaluator** keeps the team on task by critiquing poor decisions or pointing out any flaws in reasoning; tends to focus on achieving outcomes.
>
> The **plant** thinks creatively, often providing original suggestions and innovative solutions to problems; tends to stress the big picture over smaller details.
>
> The **specialist** contributes special skills and knowledge to the team; masters a specific topic or area of research, adding depth to the team's discussions.

As the team begins norming, you and the other members of the team might take some time to identify the roles each of you is playing. By identifying team roles, you can take advantage of each member's natural strengths and interests.

Not all the roles will be filled, especially in a smaller team. Instead, each team member might take on two or three roles, depending on the project. Roles may change and evolve as the project moves forward.

In other words, let the "completer/finisher" in the team worry about the deadlines and quality issues. Let the "team workers" concentrate on achieving specific tasks. Encourage the "coordinators" and "shapers" to keep an eye on the overall objectives and mission of the team.

AT A GLANCE

Norming

- Revise objectives and outcomes.
- Identify team roles.
- Use groupware to facilitate work.

Using Groupware to Facilitate Work

When working in a team, you might need to use *groupware*, a kind of software that helps move information and documents around.

Groupware allows team members to communicate and work collaboratively through a local area network (LAN) or an intranet.

Perhaps the most common use of groupware is sharing documents and sending messages. The two most popular groupware packages include IBM Lotus Notes® (Figure 4.8) and Microsoft Outlook®. These software packages and ones like them support the following kinds of activities:

Using Groupware

Monthly calendar is shown here.

Meetings are shown here.

Notes are listed here.

Figure 4.8: With groupware like Lotus Notes, shown here, members of a group can keep a common calendar, post notes, and send e-mails to each other. In this screen, the user can do a variety of functions using the menu at the top.

Source: Reprint courtesy of International Business Machines Corporation, copyright 2008 © International Business Machines Corporation.

GO TO THE NET

To learn more about team role theory, go to
www.pearsonhighered.com/johnsonweb3/4.11

Sarah "Intellagirl" Robbins-Bell

DIRECTOR OF EMERGING TECHNOLOGIES, MEDIA SAUCE, INDIANAPOLIS, INDIANA

Media Sauce is a "fantastically hip" multi-media marketing company.

AT WORK

What is a good way for virtual teams to develop and organize a project?

Virtual worlds like Second Life (www.secondlife.com) are allowing new forms of workplace teamwork. Second Life is a 3-dimensional social virtual world in which residents use voice chat, text chat, custom-created avatars, and a flexible building system to interact with each other.

A meeting with a team in a virtual world overcomes distance and allows members to talk to each other in ways they are already familiar with. Second Life allows users to communicate through text (like instant messages) and voice (like a phone call). Both text and voice chat work in spatial and distance modes. Avatars who are within less than twenty virtual meters of each other can read each other's text or hear each other's voices. Users who are separated by distance in the Second Life world can communicate as a group using voice or text.

Avatars, digital personifications of the user, are perhaps the most powerful element of a virtual world for teamwork. Avatars in the same space can use nonverbal gestures to convey messages. In a meeting, if I'm not interested in what is being said, my avatar can yawn, shrug, even slump over asleep to convey how I feel about how the meeting is going. If I like someone, I can maneuver my avatar to stand near another person's avatar, face it, and allow my avatar to look the other avatar in the eye—even offer it a hug.

The world of Second Life is filled with exotic landscapes and custom buildings: amphitheaters, meeting spaces, beaches, mountaintops, and so forth. Even in a virtual world, the space in which people meet has a profound impact on how people behave. Build a room with low ceilings, tight walls, and nowhere for avatars to sit down, and you'll find that users find it difficult to communicate and collaborate because they can't pan their camera view around to see the faces of other avatars with whom they're meeting. The building tools in Second Life also contribute to team projects by allowing users to work collaboratively while creating content.

Second Life is but one example of virtual worlds being used for teams to work and communicate. This new form of teamwork has advantages and can create more cohesive, productive teams that many digital communication forms lack.

Scheduling and calendaring—One of the more powerful features of groupware is the ability to schedule meetings and keep a common calendar for the team. With this feature, team members can regularly check the calendar to stay on task.

Discussion lists and instant messaging—Groupware offers easy access to discussion lists (usually using e-mail) and instant messaging. With these tools, team members can post notes to each other on a discussion list or have real-time discussions through instant messaging.

Document posting and commenting—Files can be posted to a common site, allowing team members to view documents, comment on them electronically, or download them.

Using groupware effectively takes some practice. Eventually, the team will begin using the groupware as a meeting place and a posting board. It will become an integral part of the project.

Performing: Improving Quality

Your team is *performing* when members are comfortable with the project and their roles in it. Team members will recognize the other members' talents and weaknesses. They also begin to anticipate each other's needs and capabilities.

When your team is performing, you can start looking for ways to improve the quality of your work. One of the gurus of quality was W. Edwards Deming, who developed many of the principles behind Total Quality Management (TQM) and Continuous Quality Improvement (CQI), which are widely used in technical workplaces. Deming argued that teams should put an emphasis on improving the *process* rather than simply exhorting people to improve the product (Deming, 2000).

How can you improve quality in your team? While performing, a helpful technique is to develop *quality feedback loops* in which your team regularly compares outcomes to the project objectives.

Link

For more ideas about improving quality through document cycling, see Chapter 12, page 322.

An effective quality feedback loop should include methods to collect feedback on the outcomes of the project. To collect feedback, perhaps "focus groups" of customers might be consulted. Perhaps the product might be thoroughly user tested. Perhaps supervisors or outside experts might be called in to offer suggestions for improvement. The type of feedback you need depends on the kind of product your team is being asked to produce.

Teams also need to regularly review the performance of their own members. Figure 4.9 shows a "Team Performance Review" form that is similar to ones found in the workplace. Your team and professor can use this form to assess performance of your team and look for places to improve. A performing team will have its ups and downs. There may even be times when the team regresses into the norming or even the storming stages. Eventually, though, the performing team usually regroups and puts the focus back on quality.

Team Performance Review

Date: _____

Name: _____

Name of Project: _____

The purpose of this performance review is to evaluate the participation of you and your team members on the project we just completed. Your feedback on this form will be taken into account when determining each member's "participation" part of the grade for this assignment.

Describe your role on the team and list your top four contributions to the project.

My role:

My contributions to the project:

1.

2.

3.

4.

Describe the role and list the contributions of your team members.

Team Member	a.	b.	c.
Role on team			
Contributions			

Rate the participation (1–10 with 10 being the highest) of your team members.

Team Member	a.	b.	c.
Attended team meetings			
Did her/his share of the project			
Contributed good ideas			
Respected the ideas of others			
Handled conflict and stress			
Communicated with team			
Overall Participation Rating (1–10)			

On the back of this sheet, write detailed comments about how well your team worked together. What were its strengths? What could have been improved? Which team members were essential to the project and why were they so important? Which team members should have done more?

The Keys to Teaming

The keys to good teaming are good planning and effective communication. The planning strategies discussed in this chapter might seem like extra work, especially when time is limited and your team is eager to start working on the project. But good planning will save your team time in the long run. Each person needs a clear understanding of the mission and the steps in the project. Then, you need to keep the communication lines open. Plan to communicate regularly by phone, e-mail, and instant messaging.

Telecommuting, or teleworking, is becoming much more common in today's technical workplace. Sometimes your team members will be working a few days per week at home. Or, they may be working while they are on the road. Some telecommuters work almost exclusively from home, going to the office only when absolutely necessary. Good planning and effective communication are the keys to success in these virtual workplace environments.

- Teams have several advantages. They can concentrate strengths, foster creativity, share the workload, and improve morale. Disadvantages include conflict with other members and disproportionate workloads.

- Tuckman's four stages of teaming are forming, storming, norming, and performing.

- In the forming stage, use strategic planning to define the mission, set objectives, define roles, and establish a project schedule.

- In the storming stage, use conflict management techniques to handle emerging disagreements, tension and anxiety, leadership challenges, and frustration.

- In the norming stage, revise the work plan to form consensus, refine the team's objectives and outcomes, and solidify team roles.

- In the performing stage, pay attention to improving the process in ways that improve the quality of the work.

- W. Edwards Deming developed many of the principles behind Total Quality Management (TQM) and Continuous Quality Improvement (CQI), which are common in technical workplaces today.

- Virtual teaming, or working together from a distance, requires good planning and effective communication.

Individual or Team Projects

1. When working with a team, keep a journal that describes your experiences working with others. In a memo to your instructor, discuss whether your team went through the four stages of teaming (forming, storming, norming, performing). Describe how each stage brought up new issues to be resolved. How did you and your team members handle them? In your memo to your instructor, discuss how you might use awareness of Tuckman's four stages to improve your ability to work with a team.

2. If you have a job now, write a report to your instructor in which you talk about how your co-workers fluctuate among Tuckman's four stages. What are some of the indications that the team is forming, storming, norming, and performing? How does the team tend to react during each of these stages? Does your team aid in forming strategic planning? How do you mediate conflicts during storming? How is norming achieved at your workplace? What does performing look like?

3. Imagine that your class is a workplace. What are the objectives and outcomes of this course? How are conflicts resolved in the classroom? Do members of your class take on various team roles in the classroom? How could you and your instructor create quality feedback loops to improve your learning experience? In a class, discuss how teaming strategies might be helpful in improving how the class is managed.

4. On the Internet, research the theories and writings of Tuckman, Belbin, or Deming. Write a report to your instructor in which you summarize the principles of one of their theories. Then, discuss the ways in which you might apply this theory to your own life and work.

Collaborative Project

With a team, try to write a report in one hour on a topic of interest to all of you. For example, you might write about a problem on campus or at your workplace. While you are writing the report, pay attention to how your group forms, norms, storms, and performs. Pay attention to how the group plans and divides up the responsibilities. Then, as the project moves forward, pay attention to the ways conflict is resolved. Finally, identify the different roles that group members tend to play as the group develops norms for the project.

When your one-hour report is finished, talk among yourselves about how the group project went. Did the group form and plan properly? Did you handle conflict well? Did the group members take on identifiable roles? Do you think you ever reached the performing stage? If you were going to do the project over, how might your group do things differently?

For additional technical writing resources, including interactive sample documents, document design tutorials and guidelines, and more, go to **www.mytechcommlab.com**.

Not a Sunny Day

Veronica Norton liked working on teams. She liked interacting with others, and she liked how working with others allowed her to accomplish projects that were too big for one person. Veronica also had a strong interest in building solar motors, like the ones that went into solar-powered cars. She hoped to find a job in the automotive industry building solar vehicles.

So, she decided to join her university's team in the American Solar Challenge race competition. She and a group of ten other students were going to build a solar car to race across the country against cars from other universities.

The first meeting of the team went great. Everyone was excited, and they found out there were funds to support the project. Several engineering departments were contributing money, and a local aerospace engineering firm was matching university funds dollar for dollar. The firm was also letting the team use its wind tunnel to help streamline the car. There were other local sponsors who wanted to participate.

The team rushed right into the project. They sketched out a design for the car, ordered supplies, and began welding together a frame for the car. Veronica began designing a solar motor that would be long lasting and have plenty of power. She noticed that the overall plan for the car was not well thought out, but she went along with it anyway. At least they were making progress.

One of the team members, George Franks, began emerging as the leader. He was finishing up his engineering degree, so he had only a couple of classes left to take. He had plenty of time to work on the solar car.

Unfortunately, he didn't like to follow plans. Instead, he just liked to tinker, putting things together as he saw fit. It wasn't long before the team started running into problems. George's tinkering approach was creating a car that would be rather heavy. Also, each time Veronica visited the shop, the dimensions of the car had changed. So, the motor she was building needed to be completely redesigned.

Perhaps the worst problem was that they were running out of money. George's tinkering meant lots of wasted materials. As a result, they had almost used up their entire budget. Sally, who was in charge of the finances for the project, told everyone they were going to be out of money in a month. Everyone was getting anxious.

Finally, at one of the monthly meetings, things fell apart. People were yelling at each other. After being blamed for messing up the project, George stormed out of the meeting. Things looked pretty hopeless. Everyone left the meeting unsure what to do.

Veronica still wanted to complete the project, though. If you were her, how might you handle this situation? What should she do to get the project going again?

For more information on the American Solar Challenge, go to
www.pearsonhighered.com/johnsonweb3/4.15

CHAPTER

5

Ethics in the Technical Workplace

In this chapter, you will learn:

- A working definition and understanding of ethics.

- To identify three ethical systems: personal, social, and conservation.

- To consider social ethics as issues of rights, justice, utility, and care.

- How copyright law affects technical communication.

- Strategies for resolving ethical conflicts in technical workplaces.

- How to balance the many issues involved in an ethical dilemma.

- About the new ethical challenges that face the computer-centered workplace.

Ethics have taken a beating recently. Enron, Exxon, Adelphia, Worldcom, and even Hewlett-Packard (for spying) have all added their names to a growing list of corporate scandals. Elsewhere, researchers studying human cloning have admitted following unethical research practices. Hurricane Katrina in 2005 exposed inept political appointees who had minimal experience with natural disasters. The reckless practices of the credit and mortgage industries rocked the world economy in 2008.

We could blame these scandals on unethical, greedy, unscrupulous people—and certainly they deserve blame—but the newness of the computerized workplace is probably also partly to blame. As we evolve into an electronic culture, the ethical boundaries are not as clear as they were only a few decades ago. Unfortunately, some people are willing to exploit these ethical gray areas for their own financial advantage.

In the technical workplace, you will run into ethical dilemmas regularly. In these situations, you need to be able to identify what is at stake and make an informed decision. Ethical behavior is more than a matter of personal virtue—it is good business. The scandals and failures of the early twenty-first century have quite clearly proved the importance of ethics in the technical workplace.

What Are Ethics?

For some people, ethics are about issues of morality. For others, ethics are a matter of law. Actually, ethics bring together many different ideas about appropriate behavior in a society.

Ethics are systems of moral, social, or cultural values that govern the conduct of an individual or community. For many people, acting ethically simply means "doing the right thing," a phrase that actually sums up ethics quite well. The hard part, of course, is figuring out the right thing to do. Ethical choices, after all, are not always straightforward.

Every decision you make has an ethical dimension, whether it is apparent or not. In most workplace situations, the ethical choice is apparent, so you do not pause to consider whether you are acting ethically. Occasionally, though, you will be presented with an *ethical dilemma* that needs more consideration. An ethical dilemma offers a choice among two or more unsatisfactory courses of action. At these decision points, it is helpful to ponder the ethics of each path so you can make the best choice.

In technical workplaces, resolving ethical dilemmas will be a part of your job. Resources, time, and reputations are at stake, so you will feel pressure to overpromise, underdeliver, bend the rules, cook the numbers, or exaggerate results. Technical fields are also highly competitive, so people sometimes stretch a little further than they should. Ethical dilemmas can force us into situations where all choices seem unsatisfactory.

Why do some people behave unethically? People rarely set out to do something unethical. Rather, they find themselves facing a tough

Want to see some other definitions of ethics? Go to
www.pearsonhighered.com/johnsonweb3/5.1

decision in which moving forward means taking risks or treating others unfairly. In these situations, they may be tempted to act unethically due to a fear of failure, a desire to survive, the pressure from others, or just a series of bad decisions. Small lies lead to bigger lies until the whole house of cards collapses on them.

Keep in mind, though, that ethics are not always about deception or fraud. A famous decision involving Albert Einstein offers an interesting example. Figure 5.1 shows a letter that Einstein wrote to President Franklin Roosevelt encouraging research into the development of the atom bomb. Throughout the rest of his life, Einstein, who was an ardent pacifist, was troubled by this letter. Five months before his death, he stated:

> I made one great mistake in my life...when I signed the letter to President Roosevelt recommending that atom bombs be made; but there was some justification—the danger that the Germans would make them. (Clark, p. 752)

In this quote, you see the ethical dilemma weighing on Einstein. He deeply regretted the atom bomb's development and use on Japan. However, he also recognized that his letter may have alerted Roosevelt to a very real danger. Historians have pointed out that Einstein's letter may have helped prevent the Nazis from creating an atom bomb. Ethical dilemmas put people in these kinds of quandaries.

Einstein with Robert Oppenheimer

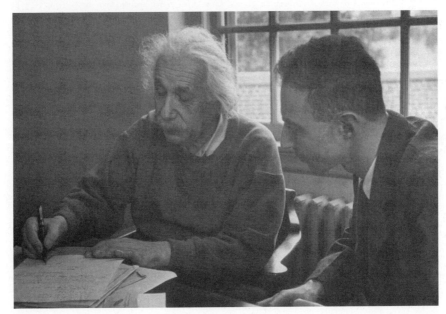

Einstein meets with Robert Oppenheimer, the leader of the U.S. efforts to develop an atom bomb. Later, Einstein regretted his involvement, though minimal, with its development.

Einstein's Letter to Roosevelt About the Atom Bomb

Albert Einstein
Old Grove Rd.
Nassau Point
Peconic, Long Island

August 2nd 1939

F.D. Roosevelt
President of the United States
White House
Washington, D.C.

Sir:

Here is Einstein's main point.

Some recent work by E.Fermi and L. Szilard, which has been communicated to me in manuscript, leads me to expect that the element uranium may be turned into a new and important source of energy in the immediate future. Certain aspects of the situation which has arisen seem to call for watchfulness and, if necessary, quick action on the part of the Administration. I believe therefore that it is my duty to bring to your attention the following facts and recommendations:

Einstein expresses the imminent problem.

In the course of the last four months it has been made probable - through the work of Joliot in France as well as Fermi and Szilard in America - that it may become possible to set up a nuclear chain reaction in a large mass of uranium, by which vast amounts of power and large quantities of new radium-like elements would be generated. Now it appears almost certain that this could be achieved in the immediate future.

He points out the possible threat of these new kinds of weapons.

This new phenomenon would also lead to the construction of bombs, and it is conceivable - though much less certain - that extremely powerful bombs of a new type may thus be constructed. A single bomb of this type, carried by boat and exploded in a port, might very well destroy the whole port together with some of the surrounding territory. However, such bombs might very well prove to be too heavy for transportation by air.

Figure 5.1:
In 1939, Einstein wrote this letter to President Franklin Roosevelt. The atomic bomb would have been built without Einstein's letter, but his prodding jump-started the U.S. nuclear program.

Source: Argonne National Laboratory, http://www.anl.gov/OPA/frontiers96arch/aetofdr.html.

-2-

The United States has only very poor ores of uranium in moderate quantities. There is some good ore in Canada and the former Czechoslovakia, while the most important source of uranium is Belgian Congo.

Einstein offers a potential solution.

In view of the situation you may think it desirable to have more permanent contact maintained between the Administration and the group of physicists working on chain reactions in America. One possible way of achieving this might be for you to entrust with this task a person who has your confidence and who could perhaps serve in an inofficial capacity. His task might comprise the following:

a) to approach Government Departments, keep them informed of the further development, and put forward recommendations for Government action, giving particular attention to the problem of securing a supply of uranium ore for the United States;

b) to speed up the experimental work, which is at present being carried on within the limits of the budgets of University laboratories, by providing funds, if such funds be required, through his contacts with y private persons who are willing to make contributions for this cause, and perhaps also by obtaining the co-operation of industrial laboratories which have the necessary equipment.

He points out that the Nazis may already be working on nuclear technology, potentially a bomb.

I understand that Germany has actually stopped the sale of uranium from the Czechoslovakian mines which she has taken over. That she should have taken such early action might perhaps be understood on the ground that the son of the German Under-Secretary of State, von Weizsäcker, is attached to the Kaiser-Wilhelm-Institut in Berlin where some of the American work on uranium is now being repeated.

Yours very truly,

A. Einstein

(Albert Einstein)

Where Do Ethics Come From?

How can you identify ethical issues and make appropriate choices? To begin, consider where values come from:

Personal ethics—Values derived from family, culture, and faith

Social ethics—Values derived from constitutional, legal, utilitarian, and caring sources

Conservation ethics—Values that protect and preserve the ecosystem in which we live

These ethical systems intertwine, and sometimes they even conflict with each other (Figure 5.2).

Intertwined Ethical Systems

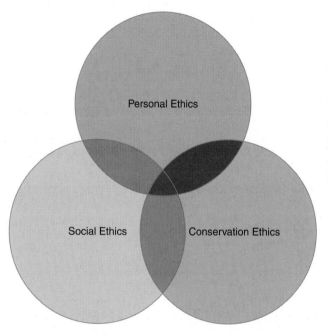

Personal Ethics

Social Ethics   Conservation Ethics

Figure 5.2: Ethics come from a variety of sources that overlap. Your personal sense of ethics guides the majority of your daily decisions. Social and conservation ethics play significant roles in the technical workplace.

Personal Ethics

By this point in your life, you have developed a good sense of right and wrong. More than likely, your personal ethics derive from your family, your culture, and your faith. Your family, especially your parents, taught you some principles to live by. Meanwhile, your culture, including the people in your neighborhood or even the people you watch on television, has shaped how you make decisions. And for many people, faith gives them specific principles about how they should live their lives.

GO TO THE NET

For websites that discuss sources of personal ethics, go to
www.pearsonhighered.com/johnsonweb3/5.3

In the technical workplace, a strong sense of personal values is essential, because these values offer a reliable touchstone for ethical behavior. A good exercise is making a list of values that you hold dear. Perhaps some of these values are honesty, integrity, respect, candor, loyalty, politeness, thoughtfulness, cautiousness, thriftiness, and caring. By articulating these values and following them, you will likely find yourself acting ethically in almost all situations.

Social Ethics

In technical workplaces, the most difficult ethical dilemmas are usually found in the social realm. Social ethics require you to think more globally about the consequences of your or your company's actions.

Ethics scholar Manuel Velasquez (2002) offers a helpful four-part categorization of social ethical situations:

Rights—Rights are fundamental freedoms that are innate to humans or granted by a nation to its citizens. *Human rights,* like those mentioned in the *U.S. Declaration of Independence* (life, liberty, and the pursuit of happiness), are innate to humans and cannot be taken away. *Constitutional rights* (freedom of speech, right to bear arms, protection against double jeopardy) are the rights held in common by citizens of a nation.

Justice—Justice involves fairness among equals. Justice takes its most obvious form in the laws that govern a society. Our laws are a formalized ethical system that is designed to ensure that people are treated equally and fairly. Similarly, *corporate policies* are the rules that ensure fairness within a company.

Utility—Utility suggests that the interests of the majority should outweigh the interests of the few. Of paramount importance to utilitarianism is *the greatest good for the greatest number of people.*

Care—Care suggests that tolerance and compassion take precedence over rigid, absolute rules. Ethics of care suggest that each situation should be judged on its own, putting heightened attention on concern for the welfare of people and preserving relationships. It also recognizes that some relationships, like those involving friends and family, will often lead to ethical choices that transcend rights, justice, and utility.

Legal issues, usually involving rights and justice, are especially important in technical communication, because the temptation to break the law to gain a competitive edge can be great. Legal issues of copyright law, patent law, liability, privacy, and fraud (which are all discussed later in this chapter) are crucial concerns that affect how individuals and companies conduct themselves. You should be aware of the laws that apply to your discipline.

When facing an ethical dilemma or controversy involving ethics, you should first identify which of these four ethical categories applies to your situation. Ethical issues that involve human or constitutional rights are usually given the most gravity (Figure 5.3). Issues involving care are still important, but they have the least gravity. In other words, if an ethical decision involves human or constitutional rights, it will take on much more importance than a decision that involves issues of justice, utility, or care.

Four Categories of Social Ethics

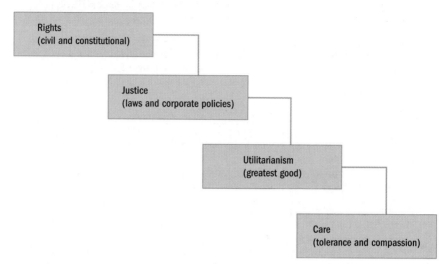

Figure 5.3:
Social ethics can be ranked. Concerns about rights usually have more gravity than justice issues, and so on.

By sorting ethical dilemmas into these four categories, you can often decide which course of action is best. For example, consider the following case study:

Case Study: Your company makes a popular action figure toy with many tiny accessories. Countless children enjoy the toy, especially all those tiny boots, hats, backpacks, weapons, etc. However, a few children have choked on some of the small pieces that go with the toy. How would the four levels of ethics govern how to handle this situation?

Answer: In this case, a human right (life) has more weight than utility (thousands of children versus a few). So, the ethical choice would be to alter the toy or stop selling it.

Link

For more information on working with international readers, see Chapter 3, page 53.

Social ethical issues are rarely this clear cut, though. After all, rights, justice, utility, and care are open to interpretation and debate. A union organizer, for example, may see a company's resistance to unionizing as a violation of the "right to free assembly" provided by the U.S. Constitution. The company, on the other hand, may point to federal laws that allow it to curb union activities. This kind of debate over rights and justice happens all the time.

Another problem is that people miscategorize their ethical issue.

Case Study: Your town's city council has decided to implement a no-smoking policy that includes all public property and restaurants. Many smokers now find it impossible to have their smoke break around public buildings and in restaurants. They argue that their "right to smoke" is being violated. How might you resolve this ethical issue?

Answer: Actually, there is no such thing as a right to smoke. Smoking is a legal issue (a matter of justice) and a utility issue (the health interests of the nonsmoking majority

Want to see the U.S. Constitution? Go to
www.pearsonhighered.com/johnsonweb3/5.5

versus the interests of a smoking minority). So, if the city chooses to ban smoking on public property or in restaurants, it can do so legally as long as it applies the law fairly to all. It is not violating anyone's human or constitutional rights.

Of course, defenders of smoking may point out that restricting smoking may hurt businesses like bars and restaurants (a utility argument). Advocates of nonsmoking places might counter that secondhand smoke may cause cancer in patrons and employees of these establishments (also a utility argument). The city council may take these arguments into consideration before passing a new law.

When making decisions about social ethical issues, it is important to first decide which ethical categories fit the ethical dilemma you are pondering. Then, decide which set of ethics has more significance, or gravity. In most cases:

- issues involving *rights* will have more gravity than issues involving *laws*.
- issues involving *laws* will have more gravity than issues involving *utility*.
- issues involving *utility* will have more gravity than issues involving *care*.

AT A GLANCE

Four Categories of Social Ethics

- Rights—Civil rights and constitutional rights
- Justice—Laws and corporate policies
- Utility—Greatest good (majority rules)
- Care—Tolerance and compassion for others

There are, of course, exceptions. In some cases, utility may be used to argue against laws that are antiquated or unfair. For example, it was once legal to smoke just about anywhere, including the workplace (and the college classroom). By using utility arguments, opponents of smoking have successfully changed those laws. Today, smoking is evermore restricted in public.

Conservation Ethics

Increasingly, issues involving our ecosystems are becoming sources for ethical dilemmas. With issues such as global warming, nuclear waste storage, toxic waste disposal, and overpopulation, we must move beyond the idea that conservation is a personal virtue. We are now forced to realize that human health and survival are closely tied to the health and survival of the entire ecosystem in which we live. Conservation ethics involve issues of water conservation, chemical and nuclear production and waste, management of insects and weeds in agriculture, mining, energy production and use, land use, pollution, and other environmental issues.

One of America's prominent naturalists, Aldo Leopold, suggested that humans need to develop a *land ethic*. He argued:

All ethics so far evolved rest upon a single premise: that the individual is a member of a community of interdependent parts.... The land ethic simply enlarges the boundaries of the community to include soils, waters, plants, and animals, or collectively: the land. (p. 239)

In other words, your considerations of ethics should go beyond the impact on humans and their communities. The health and welfare of the ecosystem around you should also be carefully considered.

People who work in technical fields need to be especially aware of conservation ethics, because we handle so many tools and products that can damage the ecosystem. Without careful concern for use and disposal of materials and wastes, we can do great harm to the environment.

Aldo Leopold and the Land Ethic

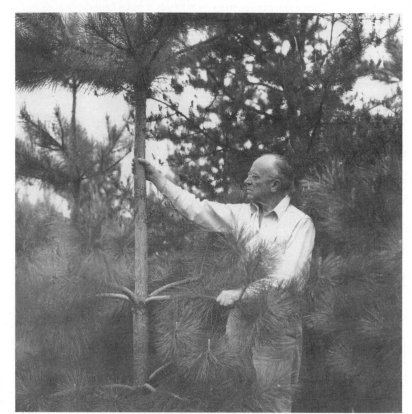

Aldo Leopold, a naturalist and conservationist, developed the concept of a "land ethic," which defines a sustainable relationship between humans and nature.

Source: Aldo Leopold Foundation.

Ultimately, conservation ethics are about *sustainability*. Can humans interact with their ecology in ways that are sustainable in the long term? Conservation ethics recognize that resources must be used. They simply ask that people use resources in sustainable ways. They ask us to pay attention to the impact our decisions have on the air, water, soil, plants, and animals on this planet.

Conservation ethics are becoming increasingly important. The twenty-first century has been characterized as the "Green Century," because humans have reached a point where we can no longer ignore the ecological damage caused by our decisions. For example, within this century, estimates suggest that human-caused climate change will raise global temperatures between 2 and 10 degrees. Such a rise would radically alter our ecosystem.

Moreover, as emerging markets such as China and India continue to grow, the world economy will need to learn how to use its limited resources in ways that are fair and conscientious.

GO TO THE NET

For more information on conservation ethics, go to
www.pearsonhighered.com/johnsonweb3/5.6

Stopping Cyberbullying and Computer Harassment

Have you ever been cyberbullied or harassed through a computer? Cyberbullying is the use of a computer to harm or threaten others psychologically, economically, or in a way that damages their careers or personal lives. Computer harassment involves using a computer to disturb or threaten others because of their race, gender, sexual orientation, disability, ancestry, religious affiliation, or other inherent characteristics. People also use computers to sexually harass others by making unwanted sexual advances and using demeaning or humiliating sexual comments.

To combat cyberbullying and computer harassment, anti-harassment policies are being developed by schools, workplaces, video games, Internet service providers, and social networking sites. Figure A, for example, shows the World of Warcraft®'s policies for dealing with harassment.

World of Warcraft's Anti-Harassment Policies

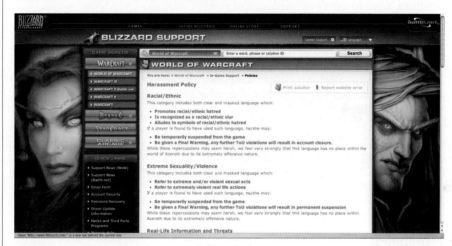

Source: Blizzard Entertainment, http://us.blizzard.com/support/article.xml?articleId=20226&categoryId=2415&parentCategoryId=2318&pageNumber=1

In some cases, people don't even realize they are bullying or harassing others. When e-mailing or texting others, they use aggressive language such as "I'll have you fired for this," or "I'm going to kill myself," or "Tomorrow, I think I'll bring my assault rifle to the meeting." Or, they might think it's funny to e-mail dirty jokes or pornographic cartoons or images to their co-workers.

(continued)

HELP

Figure A: Bullying and harassment are not new in school or the workplace, but computers add some important new elements that make these activities especially harmful. Bullies and harassers can mask their identities or pretend to be someone else. Video games often allow people to do things they would not do in real life. Meanwhile, people can spread rumors and harmful information quickly through texting, games, Listservs, and social networking.

Unfortunately, these kinds of behaviors are not uncommon on campuses and in today's workplace. So, what should you do if someone is bullying or harassing you electronically? Here are some tips for preventing and stopping cyberbullying and harassment:

Prevention

NEVER GIVE OUT PERSONAL INFORMATION ONLINE—This includes social networking sites. You might think only your friends can access your Facebook or MySpace page, but it is not that difficult for someone else to gain access to that information through one of your "friends." Your e-mails and text messages can also be easily forwarded, so any personal information may fall into the wrong hands.

DON'T PUT COMPROMISING CONTENT ONLINE—Compromising pictures, video, or statements have a way of being copied, leaked, and forwarded. If you have done or written something that would look bad to your classmates, parents, professors, or a future employer, you should not put it online or send it through your phone. Once it's out there, it can be saved and used against you.

REFUSE TO PASS ALONG MESSAGES FROM CYBERBULLIES AND HARASSERS—Instead, tell the cyberbully or harasser to stop sending the messages and report the incident to your professors, the Dean of Students, or your supervisor at work. Even if you are just making a victim aware that harmful things are being said about him or her, your forwarding of the message to the victim or others is really doing work for the bully or harasser.

KNOW WHO YOU ARE TALKING TO—Cyberbullies, harassers, and predators often try to build relationships with their victims over time. Then, once a "friendship" is established, they try to exploit their victim's trust. As soon as your "friend" does or says something strange, offensive, or makes an inappropriate request, you should end the relationship.

Stopping It

TELL THE PERSON TO STOP—The person who is bullying or harassing you may not realize he or she is being intimidating or offensive. Firmly tell that person that his or her messages are not welcome and that they should cease.

BLOCK MESSAGES—Social networking sites, video games, Internet service providers, and mobile phones almost always give you the ability to block messages from people who you don't want to hear from.

SAVE AND PRINT MESSAGES—By saving and printing messages, you are collecting evidence that can be used against the cyberbully or harasser. Even if you don't want to do something now or don't know who is doing it, you should keep any messages in case the problem escalates or the person is identified.

For more information on cyberbullying, go to
www.pearsonhighered.com/johnsonweb3/5.7

FILE A COMPLAINT—Most networking and game sites have a procedure for filing a complaint that will warn, suspend, or ban the person who is bullying or harassing you. If you are being bullied or harassed in college, you can file a complaint with your university's Dean of Students office or the Equal Opportunity Office at your campus. At a workplace, you should contact the Human Resources office at your company.

Networking sites and video games are going to challenge ethics in new ways. These games can be problematic because players earn virtual money that can be converted into real money. So, we are already seeing instances of cybermuggings, fraud, and even virtual rape. Take steps to avoid being a victim.

Resolving Ethical Dilemmas

No doubt, you will be faced with numerous ethical dilemmas during your career. There is no formula or mechanism you can use to come up with the right answer. Rather, ethical dilemmas usually force us to choose among uncomfortable alternatives.

Doing the right thing can mean putting your reputation and your career on the line. It might mean putting the interests of people above profits. It also might mean putting the long-term interests of the environment above short-term solutions to waste disposal and use of resources.

Confronting an Ethical Dilemma

When faced with an ethical dilemma, start considering it from all three ethical perspectives: personal, social, and conservation (Figure 5.4).

> **Personal ethics**—How does my upbringing in a family, culture, and faith guide my decision? How can I do unto others as I would have them do unto me?
>
> **Social ethics**—What rights or laws are involved in my decision? What is best for the majority? How can I demonstrate caring by being tolerant and compassionate?
>
> **Conservation ethics**—How will my decision affect the ecosystem? Will my choice be ecologically sustainable in the long term?

With most ethical dilemmas, you will find that ethical stances conflict. To resolve the dilemma, it helps to first locate the "ethical tension"—the point where two or more ethical stances are not compatible. For example:

- An emergency room doctor who treats gunshot victims believes gun ownership should be highly restricted. Constitutional law, however, makes gun ownership a right. *Here, rights are in tension with utility.*
- Someone offers you a draft of your competitor's proposal for an important project. With that information, your company would almost certainly win the contract. However, your industry's code of ethics regarding proprietary information forbids you from looking at it. *Here, justice is in tension with utility.*

- Your company owns the rights to the timber in a forest, but an endangered species of eagle lives there and its habitat, by law, should be protected. *Here, justice, rights, and conservation are in tension.*
- A legal loophole allows your company to pump tons of pollution into the air even though this pollution harms the health of the residents in a small town a few miles downwind. *Here, personal and conservation ethics are in tension with justice.*

Balancing the Different Issues in an Ethical Dilemma

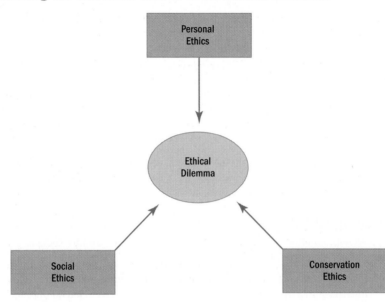

Figure 5.4: Resolving an ethical dilemma requires you to consider it from various ethical perspectives.

Link

For more information on liability in technical documents, see Chapter 20, page 562.

Product liability is a place where these kinds of conflicts become especially important. If your company produces a product that harms people unintentionally, the company may still be found liable for damages. It is not enough to simply cover your documentation with warnings. Even if warnings are provided, your company might be found negligent by the courts and ordered to pay damages.

Resolving an Ethical Dilemma

When faced with an ethical dilemma, you can use the following five questions to help you resolve it. These questions are a variation of the ones developed by Professor Sam Dragga in an article on ethics in technical communication (1996).

> **Do any laws or rules govern my decision?** — In many cases, laws at the federal, state, and local levels will specify the appropriate action in an ethical case. You can look to your company's legal counsel for guidance in these matters. Otherwise, companies often have written rules or procedures that address ethical situations.

Do you want to try solving some other ethical dilemmas? Go to
www.pearsonhighered.com/johnsonweb3/5.8

AT A GLANCE

Resolving Ethical Dilemmas

- Do any laws or rules govern my decision?
- Do any corporate or professional codes of ethics offer guidance?
- Are there any historical records to learn from?
- What do my colleagues think?
- What would moral leaders do?

Do any corporate or professional codes of ethics offer guidance?—Most companies and professional organizations have published codes of ethics. They are usually rather abstract, but they can help frame ethical situations so you can make clearer decisions. Figure 5.5 shows the code of ethics for the Institute of Electrical and Electronics Engineers (IEEE).

Are there any historical records to learn from?—Look for similar situations in the past. Your company may keep records of past decisions, or you can often find ethical cases discussed on the Internet. By noting successes or failures in the past, you can make a more informed decision.

What do my colleagues think?—Your co-workers, especially people who have been around for a while, may have some insight into handling difficult ethical situations. First, they can help you assess the seriousness of the situation. Second, they may be able to help you sort out the impact on others. At a minimum, talking through the ethical dilemma may help you sort out the facts.

What would moral leaders do?—You can look for guidance from moral leaders that you respect. These might include spiritual leaders, civil rights advocates, business pioneers, or even your friends and relatives. In your situation, what would they do? Sometimes their convictions will help guide your own. Their stories may give you the confidence to do what is right.

Facing an ethical dilemma, you will probably need to make a judgment call. In the end, you want to make an informed decision. If you fully consider the personal, social, and conservation perspectives, you will likely make a good decision.

When You Disagree with the Company

Ethical conflicts between you and your company need to be handled carefully. If you suspect your company or your supervisors are acting unethically, there are a few paths you can take:

Persuasion through costs and benefits—After you have collected the facts, take some time to discuss the issue with your supervisors in terms of costs and benefits. Usually, unethical practices are costly in the long term. Show them that the ethical choice will be beneficial over time.

Seek legal advice—Your company likely has an attorney who can offer legal counsel on some issues. You may visit legal counsel to sort out the laws involved in the situation. If your company does not have legal counsel or you don't feel comfortable using it, you may need to look outside the company for legal help.

Mediation—Companies often offer access to mediators who can facilitate meetings between you and others. Mediators will not offer judgments on your ethical case, but they can help you and others identify the issues at stake and work toward solutions.

Memos to file—In some cases, you will be overruled by your supervisors. If you believe a mistake is being made, you may decide to write a *memo to file* in which you express your concerns. In the memo, write down all the facts and your concerns. Then, present the memo to your supervisors and keep a copy for yourself. These memos are usually filed for future reference to show your doubts.

Link

For more on persuasion strategies, see Chapter 6, page 123.

Link

For more information on resolving conflicts, see Chapter 5, page 103.

IEEE CODE OF ETHICS

WE, THE MEMBERS OF THE IEEE, in recognition of the importance of our technologies in affecting the quality of life throughout the world and in accepting a personal obligation to our profession, its members and the communities we serve, do hereby commit ourselves to the highest ethical and professional conduct and agree:

1. to accept responsibility in making decisions consistent with the safety, health and welfare of the public, and to disclose promptly factors that might endanger the public or the environment;

2. to avoid real or perceived conflicts of interest whenever possible, and to disclose them to affected parties when they do exist;

3. to be honest and realistic in stating claims or estimates based on available data;

4. to reject bribery in all its forms;

5. to improve the understanding of technology, its appropriate application, and potential consequences;

6. to maintain and improve our technical competence and to undertake technological tasks for others only if qualified by training or experience, or after full disclosure of pertinent limitations;

7. to seek, accept, and offer honest criticism of technical work, to acknowledge and correct errors, and to credit properly the contributions of others;

8. to treat fairly all persons regardless of such factors as race, religion, gender, disability, age, or national origin;

9. to avoid injuring others, their property, reputation, or employment by false or malicious action;

10. to assist colleagues and co-workers in their professional development and to support them in following this code of ethics.

Approved by the IEEE Board of Directors | February 2006

Source: Institute of Electrical and Electronics Engineers. © 2006 IEEE. Reprinted with permission of the IEEE.

Figure 5.5:
In the IEEE Code of Ethics, just about every established field has a code of ethics you can turn to for guidance.

Want to see other codes of ethics? Go to
www.pearsonhighered.com/johnsonweb3/5.10

Whistle-blowing—In serious cases, especially where people's lives are at stake, you may even choose to be a whistle-blower. Whistle-blowing usually involves going to legal authorities, regulatory agencies, or the news media. Being a whistle-blower is a serious decision. It will affect your career and your company. Federal laws exist that protect whistle-blowers, but there is always a personal price to be paid.

Ethical situations should be carefully considered, but they should not be ignored. When faced with an ethical dilemma, it is tempting to walk away from it or pretend it isn't there. In any ethical situation, you should take some kind of action. Inaction on your part is both ethically wrong and might leave you or your company vulnerable to liability lawsuits. At a minimum, taking action will allow you to live with your conscience.

Websites exist that can help you make your decision by considering ethical case studies. For example, the Online Ethics Center at the National Academy of Engineering offers many case studies that are discussed by ethics experts (Figure 5.6). Perhaps one of these cases is similar to the one you face, and you can use the wisdom of these experts to make the ethical decision.

Online Ethics

Figure 5.6:
The Online Ethics Center at the National Academy of Engineering is a great place to learn about ethics in scientific and technical disciplines.

Source: The Online Ethics Center at the National Academy of Engineering, http://www.onlineethics.org.

GO TO THE NET

For more strategies for handling unethical situations at work, go to
www.pearsonhighered.com/johnsonweb3/5.11
Does whistle-blowing really work? Go to
www.pearsonhighered.com/johnsonweb3/5.12

Ethics in the Technical Workplace

Some legal and ethics scholars have speculated that the Information Age requires a new sense of ethics, or at least an updating of commonly held ethics. These scholars may be right. After all, our ethical systems, especially those involving forms of communication, are based on the printing press as the prominent technology. Laws and guidelines about copyright, plagiarism, privacy, information sharing, and proprietary information are all based on the idea that information is "owned" and shared on paper.

The fluid, shareable, changeable nature of electronic files and text brings many of these laws and guidelines into discussion. For example, consider the following case study:

> **Case Study:** You and your co-workers are pulling together a training package by collecting information off the Internet. You find numerous sources of information on the websites of consultants and college professors. Most of it is well written, so you cut and paste some of the text directly into your materials. You also find some great pictures and drawings on

AT WORK

Caroline Whitbeck, Ph.D.

DIRECTOR OF THE ONLINE ETHICS CENTER FOR ENGINEERING AND SCIENCE, CASE WESTERN RESERVE UNIVERSITY

The Center for Engineering and Science in Cleveland, Ohio, is an academic division that researches issues involving technology.

Why should technical professionals learn about ethics?

The practice of a profession, such as the profession of engineering, is characterized by two elements: (1) the practice directly influences one or more major aspect of human well-being and (2) it requires mastery of a complex body of knowledge and specialized skills. To become a professional in a field requires both formal education and practical experience.

The responsibility to achieve certain ends characterizes the core of professional ethics. For example, engineers have a responsibility for the public safety, and research investigators have a responsibility for the integrity of research. Achieving ends requires judgment in the application of professional knowledge.

Following moral rules, such as "Do not offer or accept bribes," important as they are, does not demand the exercise of judgment that is required to fulfill responsibilities. Because the judgment needed to fulfill responsibilities requires professional knowledge, those without professional knowledge cannot judge whether a professional is making responsible judgments, that is, behaving both competently and with due concern. This is the reason why professions establish standards of responsible practice for their practitioners.

Of course, individual practitioners and sometimes the professions themselves may prove untrustworthy, but when they do, everyone loses. When professionals prove irresponsible, they may be monitored more closely. Monitoring may work to see if the rules are being followed, but it does not readily check on the trustworthiness of professional judgments.

If society loses trust in a profession, people avoid relying on the service of members of that profession.

GO TO THE NET

To access the Online Ethics Center and similar sites, go to
www.pearsonhighered.com/johnsonweb3/5.13
Want to learn more about wired ethics? Go to
www.pearsonhighered.com/johnsonweb3/5.14

the Internet to add to your presentation. At what point does your cutting and pasting of text become a violation of copyright law? How can you avoid any copyright problems?

In the past, these kinds of questions were easier to resolve, because text was almost exclusively paper-based. Printed text is rather static, so determining who "owns" something is a bit easier. Today, the flexibility and speed of electronic media make these questions much more complicated.

At this time, many of our laws governing the use of information and text are evolving to suit new situations.

Copyright Law

Today, copyright law is being strained by the electronic sharing of information, images, and music. In legal and illegal forms, copies of books, songs, and software are all available on the Internet. According to the law, these materials are owned by the people who wrote or produced them; however, how can these materials be protected when they can be shared with a few clicks of a mouse? You can find out more about copyright law at the website of the U.S. Copyright Office (www.copyright.gov). Later in this chapter, U.S. copyright law is discussed in depth.

Trademarks

People or companies can claim a symbol, word, or phrase as their property by trademarking it. Usually, a trademark is signaled with a ™ symbol. For example, the Internet search engine Google™ is a trademarked name. The trademark signals that the company is claiming this word for its use in the area of Internet search engines.

To gain further protection, a company might register its trademark with the U.S. Patent and Trademark Office, which allows the company to use the symbol ® after the logo, word, or phrase. Once the item is registered, the trademark owner has exclusive rights to use that symbol, word, or phrase. For example, IBM's familiar blue symbol is its registered trademark, and it has the exclusive right to use it.

There are exceptions, though. The First Amendment of the U.S. Constitution, which protects free speech, has allowed trademarked items to be parodied or critiqued without permission of the trademark's owner.

Patents

Inventors of machines, processes, products, and other items can protect their inventions by patenting them. Obtaining a patent is very difficult because the mechanism being patented must be demonstrably unique. But once something is patented, the inventor is protected against others' use of his or her ideas to create new products.

Privacy

Whether you realize it or not, your movements are regularly monitored through electronic networks. Surveillance cameras are increasingly prevalent in our society, and your online movements are tracked. Websites will send or ask for *cookies* that identify your computer. These cookies can be used to build a profile of you. Meanwhile, at a workplace, your e-mail and phone conversations can be monitored by your supervisors. Privacy laws are only now being established to cover these issues.

Privacy in an Electronic World

Electronic networks offer many more opportunities for surveillance.

Information Sharing

Through electronic networks, companies can build databases with information about their customers and employees that can be shared with other companies. Information sharing, especially involving medical information, is an important issue that will probably be resolved in the courts.

Proprietary Information

As an employee of a company, you have access to proprietary information that you are expected not to share with others outside the company. In government-related work, you may even have a *security clearance* that determines what kinds of information you have access to. When you leave that company, you cannot take copies of documents, databases, or software with you. In some cases, you may even be asked to sign papers that prevent you from sharing your previous employer's secrets with your new employer.

Libel and Slander

You or your company can be sued for printing falsehoods (libel) or speaking untruths (slander) that damage the reputation or livelihood of another person or company. With the broadcasting capabilities of the Internet, libel and slander have much greater reach.

110   Chapter 5
**Ethics in the
Technical
Workplace**

Want to learn more about Internet libel, slander, and fraud? Go to
www.pearsonhighered.com/johnsonweb3/5.16

For example, a website that libels another person or company may be a target for legal retaliation. If you use e-mail to libel others, these messages could be used against you.

Fraud

The Internet is also opening whole new avenues for fraud. Con artists are finding new victims with classic fraud schemes. Websites, especially, can sometimes give the appearance of legitimacy to a fraudulent operation. Meanwhile, con artists use e-mail to find victims. (Have you received any e-mails from the widows of wealthy Nigerian dictators recently?)

Copyright Law in Technical Communication

An interesting flash point today is copyright law. A copyright gives someone an exclusive legal right to reproduce, publish, or sell his or her literary, musical, or artistic works. Copyright law in the United States was established by Article 1 of the U.S. Constitution. The U.S. law that governs copyright protection is called "Title 17" of the U.S. Code. You can find this code explained on the U.S. Copyright Office website at www.copyright.gov (Figure 5.7).

The U.S. Copyright Office Website

Figure 5.7:
You can visit the U.S. Copyright Office website to learn more about copyright law or to protect your own work.

Source: United States Copyright Office, http://www.copyright.gov.

Essentially, a copyright means creative work is someone's property. If others would like to duplicate that work, they need to ask permission and possibly pay the owner. Authors, musicians, and artists often sign over their copyrights to publishers, who pay them royalties for the right to duplicate their work.

New electronic media, however, have complicated copyright law. For example,

- When you purchase something, like a music CD, you have the right to duplicate it for your own personal use. What happens if you decide to copy a song off a CD and put it on your website for downloading? You might claim that you put the song on your website for your personal use, but now anyone else can download the song for free. Are you violating copyright law?
- According to Title 17, section 107, you can reproduce the work of others "for purposes such as criticism, comment, news reporting, teaching (including multiple copies for classroom use), scholarship, or research." This is referred to as "fair use." So, is it illegal to scan whole chapters of books for "teaching purposes" and put them on a CD for fellow students or co-workers?
- New technology like webcasting (using digital cameras to broadcast over the Internet) allows people to produce creative works. If you decided to webcast your and your roommates' dorm room antics each evening, would you be protected by copyright law?
- Blogs, or web logs, have become a popular way to broadcast news and opinions. Are these materials copyrighted?

The answer to these questions is "yes," but the laws are still being worked out. It is illegal to allow others to download songs off your website. It would be illegal to scan large parts of a book, even if you claimed they were being used for educational purposes. Meanwhile, you can protect webcasting and blogs through the copyright laws.

The problem is the ease of duplication. Before computers, copyrights were easier to protect because expensive equipment like printing presses, sound studios, and heavy cameras were required to copy someone else's work. Today, anyone can easily duplicate the works of others with a scanner, CD/DVD recorder, or digital video recorder.

Ultimately, violating copyright is like stealing someone else's property. The fact that it is easier to steal today does not make it acceptable. Nevertheless, a few scholars have argued that copyright law is antiquated and that this kind of electronic sharing is how people will use text and music in the future.

Asking Permission

To avoid legal problems, it is best to follow copyright law as it is currently written. You need to ask permission if you would like to duplicate or take something from someone else's work. You can ask permission by writing a letter or e-mail to the publisher of the materials. Publishers can almost always be found on the Internet. On their websites, they will often include a procedure for obtaining permissions. Tell them exactly what you want to use and how it will be used.

In some cases, especially when you are a student, your use may fall under the "fair use clause" of the Copyright Act. Fair use allows people to copy works for purposes of "criticism, comment, news reporting, teaching (including multiple copies for classroom use), scholarship, or research" (17 U.S. Code, sec. 107). If your use of the materials falls under these guidelines, you may have a *limited* right to use the materials without asking permission.

For example, fair use would likely allow you to use a song legally downloaded from the Internet as background music in a presentation for your class. However, it does not allow you to distribute that song freely to your friends, even if you claim you are doing so for educational purposes.

Copyrighting Your Work

What if you write a novel, take a picture, produce a movie, or create a song? How do you copyright it? The good news is that you already have. In the United States, a work is copyrighted as soon as it exists in written form. If you want, you can add the copyright symbol "©" to your work to signal that it is copyrighted. The copyright symbol, however, is no longer necessary to protect a work.

If you want to formally protect your work from copyright infringement (i.e., so you can sue someone who uses your work without your permission), you should register your copyright with the U.S. Copyright Office. This step is not necessary to protect your work, but it makes settling who owns the material much easier.

Plagiarism

One type of copyright infringement is plagiarism. In Chapter 7 of this book, plagiarism is discussed in depth, but the subject is worth briefly mentioning here. Plagiarism is the use of someone else's text or ideas as your own without giving credit. Plagiarism is a violation of copyright law, but it is also a form of academic dishonesty that can have consequences for your education and career.

For example, cutting and pasting words and images off the Internet and "patchwriting" them into your documents is a form of plagiarism, unless those materials are properly cited. To avoid questions of plagiarism, make sure you cite your sources properly and, when needed, ask permission to use someone else's work.

CHAPTER REVIEW

- Ethics are systems of moral, social, or cultural values that govern the conduct of an individual or community.

- Ethical dilemmas force us to choose among uncomfortable alternatives.

- When you are faced with an ethical dilemma, consider it from all three ethical perspectives: personal, social, and conservation.

- You can turn to sources like laws, professional codes of ethics, historical records, your colleagues, or moral leaders to help you make ethical choices.

- When you disagree with the company, use persuasion first to discuss costs and benefits. You may turn to legal avenues if persuasion doesn't work.

- Ethical guidelines are evolving to suit the new abilities of computers.

- Copyright law and plagiarism are two rapidly evolving areas of ethics in this computer-centered world.

- Privacy and information sharing are also becoming hot topics, because computer networks facilitate the collection of so much information.

Individual or Team Projects

1. At www.pearsonhighered.com/johnsonweb3/5.11, you will find codes of ethics that apply to a number of career paths. If you are an engineer, you might consider looking at the IEEE Code of Ethics. If you are going into medicine, you might look at the American Medical Association (AMA) Code of Ethics. Write a memo to your instructor in which you summarize the different codes of ethics, and highlight their important features. Discuss which specific ethical issues seem to pertain to your field.

2. Describe a real or fictional situation that involves a communication-related ethical dilemma. As you describe the situation, try to bring personal, social, and conservation ethics into conflict. At the end of the situation, leave the readers with a difficult question to answer.

   In a memo to your instructor, identify the ethical issues at stake in the situation you described and offer a solution to the ethical dilemma. Then, give your description to someone else in your class. He or she should write a memo to you and your instructor discussing the ethical issue at stake and offering a solution to the problem. Compare your original solution to your classmate's solution.

3. Visit the U.S. Copyright Office at http://www.copyright.gov and write a brief synopsis of copyright law. What are some copyright issues that seem to be changing? What are some issues that will likely stay the same? Do you believe that the current copyright law will stand the test of time? Do you think an alternative to copyright law is available, specifically an alternative that allows authors, musicians, and artists to be paid for their creative work?

4. Find examples of advertising that seem to stretch ethics by making unreasonable claims. Choose one of these examples and write a short report to your instructor in which you discuss why you find the advertisement unethical. Use the terminology from this chapter to show how the advertisement challenges your sense of personal ethics, social ethics, or conservation ethics.

5. After researching conservation ethics on the Internet (see www.pearsonhighered.com/johnsonweb3/5.6), develop a Conservation Code of Ethics for your campus. In your code, you might discuss issues of recycling, water usage, chemical usage, testing on animals, or release of pollution. If you were an administrator at your university, how might you go about putting your code of ethics into action?

114   Chapter 5
**Ethics in the
Technical
Workplace**

Collaborative Project

The case study at the end of this chapter discusses a difficult case in which a company might be doing something unethical. Read and discuss this case with a group of others. Sort out the ethical issues involved by paying attention to the personal, social, and conservation factors that shape the ethical dilemma.

If you were Hanna, how would you react to this ethical dilemma? How would you use writing to turn your reaction into action? Would you write a memo to your supervisor? Would you contact a government agency? Would a memo to file be appropriate? Would you blow the whistle?

Whichever path you choose, write a letter or memo to a specific reader (e.g., your supervisor, corporate management, the human resources office, a newspaper journalist) that urges the reader to take action. Summarize the situation for the reader of your document, and then suggest an appropriate course of action. Support your recommendation by highlighting the ethical issues involved and discussing the ramifications of inaction.

> For additional technical writing resources, including interactive sample documents, document design tutorials and guidelines, and more, go to **www.mytechcommlab.com.**

This Company Is Bugging Me

Last week, Hanna Simpson's employer, Ventron United, sent out an identical memo and e-mail to all 50 employees. The memo/e-mail announced that all employees would be required to have a Radio Frequency Identification (RFID) chip inserted into their left or right hand. The RFID would allow employees to open secure doors, use equipment, and log onto computers throughout the company's campus in Portland, Oregon.

The announcement explained that the company would be working on some projects that required high security. So, management wanted to ensure that only employees who were supposed to enter secure areas or use secure computers would have access. Also, the company experienced a break-in a couple months ago. The thieves took a few laptop computers, but the rumor was that the thieves were really looking for information on Ventron United's new products. They had used an employee's access card, which was reported missing the next day.

Hanna's two closest co-workers, Jim Peters and Georgia Miller, had very different reactions to the memo. Jim thought it was a great idea. He already had a personal RFID implanted in one of his hands. "It's about the size of a grain of rice. I don't notice it at all. But, now I don't need to carry keys anymore. It opens the doors to my house, and I can turn on my car with just a button. I'm looking forward to the day when I can pay for my groceries with the wave of my hand."

Georgia was less enthusiastic, even paranoid. She confided to Hanna, "I think they're just trying to track our movements around here. Besides, my body is my personal space. These RFIDs are an invasion of my privacy."

"But would you quit your job over this?" Hanna asked Georgia.

"I hope it doesn't come to that, but yes I would," Georgia replied. "I'm not some dog that needs a chip implanted in me to keep track of me."

Other employees were worried about the health implications of the chips. Some just felt uncomfortable about the idea of a chip being injected into their hands. Some were considering a lawsuit if the company didn't back down on the issue.

Do some research on the Internet about RFIDs. What ethical issues are in conflict here?

GO TO THE NET

To learn more about the ethics of RFIDs, go to
www.pearsonhighered.com/johnsonweb3/5.17

PART

2

Communicating in the Technical Workplace

CHAPTER

6

Planning and Persuasion

In this chapter, you will learn:

- To plan a project in an organized way.

- Strategies for identifying your objectives and "top-rank objective."

- To develop a list of tasks and a timeline for your project.

- How to use the technical writing process as a planning tool.

- How people are persuaded by reasoning and values.

- How to reason with logic, examples, and evidence to support your views.

- How to use values to appeal to common goals and ideals while using language familiar to your readers.

- About persuasion in high-context cultures.

The writing process for any technical document should start with a "planning phase" in which you think about how you are going to persuade your readers. Solid planning will save you time while helping you write a much more effective document.

While planning, you should also develop a *persuasion strategy*. All technical documents are persuasive in some way. Some kinds of documents, like proposals and recommendation reports, are intentionally persuasive. The aim of these documents is to convince the readers to say yes to an idea or take a specific action.

Other documents, like instructions or technical descriptions, may not seem persuasive in an obvious way. Nevertheless, these documents need to build the readers' confidence in you, your company, and your company's products and services. Even strictly "informational" documents still need to be persuasive, if only in a minimal way.

Planning and Persuasion Are Critical Skills

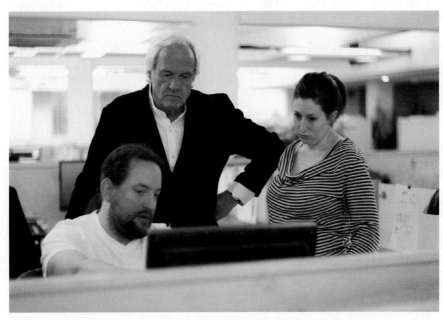

Effective planning and persuasion are important skills in today's technical workplace.

Using Strategic Planning

Strategic planning will save you time, while helping you produce higher quality documents and presentations that are persuasive to your readers. A time-tested method for strategic planning includes three steps: (1) setting objectives, (2) creating a list of tasks, and (3) setting a timeline (Figure 6.1).

Steps of Strategic Planning

Set Objectives

Create a List of Tasks

Set a Timeline

Figure 6.1: Good project planning involves identifying your objective and then breaking the project down into tasks that are set on a timeline.

Step 1: Set Your Objectives

To begin the planning process, you first need to figure out the objectives of your project.

LIST PROJECT OBJECTIVES   On your computer or a sheet of paper, make a list of the objectives of your project. For a smaller project, you may list only a few objectives. For a larger project, your list of objectives might include a variety of items, both significant and less significant.

IDENTIFY THE TOP-RANK OBJECTIVE   Rank your objectives from the most important to the least important. Identify your "top-rank objective" (or what a marketing guru would call your "TRO"). More than likely, your top-rank objective is going to be almost identical to the *purpose* of your project.

Then, express your project's top-rank objective in one sentence:

> Our main objective is to persuade the vice president of information technology to upgrade the wireless network on campus.

> The primary goal of this project is to develop a solar car that will be competitive in the American Solar Challenge race.

If you are having trouble expressing your top-rank objective or purpose in one sentence, you probably need to narrow the scope of your project. A top-rank objective that requires more than one sentence is probably too complicated to guide your strategic planning.

Step 2: Create a List of Tasks (or Task List)

Convert your objectives into tasks that you will need to perform. Logical mapping and developing a "task list" are helpful ways to make this conversion from objectives to tasks.

MAP OUT THE PROJECT TASKS   Put your top-rank objective (purpose) in the middle of your screen or a piece of paper, and ask yourself, "What are the two to five major steps necessary to achieve this goal?" Once you have identified your major steps, then identify the two to five minor steps that will help you achieve each major step.

You shouldn't reinvent the wheel with every new project. Instead, use a writing process that you can follow consistently. When mapping out your project tasks, you can use the "technical writing process" described in Chapter 2 to help you figure out the major steps of your project (Figure 6.2). Here are the stages of the writing process again:

Link

For more tips on using logical maps, go to Chapter 7, pages 146–148.

Planning and researching

Organizing and drafting

Improving the style

Designing

Revising and editing

Link

For more information about the technical writing process, go to Chapter 2, pages 18–34.

CREATE A TASK LIST   When the logical map is finished, you can transform it into a list of tasks, or a *task list* (Figure 6.2). The major steps in your map will become the larger stages of the project. Meanwhile, the minor steps will become individual tasks that you need to complete in each stage.

AT A GLANCE

Project Planning Steps
• Step 1: Set your objectives
• Step 2: Create a list of objectives
• Step 3: Set a timeline

Mapping Out a Plan

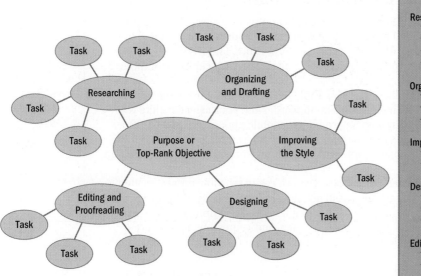

Project Task List

Researching
  Task
  Task
  Task
  Task

Organizing and Drafting
  Task
  Task
  Task

Improving the Style
  Task
  Task

Designing
  Task
  Task
  Task

Editing and Proofreading
  Task
  Task
  Task

Figure 6.2: To create a project plan, map out the two to five major steps. Then, add two to five minor tasks for each major step. Your plan can then be converted into a task list.

Step 3: Set a Timeline

In technical workplaces, setting timelines is essential. A timeline allows you to keep track of your progress toward completing the project. If you are working alone, the timeline will help you avoid procrastination (and a mad rush to the finish). If you are working with a team, the timeline will help everyone work together to reach the same milestones and deadlines.

ASSIGN A DATE TO EACH TASK   Working backward from your project's deadline, identify how much time each part of the project will require. Then, on your task list, write down specific dates when each part of the project should be completed. To help you fill out your timeline, online calendars and project planning software are available (see the Help box below). These calendering programs are widely used in technical workplaces.

Link

For more information on creating a project calendar, go to Chapter 4, pages 70–71.

SCHEDULE MEETINGS OR CHECKPOINTS   At regular intervals on your timeline (each week, every two weeks, or each month), schedule meetings with your team. Meetings can be boring, but people tend to use them as deadlines to get their tasks completed. If you are working alone, you can use the major steps in your project as "checkpoints." By putting project checkpoints into your schedule, you will ensure that you are making steady progress toward finishing the project.

HELP

Planning with Online Calendars

An online calendar is a helpful tool for planning out your project and coordinating with team members. Until recently, the best online calendars have been tied to large software suites like Microsoft Outlook, Corel Office, and IBM Lotus. These calendar programs are closely linked to the e-mail services included with these software suites.

When search engines like Google and Yahoo! jumped into the e-mail game, online calendars took an interesting leap forward. Now, your calendar (like your e-mail) can be accessed through any computer, mobile phone, or PDA that gives you access to Google or Yahoo! portals (Figure A). So, your calendar is no longer tied to your personal computer. You can access it anywhere. That's a significant advantage.

Online calendars are helpful because you can easily schedule events, including any deadlines and meetings (and social activities). Then, you can set up your preferences, and the calendar will send you reminder messages through your e-mail. That way, you won't miss an important meeting, and you will be fully aware when you blow past the deadline for a project—even if there's nothing you can do about it.

You can set up your calendar to let others check your schedule to find times when they can meet with you. If you scheduled a meeting or an after-work gathering, your calendar program can remind people with e-mail messages ("Hey, don't forget we're meeting at Cy's Roost on Thursday at 5:30").

Online calendars are especially useful for project planning. Once you have created your list of project tasks, you should enter the items into your calendar. Work

Want to learn more about creating a schedule? Go to
www.pearsonhighered.com/johnsonweb3/6.4

Scheduling with an Online Calendar

Figure A:
Online calendars, like this one from Yahoo!, are widely used in technical workplaces. They are usually linked to e-mail programs.

Source: Reproduced with permission of Yahoo! Inc. © 2006 by Yahoo! Inc. YAHOO! and the YAHOO! logo are trademarks of Yahoo! Inc.

backward from the project deadline, as shown in Figure A. Put something like "Proposal Due" on the deadline. From last to first, start entering the other tasks into the calendar. Once all the tasks are entered, you can move them around to create a project timeline.

Finally, set up your preferences so your online calendar sends you reminders about when parts of the project need to be completed. The reminders should keep you on schedule.

Online calendaring is easy, and it's a great way to keep yourself organized. Give it a try.

Developing a Persuasion Strategy

When you have finished your strategic planning, you should spend some time developing your "persuasion strategy." Persuasion, is more than a matter of changing other people's minds. Persuasion is also about giving people good reasons to do things they already want to do. In some cases, persuasion is about building someone else's confidence in you, your company, or your company's products and services.

Use a free online calendar program. Go to
www.pearsonhighered.com/johnsonweb3/6.5

Developing a Persuasion Strategy   123

So, how can you be more persuasive? In technical workplaces, people tend to respond to two types of persuasion:

Reasoning-based persuasion—The use of logic, examples, and evidence to support claims and ideas

Values-based persuasion—The use of shared ideals, mutual values, common goals, and earned credibility to build and strengthen relationships

Effective technical documents typically include a blend of reasoning-based and values-based persuasion strategies (Figure 6.3).

Types of Persuasion

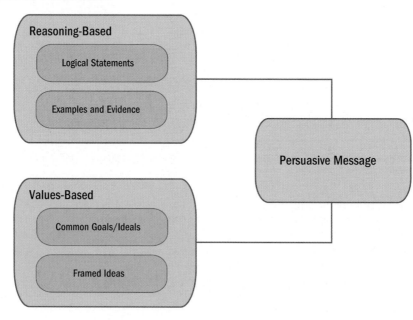

Figure 6.3: Technical documents and presentations typically include a blend of reasoning-based and values-based statements.

Persuading with Reasoning

Let's look at reasoning first. Reasoning is the use of logic and examples to show others the strengths and merits of your ideas. Reasoning has two basic forms:

Reasoning with logic—Using logically constructed statements such as *if . . . then, either . . . or, cause and effect, effect . . . because, costs and benefits,* and *better and worse*

Reasoning with examples and evidence—Using real or realistic statements, such as examples, experiences, facts, data, observations, and quotes from experts

Figure 6.4 shows a briefing from NASA that uses an assortment of logical statements, examples, and evidence to support its points. In this example, notice

To learn more about rhetoric and persuasion, go to
www.pearsonhighered.com/johnsonweb3/6.6
For websites that discuss using logic, go to
www.pearsonhighered.com/johnsonweb3/6.7

A Document That Uses Reasoning

Figure 6.4: In this letter, scientists at NASA Goddard Institute try to explain their findings about global warming. Due to political resistance, scientists often need to be persuasive to explain the results of their research.

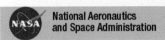
National Aeronautics and Space Administration

Scientists Confirm Earth's Energy Is Out of Balance
04.28.05

Scientists have concluded more energy is being absorbed from the sun than is emitted back to space, throwing the Earth's energy "out of balance" and warming the globe.

Scientists from NASA, Columbia University, New York, and Lawrence Berkeley National Laboratory, Berkeley, Calif., used satellites, data from buoys and computer models to study the Earth's oceans. They confirmed the energy imbalance by using precise measurements of increasing ocean heat content over the past 10 years.

These images are visual observations that show the changes in the Earth's climate.

Image above: *Clouds and the Earth's Radiant Energy System (CERES) measurements show the reflected solar radiation (left) and emitted heat radiation (right) for January 1, 2002. In both images, the lightest areas represent thick clouds, which both reflect radiation from the Sun and block heat rising from the Earth's surface. Notice the clouds above the western Pacific Ocean, where there is strong uprising of air, and the relative lack of clouds north and south of the equator.*

Data is used to support factual claims.

The study reveals Earth's energy imbalance is large by standards of the planet's history. The imbalance is 0.85 watts per meter squared. That will cause an additional warming of 0.6 degrees Celsius (1 degree Fahrenheit) by the end of this century.

The main point is expressed as a cause and effect statement.

To understand the difference, think of a one-watt light bulb shining over an area of one square meter (10.76 square feet). Although it doesn't seem like much, adding up the number of feet around the world creates a big effect. To put this number into perspective, an imbalance of one-watt per square meter, maintained for the past 10,000 years is enough to melt ice equivalent to one kilometer (0.6 mile) of sea level, if there were that much ice.

Source: Rani Chohan, Goddard Space Flight Center, NASA, http://www.nasa.gov/vision/ earth/environment/earth_energy.html.

(continued)

Image above: *The Earth's energy balance refers to the amount of energy received from the sun (yellow arrows) minus the energy reflected and emitted from the Earth (red arrows). Clouds play an important role in regulating this balance. Thin cirrus clouds permit sunlight to pass through them, while blocking a significant amount of the heat radiating from the surface. Thick cumulus clouds reflect most sunlight, and block the majority of heat radiating from the surface. Credit: NASA*

"The energy imbalance is an expected consequence of increasing atmospheric pollution, especially carbon dioxide, methane, ozone, and black carbon particles. These pollutants block the Earth's heat radiation from escaping to space, and they increase absorption of sunlight," said Jim Hansen of NASA's Goddard Institute for Space Studies, New York. He is the lead author of the new study, which is in this week's Science Magazine Science Express.

As the Earth warms it emits more heat. Eventually the Earth will be back in balance, if the greenhouse gas emissions are kept at the same level of today. Scientists know it takes the ocean longer to warm than the land. The lag in the ocean's response has practical consequences. It means there is an additional global warming of about one degree Fahrenheit that is already in the pipeline. Even if there were no further increase of human-made gases in the air, climate would continue to warm that much over the next century.

Warmer world-wide water temperatures also affect other things. "Warmer waters increase the likelihood of accelerated ice sheet disintegration and sea level rise during this century," Hansen said. Since 1993, sea levels have been measured by satellite altimeters. Data has shown they have risen by approximately 3.1 centimeters or 1.26 inches per decade.

Although 3.1 centimeters is a small change, the rate of increase is twice as large as in the preceding century. There are positive feedbacks that come into play, as the area of ice melt increases. The researchers agree monitoring ice sheets and sea level is necessary to best ensure the system is in balance.

This quote from an expert supports and clarifies the message.

Here are "if . . . then" statements.

These are cause and effect statements.

The data used here helps explain the magnitude of the problem.

how the writer from NASA weaves logic, examples, and evidence together to present a persuasive document.

Reasoning with Logic

Logic offers you a variety of ways to persuade your readers. When using logic, you are trying to appeal to your readers' common sense or beliefs. Logic allows you to build

More information about using examples is available at
www.pearsonhighered.com/johnsonweb3/6.8

up more complex ideas from simpler facts. When using logical reasoning, you are essentially saying things like, "If you believe X, then you should also believe Y," or "Either you believe X, or you believe Y," or "X is happening because Y happened."

IF…THEN   Perhaps the most common logical pattern is the *if…then* statement. When using *if…then* statements, you persuade your readers by demonstrating that something they already believe leads logically to something else they should also accept.

> If we are going to be ready for next summer's hurricane season, then we cannot wait until next spring to begin planning.

> Internet thieves will be able to steal your identity if you don't take steps to protect yourself.

EITHER…OR   When using *either…or* statements, you are offering your readers a choice between two paths.

> Either we take steps to control crime in our area or we risk handing over our streets to criminals.

> We need to either start redesigning the car to use a hybrid engine or take the risky path of hoping oil prices drop dramatically over the next few years.

Either…or statements can be risky because the readers may choose the wrong path, or they may reject both choices. So, you need to make the "correct" path obvious to them if you want them to go in a particular direction.

CAUSE AND EFFECT   When using *cause and effect* statements, you are demonstrating to your readers how specific causes lead to specific effects.

> Gradually, desertification causes a dryland, such as the Sonoran Desert, to lose its ability to support plants and animals.

> The effects of this problem can be sobering. Last year, intoxicated drivers caused 83 accidents in Holt County, killing four people and costing taxpayers $1.1 million.

Similarly, the word *because* can also signal a cause and effect relationship:

> The Stonyridge Windfarm project should be approved because it will generate electricity and revenue without further polluting our area.

> The Internet went down repeatedly over the summer because the server kept crashing.

COSTS AND BENEFITS   When trying to persuade readers, you might find it helpful to show them the *costs and benefits* of your ideas. In most cases, you will want to show them that the benefits outweigh the costs. In some cases, though, you might point out that the costs would be too high for the few benefits gained.

> Building a wireless network in the fraternity house will require an up-front investment, but we would save money because each member won't need to pay the phone company for a separate DSL line.

> Since St. Elizabeth Hospital's main building has become obsolete in almost every way, the benefits of remodeling it would not justify the costs.

Link

For more on using if…then and either…or statements, go to Chapter 8, pages 191–193.

Link

For more help using cause and effect and costs and benefits, go to Chapter 8, pages 188–191.

Link

For more help with using empirical observations, go to Chapter 7, page 153.

BETTER AND WORSE Another persuasive strategy is to show that your ideas are better than the alternatives.

> In 2008, we decided to go with AMD's Athlon 64 FX-57 microprocessor for our gaming-dedicated computers. The other chips on the market just couldn't match its balance of speed and reliability.

> In the long run, we would be better off implementing our automation plan right now, while we are retooling the manufacturing plant. If we wait, automating the lines will be almost impossible as we return to full capacity.

Reasoning with Examples and Evidence

Examples and evidence allow you to reason with your readers by showing them real-world or realistic support for your claims.

FOR EXAMPLE Using an example is a good way to clarify and support a complex idea while making it seem more realistic to the readers.

> For example, some parasitoid wasps inject polydnaviruses into the egg or larva of the moth host. The wasp eggs survive in the host's body because the virus suppresses the immunity of the host.

> If, for instance, a high-speed railway were to be built between Albuquerque and Santa Fe, commuters would cut their commuting times in half, while avoiding the dangerous 50-mile drive.

Phrases like *for instance, in a specific case,* and *such as* signal that you are using an example.

EXPERIENCES AND OBSERVATIONS Personal experiences and observations can often be persuasive as long as the readers trust the credibility of the source.

> While we were in the Arctic Circle, we observed a large male polar bear kill and devour a cub, turning to cannibalism to survive.

> Once our team began closely monitoring the medications of schizophrenic prisoners at the Oakwood Correctional Facility, we observed a dramatic decline in the number of hallucinations and delusional episodes among the prisoners.

FACTS AND DATA Empirically proved facts and data generated from experiments and measurements can offer some of the strongest forms of evidence. People trust observed facts and numbers.

> The facts about influenza epidemics on college campuses are amazing. A 2004 study on college campuses showed that 5 out of 10 students will become infected when the virus finds its way into a dormitory. (Venn, p. 15)

> Recently published data shows that for every child who may experience a prolonged benefit from the hemophilus vaccine, 2 to 3 children may develop vaccine-induced diabetes. (Akers & Wilson, p. 126)

Remember always to cite any sources when referring to facts and data.

AT A GLANCE

Reasoning-Based Persuasion

- If...then
- Either...or
- Cause and effect
- Costs and benefits
- Better and worse
- For example
- Experiences and observations
- Facts and data
- Quotes from experts

QUOTES FROM EXPERTS   A recognized authority on a subject can also be used to support your points. You can use quotes as evidence to back up your claims.

> Dr. Jennifer Xu, a scientist at Los Alamos National Laboratory, recently stated, "Our break-throughs in research on edge localized modes (ELMs) demonstrate that we are overcoming the hurdles to fusion nuclear power." (Xu, 2006)

> The lead biologist for the study, Jim Filks, told us that mercury levels in fish from the Wildcat Creek had dropped 18 percent over the last ten years, but levels were still too high for the fish to be safely consumed.

Link

For more ideas about using interviews and observations to collect information, go to Chapter 7, page 153.

Persuading with Values

Values-based persuasion can be more subtle than reasoning-based persuasion. Values-based persuasion is effective because people prefer to say yes to someone who holds the same values and beliefs as they do (Lakoff). Moreover, confidence and trust go a long way toward convincing people what to believe and what to do.

Values-based persuasion uses two forms:

> **Goals and ideals**—The use of goals, needs, values, and attitudes that you share with your readers

> **Frames**—The use of words, phrases, and themes that reflect your readers' point of view and values

When using values-based persuasion, you are trying to convince your readers to *identify* with you, your company, or your company's products and services. Advertisers, of course, spend billions each year appealing to consumers' sense of values (Figure 6.5). This kind of "branding" or name recognition can take time to develop. But once a company succeeds in associating itself with a particular set of positive values, their written materials will be much more effective.

Appealing to Common Goals and Ideals

To appeal to common goals and ideals, you should begin by looking closely at the profile you developed of your readers, which was discussed in Chapter 3. Your readers' profile will contain clues about your readers' goals, needs, values, and attitudes.

Link

For more information about profiling readers, turn to Chapter 3, pages 39–46.

GOALS   Almost all people have personal and professional goals they are striving to reach, and so do most companies. If possible, discuss your readers' goals with them. A good idea is to ask your readers to state their goals and tell you which ones are most important. Then, in your document, you can show them how your product or services will help them reach those goals.

> Solar power will maintain the appearance of your home while adding value. From the street, photovoltaic shingles look just like normal roofing materials, but they will give you energy independence, save you money, and increase the value of your home.

> By renovating Centennial Park, Mason City will go a long way toward reaching its goal of becoming a town that promotes a lifestyle of learning and leisure.

Appealing to Values

Readers identify with these elderly people who are still apparently active people.

Notice how words like "approved," "science," and "common sense" reflect reader values.

The fear of being unable to communicate (an attitude) is brought out here and other places in the text.

Use of "you" appeals directly to the readers.

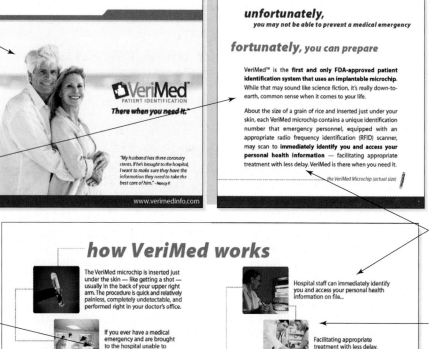

Figure 6.5: The product described in this booklet is an RFID (Radio Frequency Identification) device that is injected under the skin—a potentially troubling product. The authors appeal to values to persuade readers of the merits of the product.

The shared goal of quick service is regularly mentioned.

Images are used to put the readers in the place of a patient.

The image shows the achievement of a common goal.

Source: VeriMed, http://www.verimedinfo.com.

NEEDS   A fundamental difference exists between needs and goals. *Needs* are the basic requirements for survival.

> Your just-in-time manufacturing process requires suppliers to have parts ready to ship within hours. We can guarantee that parts will be shipped within one hour of your order.

> Equipment in the operating room must be sterilized with alcohol to ensure our patients' safety.

If you aren't sure what your readers need, go ahead and ask them. They will usually be very specific about what they require.

SHARED VALUES   Spend some time identifying any values that your readers, you, and your company have in common. For example, if your readers stress high-quality

Mary Nauman
FREELANCE WRITER-EDITOR OF GRANTS AND PROPOSALS

Mary Nauman specializes in writing grant proposals, including ones for scientific, medical, and technical research.

Can you give us some tips for writing persuasive proposals?

Two key methods for crafting a persuasive proposal are: (1) always provide your reader with a compelling yet accurate story about your project; and (2) ensure your goals are aligned with the mission of the people who will fund the project.

First, your proposal needs to be compelling if you're going to persuade someone to fund the project. Recently, I asked a client to explain why the federal government should fund his research on mammograms. He answered, "This would provide a tool that doesn't currently exist." Sorry. There are a lot of tools that don't currently exist, but that doesn't mean these tools are needed. Proposal writers must take a step back and look at the full impact of the project.

To describe the full impact, my client's proposal immediately stated that his research is a first step toward reducing the number of false-positive mammograms. False-positive results are a painful and expensive problem experienced by 49% of all women at least once within their first 10 mammograms. By providing this context without overselling his research, my client conveyed his comprehensive vision. He also paved the way for a long-term relationship with the funder.

Secondly, a proposal is only successful if it matches the funder's interests. For example, two private foundations may state that their missions are to reduce smoking. One may support lung cancer research, however, while the other may only support anti-smoking advertisements. It is important to carefully do research on your target donor and, if necessary, to call and ask specific questions. Selectively choosing your potential funding agency and placing your project within the appropriate context are both critical steps toward crafting a compelling proposal.

AT WORK

service, more than likely they will want to work with suppliers who also value high quality.

> If you have any questions or problems with your custom CAD software, we have engineers available 24/7 through our website and over the phone. When you call, a person, not a machine, will answer the phone.

> Like your organization, our company has a "People First" policy, which we believe is essential toward keeping our employees satisfied and productive.

You can usually find any shared values on your readers' websites.

ATTITUDES   A reader's attitude toward your subject, you, or your company can greatly determine whether your message is persuasive. So, you might use your reader's positive attitude to your advantage or show understanding when the reader has a negative attitude.

Link

For more strategies on identifying shared values through the Internet, go to Chapter 3, pages 45–46.

Values-Based Persuasion

- Goals
- Needs
- Shared values
- Attitudes
- Framing
- Reframing

Purchasing your first new car can be exciting and just a little bit stressful. This guide was created to help you survive your first purchase.

Like you, we're always a bit nervous when our company upgrades operating systems on all its computers. The process, though, is mostly painless, and the improved speed and new features are worth the effort.

Words alone will rarely change someone's attitude, but you can show that you empathize with your readers' point of view. Perhaps your readers will give you a chance if they think you understand how they feel about the subject.

Framing Issues from the Readers' Perspective

Being persuasive often means seeing and describing an issue from your readers' perspective. Linguists and psychologists call this *framing* an issue (Lakoff, p. 24). By properly framing or reframing an issue, you can appeal to your readers' sense of values.

FRAMING   To frame an issue, you should look closely at the profile you developed of the readers. Locate the one or two words or phrases that best characterize your readers' perspective on the issue. For example, let's say your readers are interested in *progress.* Your profile shows that they see the world in terms of *growth* and *advancement.* You can use logical mapping to develop a frame around that concept.

In Figure 6.6, for example, the word "progress" was put in the middle of the screen. Then, words and phrases that cluster around "progress" were added. As you fill it in,

Link

For more help on using style persuasively, go to Chapter 9, page 222.

Using Logical Mapping to Create a Frame

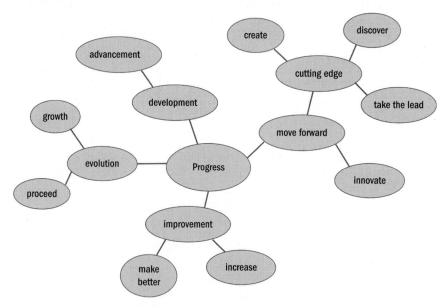

Figure 6.6: Framing is a useful way to describe an idea through the readers' point of view. Here, a logical map is showing how the word "progress" would be framed.

Want to learn more about framing? Go to
www.pearsonhighered.com/johnsonweb3/6.12

your logical map should show you the frame from which your readers understand the issue you are discussing. Knowing their frame will help you choose content and phrasing that support or reinforce their point of view.

REFRAMING   In some cases, your readers might not see an issue in a way that is compatible with your or your company's views. In these situations, you may need

Reframing by Using Common Values

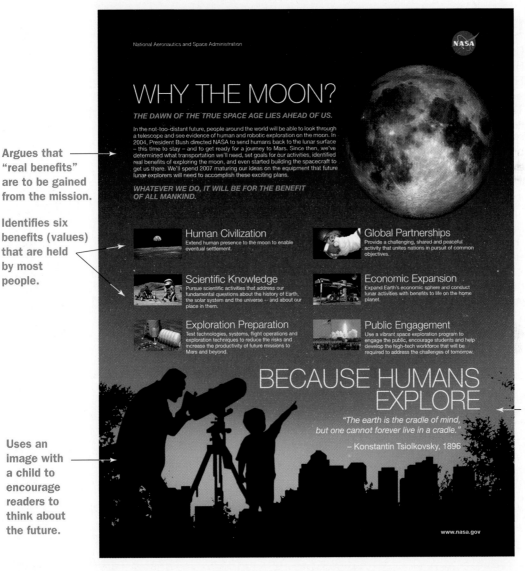

Argues that "real benefits" are to be gained from the mission.

Identifies six benefits (values) that are held by most people.

Uses an image with a child to encourage readers to think about the future.

Figure 6.7:
In this poster, NASA uses several common values to argue that travel to the moon is important. NASA's goal is to overcome public skepticism about the cost and necessity of the program.

The answer to the "Why the Moon?" question is answered by appealing to an assumed trait of humanity—a desire to explore.

to "reframe" the issue for them. To reframe an issue, look a little deeper into your reader profile to find a value that you and your readers share. Once you locate a common value, you can use logical mapping to reframe the issue in a way that appeals to the readers. For example, the poster in Figure 6.7 shows how NASA is trying to reframe the debate over humans traveling to the moon. They recognize that some people may be skeptical about the need to go to the moon. So, they reframe the argument around common values that most readers will share.

Persuasion in High-Context Cultures

Cross-cultural communication can be difficult, but persuasion is especially challenging when the audience or readers are from a "high-context" culture. High-context cultures, which include many in Asia, the Middle East, and sub-Saharan Africa, tend to put more emphasis on community than individuals. These cultures also tend to put a high value on consensus, interpersonal harmony, hierarchy, and rituals. They tend to stress long-term goals over short-term gains.

What is not said in a conversation is often very significant, because people from high-context cultures rely on *contextual cues* to interpret what a speaker means. As a result, negotiators from low-context cultures, such as Europe, the Americas, and Australia, tend to mistakenly believe that high-context negotiators are inefficient, indirect, vague, and ambiguous. Meanwhile, negotiators from high-context cultures, such as Japan, China, Korea, Indonesia, and parts of Africa, often find their low-context counterparts, especially Americans, to be abrupt, aggressive, and far too emotional.

Persuasion and negotiation are still important in high-context cultures, but the interactions are more subtle and indirect. Here are some guidelines to help you out:

Guideline 1: Develop long-term relationships — High-context cultures put a high value on existing relationships and reputations. Spend time familiarizing your readers with your company and explaining its reputation in the field. This will take time, because companies and goverments from high-context cultures tend to think long term. Your company may need years of effort to build a strong relationship with a company from a high-context culture.

Guideline 2: Use intermediaries to build relationships — In high-context cultures, strangers tend to be treated skeptically. So, you might look for a trusted intermediary who knows both parties to help you make connections.

Guideline 3: Rely on facts and reasoning — In high-context cultures, attempts to persuade directly are often viewed negatively. So, a fact-based presentation with solid reasoning tends to be much more effective.

Working in High-Context Cultures

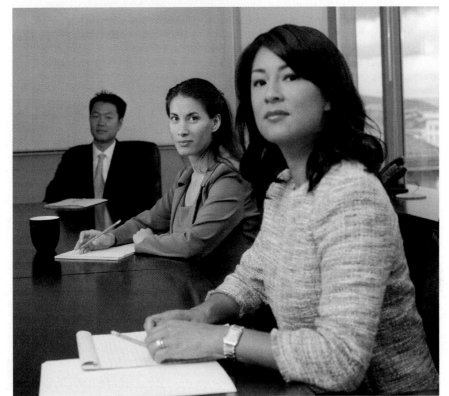

Readers from high-context cultures often use different forms of persuasion.

Guideline 4: Avoid arguing strongly for or against—Outward argumentativeness can be viewed as threatening and disrespectful and is often counterproductive. Moreover, arguing directly against someone else's position might be perceived as an attack on that person. Instead, if you disagree with someone, restate the facts of your ideas and use reasoning to explain them.

Guideline 5: Strive to reach consensus—Social harmony is greatly valued in high-context cultures. So, you should strive for consensus with your high-context counterparts. They will tend to react skeptically to plans in which one side seems to benefit more than another. Instead, your ideas will be most persuasive if you present them as a win-win for both sides.

Guideline 6: Speak collectively, not individually—To maintain social harmony, you should tend toward speaking collectively ("we" or "us") rather than individually ("I" or "me"). Avoid saying something like, "Here's my opinion." Instead, strive to characterize what is best for all.

Guideline 7: Be patient and wait for the "point"—People from high-context cultures often approach complex issues holistically, discussing all issues at the same time. To a low-context person, it can seem as though nothing is being decided. People from low-context cultures tend to be eager to "get to the point," while people from high-context cultures usually assume the point is obvious and need not be stated. The point of a discussion or document will tend to arrive near the end, so you should be patient and wait for it.

Guideline 8: Remember that "no" is rarely used—Direct refusal, rejection, and the answer "no" tend to be avoided in high-context cultures, especially in professional settings. A direct refusal can be considered an insult. So, refusals are handled with some care. If you are being refused, you might hear a deferral like, "We will consider your ideas." In some cases, as in Indonesia, you may receive an initial "yes" with a later "but" that signals the refusal. When a "no" needs to be conveyed, often an intermediary will be asked to deliver the bad news.

Guideline 9: Don't be informal—In high-context cultures, hierarchy and rituals are important, and respect for social status is expected. In professional settings, people from high-context cultures tend to address each other formally. The American tendency to quickly become informal and familiar (e.g., "You can just call me Jim") can be viewed as disrespectful or aggressive. Even when you know someone well, you should address that person by title and demonstrate respect for his or her position.

Guideline 10: Defer to hierarchy—High-context cultures tend to put great emphasis on hierarchy and social standing. A person of lower standing is expected to defer to someone of higher standing. Meanwhile, causing someone of higher standing to become embarrassed or agitated, even accidentally, will usually undermine or scuttle any negotiations.

Guideline 11: Minimize emotions—Smiles are welcome in high-context cultures, but obvious signs of emotion, like anger, hilarity, annoyance, or bemusement, will usually be taken far more seriously than they would in low-context cultures. Losing your composure is almost a sure way to end a professional relationship.

Link

For more strategies on communicating with cross-cultural readers, go to Chapter 3, pages 53–58.

Research is always helpful if you need to communicate with people from another culture. A surprising amount of helpful information can be found on the web.

As technical fields globalize, North Americans are finding it helpful to learn how other cultures operate. To be persuasive, you should learn how people from the target culture tend to think and negotiate. Listen carefully and don't become too frustrated when you make mistakes—because you will.

Want to learn more about etiquette in Asian cultures? Go to
www.pearsonhighered.com/johnsonweb3/6.13

- Developing a project plan is a process of identifying tasks and setting a timeline for completing them.

- All technical documents are persuasive in some way, even ones that are intended to be strictly "informational."

- Persuasion is about giving people good reasons to do things, while building their confidence in you and your company.

- Technical workplaces tend to rely on reasoning-based persuasion and values-based persuasion.

- Reasoning-based persuasion relies on logic, examples, and evidence to support claims.

- Values-based persuasion uses shared ideals, mutual values, common goals, and credibility to build and strengthen relationships.

- Cross-cultural communication can be difficult, but persuasion is especially challenging when the audience or readers are from a "high-context" culture.

Individual or Team Projects

1. In a magazine, find some advertisements that are intended to be persuasive. Using the vocabulary in this chapter, write an analysis of an advertisement for your instructor in which you discuss how the advertisement is persuasive or not. Tell your instructor why you think the advertiser's strategies work (or don't work).

2. Compare and contrast two technical documents. In a PowerPoint presentation, show why one document is more persuasive than the other. How does the more persuasive document use reasoning and values to persuade the readers? What would make the other document more persuasive?

3. Find websites that use images and graphics to set a particular tone. How persuasive are the images? Why do you think they persuade the readers, or don't persuade the readers? If you were the intended reader of the website, what would make the website more persuasive to you? What would make you do what the authors of the website want you to do?

4. For your next writing project, use the three-step stragetic planning process described in this chapter. Start out by listing your objectives. Then, rank those objectives and identify your top-rank objective. Use mapping to sketch out your plan and then convert it to a task list. Finally, develop a timeline for completing the project. Write a memo to your instructor that describes your work plan, and provide a calendar that shows the milestones or checkpoints in your project.

5. Write a white paper to your class that studies the persuasion strategies of a culture other than your own. For example, you might explore persuasion strategies in China or France. What would typical people from these cultures find persuasive? And what would be the most effective way to persuade them, without offending them?

Collaborative Project

Ask one of your team members to bring in a common household product (e.g., toaster, popcorn popper, music CD, mobile phone, etc.). Using reasoning-based and values-based persuasion strategies, develop a strategy to promote this item to college students—in other words, people much like your classmates. What would be some of your best arguments in favor of the product? How would you downplay any weaknesses? How might you use the buyers' values to urge them to identify with the product?

Now, let's change directions. Imagine you need to promote this same product to people who are 60 years or older. How might your persuasion strategies change? What kinds of reasoning would be more appropriate for this audience? How might you encourage them to value the product by identifying with it?

In a five-minute presentation to your class, compare and contrast your team's strategies for marketing the product to college students and older people. How would your strategies be similar? How would they differ? How might these differences affect the organization, style, and design of the documentation that goes with the product?

For additional technical writing resources, including interactive sample documents, document design tutorials and guidelines, and more, go to www.mytechcommlab.com.

To learn more about high-context cultures, go to
www.pearsonhighered.com/johnsonweb3/6.14

Leapfrogging with Wireless

Tim Jenks knew this was going to be an interesting project. Tim was an electrical engineer who worked for Seal Beach Communications, a company that built wireless networks in southern California. Most of the company's customers lived and worked in the suburbs of Los Angeles.

Recently, his company won a contract to install wireless networks on Native American reservations in New Mexico. The tribes had received a $6 million grant to build wireless networks that would bring phone and Internet services to their members. The tribes were leapfrogging the technology, in some cases going from almost no services in some areas straight to high-speed wireless services in all areas. The grant was designed to help tribal members access educational opportunities and perhaps even work from home.

The project sounded fascinating, and Tim was eager to work with Native Americans, with whom he had almost no experience. Plus, this project was completely different from his normal work in southern California. It sounded like something he could be proud of.

His boss, Tina Guerrero, visited him at a worksite. "Tim, I have some good news. You're going to be our reconnaissance guy on the New Mexico project. Next week, we're sending you on a fact-finding mission to New Mexico."

Tina explained that Tim would be meeting with tribal leaders to talk about where towers would be placed, where wireless clouds should be created, and what public facilities would be covered (e.g., libraries, cultural centers, schools).

"There is one complication—and it's a big one," cautioned Tina. "One of my friends who has worked with reservations in New Mexico says the tribes are all different. The Navajo sometimes handle issues very differently than the Pueblo tribes along the Rio Grande. We know very little about the Apache tribes in the north and south of the state."

A bit concerned, Tim conceded he didn't know anything about these cultural differences.

Tina continued, "The last thing we need is one of our engineers offending people and making things harder on us. So, I want you to find out whatever you can about the tribes in New Mexico and how they operate in business settings."

Imagine that you are Tim. Using the Internet and any other sources, discover what he should know about the cultural customs and business practices of Native Americans in New Mexico. How should he handle the cross-cultural issues involved in this project, especially when meeting tribal leaders for the first time?

For some websites that discuss Native American culture and practices, go to **www.pearsonhighered.com/johnsonweb3/6.15**

CHAPTER

7

Researching and Managing Information

In this chapter, you will learn:

- How to define your research subject using logical mapping to define its boundaries.

- How to formulate a research question or hypothesis that will guide your research.

- To develop a research methodology and revise it as needed.

- Methods for triangulating information to ensure reliability.

- To use the many available electronic research tools.

- To find electronic, print, and empirical sources for your research.

- How to take accurate notes and use your notes appropriately in paraphrasing, summarizing, and quoting.

- How to appraise the information you gather for reliability and bias while avoiding plagiarism.

Computers and computer networks have made research both easier and more challenging. Not long ago, finding *enough* information was the hard part when doing research. Today, with access to the Internet, you will find seemingly endless amounts of information available on any given topic. If you run a search on Google.com, Ask.com, or Yahoo.com, thousands of webpages might refer to your subject. Even a traditional search at your library will unearth more information than you could ever collect.

The problem caused by this overwhelming amount of facts, data, and opinions is an *information glut.* An information glut exists when there is more information than time available to collect, interpret, and synthesize that information.

What should you do about this overwhelming access to information? You should view "research" as a form of *information management.* Research is now a process of shaping the flow of information, so you can locate and utilize the information you need. As an information manager, you need to learn how to evaluate, prioritize, interpret, and store that information so you can use it effectively.

Empirical research is a critical part of working in technical disciplines.

Of course, in addition to collecting existing information, *primary research* (empirical research) is important in the technical workplace. Primary research involves observing and/or directly experiencing the subject of your study. By conducting experiments, doing field studies, using surveys, and following other empirical methods, you can make your own observations and collect your own data. The most effective research usually blends these kinds of empirical observations with the existing information available through computer networks and libraries.

Beginning Your Research

In technical fields, researchers typically use a combination of primary and secondary sources to gain a full understanding of a particular subject.

Primary sources—Information collected from observations, experiments, surveys, interviews, ethnographies, testing

Secondary sources—Information drawn from academic journals, magazine articles, books, websites, research databases, CD-ROMs, and reference materials

Most researchers begin their research by first locating the secondary sources available on their subject. Once they have a thorough understanding of their subject, they use primary research to expand on these existing materials.

Your research with primary and secondary sources should follow a process similar to the following:

1. Define the research subject.

2. Formulate a research question and hypothesis.

3. Develop a research methodology.

4. Triangulate electronic, print, and empirical sources of information.

5. Appraise collected information to determine reliability.

A good research process begins by clearly defining the research subject. Then, it follows a research methodology in which a variety of sources are located and appraised for reliability (Figure 7.1).

A Research Process

Define your research subject.

Formulate a research question and hypothesis.

Develop a research methodology.

Triangulate electronic, print, and empirical sources of information.

Appraise collected information and determine reliability.

Figure 7.1: To ensure the collection of reliable information, it helps to follow a predictable research process.

Defining Your Research Subject

Your first task is to define your research subject as clearly as possible. You should begin by identifying what you already know about the subject and highlighting areas where you need to do more research.

A reliable way to start is to first develop a *logical map* of your research subject (Figure 7.2). To create a logical map, write your subject in the middle of your screen or a piece of paper. Then, around that subject, begin noting everything you already know or believe about it. As you find relationships among these ideas, you can draw lines to connect them together into clusters. In places where you are not sure of yourself, simply jot down your thoughts and put question marks (?) after them.

Using Mapping to Find the Boundaries of a Subject

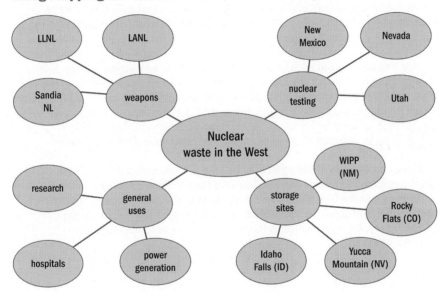

Figure 7.2:
A logical map can help you generate ideas about your subject. It can also show you where you need to do research.

As you make your logical map, you will notice that some ideas will lead to other, unexpected ideas—some seemingly unrelated to your subject. When this happens, just keep writing them down. Don't stop. These unexpected ideas are evidence that you are thinking creatively by tapping into your visual-spatial abilities. You may end up crossing out many of these ideas, but some may offer you new insights into the subject.

Mapping is widely used in technical disciplines, and it is gaining popularity in highly scientific and technical research (see the At Work box on page 144). You might find it strange to begin your research by drawing circles and lines, but mapping will reveal relationships that you would not otherwise discover. It is a great way to tap into your creativity through your visual abilities.

Want to learn more about being creative? Go to
www.pearsonhighered.com/johnsonweb3/7.2
**Defining Your
Research Subject**
143

AT WORK

Greg Wilson

RESEARCHER, SYSTEMS ETHNOGRAPHY AND QUALITATIVE MODELING TEAM AT LOS ALAMOS NATIONAL LABORATORY

Los Alamos National Laboratory is a U.S. government research center in Los Alamos, New Mexico.

Do researchers in the technical workplace *really* use logical mapping?

Mapping is a regular part of our research at Los Alamos National Laboratory (LANL). The mission of LANL is to address big science problems related to the U.S. nuclear stockpile, terrorism and weapons of mass destruction, defense, energy, environment, and infrastructure. These types of problems involve complex technical systems, political and social concerns, and interdisciplinary teams of experts.

We use mapping methods to create representations that help the teams of experts understand the important elements of a problem and how they relate to each other. I do textual research, interviewing, and ethnographies to identify the ways that different experts understand the problem. Then, I create maps so the experts can understand each other's perspectives and communicate about how to solve the problem. For example, if you have a biologist, epidemiologist, physician, emergency response planner, police chief, and political decision maker all trying to plan how a major city should respond in the event of a bioterrorist attack, each of them understands the problem in a different way. They talk about the problem in different ways, making efficient communication and problem solving difficult.

Using mapping as a centerpiece of research, I work with teams like this to build a common graphical representation of the problem. This representation also serves as a framework for identifying what data and information exist at each node in the graph, so that all relevant information is available to the team. Usually, a simple picture is the best way to begin to understand a complex problem.

Narrowing Your Research Subject

After defining your subject, you also need to look for ways to narrow and focus your research. Often, when people start the research process, they begin with a very broad subject (e.g., nuclear waste, raptors, lung cancer). Your logical map and a brief search on the Internet will soon show you that these kinds of subjects are too large for you to handle in the time available.

To help narrow your subject, you need to choose an *angle* on the subject. An angle is a specific direction that your research will follow. For example, "nuclear waste" may be too large a subject, but "the hazards of transporting nuclear waste in the western United States" might be a good angle for your research. Likewise, research on raptors

Want to see other workplaces that use mapping? Go to
www.pearsonhighered.com/johnsonweb3/7.3

GO TO THE NET

General Subject (too broad)	Angled Research Area (narrowed)
Nuclear waste	Transportation of nuclear waste in western states
Eagles	Bald eagles on the Mississippi
Lung cancer	Effects of secondhand smoke
Water usage	Water usage on the TTU Campus
Violence	Domestic abuse in rural areas

is probably too large a subject, but "the restoration of bald eagles along the Mississippi River" might be a manageable project.

By choosing an angle, you will help yourself narrow your research subject into a manageable size.

Formulating a Research Question or Hypothesis

Once you have narrowed your subject, you should then formulate a *research question* and *hypothesis.*

The purpose of a research question is to guide your empirical or analytical research. Your research question does not need to be very specific when you begin your research. It simply needs to give your research a direction to follow.

Try to devise a research question that is as specific as possible:

Why do crows like to gather on our campus during the winter?

What are the effects of violent television on boys between the ages of 10 and 16?

Is solar power a viable energy source for South Dakota?

Your hypothesis is your best guess about an answer to your research question:

Hypothesis: The campus is the best source of available food in the wintertime, because students leave food around. Crows naturally congregate because of the food.

Hypothesis: Boys between the ages of 10 and 16 model what they see on violent television, causing them to be more violent than boys who do not watch violent television.

Hypothesis: Solar power is a viable energy source in the summer, but cloudiness in the winter makes it less economical than other forms of renewable energy.

As you move forward with your research, you will probably need to refine or sharpen your original research question and hypothesis. For now, though, ask the question that you would most like to answer. Then, to form your hypothesis, answer this question to the best of your knowledge. Your hypothesis should be your best guess for the moment.

Having trouble refining your hypothesis? Go to
www.pearsonhighered.com/johnsonweb3/7.4

Formulating a Research Question or Hypothesis

145

Developing a Research Methodology

With your research question and hypothesis formed, you are ready to start developing your *research methodology*. A research methodology is a step-by-step procedure that you will use to study your subject. As you and your research team consider how to do research on your subject, begin thinking about all the different ways you can collect information.

Mapping Out a Methodology

Logical mapping can help. Put the purpose of your research in the middle of your screen or a piece of paper. Ask, "*How* are we going to achieve this purpose?" Then, answer this question by formulating the two to five major steps you will need to take in your research. Each of these major steps can then be broken down into minor steps (Figure 7.3).

Using the map in Figure 7.3, for example, a team of researchers might devise the following methodology for studying their research question:

Mapping Out a Methodology

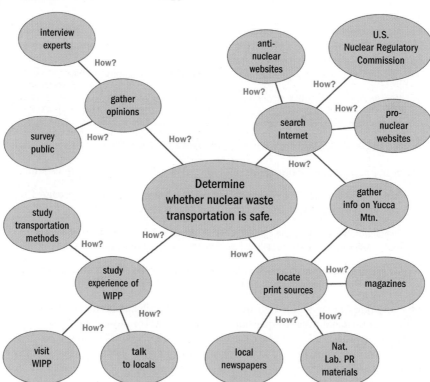

Figure 7.3: Logical mapping can help you sketch out a methodology. Keep asking the *How?* question as you consider the steps needed to complete your project.

Need help developing a methodology? Go to
www.pearsonhighered.com/johnsonweb3/7.5

GO TO THE NET

Methodology for Researching Nuclear Waste Transportation:

- Collect information off the Internet from sources for and against nuclear waste storage and transportation.

- Track down news stories in the print media and collect any journal articles available on nuclear waste transportation.

- Interview experts and survey members of the general public.

- Study the Waste Isolation Pilot Plant (WIPP) in New Mexico to see if transportation to the site has been a problem.

Note that these researchers are planning to collect information from a range of electronic, print, and empirical sources.

Describing Your Methodology

After mapping out your methodology, begin describing your methodology in outline form (Figure 7.4).

Sometimes, as shown in Figure 7.4, it is also helpful to identify the kinds of information you expect to find in each step. By clearly stating your *expected findings* before you start collecting information, you will know if your research methodology is working the way you expected.

Outlining a Research Methodology

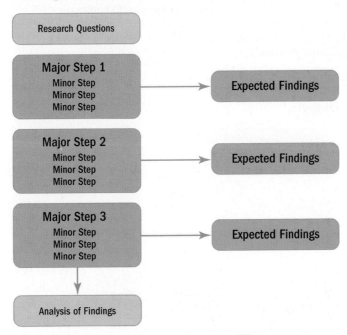

Figure 7.4: The major and minor steps in the research methodology should result in specific kinds of findings. At the end of the methodology, leave time for analyzing your findings.

At the end of your methodology, add a step called "Analysis of Findings." If you collected data, you will need to do some statistical analysis. If you conducted interviews or tracked down information on the Internet, you will need to spend some time checking and verifying your sources.

Using and Revising Your Methodology

A good methodology is like a treasure map. You and your research team can use it as a guide to uncover answers to questions that intrigue you.

Almost certainly, you will deviate from your methodology while doing your research. Sometimes you will find information that takes you down an unexpected path. Sometimes information you expected to find is not available. In other cases, experiments and surveys return unexpected findings.

When you deviate from your methodology, note these changes in direction. A change in the methodology is not a failure. It is simply a recognition that research is not formulaic and can be unpredictable. Research is a process of discovery. Sometimes the most important discoveries are made when we deviate from the plan.

Triangulating Materials

To ensure that your methodology is reliable, you should draw information from a variety of sources. Specifically, you should always try to *triangulate* your materials by collecting information from electronic, print, and empirical sources (Figure 7.5). Triangulation allows you to compare and contrast sources, thereby helping you determine which information is reliable and which information is not. Plus, triangulation gives your readers confidence in your research, because you will have collected information from a variety of sources.

The Research Triangle

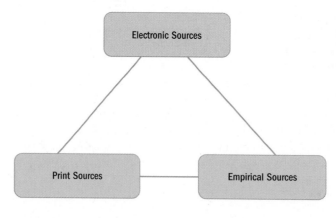

Figure 7.5:
In any research project, try to draw information from electronic, print, and empirical sources.

Tools for helping you do research electronically
are available at
www.pearsonhighered.com/johnsonweb3/7.6

Solid research draws from three kinds of information:

- **Electronic sources:** Websites, CD-ROMs, listservs, research databases television and radio, videos, podcasts, blogs
- **Print sources:** Books, journals, magazines and newspapers, government publications, reference materials, microform/microfiche
- **Empirical sources:** Experiments, surveys, interviews, field observations, ethnographies, case studies

By drawing information from all three kinds of sources, you will be able to verify the facts you find:

- If you find similar facts in all three kinds of sources, you can be reasonably confident that the information is reliable.
- If you find the information in two of three kinds of sources, the information is probably still reliable, though you should be less confident.
- If, however, you find the information in only one kind of source, it might not be reliable and needs further confirmation.

Something to remember is that "truth" and "facts" are more slippery than we want to admit. A source may claim it is providing you the truth, but until you can confirm that source's facts through triangulation, you should always treat any claims skeptically.

Also keep in mind that there are always two sides to every issue. So, do not restrict your research to only one side. If you look only for sources that confirm what you already believe, you will probably not gain a deeper understanding of the subject. After all, even when you absolutely disagree with someone else, his or her argument may give you additional insight into the issue you are researching. Keep an open mind.

Using Electronic Sources

Because electronic sources are so convenient, a good place to start collecting information is through your computer.

Websites—Websites are accessible through browsers like Firefox, Explorer, or Safari. When using search engines like MetaCrawler.com, Google.com, Yahoo.com, and Ask.com, among many others, you can run keyword searches to find information on your subject (Figures 7.6 and 7.7).

CD-ROMs—A compact disc (CD-ROM) can hold a library worth of text and images. Often available at your local or campus library, CD-ROMs are usually searchable through keywords or subjects. Encyclopedias and databases are also available on CD-ROM.

Listservs—Listservs are ongoing e-mail discussions, usually among specialists in a field. Once you find a listserv on your subject, you can usually subscribe to the discussion. A politely phrased question may return some helpful answers from other subscribers to the listserv.

Link

To improve your use of search engines, go to Chapter 3, page 45.

Television and radio—You can locate television and radio documentaries or news programs that address your subject. In some cases, copies of these materials will be available at your library.

Research databases—If you are looking for scientific and technical articles on your subject, you might first locate a research database that collects materials about your subject (Figure 7.7). Your campus library likely subscribes to a variety of databases that can be searched electronically.

Podcasts—Increasingly, information is being put on websites as podcasts. Podcasts can be played on your computer or an MP3 player (not only on an iPod). They often sound like radio broadcasts. Video podcasts are also becoming more popular.

Videos—Increasingly, documentaries and training videos are available on digital videodiscs (DVDs) or videotapes (VHS). Your library or even video rental stores may have these kinds of materials available.

Internet Search Engine

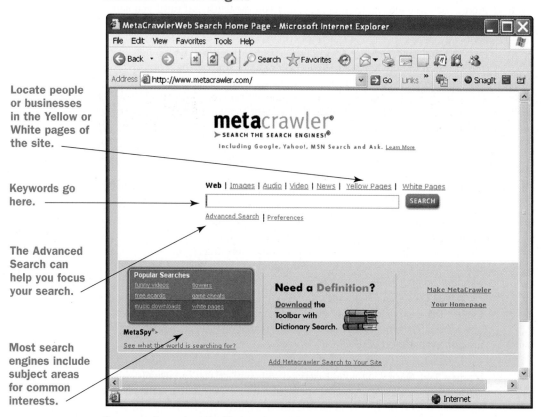

Locate people or businesses in the Yellow or White pages of the site.

Keywords go here.

The Advanced Search can help you focus your search.

Most search engines include subject areas for common interests.

Figure 7.6: MetaCrawler is one of the more useful Internet search engines. By typing in some keywords, you can locate limitless amounts of information.

Source: metacrawler, http://www.metacrawler.com.

For links to solid electronic sources of information, go to
www.pearsonhighered.com/johnsonweb3/7.8

Research Databases

Figure 7.7: A research database can help you target information in specialized fields.

Here you can enter the terms you want searched.

You can browse journals on the subject here.

Source: American Society of Civil Engineers, http://www.ascelibrary.org.

Blogs—Blogs are Internet sites where a commentator or group of commentators often "publish" raw information, opinions, and hearsay. Blogs can be a good source for keeping up with the cutting edge of a research area.

Using Print Sources

With easy access to information through the Internet, you may be tempted to forego using the print sources available at your library. Researchers who neglect print sources are making a serious mistake.

Printed documents are still the most abundant and reliable sources of information. In the rush to use electronic sources, many people have forgotten that their nearby library is loaded with books and periodicals on almost any subject. These print sources can usually be located by using your computer to access the library's website (Figure 7.8).

Want to know more about blogs? Go to
www.pearsonhighered.com/johnsonweb3/7.9

GO TO THE NET

Triangulating Materials   151

A Library's Search Engine

Figure 7.8: Your library likely has a website for finding print sources. This is the University of Minnesota's page for finding a variety of different kinds of materials. Your campus has trained librarians who are there to help you. Don't be afraid to ask.

Source: The University of Minnesota, http://www.lib.umn.edu.

Here are a few of the many kinds of print materials that you can use:

Books—Almost all libraries have electronic cataloging systems that allow you to use author name, subject, title, and keywords to search for books on your subject. Once you have located a book on your subject, look at the books shelved around it to find other useful materials.

Journals—Using a *periodical index* at your library, you can search for journal articles on your subject. Journal articles are usually written by professors and scientists in a research field, so the articles can be rather detailed and hard to understand. Nevertheless, these articles offer some of the most exact research on any subject. Periodical indexes for journals are usually available online at your library's website, or they will be available as printed books in your library's reference area.

Magazines and newspapers—You can also search for magazine and newspaper articles on your subject by using the *Readers' Guide to Periodical*

For access to online news and legal and business information, go to **www.pearsonhighered.com/johnsonweb3/7.10**

Literature or a newspaper index. The *Readers' Guide* and newspaper indexes are likely available online at your library's website or in print form. Recent editions of magazines or newspapers might be stored at your library. Older magazines and newspapers have usually been stored on microform or microfiche.

Government publications—The U.S. government produces a surprising amount of useful books, reports, maps, and other documents. You can find these documents through your library or through government websites. A good place to start is *The Catalog of U.S. Government Publications* (www.catalog.gpo.gov), which offers a searchable listing of government publications and reports.

Reference materials—Libraries contain many reference tools like almanacs, encyclopedias, handbooks, and directories. These reference materials can help you track down facts, data, and people. Increasingly, these materials can also be found online in searchable formats.

Microform/microfiche—Libraries will often store copies of print materials on microform or microfiche. Microform and microfiche are miniature transparencies that can be read on projectors available at your library. You will usually find that magazines and newspapers over a year old have been transferred to microform or microfiche to save space in the library. Also, delicate and older texts are available in this format to reduce the handling of the original documents.

Using Empirical Sources

You should also generate your own data and observations to support your research. Empirical studies can be *quantitative* or *qualitative,* depending on the kinds of information you are looking for. Quantitative research allows you to generate data that you can analyze statistically to find trends. Qualitative research allows you to observe patterns of behavior that cannot be readily boiled down into numbers.

Experiments—Each research field has its own experimental procedures. A controlled experiment allows you to test a hypothesis by generating data. From that data, you can confirm or dispute the hypothesis. Experiments should be *repeatable,* meaning their results can be replicated by another experimenter.

Field observations—Researchers often carry field notebooks to record their observations of their research subjects. For example, an ornithologist might regularly note the birds she observes in her hikes around a lake. Her notebook would include her descriptions of birds and their activities.

Interviews—You can ask experts to answer questions about your subject. On almost any given college campus, experts are available on just about any subject. Your well-crafted questions can draw out very useful information

Empirical research requires you to observe your subject directly.

and quotes. Figure 7.9, for example, shows a classic interview with Bill Gates and Steve Jobs.

Surveys and Questionnaires—You can ask a group of people to answer questions about your subject. Their answers can then be scored and analyzed for trends. Survey questions can be *closed-ended* or *open-ended*. Closed-ended questions ask respondents to choose among preselected answers. Open-ended questions allow respondents to write down their views in their own words. Figure 7.10 shows pages from a survey with both closed-ended and open-ended questions.

Ethnographies—An ethnography is a systematic recording of your observations of a defined group or culture. Anthropologists use ethnographies to identify social or cultural trends and norms.

Case studies—Case studies typically offer in-depth observations of specific people or situations. For example, a case study might describe how a patient reacted to a new treatment regimen that manages diabetes.

For more information on conducting qualitative empirical studies, go to
www.pearsonhighered.com/johnsonweb3/7.12

GO TO THE NET

Interviewing People

Scripted question

Uses follow-up prompt for more detail.

Scripted question

Following is a transcript of the interview Kara Swisher and Walt Mossberg conducted with Microsoft Chairman Bill Gates and Apple CEO Steve Jobs at the Wall Street Journal's D: All Things Digital *conference on May 30, 2007.*

Kara: So let's get started. I wanted to ask, there's been a lot of mano-a-mano/catfight kind of thing in a lot of the blogs and the press and stuff like that, and we wanted to–the first question I was interested in asking is what you think each has contributed to the computer and technology industry, starting with you, Steve, for Bill, and vice versa.

Steve: Well, you know, Bill built the first software company in the industry and I think he built the first software company before anybody really in our industry knew what a software company was, except for these guys. And that was huge. That was really huge. And the business model that they ended up pursing turned out to be the one that worked really well, you know, for the industry. I think the biggest thing was, Bill was really focused on software before almost anybody else had a clue that it was really the software.

Kara: Was important?

Steve: That's what I see. I mean, a lot of other things you could say, but that's the high order bit. And I think building a company's really hard, and it requires your greatest persuasive abilities to hire the best people you can and keep them at your company and keep them working, doing the best work of their lives, hopefully. And Bill's been able to stay with it for all these years.

Walt: Bill, how about the contribution of Steve and Apple?

Bill: Well, first, I want to clarify: I'm not Fake Steve Jobs. What Steve's done is quite phenomenal, and if you look back to 1977, that Apple II computer, the idea that it would be a mass-market machine, you know, the bet that was made there by Apple uniquely–there were other people with products, but the idea that this could be an incredible empowering phenomenon, Apple pursued that dream. Then one of the most fun things we did was the Macintosh and that was so risky. People may not remember that Apple really bet the company. Lisa hadn't done that well, and some people were saying that general approach wasn't good, but the team that Steve built even within the company to pursue that, even some days it felt a little ahead of its time–I don't know if you remember that Twiggy disk drive and...

Steve: One hundred twenty-eight K.

Kara: Oh, the Twiggy disk drive, yes.

Bill: Steve gave a speech once, which is one of my favorites, where he talked about, in a certain sense, we build the products that we want to use ourselves. And so he's really pursued that with incredible taste and elegance that has had a huge impact on the industry. And his ability to always come around and figure out where that next bet should be has been phenomenal. Apple literally was failing when Steve went back and re-infused the innovation and risk-taking that have been phenomenal. So the industry's benefited immensely from his work. We've both been lucky to be part of it, but I'd say he's contributed as much as anyone.

Steve: We've also both been incredibly lucky to have had great partners that we started the companies with and we've attracted great people. I mean, so everything that's been done at Microsoft and at Apple has been done by just remarkable people, none of which are sitting up here today.

Figure 7.9:
In this interview transcript, notice how the interviewers are asking scripted questions and then following up with unscripted questions.

Source: D: All Things Digital, http://d5.allthingsd.com/200705311/d5-gates-jobs-transcript/.

 For websites that discuss empirical research methods, go to
www.pearsonhighered.com/johnsonweb3/7.13

Triangulating Materials

Pages from a Questionnaire on Campus Safety

Campus Survey 75

Campus Perception Survey

The following questions are about how safe you feel or don't feel on campus. For each situation please tell us if you feel: very safe, reasonably safe, neither safe nor unsafe, somewhat unsafe, very unsafe, or if this situation does not apply to you. (Please circle the number that best represents your answer or **NA** if the situation does not apply to you.)

How safe do you feel...

	Very Unsafe	Somewhat Unsafe	Neither Safe Nor Unsafe	Reasonably Safe	Very Safe	
walking alone on campus during daylight hours?	1	2	3	4	5	NA
waiting alone on campus for public transportation during daylight hours?	1	2	3	4	5	NA
walking alone in parking lots or garages on campus during daylight hours?	1	2	3	4	5	NA
walking alone on campus after dark?	1	2	3	4	5	NA
waiting alone on campus for public transportation after dark?	1	2	3	4	5	NA
walking alone in parking lots or garages on campus after dark?	1	2	3	4	5	NA
working in the library stacks late at night?	1	2	3	4	5	NA
while alone in classrooms?	1	2	3	4	5	NA
Student Activity Center during the day?	1	2	3	4	5	NA
Student Activity Center at night?	1	2	3	4	5	NA

Are there any specific areas on campus where you do not feel safe? Please specify which areas, which campus, and when; for example, evenings only or any time. _____

_____

Do you have any special needs related to safety on campus? _____

Have you ever used services related to safety issues, sexual harassment, or sexual assault that are provided on campus by the following?

How satisfied were you with help from this source?

			Very Dissatisfied	Somewhat Dissatisfied	Neither Satisfied Nor Dissatisfied	Somewhat Satisfied	Very Satisfied
Campus Police	YES	NO	1	2	3	4	5
Women's Center	YES	NO	1	2	3	4	5
Campus ministry	YES	NO	1	2	3	4	5
Campus counseling (Belknap)	YES	NO	1	2	3	4	5
Student Health Services (Belknap)	YES	NO	1	2	3	4	5
Campus counseling (Health Science)	YES	NO	1	2	3	4	5
Student Health Services (Health Science)	YES	NO	1	2	3	4	5
Psychological Services Ctr (Psychology Clinic)	YES	NO	1	2	3	4	5
Affirmative Action Office	YES	NO	1	2	3	4	5
Security escort services after dark	YES	NO	1	2	3	4	5
Residence Hall staff	YES	NO	1	2	3	4	5
Office of Student Life	YES	NO	1	2	3	4	5
Disability Resource Center	YES	NO	1	2	3	4	5
Access Center	YES	NO	1	2	3	4	5
Faculty member	YES	NO	1	2	3	4	5
Other _____	YES	NO	1	2	3	4	5
(Please specify.)							

These closed-ended questions yield numerical data.

Figure 7.10: A survey is a good way to generate data for your research. In this example, both closed-ended and open-ended questions are being used to solicit information.

Open-ended questions give participants an opportunity to elaborate on their answers.

Source: Bledsoe & Sar, 2001.

Campus Survey 78

The following are some beliefs that may be held about the role of women and men in today's society. There are no right or wrong answers.   (Please circle the response that best describes your opinion.)

	Strongly Disagree		Somewhat Agree		Strongly Agree

A man's got to show the woman who's boss right from the start.   1   2   3   4   5

Women are usually sweet until they've caught a man, but then they let their true self show.   1   2   3   4   5

In a dating relationship a woman is largely out to take advantage of a man.   1   2   3   4   5

Men are out for only one thing.   1   2   3   4   5

A lot of women seem to get pleasure from putting a man down.   1   2   3   4   5

A woman who goes to the home or apartment of a man on their first date implies that she is willing to have sex.   1   2   3   4   5

Any female can get raped.   1   2   3   4   5

Any healthy woman can successfully resist a rapist if she really wants to.   1   2   3   4   5

Many women have an unconscious wish to be raped, and may then unconsciously set up a situation in which they are likely to be attacked.   1   2   3   4   5

If a woman gets drunk at a party and has intercourse with a man she's just met there, she should be considered "fair game" to other males at the party who also want to have sex with her whether she wants to or not.   1   2   3   4   5

What **percentage of women** who report a rape would you say are lying because they are angry and want to get back at the man they accuse?   _____ %

What **percentage of reported rapes** would you guess were merely invented by women who discovered they were pregnant and wanted to protect their own reputation?   _____ %

Did you attend any type of student orientation conducted by University of Louisville during Summer 2000 or at the beginning of this term? (Please circle your answer).   **YES NO**

If your answer was "No, I did not attend orientation", please continue on next page.

At the orientation you attended, how much information about violence against women issues did you receive? (Please circle your answer).   None 1   2   Some 3   4   A Lot 5

The survey uses statements to measure the participants' reactions to specific situations or opinions.

Using the Scientific Method

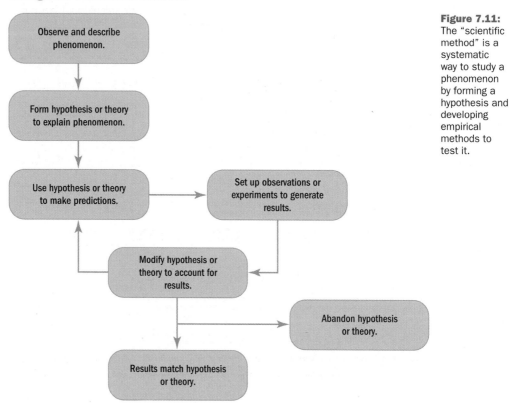

Figure 7.11:
The "scientific method" is a systematic way to study a phenomenon by forming a hypothesis and developing empirical methods to test it.

When conducting empirical research, you should follow the *scientific method* (Figure 7.11). The concept of a scientific method was first conceived by Francis Bacon, a seventeenth-century English philosopher. Later in the seventeenth century, the London Royal Society, a club of scientists, gave the scientific method the form we recognize now.

The Scientific Method:

1. Observe and describe a phenomenon.

2. Formulate a hypothesis or theory that explains the phenomenon.

3. Use the hypothesis or theory to make predictions.

4. Use observations and experiments to generate results that confirm or deny your predictions.

5. Modify the hypothesis or theory to account for your results.

6. Repeat steps 3 through 5 until results match your hypothesis or theory OR you abandon the hypothesis or theory.

To learn more about the scientific method, go to
www.pearsonhighered.com/johnsonweb3/7.14

Triangulating Research

Solid research draws from three kinds of information:

• Electronic sources—Internet, CD-ROMs, listservs, television and radio, videos, blogs

• Print sources—books, journals, magazines and newspapers, government publications, reference materials, microform/microfiche

• Empirical sources—experiments, surveys, interviews, field observations, ethnographies, case studies

The scientific method can be used with quantitative or qualitative forms of empirical research. Whether you are doing an experiment in a laboratory or making field observations, following the scientific method will help you focus and streamline your research. It should help you produce the kinds of results that will provide a solid empirical foundation for your work.

Managing Information and Taking Notes

On almost any subject, you are going to find a wealth of information. At this point, you need to start thinking like an information manager. After all, only some of the information that you collected will be important to your readers. Most of the information you find will not be needed by your readers to take action or make a decision (Figure 7.12).

Managing Information

As you decide what to include in the document you are writing, you need to distinguish between *need-to-know* information and *want-to-tell* information.

• *Need-to-know information* includes material that your readers require to take action or make a decision.

• *Want-to-tell information* includes material that you would like to tell your readers but that is not necessary for them to take action or make a decision.

After you have gone through all the effort to collect information, you will want to tell the readers about everything you found. But your readers don't need (or want) all

Need-to-Know Versus Want-to-Tell Information

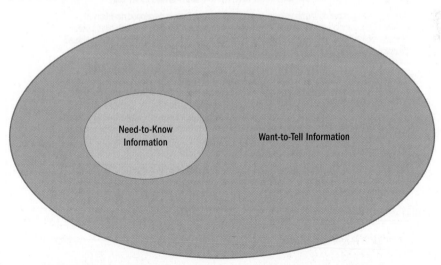

Need-to-Know Information

Want-to-Tell Information

Figure 7.12: While researching, you will find much more information than your readers need. Give them only the information they need to make a decision.

that information. They want only the information they need to take action or make an informed decision. Any extra want-to-tell information will just cloud their ability to understand your document.

Careful Note Taking

Reliable note taking is essential when you do research. If you are organized when you take notes, you will find the information you collected easy to use in the document you are writing.

Note-organizing software and database programs can help you keep track of the information you find. Many researchers write their notes exclusively on laptops or their personal digital assistants (PDAs). A pen and pad of paper is also still a good way to keep track of information.

What is most important, though, is to have a workable system for taking notes. Here are some note-taking strategies you might consider using:

RECORD EACH SOURCE SEPARATELY   Make sure you clearly identify the author, title of the work, and the place where you found the information (Figure 7.13). For information off the Internet, write down the webpage address (URL) and the date and time you found the information. For a print document, write down where the information was published and who published it. Also, record the library number of the document.

Keeping Notes on Your Computer

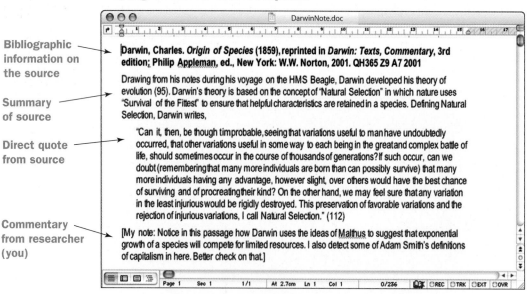

Bibliographic information on the source

Summary of source

Direct quote from source

Commentary from researcher (you)

Figure 7.13: Most notes include a combination of summaries, paraphrases, direct quotes, and personal comments.

Source: © 2008 Regents University of Minnesota. All rights reserved.

For some great ideas on note taking, go to
www.pearsonhighered.com/johnsonweb3/7.15

For large research projects, you might consider making a separate word-processing file for each of your authors or sources, like the one shown in Figure 7.13. That way, you can more easily keep your notes organized.

TAKE DOWN QUOTATIONS   When taking down a quotation from the source, you need to be sure that you copy the exact wording of the author. If you are taking a quote from a website, you might avoid errors by using the Copy and Paste functions of your computer to copy the statement directly from your source into your notes.

In your notes, you should put quotation marks around any material you copied word for word from a source.

> According to Louis Pakiser and Kaye Shedlock, scientists for the Earthquake Hazards Program at the U.S. Geological Survey, "the assumption of random occurrence with time may not be true." (1997, para. 3)

If the quoted material runs more than three lines in your text, you should set off the material by indenting it in the text.

> Louis Pakiser and Kaye Shedlock, scientists for the Earthquake Hazards Program at the U.S. Geological Survey, make the following point:
>
>> When plate movements build the strain in rocks to a critical level, like pulling a rubber band too tight, the rocks will suddenly break and slip to a new position. Scientists measure how much strain accumulates along a fault segment each year, how much time has passed since the last earthquake along the segment, and how much strain was released in the last earthquake. (1997, para. 4)
>
> If we apply this rubber band analogy to the earthquake risk here in California...

When you are quoting a source, you also need to include an in-text citation at the end of the quote. In these two examples above, the in-text citation is the information in the parentheses.

Overall, you should use direct quotes sparingly in your technical writing. You might be tempted to use several quotes from a source, because the authors "said it right." If you use too many quotes, though, your writing will sound fragmented and patchy, because the quotes disrupt the flow of your text.

PARAPHRASE IDEAS   A better way to incorporate someone else's ideas into your writing is to paraphrase them. When paraphrasing, you are presenting another person's ideas in your own words. You still need to give the original author credit for the ideas, but you do not need to use quotation marks around the text. To paraphrase something, you should:

- reorganize the information to highlight important points.
- use plain language, replacing jargon and technical terms with simpler words.
- include an in-text citation.

Link

For more information on citing sources, go to Appendix C, which begins on page A-23.

In the following example, a quote from an original document is paraphrased:

Original Quote

"But in many places, the assumption of random occurrence with time may not be true, because when strain is released along one part of the fault system, it may actually increase on another part. Four magnitude 6.8 or larger earthquakes and many magnitude 6–6.5 shocks occurred in the San Francisco Bay region during the 75 years between 1836 and 1911. For the next 68 years (until 1979), no earthquakes of magnitude 6 or larger occurred in the region. Beginning with a magnitude 6.0 shock in 1979, the earthquake activity in the region increased dramatically; between 1979 and 1989, there were four magnitude 6 or greater earthquakes, including the magnitude 7.1 Loma Prieta earthquake. This clustering of earthquakes leads scientists to estimate that the probability of a magnitude 6.8 or larger earthquake occurring during the next 30 years in the San Francisco Bay region is about 67 percent (twice as likely as not)."

Effective Paraphrase

In-text citation →

Simple language is used. →

Pakiser and Shedlock (1997) report that large earthquakes are mostly predictable, because an earthquake in one place usually increases the likelihood of an earthquake somewhere nearby. They point out that the San Francisco area–known for earthquakes–has experienced long periods of minor earthquake activity (most notably from 1836 to 1911, when no earthquakes over magnitude 6 occurred). At other times in San Francisco, major earthquakes have arrived with more frequency, because large earthquakes tend to trigger other large earthquakes in the area.

← Some of the more technical details have been removed to enhance understanding.

Improper Paraphrase

Much of the original wording is retained. →

Pakiser and Shedlock (1997) report the assumption of random occurrence of earthquakes may not be accurate. Earthquakes along one part of a fault system may increase the frequency of earthquakes in another part. For example, the San Francisco Bay region experienced many large earthquakes between 1836 and 1911. For the next six decades until 1979, only smaller earthquakes (below magnitude 6) occurred in the area. Then, there was a large rise in earthquakes between 1979 and 1989. Scientists estimate that the probability of an earthquake of magnitude 6.8 or larger is 67 percent in the next 30 years in the Bay area.

← Language is still overly technical for the readers.

Link

For more information on plagiarism, see page 170 in this chapter.

The "effective" paraphrase shown here uses the ideas of the original quote, while reordering information to highlight important points and simplifying the language. The "improper" paraphrase above duplicates too much of the wording from the original source and does not effectively reorder information to highlight important points. In fact, this improper paraphrase is so close to the original, it could be considered plagiarism.

In many ways, paraphrasing is superior to using direct quotes. A paraphrase allows you to simplify the language of a technical document, making the information easier for the readers to understand. Also, you can better blend the paraphrased information into your writing because you are using your writing style, not the style of the source.

As a warning, make sure you are paraphrasing sources properly. Do not use the author's original words and phrases. Otherwise, when you draft your document, you may forget that you copied some of the wording from the original text. These duplications may leave you vulnerable to charges of plagiarism or copyright violation.

SUMMARIZE SOURCES   When summarizing, your goal is to condense the ideas from your source into a brief passage. Summaries usually strip out many of the examples, details, data, and reasoning from the original text, leaving only the essential information that readers need to know. Like a paraphrase, summaries should be written in your own words. When you are summarizing a source for your notes:

- Read the source carefully to gain an overall understanding.
- Highlight or underline the main point and other key points.
- Condense key points into lists, where appropriate.
- Organize information from most important to least important.
- Use plain language to replace any technical terms or jargon in the original.
- Use in-text citations to identify important ideas from the source.

To see an example of summarizing, consider the passage about predicting earthquakes shown in Figure 7.14. When summarizing this text, you would first need to identify the main point and key points in the text. The main point is that scientists

Original Text to Be Summarized

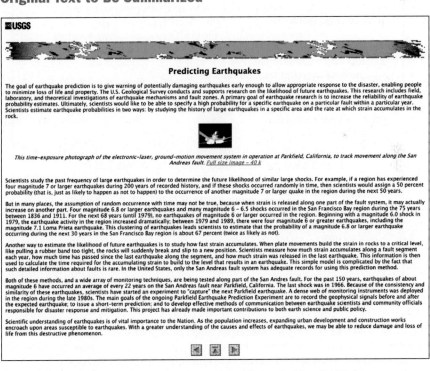

Figure 7.14: The original text contains many details that can be condensed into a summary.

Source: U.S. Geological Survey, http://pubs.usgs.gov/gip/earthg1/predict.html.

are increasingly able to estimate the probability of an earthquake in a specific area in the near future.

Now, locate the other key points in the text, of which there are three: (1) the frequency of earthquakes in the past helps scientists predict them in the future; (2) earthquakes are not random events, and they tend to occur in clusters; and (3) measurements of the strain on the earth can help scientists measure the probability of a future earthquake.

In the summary shown in Figure 7.15, the details in the original text have been stripped away, leaving only a condensed version that highlights the main point and a few other key issues. As shown here, the summary uses the writer's own words, not the words from the original source.

Summary of Original Text

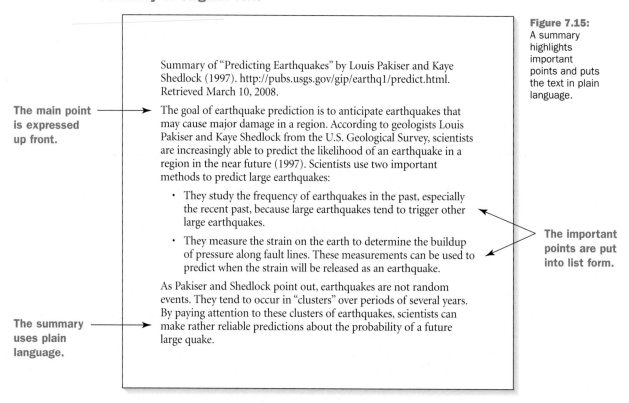

Figure 7.15: A summary highlights important points and puts the text in plain language.

Summary of "Predicting Earthquakes" by Louis Pakiser and Kaye Shedlock (1997). http://pubs.usgs.gov/gip/earthq1/predict.html. Retrieved March 10, 2008.

The main point is expressed up front.

The goal of earthquake prediction is to anticipate earthquakes that may cause major damage in a region. According to geologists Louis Pakiser and Kaye Shedlock from the U.S. Geological Survey, scientists are increasingly able to predict the likelihood of an earthquake in a region in the near future (1997). Scientists use two important methods to predict large earthquakes:

- They study the frequency of earthquakes in the past, especially the recent past, because large earthquakes tend to trigger other large earthquakes.

- They measure the strain on the earth to determine the buildup of pressure along fault lines. These measurements can be used to predict when the strain will be released as an earthquake.

The important points are put into list form.

The summary uses plain language.

As Pakiser and Shedlock point out, earthquakes are not random events. They tend to occur in "clusters" over periods of several years. By paying attention to these clusters of earthquakes, scientists can make rather reliable predictions about the probability of a future large quake.

WRITE COMMENTARY   In your notes, you might offer your own commentary to help interpret your sources. Your commentary might help you remember why you collected the information and how you thought it could be used. To avoid plagiarism, it is important to visually distinguish your commentary from summaries, paraphrases, and

For more tips on paraphrasing and summarizing, go to
www.pearsonhighered.com/johnsonweb3/7.17

quotations drawn from other sources. You might put brackets around your comments or use color, italics, or bold type to set them off from your other notes.

Documenting Sources

As you draft your text, you will need to *document* your sources. Documentation involves (1) naming each source with an *in-text citation* and (2) recording your sources in the *References* list at the end of the document. Documenting your sources offers the advantages of:

- Supporting your claims by referring to the research of others.
- Helping build your credibility with readers by showing them the support for your ideas.
- Reinforcing the thoroughness of your research methodology.
- Allowing your readers to explore your sources for more information.

When should you document your sources? Any ideas, text, or images that you draw from another text need to be properly acknowledged. If you are in doubt about whether you need to cite someone else's work, you should go ahead and cite it. Citing sources will help you avoid any questions about the integrity and soundness of your work.

In Appendix C at the end of this book, you will find a full discussion of three documentation systems (APA, CSE, and MLA) that are used in technical fields. Each of these systems works differently.

The most common documentation style for technical fields is offered by the American Psychological Association (APA). The APA style, published in the *Publication Manual of the American Psychological Association,* is preferred in technical fields because it puts emphasis on the year of publication. As an example, let us briefly look at the APA style for in-text citations and full references.

APA IN-TEXT CITATIONS   In APA style, in-text citations can include the author's name, the publication year, and the page number where the information was found.

> One important study showed that physicians were regularly misusing antibiotics to treat viruses (Reynolds, 2003, p. 743).

> According to Reynolds (2003), physicians are regularly misusing antibiotics to treat viruses.

> According to Reynolds, "Doctors are creating larger problems by mistakenly treating viruses with antibiotics" (2003, p. 743).

These in-text citations are intended to refer the readers back to the list of full references at the end of the document.

APA FULL REFERENCES   The full references at the end of the document provide readers with the complete citation for each source (Figure 7.16).

> Friedman, S. M. (2005). *Einstein online*. Retrieved from http://www.westegg.com/einstein

> Pauling, L., & Wilson, E. B. (1935). *Introduction to quantum mechanics*. New York: Dover Publications.

Link

For a full discussion of documentation, including models for documenting references, turn to Appendix C, page A-23.

Need help citing sources? Go to
www.pearsonhighered.com/johnsonweb3/7.18

Managing
Information and
Taking Notes

165

Elements of an APA Full Reference

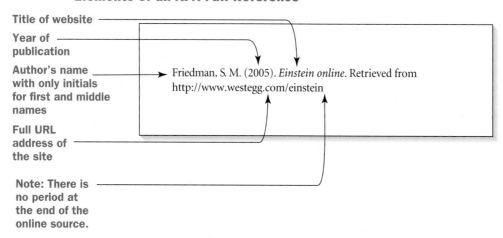

Title of website

Year of publication

Author's name with only initials for first and middle names

Full URL address of the site

Note: There is no period at the end of the online source.

Friedman, S. M. (2005). *Einstein online*. Retrieved from http://www.westegg.com/einstein

Figure 7.16: The full reference for an APA citation contains some standard elements. Here is an example of a reference for a website.

As you take notes, you should keep track of the information needed to properly cite your sources. That way, when you draft the document and create a references list, you will have this important information available. It's very difficult to locate the sources of your information after you finish drafting the document.

HELP

Avoiding Junk Science on the Internet

The Internet has been a boon for the "junk science" industry. Junk science is really not science at all. It is a public relations tool that corporations, lawyers, and special-interest groups use to confuse the public or cast doubt on the findings of legitimate scientists. Junk science is the selective use of data, scientific style, or scientific-like methods to advance a political or economic agenda.

The tobacco industry has been the most flagrant user of junk science. For decades, it employed scientists to confuse and cast doubt on findings that smoking causes cancer and other health problems. The tobacco industry's deceptions were exposed only when their own scientists, most notably chemist Jeffrey Wigand, turned over key documents to the government and media.

One of the most infamous uses of junk science was the response to biologist Rachel Carson's book *Silent Spring*. In the book, Carson exposed DDT, an insecticide, as the cause of a variety of illnesses. Designed to kill insects, DDT was widely used in the United States, and ended up killing many forms of wildlife, especially birds. Humans also became dangerously ill. Whole neighborhoods and wildlife

GO TO THE NET

Want to learn more about junk science? Go to
www.pearsonhighered.com/johnsonweb3/7.19

areas would be sprayed with DDT, leaving a path of dead birds, sickened animals, and poisoned people.

Instead of immediately pulling DDT off the market, Carson's opponents used junk science tactics to attack her credibility and her book. John Stauber and Sheldon Rampton, in their book *Toxic Sludge Is Good for You,* show how public relations agents did everything possible to undermine Carson and her findings.

The attacks were vicious, and Carson's reputation was compromised. But Carson was right: DDT is a very dangerous chemical. Eventually, it was banned in the United States as the scientific evidence mounted against it. Unfortunately, the junk science attacks on her reputation did great damage and allowed DDT to stay on the market longer than it should have.

Junk science can be used on all sides of a debate to support various causes. You will find junk science actively used by chemical companies, extreme environmental groups, antienvironmental groups, the diet industry, the oil industry, the tobacco industry, and proponents and opponents of genetically modified foods.

How do you distinguish real science from junk when you are doing research on the Internet? It's actually very difficult, but here are some pointers:

Follow the money—If the scientists behind a study receive all of their funding from a corporation or group that benefits from their results, there is a good chance that these scientists are being paid to generate specific results. Their results could be biased.

Check the reputations—Some "experts" on scientific issues are not really scientists at all. They are public relations consultants whose real job is to spin science for the media, casting doubt on the work of reliable scientists. So, look into the education and work experience of these "experts" to determine whether they are scientists.

Check the source—Reliable science is typically published in *peer-reviewed journals.* If the only "scientific" evidence you find for an argument is from a company or special-interest group, it might be junk science. Look for impartial sources to confirm that information.

Check the science—Reliable science follows the scientific method. The results should be repeatable or publicly available. If the methodology looks flimsy or the data are not available, the results are probably unreliable.

Remember that if it sounds too extreme, it probably is—Special-interest groups like to scare people by exaggerating the results of scientific studies. Be skeptical of studies that seem to contradict common sense (e.g., global warming is good for the planet) or seem too far-fetched (e.g., AIDS was a CIA experiment gone awry).

Junk science is a serious problem on the Internet. As you are researching an issue, you need to be skeptical and cross-check your sources. Always rely on a broad base of sources to confirm your information.

For help evaluating electronic sources, go to
www.pearsonhighered.com/johnsonweb3/7.20

Managing
Information and
Taking Notes

167

Appraising Your Information

All information is not created equal. In fact, some information is downright wrong or misleading. Keep in mind that even the most respected authorities usually have agendas that they are pursuing with their research. Even the most objective experiment will include some tinge of bias.

To avoid misleading information and researcher biases, you need to appraise the information you have collected and develop an overall sense of what the truth might be (Figure 7.17).

Questions for Appraising Your Information

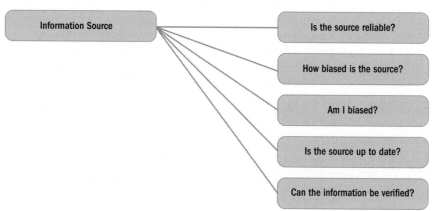

Figure 7.17: Challenge your sources by asking questions about their biases and validity.

Is the Source Reliable?

Usually, the most reliable sources of information are sources that have limited personal, political, or financial stakes in the subject. For example, claims about the safety of pesticides from a company that sells pesticides need to be carefully verified. Meanwhile, a study on pesticides by a university professor should be less biased, because the professor is not selling the product.

Link

For more information on finding people on the Internet, go to Chapter 3, page 45.

To ensure that your sources are reliable, you should always do some checking on their authors. Use an Internet search engine like Ask.com or Google.com to check out the authors, company, or organization that produced the materials. If the researchers have a good reputation, the information is probably reliable. If you can find little or no information about the researchers, company, or organization, you should be skeptical about their research.

How Biased Is the Source?

It is safe to say that all sources of information have some bias. There is no such thing as a completely objective source. So, you need to assess the amount of bias in your source. For example, facts on creation science websites that are used to dispute Darwinian evolution are usually biased toward theories that reinforce the biblical creation story. Their information is still usable in some situations—you might even accept it as true—but you need to recognize the inherent bias in such material.

Even the most reliable sources have some bias. Researchers, after all, very much want their hypotheses to be true, so irregularities in their results might be overlooked. Bias is a natural part of research. So, when you are assessing bias, consider how much the researchers want their results to be true. If the researchers indicate that at the beginning of their research, they were open to a range of answers, then the bias of the material is probably minimal. If only one answer was acceptable to the researchers (e.g., smoking does not cause lung cancer), then the material should be considered heavily biased.

Am I Biased?

As a researcher, you need to carefully examine your own biases. We all go into a research project with our own beliefs and expectations of what we will find. Our own biases can cause us to overlook evidence that contradicts our beliefs or expectations. For example, our beliefs about gender, race, sexuality, poverty, or religion, among other social issues, can strongly influence the way we conduct research and interpret our findings. These influences cannot be completely avoided, but they can be identified and taken into consideration.

To keep your own biases in check, consider your research subject from an alternative or opposing perspective. At a minimum, considering alternative views will only strengthen your confidence in your research. But, in some cases, you may actually gain a new perspective that can help you further your research.

Is the Source Up to Date?

Depending on the field of study, results from prior research can become obsolete rather quickly. For instance, three-year-old research on skin cancer might already be considered outdated. On the other hand, climate measurements that are over 100 years old are still usable today.

Try to find the most recent sources on your subject. Reliable sources will usually offer a *literature review* that traces research on the subject back at least a few years. These literature reviews will show you how quickly the field is changing, while allowing you to judge whether the information you have located is current.

Can the Information Be Verified?

You should be able to locate more than one independent source that verifies the information you find. If you locate the same information from a few different independent sources, chances are good that the information is reliable. If you find the information in only one or two places, it is probably less reliable.

Triangulation is the key to verifying information. If you can find the information in diverse electronic and print sources, it is probably information you can trust. You might also use empirical methods to confirm or challenge the results of others.

Avoiding Plagiarism

One thing to watch out for is plagiarism in your own work, whether it is intentional or unintentional.

Plagiarism is the use of others' words, images, or ideas without acknowledgment or permission. In most cases, plagiarism is unintentional. While researching, a person might cut and paste information off websites or duplicate passages from a book. Later, he or she might use the exact text, forgetting that the information was copied directly from a source.

In rare cases, plagiarism is intentional and therefore a form of academic dishonesty. In these cases, teachers and colleges will often punish plagiarizers by having them fail the course, putting them on academic probation, or even expelling them. Intentional plagiarism is a serious form of dishonesty.

To avoid plagiarizing, keep careful track of your sources and acknowledge where you found your information.

Keep track of sources—Whenever you are gathering information from a source, carefully note where that information came from. If you are cutting and pasting information from an online source, make sure you put quotation marks around that material and clearly identify where you found it.

Acknowledge your sources—Any words, sentences, images, data, or unique ideas that you take from another source should be properly cited. If you are taking a direct quote from a source, use quotation marks to set it off from your writing. If you are paraphrasing the work of others, make sure you cite them with an in-text citation and put a full-text citation in a references list.

Ask permission—If you want to include others' images or large blocks of text in your work, write them an e-mail to ask permission. Downloading pictures and graphics off the Internet is really easy. But those images are usually someone's property. If you are using them for educational purposes, you can probably include them without asking permission. But, if you are using them for any other reason, you likely need to obtain permission from their owners.

Link

For more information on obtaining permission, go to Chapter 5, page 170.

You do not need to cite sources that offer information that is "common knowledge." If you find the same information in a few different sources, you probably do not need to document that information. But, if you have any doubts, you might want to cite the sources anyway to avoid any plagiarism problems.

Unfortunately, cases of plagiarism are on the rise. One of the downsides of online texts, such as websites, is the ease of plagiarism. Some students have learned techniques of "patchwriting," in which they cut and paste text from the Internet and then revise it into a document. This kind of writing is highly vulnerable to charges of plagiarism, so it should be avoided.

In the end, plagiarism harms mostly the person doing it. Plagiarism is kind of like running stoplights. People get away with it for only so long. Then, when they are caught, the penalties can be severe. Moreover, whether intentional or unintentional, plagiarizing reinforces some lazy habits. Before long, people who plagiarize find it difficult to do their own work, because they did not learn proper research skills. Your best approach is to avoid plagiarism in the first place.

Assessing Your Information

- Is the source reliable?
- How biased is the source?
- Am I biased?
- Is the source up to date?
- Can the information be verified?

AT A GLANCE

To learn more about plagiarism, go to
www.pearsonhighered.com/johnsonweb3/7.21

- Research today involves collecting information from diverse sources that are available in many media, including the Internet.

- Effectively managing existing information is often as important as creating new information.

- Logical mapping can be used to define a subject and highlight places where information needs to be found.

- A research methodology is a planned, step-by-step procedure that you will use to study the subject. Your research methodology can be revised as needed as your research moves forward.

- Triangulation is a process of using electronic, print, and empirical sources to obtain and evaluate your findings and conclusions.

- After collecting information, you should carefully assess whether it is biased or outdated. Also, be aware of your own biases.

- Careful note taking is essential for research. You should keep close track of your sources and use your research carefully in summarizing, paraphrasing, and quoting sources.

Individual or Team Projects

1. Think of a technical subject that interests you. Then, collect information from electronic and print sources. Write a progress report to your instructor in which you highlight themes in the materials you've found. Discuss any gaps in the information that you might be able to fill with more searching or empirical study. Some possible topics might include the following:

   Wildlife on campus
   Surveillance in America
   Hybrid motor cars
   The problems with running red lights on or near campus
   Safety on campus at night
   The effects of acid rain in Canada
   Migration of humpback whales

2. On the Internet, find information on a subject that you think is junk science or is influenced by junk science. Pay close attention to the reputations of the researchers and their results. Can you find any information to back up their claims? Pay special attention to where they receive their funding for the research. When you are finished searching the web, make a report to your class on your findings. Show your audience how junk science influences the debate on your subject.

   Here are a few possible topics:

   Evolution versus creation science
   Genetically engineered foods
   Managing forests to prevent fires

Cell phones and cancer
Experimentation on animals
Herbicides and insecticides
Diets and dietary supplements
Global warming
Smoking and secondhand smoke
Air and water pollution
Transporting nuclear waste
Welfare abuse

3. Survey your class on a campus issue that interests you. Write five questions and let your classmates select among answers like "strongly agree," "agree," "disagree," and "strongly disagree." Then, tabulate the results of the survey. Write a memo to your instructor in which you discuss the trends you found in your findings. In your memorandum, also point out places where your methodology might be challenged by someone who doubts your findings. Discuss how you might strengthen your survey if you wanted to do a larger study on this subject.

Collaborative Project

With a group, develop a methodology for studying substance abuse (alcohol abuse or abuse of prescription drugs or illegal drugs) on campus. First, use logical mapping to identify what you already know or believe about substance abuse on your campus. Second, formulate a research question that your research will answer. Third, use logical mapping to sketch out a methodology that would help you generate results to answer your research question.

Your methodology should use triangulation to gather information from a broad range of sources. In other words, you should plan to gather information from electronic, print, and empirical sources.

Finally, write up your methodology, showing the step-by-step procedures you will use to study substance abuse on campus. Your methodology should be written in such a way that others can duplicate it. It should also clearly identify the kinds of results you expect your research to generate.

Give your methodology to your instructor. At this point, your instructor may ask you to continue your research, following your methodology. As you do your research, note places where you changed your methodology or found information you did not expect.

For additional technical writing resources, including interactive sample documents, document design tutorials and guidelines, and more, go to **www.mytechcommlab.com**.

For websites that discuss substance abuse on college campuses, go to
www.ablongman.com/johnsonweb3/7.22

The Life of a Dilemma

Jen Krannert was a third year student in the biomedical engineering program at North Carolina State University. She was looking forward to doing a co-op semester in the spring, so she could gain some valuable experience (and earn a little money).

She applied to several co-op programs, including programs at Baxter, Biogen, and Boston Scientific. The one that caught her interest most, though, was at GenBenefits, a small biotech laboratory in California. At the interview, GenBenefit's recruiters talked with her about their work on embryonic stem cell research and how they were on the cusp of some major breakthroughs. She was very excited about the possibilities of being part of that kind of research.

Two weeks after the interview, Allen Marshall, GenBenefits' Vice President of Research, called Jen personally to offer her the co-op position. She accepted right away and spent the next hour e-mailing her other co-op opportunities to tell them she had accepted a position. She also called her parents, who were thrilled for her. They were very excited about her being part of this cutting-edge research.

The next day, she called her best friend Alice Cravitz, who was a student at Duke University. At first, Alice was enthusiastic, but when Jen told her that she would be working with embryonic stem cells, Alice grew quiet.

"What's wrong?" Jen asked.

Alice said, "I just think doing research on embryonic stem cells is unethical. Those are human lives you will be messing around with and ultimately destroying."

Jen became a little defensive. "First, they are embryos, not people. Second, these embryos are the leftover products of fertilization clinics. They will never be implanted and will probably be destroyed anyway. Third, we could save many, many lives with this research."

Sensing Jen's defensiveness, Alice changed the subject. After Jen hung up the phone, she wasn't sure how she felt about her co-op now. She realized she wasn't sure how she felt about embryonic stem cell research, so she didn't know if she believed it was ethical or not. The co-op was a great opportunity. If she turned it down, she would likely not find another for the spring. Plus, she really believed that this research could lead to some incredible medical breakthroughs.

If you were Jen, how might you use the research methods described in this chapter to sort out this ethical dilemma?

Want to learn more about embryonic stem cell research? Go to www.pearsonhighered.com/johnsonweb3/7.23

GO TO THE NET

Case Study 173

CHAPTER

8

Organizing and Drafting

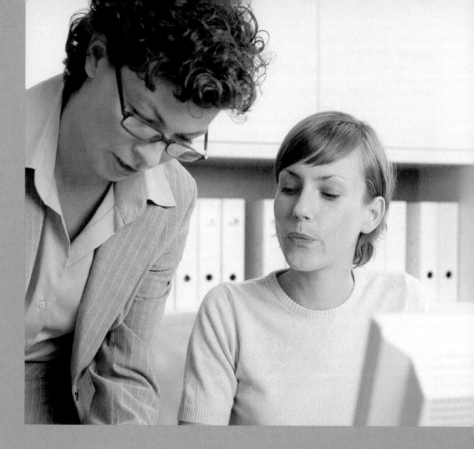

In this chapter, you will learn:

- A basic organizational pattern that any document can follow.
- How genres offer patterns for organizing documents.
- How to outline the organization of a document.
- How to use presentation software to organize your information.
- How to organize and draft a document's introduction.
- Patterns of arrangement for organizing and drafting sections of a document's body.
- Basic moves used in a document's conclusion.
- Strategies for organizing documents for international readers.

A Home Page

Figure 8.7:
Readers expect a home page to make the same moves as an introduction in a print document.

Subject

Purpose

Main point

Forecasting

Background information

Importance of subject

Source: Pacific Northwest Search and Rescue, http://www.pnwsar.org.

As discussed more thoroughly in Chapter 14, home pages in websites are also introductions. For example, the home page shown in Figure 8.7 also makes the typical moves found in an introduction.

Organizing and Drafting the Body

The body of the document is where you are going to provide the *content* that readers need to know. Here is where you will give them the information they need (facts, details, examples, and reasoning) to understand your subject and/or take action.

Carving the Body into Sections

With the exception of small documents (e.g., small memos or letters), the bodies of technical documents are typically carved into *sections*. In many ways, sections are like

For more help with writing introductions, go to
www.pearsonhighered.com/johnsonweb3/8.8

Two Versions of an Introduction

Version One

Solving the Flooding Problems in Darbey, Missouri: A Recommendation Report

Subject → Flooding has become a recurring problem in Darbey, our small community set in the Curlew Valley south of St. Louis.

Background information → As we mentioned in our presentation to the Darbey City Council last month, the town has been dealing with flooding since it was founded. Unfortunately, the problem seems to be growing worse. Previously, the downtown was flooded three times (1901, 1922, and 1954). In recent years, though, Darbey has experienced flooding with much more frequency. The downtown has flooded in 1995, 1998, 2000, and 2003.

Purpose → This proposal offers some strategies for managing flooding in the Darbey area.

Main point → We argue that the only long-term way to control flooding around Darbey is to purchase and restore the wetlands around the Curlew River, while enhancing some of the existing flood control mechanisms like levees and diversion ditches.

Importance → Otherwise, if development continues to expand between the town and the river, the flooding around Darbey will only continue to worsen, potentially causing millions of dollars in damage.

Forecasting → In this proposal, we will first identify the causes of Darbey's flooding problems. Then, we will offer some solutions for managing future flooding. And finally, we will discuss the costs and benefits of implementing our solutions.

Version Two

Solving the Flooding Problems in Darbey, Missouri: A Recommendation Report

Purpose and subject → This proposal offers strategies for managing flooding in the Darbey area.

Main point → The only long-term way to control flooding around Darbey is to purchase and restore the wetlands around the Curlew River, while enhancing some of the existing flood control mechanisms like levees and diversion ditches.

Importance → Otherwise, if development expands between the town and the river, the flooding around Darbey will only continue to worsen, potentially causing millions of dollars in damage.

Background information → We cannot ignore this problem. Flooding has become a recurring problem in Darbey, our small community set in the Curlew Valley south of St. Louis. As we mentioned in our presentation to the Darbey City Council last month, this town has been dealing with flooding since it was founded. Previously, the downtown was flooded three times (1901, 1922, and 1954). In recent years, however, Darbey has experienced flooding with much more frequency. The downtown has flooded in 1995, 1998, 2000, and 2003.

Forecasting → In this proposal, we will first identify the causes of Darbey's flooding problems. Then, we will offer some solutions for managing future flooding. And finally, we will discuss the costs and benefits of implementing our solutions.

Figure 8.6: As shown in these two sample introductions, the six introductory moves can be arranged in just about any order. Some arrangements, however, are more effective than others, depending on the subject and purpose of the document.

Brian Burridge
INTERNET SOFTWARE ARCHITECT, DESIGNER, AND DEVELOPER

What are some ways to succeed as a telecommuter in technical careers?

Although we've had the technology to telecommute for many years, it has taken many more years for companies to adapt their methodologies and workplace paradigms. Human psychology often changes much slower than technology. Even today, we continue to have managers who think and manage as though it were the industrial age. They expect to see employees in their seats, arriving and leaving at a specified time. They believe that if they can't see what the employee is doing, he or she must not be working.

But this is the information age and, beyond that, the age of creativity. Employee expectations have changed, as well as increasing fuel costs, extended commutes, concerns about carbon footprints, and the ever-increasing cost of real estate for business. So, more and more companies are turning to telecommuting as a solution.

Here are the keys to being as successful telecommuter:

- Schedule a trip to the facility at the beginning of the project. A few shared lunches, dinners, and perhaps other activities (catch a movie with fellow co-workers) can help break the ice and open the door of opportunity for relationship building.
- Get to know people by finding out what they like, whether they have families, what they do for fun, what they read, what they like to eat, career goals, and so forth.
- Use social websites, like LinkedIn, MySpace, and FaceBook to blog and share your personal and professional experiences. This can help build relationships on a deeper level.
- Send daily reports that provide task status and offer updates any time a problem impacts the schedule. An end-of-day report should be sufficient in most cases.
- Be available via multiple forms of communication so that people feel they can easily get ahold of you, especially during normal working times.
- Look for ways to be helpful to other teammates on their tasks. Take on new responsibilities when possible.
- Finish tasks in advance of scheduled completion times to help build trust.

The successful telecommuter needs to put a special effort toward building trust with managers and team members. They cannot see you working, so you need to show them evidence by communicating with them regularly.

Finally, quality communication depends on the telecommuter's ability to use the written word (in e-mail and instant messages) to convey the point in a succinct and yet effective manner. This is the number one problem with telecommuting; simply that so many workers are really very bad at communicating their thoughts, defining problems, and proposing solutions with the written word. This is what frustrates team members and then makes them complain to their boss about the off-site employee who can't communicate. To communicate clearly, get to the point; leave out too many side notes and details that don't directly support the purpose of your communication.

The only long-term way to control flooding around Darbey is to purchase and re-store the wetlands around the Curlew River, while enhancing some of the existing flood control mechanisms like levees and diversion ditches.

Are you giving away the ending by telling the readers your main point up front? Yes. But technical documents are not mystery stories. Just tell readers your main point in the introduction. That way, as they read the document, they can see how you came to your decision.

MOVE 4: STRESS THE IMPORTANCE OF THE SUBJECT   Make sure you give your readers a reason to care about your subject. You need to answer their "So what?" questions if you want them to pay attention and continue reading.

If development continues to expand between the town and the river, the flooding around Darbey will only continue to worsen, potentially causing millions of dollars in damage.

MOVE 5: PROVIDE BACKGROUND INFORMATION   Typically, background information includes material that readers already know or won't find controversial. This material could be historical, or it could stress a connection with the readers.

As we mentioned in our presentation to the city council last month, Darbey has been dealing with flooding since it was founded. Previously, the downtown was flooded three times (1901, 1922, and 1954). In recent years, Darbey has experienced flooding with much more frequency. The downtown was flooded in 1995, 1998, 2000, and 2003.

MOVE 6: FORECAST THE CONTENT   Forecasting describes the structure of the document for your readers by identifying the major topics it will cover.

In this proposal, we will first identify the causes of Darbey's flooding problems. Then we will offer some solutions for managing future flooding. And finally, we will discuss the costs and benefits of implementing our solutions.

Forecasting helps readers visualize the organization of the rest of the document. It gives them a map to anticipate the topics the document will cover.

Drafting with the Six Moves

In an introduction, these moves can be made in just about any order. Figure 8.6, for example, shows how the introductory moves can be used in different arrangements.

Any information that goes beyond these six moves should be removed from the introduction. After all, this extra information will only make it more difficult for the readers to locate the subject, purpose, and main point of your document.

Organizing and Drafting the Introduction

When organizing and drafting your document, you should put yourself in your readers' place. If you were the reader, what information would you want to know up front?

What is this document about?

Why did someone write this for me?

What is the main point?

Is this information important?

How is this document organized?

When an introduction answers these questions up front, readers are better able to understand the rest of the document.

Six Opening Moves in an Introduction

These kinds of questions translate into six opening "moves" made in an introduction:

Link

For more Information on crafting a purpose statement, go to Chapter 2, page 21.

MOVE 1: DEFINE YOUR SUBJECT   Tell readers what your document is about by defining the subject.

> Flooding has become a recurring problem in Darbey, our small town nestled in the Curlew Valley south of St. Louis.

In some cases, to help define the boundaries of your subject, you might also tell readers what your document is *not* going to cover.

MOVE 2: STATE YOUR PURPOSE   Tell readers what you are trying to achieve. Your purpose statement should be clear and easy to find in the introduction. It should plainly tell your readers what the document will do.

> This proposal offers some strategies for managing flooding in the Darbey area.

You should be able to articulate your purpose in one sentence. Otherwise, your purpose may not be clear to your readers—and perhaps not even to yourself.

MOVE 3: STATE YOUR MAIN POINT   Tell your readers the key idea or main point that you would like them to take away from the document. Usually your main point is your overall decision, conclusion, or solution that you would like the readers to accept.

To learn more about how people read introductions, go to
www.pearsonhighered.com/johnsonweb3/8.6

4. Create a new page for each section of your document, and give it a unique title.
5. Create a conclusion page that restates your main point and looks to the future.

Then, working section by section, fill in your information using only bulleted lists. For each slide, use a new title and limit yourself to two to five bullets (Figure A). If one of your slides has only one bullet of information, merge its contents with another neighboring slide. If a slide has more than five bullets, divide the slide into two separate slides with unique titles.

As you fill in your information, you will find that the presentation software forces you to think about

- whether information is need-to-know content.
- how smaller pieces of information fit into larger sections.
- where information that your readers will need to know is missing.
- how you can express the information you gathered in concise ways.

When you have finished filling in the slides, you will be able to review the structure of your whole document. Now you can sharpen and clarify the materials by editing your slides. You might move some slides around, consolidate a few, and expand others. You should also pay attention to the titles of the slides, because they will eventually become the headings in the written document.

When you are done filling and editing the slides, you will have created a very helpful outline for writing your document. Each of the slides will probably become a section of your written document. Each of the bullet points will probably need a paragraph or two of coverage.

Ultimately, you are using the presentation software as an outlining tool. But the presentation software is much more flexible and visual than a written outline. Try out this technique. You will find it very effective.

Organizing Content on a Presentation Slide

Major heading of section in report

Topics to be discussed in this section

Figure A: Presentation software can be a very helpful tool for organizing your information, even if you are not going to present the material orally.

Want more advice on using presentation software? Go to
www.pearsonhighered.com/johnsonweb3/8.5

Outlining the Document 181

Outline View

The toolbar allows you to create levels in the text.

Information can be organized into hierarchies.

These icons allow you to "collapse" or "expand" parts of the outline.

These numbers allow you to view the text at different levels.

Figure 8.5: Outline View is especially helpful for gaining an overview of your document. It is also helpful when you are brainstorming with a team of co-workers on a large project.

Organizing with Presentation Software

Not everyone likes using outlines, but as a document gets larger, it becomes increasingly difficult to organize all that information without some kind of outline. One clever way to organize your information is to use presentation software, like Corel Presentations or Microsoft PowerPoint, to frame and arrange your ideas. Even if you don't need to present your materials orally, the presentation software will help you structure and clarify your thoughts.

Let's say you need to write a report. You have collected a large amount of information from the Internet, empirical studies, interviews, and trips to the library. You are almost overwhelmed by the amount of information you need to put into your report. Presentation software will force you to boil all that information down into headings and bullet points. Before too long, you will have created a very effective outline with which to start drafting.

To organize your materials with presentation software, follow these steps:

1. Create a title page and write a title.
2. State your document's main point as a subtitle on the title page.
3. Create an introduction page that states your subject, purpose, and the importance of the subject.

For more help using Outline View, go to
www.pearsonhighered.com/johnsonweb3/8.4

A Rough Outline

This outline follows the proposal genre for organizing the text.

Major headings sketch out the basic structure of the document.

Subheadings begin filling in the content.

Figure 8.4:
An outline doesn't need to be formal, and it should always be open to change. Here is a rough outline with some guesses about what kinds of topics will be discussed in the proposal.

When outlining, you might also consider using the Outline View in your word processor. In Outline View, you can begin arranging your information by listing the headings for the document's major sections and subsections (Figure 8.5). Then, you can decide if:

- smaller sections should be merged or incorporated into larger sections.
- larger sections should be divided into separate sections.
- any need-to-know information is missing from sections.
- there is too much want-to-tell information.

While you are revising your document, Outline View also helps you quickly move information around. You can "collapse" a whole part of a document under a heading. Then, you can cut and paste to move that part somewhere else in the document.

Outline View is a great tool when you are working with a team. When you and your co-workers are brainstorming about a project or document, just toss your ideas into Outline View. Then, when you're finished brainstorming, you can start organizing all that information into a more structured outline. At the end of the meeting, you can print out the outline or e-mail it to the whole group to guide the drafting of the document.

Overall, an outline should be as flexible as the document itself. A computer-generated outline can be a helpful tool for planning, drafting, and revising your work.

Link

For more information on working with a team, see Chapter 4, page 66.

For websites that offer outlining strategies, go to
www.pearsonhighered.com/johnsonweb3/8.3

Both genres have an introduction and conclusion, but otherwise they are very different in organization.

Indeed, each genre achieves a different purpose and therefore requires a different organizational pattern. Once you know your purpose, you should be able to easily figure out what kind of document you are being asked to write and how it should be organized (Figure 8.3).

Keep in mind that genres are patterns, not formulas. There are countless ways to organize analytical reports and procedures, depending on the needs of your readers and the context in which the document will be used. You can use these genres as starting places from which to begin organizing and drafting your documents.

Chapters 17 through 23 discuss a range of workplace genres in depth. When you begin writing one of these documents, you should turn to the chapter that describes it.

Choosing a Genre

Figure 8.3:
Once you know the purpose of your document, you can decide which genre fits your need. Each genre provides a pattern for you to follow as you organize and draft your document.

Which genre fits my writing task?

Purpose	Genre	Type of Document
I need to tell others about a decision or event.	Correspondence (Chapters 13 and 17)	E-mail, Letter, or Memo
I need to define a concept or set of terms.	Definition (Chapter 18)	Technical Definition, Glossary
I need to describe an item, product, or service.	Description (Chapter 19)	Technical Description, White Paper
I need to explain how to do something.	Instructions (Chapter 20)	Instructions, Specifications, Procedures
I need to make a suggestion or propose a new project.	Proposal (Chapter 21)	Research Proposal, Planning Proposal, Implementation Proposal
I need to present information or make a recommendation.	Report (Chapters 22 and 23)	Research Report, Feasibility Report, Recommendation Report, Lab Report, Progress Report
I need to provide on-line information to others.	Hypertext (Chapter 14)	Website, Multimedia Document, CD-ROM
I need a job.	Résumé (Chapter 15)	Portfolio, Résumé, Scannable Résumé

Outlining the Document

Once you have identified the genre of your document, you can begin developing a rough outline of its organization.

Outlining may seem a bit old-fashioned, but it is very helpful when you are trying to sort out your ideas, especially as you prepare to write a large technical document. In the workplace, most people sketch out a rough outline to help them organize their ideas. In their outline, they type the document's main headings on the screen (Figure 8.4). Then, they list the contents of each section separately. The outline will usually change as new ideas, evidence, or issues emerge.

If you are unsure how to outline your document, let the genre you are following guide your thought process. The genre should give you a good sense of the larger sections of the document. From there, you can begin filling in the smaller topics in each section.

Want to learn more about genres? Go to
www.pearsonhighered.com/johnsonweb3/8.2

The "Smoking Gun" *Challenger* Memo

MORTON THIOKOL, INC

Wasatch Division Interoffice Memo

July 31, 1985

2870:FY86:073

TO: R. K. Lund
              Vice President, Engineering
CC: B. C. Brinton, A. J. McDonald, L. H. Sayer, J. R. Kapp
FROM: R. M. Boisjoly
              Applied Mechanics - Ext. 3525
SUBJECT: SRM O-Ring Erosion/Potential Failure Criticality

The introduction contains the purpose and main point.

This letter is written to insure that management is fully aware of the seriousness of the current O-ring erosion problem in the SRM joints from an engineering standpoint.

The mistakenly accepted position on the joint problem was to fly without fear of failure and to run a series of design evaluations which would ultimately lead to a solution or at least a significant reduction of the erosion problem. This position is now drastically changed as a result of the SRM 16A nozzle joint erosion which eroded a secondary O-ring with the primary O-ring never sealing.

The body supports the argument.

If the same scenario should occur in a field joint (and it could), then it is a jump ball as to the success or failure of the joint because the secondary O-ring cannot respond to the clevis opening rate and may not be capable of pressurization. The result would be a catastrophe of the highest order—loss of human life.

An unofficial team (a memo defining the team and its purpose was never published) with leader was formed on July 19, 1985 and was tasked with solving the problem for both the short and long term. This unofficial team is essentially nonexistent at this time. In my opinion, the team must be officially given the responsibility and the authority to execute the work that needs to be done on a non-interference basis (full time assignment until completed.)

The conclusion restates the main point.

It is my honest and very real fear that if we do not take immediate action to dedicate a team to solve the problem with the field joint having the number one priority, then we stand in jeopardy of losing a flight along with all the launch pad facilities.

R. M. Boisjoly

Concurred by: J. R. Kapp, Manager

Applied Mechanics

Source: Report of the Presidential Commission on the Space Shuttle Challenger Accident, 1986.

Standard Organization of a Document

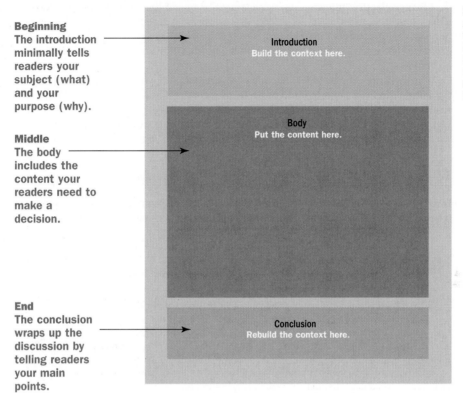

Beginning
The introduction minimally tells readers your subject (what) and your purpose (why).

Middle
The body includes the content your readers need to make a decision.

End
The conclusion wraps up the discussion by telling readers your main points.

Introduction
Build the context here.

Body
Put the content here.

Conclusion
Rebuild the context here.

Figure 8.1:
An effective technical text will have a beginning (introduction), middle (body), and end (conclusion).

Using Genres to Organize Information

Once you move beyond the introduction-body-conclusion pattern, technical documents follow a variety of different organization patterns. An analytical report, for example, is very different from a set of instructions, and a proposal shows little resemblance to a technical specification. These documents follow different patterns of organization, which are called *genres*.

A genre is a predictable pattern for organizing information to achieve specific purposes. For instance, here are two genres, for the report and a procedure, side by side:

Analytical Report
- Introduction
- Methods
- Results
- Discussion
- Conclusion

Procedure
- Introduction
- Materials
- List of parts/tools
- Sequentially ordered steps
- Conclusion

Computers are great tools for helping us manage large amounts of information. Today, effective "information management" is one of the most significant challenges to communicating in the technical workplace. After all, computer networks put amazing amounts of information at our fingertips. But, we still need to organize that information in ways that make it accessible and understandable to our readers.

When writing your technical document or presentation, you will need to organize the information you collected into patterns that are familiar to your readers. Your readers, after all, are interested in the information you have gathered. But, they need you to present that information in a predictable and usable way. Otherwise, they won't be able to take full advantage of your thoughts and research on the subject.

Basic Organization for Any Document

Despite their differences, technical documents usually have something important in common. They typically include an *introduction, body,* and *conclusion.*

> **Introduction**—The introduction of your document needs to tell readers *what* you are writing about and *why* you are writing about it.
>
> **Body**—The body of your document presents the content that your readers need to know to take action or make a decision.
>
> **Conclusion**—The conclusion of your document wraps up your argument by restating your main point(s).

Sometimes it helps to remember the familiar speechwriters' saying "Tell them what you are going to tell them. Tell them. Then, tell them what you told them."

This beginning-middle-end pattern might seem rather obvious, but people regularly forget to include distinct introductions, bodies, and conclusions in their documents. All too often, they toss readers into the details without first telling them the subject and purpose of their documents. In some cases, their documents end abruptly without summing up the major points.

Introductions and conclusions are especially important in technical documents, because they provide a *context,* or framework, for understanding the *content* in the body of the text (Figure 8.1). Without "contextual" information at the beginning and end of the document, readers find it very difficult to figure out what the author is telling them. Have you ever read a document that didn't seem to have an obvious point? More than likely, it was lacking an effective introduction and/or conclusion.

To see an example of a good introduction, body, and conclusion, consider the classic memo in Figure 8.2. In this memo, Roger Boisjoly, an engineer at Morton Thiokol, warns a vice president of the company that the space shuttle *Challenger* is in danger of blowing up. This memo is often referred to as the "smoking gun" memo that demonstrated that NASA and Morton Thiokol were ignoring erosion problems with the shuttle's O-rings. The purpose of these O-rings was to keep explosive exhaust gases from destroying the rocket boosters—and the shuttle.

The problem described in this memo is clear, and Boisjoly does his best to stress the importance of the problem in the introduction and conclusion. Unfortunately, his and other engineers' warnings were not heeded by higher-ups.

Why do documents need introductions and conclusions? For answers, see
www.pearsonhighered.com/johnsonweb3/8.1

Basic Organization for Any Document 175

A Section with an Opening, Body, and Closing

Figure 8.8:
A section includes an opening and body. The closing is optional, but it can be helpful to sum up the section's point or points.

miniature documents, needing their own beginning, middle, and end. They typically include an *opening, body,* and *closing* (Figure 8.8).

OPENING   An opening is usually a sentence or small paragraph that identifies the subject and purpose of the section. The opening usually includes a claim or set of claims that the rest of the section will support.

Results of Our Study

The results of our study allow us to draw two conclusions about the causes of flooding in Darbey. First, Darbey's flooding is mostly due to the recent construction of new levees by towns farther upriver. Second, development around the river is taking away some of the wetlands that have protected Darbey from flooding in the past. In this section, we will discuss each of these causes in depth.

BODY   The body of a section is where you will offer support for the claim you made in the opening. The body of a section can run anywhere from one paragraph to many paragraphs, depending on the purpose of the section. For example, if you are discussing the results of a research study, your Results section may require three or more paragraphs in the body—one paragraph per major result.

CLOSING (OPTIONAL)   A section that is large and/or complex might need a closing paragraph or sentence to wrap up the discussion. A closing usually restates the claim you made in the opening of the section. It might also look forward to the next section.

> In sum, these two causes will likely become only more significant over time. As upriver towns grow in population, there will be more pressure than ever to build more levees to protect them. Meanwhile, if development continues in the available wetlands around the Curlew River, Darbey will find that some of its last defenses against flooding have disappeared.

Overall, a section should be able to work as a stand-alone unit in the document. By having its own beginning, middle, and end, a well-written section feels like a miniature document that makes a specific point.

Patterns of Arrangement

When writing each section, you can usually follow a *pattern of arrangement* to organize your ideas. These patterns are based on logical principles, and they can help you organize your information so that your views will be presented in a reasoned way. Major patterns of arrangement are:

- Cause and effect
- Comparison and contrast
- Better and worse
- Costs and benefits
- If...then
- Either...or
- Chronological order
- Problem/needs/solution
- Example

You will find that each section in your document likely follows one of these patterns of arrangement. One section, for example, may discuss the causes and effects of a problem. A later section might use a discussion of the costs and benefits of doing something about the problem. So, as you are organizing and drafting each section, decide which pattern of arrangement best fits your needs.

CAUSE AND EFFECT   In a sense, all events are subject to *causes* and *effects*. For example, if a bridge suddenly collapses, investigators would immediately try to determine the causes for the collapse.

> In 2002, the I-40 bridge over the Arkansas River collapsed for a few different reasons. The actual collapse occurred when a runaway barge on the river rammed into one of the bridge's supports. But other causes were evident. The bridge was already weakened by erosion around the pilings, deteriorated concrete, and attrition due to recent seismic activity.

A cause can also have various effects.

> Leaving a wound untreated can be dangerous. The wound may become contaminated with dirt and germs, thus requiring more healing time. In some cases, the wound may grow infected or even gangrenous, requiring much more treatment at a hospital. Infections can be life-threatening.

For sample sections that you can model, go to
www.pearsonhighered.com/johnsonweb3/8.9

A Section Using Cause and Effect

Figure 8.9:
A section that
uses the cause
and effect
pattern of
arrangement.

Buckinghamshire Flooding Facts
Flooding—Cause and Effect

In Buckinghamshire, as with other inland counties, there are three main causes of flooding, which are river (also known as riparian), flash, and groundwater. In all three cases the ability of the ground to absorb rainfall, like a sponge, plays a major part, and this will vary according to weather conditions.

← Opening paragraph lists the causes of the problem.

River flooding occurs when rivers cannot cope with the amount of water draining into them off the land. In the winter months this is usually because the ground has become saturated and can no longer absorb water properly, so when rainfall is heavy and/or prolonged, runoff reaches the rivers faster and eventually they overtop their banks, as happened along the River Thames in January 2003.

Flash flooding can happen anywhere at any time, although it is more likely in the summer months when the ground is hard and dry. Sudden downpours, such as those associated with thunderstorms, cannot soak in fast enough and the water runs off quickly into drains, ditches, and culverts that cannot cope with the volume. Sometimes there is so much rain that a hillside may have the soil washed off it—often into nearby properties.

← Body paragraphs discuss each cause separately.

Groundwater flooding is rare, occurring only when the underground water table rises to an unusually high level. After the very heavy autumn and winter rainfall of 2000/01, the water table in the Chilterns was at its highest level for decades. Springs and wells that had been dry for, in some cases, 50 years or more began to flow again and did so throughout the summer, causing property flooding in many areas. Groundwater again rose significantly following the very wet winter of 2002/03, though fortunately the drier spring weather meant that levels did not reach a critical point.

Whatever the cause, the effects are much the same. If floodwater enters your property it will ruin carpets, furniture, household goods, decor, and electrics. Often floodwater will be mixed with raw sewage, as drains overflow. Even when the flood recedes, your home will take a long time to dry out and may smell for weeks. Plaster may have to be stripped off walls. If you are fully insured there should be no problem in getting the necessary repairs paid for, but the whole process will probably take months, and the experience is severely depressing. People who have been flooded will often say that it is far worse than being burgled.

← Closing paragraph discusses the effects of the problem.

Source: Buckinghamshire County Council, http://www.southbucks.gov.uk/documents/Flooding%20FACTS3.doc.

As you discuss causes and effects, show how effects are the results of specific causes. For example, the white paper shown in Figure 8.9 explains how flooding occurs when specific events (causes) occur.

Want to see examples of cause and effect sections? Go to
www.pearsonhighered.com/johnsonweb3/8.10

Figure 8.10:
Comparison and
contrast puts two
things side by
side to show their
similarities and
differences.

Comparison Between Hemyock and Bodiam Castles

Opening states the claim for the section to follow.

The plan of Hemyock Castle has similarities with Bodiam Castle in Kent, built some five years later.

Both Hemyock and Bodiam are typical of small late medieval castles: a rectangular site with high, round corner towers and central interval towers, connected by a high curtain wall; all topped with crenellations; surrounded by a water-filled moat. Both had massive fortified gatehouses. The two castles were roughly the same size. Even the detail of Hemyock's NE Tower appears similar to Bodiam's Well Tower.

Comparisons are made by noting similarities.

There are obvious differences:

- Hemyock Castle was built around the family's old manor house, whereas Bodiam Castle was built on a new site.

- Judging by the remains, Bodiam Castle appears to have been more lavish.

- Bodiam's interval towers were rectangular rather than round.

- Bodiam has a huge moat.

Contrasts are made by noting differences.

Presumably, Hemyock Castle was more functional. Much of the accommodations would have been provided by the old manor house, so the defensive outer walls could be simpler. This would also have allowed the use of stronger, more functional round towers. The rectangular towers at Bodiam were required to provide comfortable accommodations. Further accommodation was built into Bodiam's outer walls.

Bodiam had a huge, if easily drained, moat and complex entrance causeways.

Hemyock's moat was more functional but not easily drained. Hemyock's western entrance (now lost) may have included a short defensive causeway.

The massive gatehouse at Hemyock's eastern entrance was protected by a drawbridge across the moat and by outer bastions (now the "Guard Houses" holiday cottages). In recent centuries there have been great changes around Hemyock's eastern entrance, including diversion of the river (St. Margaret's Brook) and extension of St. Mary's Church. It is just possible that there was a more complex series of defenses and water obstacles.

Source: Hemyock Castle, http://www.hemyockcastle.co.uk/bodiam.htm.

COMPARISON AND CONTRAST You can *compare and contrast* just about anything. When comparing and contrasting two things, first identify all the features that make them similar. Then, contrast them by noting the features that make them different (Figure 8.10). By comparing and contrasting two similar things, you can give your readers a deeper understanding of both.

BETTER AND WORSE In technical workplaces, you are often faced with moments in which you need to choose among different paths. In these cases, you may need to play the advantages off the disadvantages.

The "better" is discussed in terms of advantages. → Automating our assembly line with robotic workstations has clear advantages. With proper maintenance, robots can work around the clock, every day of the week. They don't take vacations, and they don't require benefits. Moreover, after an initial up-front investment, they are less expensive per unit than human labor.

The "worse" is shown in a less favorable light. → Our alternative to automation is to stay with human labor. Increasingly, we will become less profitable, because our competitors are moving their operations to offshore facilities, where labor is much cheaper and environmental laws are routinely ignored. Meanwhile, if we do not automate, the increasing costs of health care and an aging workforce will eventually force us to close some of our manufacturing plants in North America.

COSTS AND BENEFITS   By directly weighing the *costs and benefits*, you can show readers that the price is ultimately worth the benefits of moving forward with a project.

Why College Education Is Still Worth It

How good an investment finishing college is depends on both earnings and costs—the earnings of college graduates relative to high school graduates and the costs of attending college (both tuition and foregone earnings). Tuition and fees for a four-year college for the 2003–2004 academic year averaged $7,091; the average net price—tuition and fees net of grants—was $5,558 (both amounts in 2003 dollars). If we assume that tuition and fees continue to rise as they did between the 1999–2000 and 2003–2004 school years, and conservatively look at sticker rather than net prices, the average full-time student entering a program in the fall of 2003 who completes a bachelor's degree in four years will pay $30,325 in tuition and fees. If we assume an opportunity cost equal to the average annual earnings of a high school graduate (from the March 2004 Current Population Survey) and a 5 percent discount rate for time preference, the total cost of attending college rises to $107,277. In other words, college is worthwhile for an average student if getting a bachelor's degree boosts the present value of her lifetime earnings by at least $107,277.

What is the boost to the present value of wages? At a 5 percent annual discount rate, it is $402,959. The net present value of a four-year degree to an average student entering college in the fall of 2003 is roughly $295,682—the difference between $402,959 in earnings and $107,277 in total costs. A student entering college today can expect to recoup her investment within 10 years of graduation.

It still pays to go to college—very much so, at least as much as ever before.

Source: L. Barrow, C. E. Rouse, (2005). "Does college still pay?" The Economists' Voice 2 (4), 1–8.

Stressing the benefits to your readers is always a good way to reason with them. By putting the costs in contrast to these benefits, you can show how the advantages ultimately outweigh the costs.

IF...THEN   Perhaps the most common way to reason is using *if...then* arguments. Essentially, you are saying, "If you believe in X, then you should do Y" or perhaps "If X happens, then Y is likely to happen also" (Figure 8.11). When using *if...then* arguments, you are leveraging something the readers already believe or consider possible to convince them that they should believe or do something further.

EITHER...OR   When using *either...or* statements, you are offering the readers a choice or showing them two sides of the issue. You are saying, "Either you believe X,

The opening sets a premise with two *if ... then* statements.

The body of the section explores the implications of those *if ... then* statements.

The closing poses *if ... then* statements that are more speculative.

Effects of Climate Change on Forests

The projected 2°C (3.6°F) warming could shift the ideal range for many North American forest species by about 300 km (200 mi.) to the north. If the climate changes slowly enough, warmer temperatures may enable the trees to colonize north into areas that are currently too cold, at about the same rate as southern areas became too hot and dry for the species to survive. If the earth warms 2°C (3.6°F) in 100 years, however, the species would have to migrate about 2 miles every year.

Trees whose seeds are spread by birds may be able to spread at that rate. But neither trees whose seeds are carried by the wind, nor nut-bearing trees such as oaks, are likely to spread by more than a few hundred feet per year. Poor soils may also limit the rate at which tree species can spread north. Thus, the range over which a particular species is found may tend to be squeezed as southern areas become inhospitably hot. The net result is that some forests may tend to have a less diverse mix of tree species.

Several other impacts associated with changing climate further complicate the picture. On the positive side, CO_2 has a beneficial fertilization effect on plants, and also enables plants to use water more efficiently. These effects might enable some species to resist the adverse effects of warmer temperatures or drier soils. On the negative side, forest fires are likely to become more frequent and severe if soils become drier. Changes in pest populations could further increase the stress on forests. Managed forests may tend to be less vulnerable than unmanaged forests, because the managers will be able to shift to tree species appropriate for the warmer climate.

Perhaps the most important complicating factor is uncertainty (see U.S. Climate in the Future Climate section) whether particular regions will become wetter or drier. If climate becomes wetter, then forests are likely to expand toward rangelands and other areas that are dry today; if climate becomes drier, then forests will retreat away from those areas. Because of these fundamental uncertainties, existing studies of the impact of climate change have ambiguous results.

Source: U.S. Environmental Protection Agency, http://yosemite.epa.gov/oar/globalwarming. nsf/content/ImpactsForests.html.

or you believe Y" or perhaps "Either X will happen, or Y will happen" (Figure 8.12). Statements using *either...or* patterns suggest that there is no middle ground, so you should use this strategy only when appropriate. When used appropriately, an either/or argument can prompt your readers to make a decision. When used inappropriately, these arguments can invite the readers to make a choice you didn't expect—or perhaps to make no choice at all.

A Section Using *Either . . . Or*

Figure 8.12:
The *either . . . or* pattern is helpful when only one belief or action is the right one.

Shapley-Curtis Debate

During the 1920s, Harlow Shapley and Heber Curtis debated whether observed fuzzy "spiral nebulae," which are today known as galaxies, exist either in our Milky Way Galaxy or beyond it. Shapley took the stance that these "spiral nebulae" observed in visible light do indeed exist in our galaxy, while the more conservative Curtis believed in the converse. While both sides of this debate seemed to either make too many suppositions or supported their claims with flawed data, the main points of both stances made the debate fairly close at the time.

The opening sets up two sides with either . . . or *statements.*

Shapley's Argument

Shapley argued that these "spiral nebulae" are actually within our galaxy's halo. He presented faulty data which seemed to confirm that M101, a large angular-diameter galaxy, changed angular size noticeably. We know now that for M101 to change angular size appreciably (0.02 arcsecond/year), it would have to recede from us at warp speeds based upon our current knowledge of its diameter and distance. To account for all the galaxies' observed recession from earth, not knowing of the universe's expansion, Shapley invented a special repulsion force to explain this strange phenomenon. He was not cogently able to explain why galaxies observed from earth are less densely distributed along the galactic equator, though.

The remainder of the section discusses the two sides.

Curtis's Argument

Curtis remained more conservative in his argument by not introducing outlandish suppositions like the special repulsion forces of Shapley. Curtis, although not able to explain galaxies' redshifts well either, made a convincing claim for the distribution of galaxies in the sky. Because observed galaxies seen edge-on often contain opaque gas lanes, and if our galaxy is similar to those observed galaxies, then we should not see extragalactic objects near our equator because of our own opaque dust lane. Hubble's observation of Cepheid variables in the Andromeda Galaxy confirmed the distance to a galaxy: millions of light-years, not thousands! Curtis was obviously correct.

Here, the section ends by saying which side won the debate.

Source: A. Aversa, 2003.

CHRONOLOGICAL ORDER   Time offers its own logic because events happen in chronological order. You can arrange information logically according to the sequence of events.

> Three things happen inside the bronchial tubes and airways in the lungs of people with asthma. The first change is inflammation: The tubes become red, irritated, and swollen. This inflamed tissue "weeps," producing thick mucus. If the inflammation persists, it can lead to permanent thickening in the airways.

Want to see other models that use chronological order? Go to
www.pearsonhighered.com/johnsonweb3/8.12

Next comes constriction: The muscles around the bronchial tubes tighten, causing the airways to narrow. This is called *bronchospasm* or *bronchoconstriction*.

Finally, there's hyperreactivity. The chronically inflamed and constricted airways become highly reactive to so-called triggers: things like allergens (animal dander, dust mites, molds, pollens), irritants (tobacco smoke, strong odors, car and factory emissions), and infections (flu, the common cold). These triggers result in progressively more inflammation and constriction.

Source: G. Shapiro, 1998.

PROBLEM/NEEDS/SOLUTION The *problem/needs/solution* pattern is a common organizational scheme in technical documents, because technical work often involves solving problems. When using this pattern, you should start by identifying the problem that needs to be solved. Then, state what is needed to solve the problem. Finally, end the section by stating the solution (Figure 8.13). The three-part structure leads the readers logically from the problem to the solution.

EXAMPLE Using an *example* is a good way to support your claims. Your readers might struggle with facts and data, but an example can help make all those technical details more realistic and familiar.

To fool predators, most butterflies have evolved colors and patterns that allow them to survive. For example, the delicious Red-spotted Purple is often not eaten by birds, because it looks like a Pipevine Swallowtail, a far less appetizing insect. Other butterflies, like anglewings, look like tree bark or leaves when seen from below. If a predator comes near, though, angle wings can surprise it with a sudden burst of color from the topside of their wings. (Pyle, p. 23)

A Section Describing Problems, Needs, and Solutions

Opening paragraph identifies the problem.

North Creek Water Cleanup Plan (TMDL)

North Creek does not meet state standards for swimming and wading because there is too much bacteria in the water. Also, the federal government has determined that Chinook salmon are threatened, and other salmon species face continuing pressure from urban development.

In the 1960s, much of the watershed was home to small ranches and hobby farms. Over the past 40 years, much of the land has been redeveloped with a trend towards more urban, commercial, and suburban residential development. The basin's hydrology, how water is stored and managed throughout the basin, has also changed.

Why Are These Waters Polluted?

Pollution in the North Creek watershed comes from thousands of sources that may not have clearly identifiable emission points; this category of pollution is called "nonpoint" pollution. These nonpoint sources can contribute a variety

Source: Washington State Department of Ecology, http://www.ecy.wa.gov/programs/wq/tmdl/watershed/north_creek/solution.html.

Figure 8.13:
The problem/needs/solution pattern is a good way to lead people logically toward taking action.

GO TO THE NET

Want to see people using examples in their documents? See
www.pearsonhighered.com/johnsonweb3/8.13

The section's body elaborates on the problem.

of pollutants that may come from failing septic systems, livestock and pet wastes, at-home car washing, lawn and garden care, leaky machinery, and other daily activities. Some of these nonpoint sources create fecal coliform bacterial pollution that indicate the presence of fecal wastes from warm-blooded animals. Ecology has confirmed that high levels of fecal coliform bacteria exist in North Creek. For this reason, Total Maximum Daily Loads for fecal coliform bacteria were subsequently established at multiple locations through each watershed.

Although wildlife can also contribute bacteria, such sources are not defined as pollution; however, when such natural sources combine with nonpoint pollution, the result can cause the kind of problems found in North Creek.

North Creek became polluted because of the way we do things, not the activities themselves. For example, having dogs, cats, horses, and other animals as part of our life is not a problem; rather, it is the way that we care for these animals. Similarly, roads and parking lots are a necessity of our modern society, but the way we build roads, neighborhoods, and shopping centers is causing our local streams and creeks to be polluted. There are solutions that can be undertaken by local governments, businesses, organizations, and citizens to solve the problem.

What Can You Do to Improve North Creek?

These questions are designed to highlight what is needed.

If cleaning up local waters is important to you, think about what you can do on your own first. Do you always pick up after your pet? Can you use organic fertilizer? Do you wash your car on your lawn or take it to a car wash? Can you reduce the amount of stormwater runoff from your property? Can you develop a farm plan to ensure your horse's manure is not reaching local streams? Do you practice good on-site septic system maintenance? The draft Action Plan includes information about current and future activities to clean up local waters.

There are many things residents can do now to reduce pollution reaching water bodies and to improve water quality. Here are some ways that you can help:

- Be responsible for proper septic tank maintenance or repair. If you have questions about your on-site septic system (exactly where it is located, how to maintain it), you can call the Snohomish Health District for technical assistance.

- Can you reduce stormwater leaving your property? To have a free survey of your property for ways to reduce potential water quality problems and

(continued)

improve stormwater management, contact Craig Young, your North Creek Basin steward.

Here are the recommended solutions to the problem.

- Keep pet and other animal wastes out of your local streams. Pick up after your pet and work with your community, association, or local government to get a pet waste collection station installed where it is needed most.

- Use landscaping methods that eliminate or reduce fertilizers and pesticides. If fertilizers are needed, organic products break down more slowly and help prevent big flushes of pollution when we have heavy rains; they also improve soil structure.

- Join local volunteers in planting trees and performing other activities that help local streams. Snohomish County, the Stilly/Snohomish Task Force and North Creek Streamkeepers, and the City of Bothell water quality volunteer program improve water quality by helping with stream restoration activities. They provide help or other opportunities to plant trees on your property or at other needed locations to help water quality (you can also volunteer to plant trees in other areas that need help too).

- Get involved in your local government's programs. Folks interested in sampling their local waters can contact Snohomish County, North Creek Streamkeepers, or your local city to explore the availability of volunteer monitoring opportunities. If you live outside of a city, be a Salmon Watcher or Watershed Keeper, or get involved in other individual or group activities to improve local waters; these are coordinated by Snohomish County staff.

The section closes by restating the main point and looking to the future.

How else can you be involved? The solution to polluting our local waters is to do some things a little differently. In this way, we can still live a normal 21st-century lifestyle, have animals as a close part of our lives, and have clean water. Citizen involvement in deciding what needs to be done is essential to making our water bodies safe places for people and fish, and you may be part of helping to design future watershed activities that haven't even been thought of yet! Check out the Related Links page for more ideas.

Organizing and Drafting the Conclusion

An effective conclusion rounds out the discussion by bringing readers back to the subject, purpose, and main point of your document. Conclusions should be concise and to the point.

Again, put yourself in your readers' place. As a document concludes, what do you want to know? More than likely, you want answers to these questions:

What is the main point?

Why is this information important to me?

Where do we go from here?

Five Closing Moves in a Conclusion

These concluding questions translate into five moves typically found in a conclusion, which will help round off your document and reestablish the context that you created in the introduction.

MOVE 1: MAKE AN OBVIOUS TRANSITION   After reading the body of your document, your readers need to wake up a bit. By using a heading such as "Final Points" or a transitional phrase such as "To sum up," you will signal to the readers that you are going to tell them your main points. Your readers will sit up and begin reading closely again. Here are some transitions that will wake up your readers:

In conclusion,	*Put briefly,*	*Overall,*
To sum up,	*In brief,*	*As a whole,*
Let us sum up,	*Finally,*	*In the end,*
In summary,	*To finish up,*	*On the whole,*
In closing,	*Ultimately,*	

Link

For more information on using transitions, see Chapter 9, page 215.

The readers' heightened attention will last only a little while, perhaps for a paragraph or a page. If your conclusion runs more than a page or two, your readers will lose interest and even become annoyed by its length.

MOVE 2: RESTATE YOUR MAIN POINT   In the conclusion, you need to restate your main point one more time to drive it home. After all, your readers now have all the facts, so they should be ready to make a final decision.

> If Darbey is to survive and thrive, we need to take action now to address its increasing flooding problem. By restoring wetlands, developing greenways, and building levees, we can begin preparing for the flooding problems that are almost certainly a risk in the future.

Your main point should be absolutely obvious to readers as they finish the document. You may need to say something direct, such as, "Our main point is..." to ensure your readers don't miss it.

MOVE 3: RESTRESS THE IMPORTANCE OF THE SUBJECT   Sometimes readers need to be reminded of why the subject of your document is important to them.

> If we can reduce or eliminate flooding in Darbey, we will save our citizens millions of dollars in lost revenues and reconstruction. Moreover, Darbey will be viewed as a place with a future, because flooding will not continually undo all our hard work.

Don't try to scare the readers at this point. Stay positive. Stress the benefits of accepting your ideas or agreeing to your recommendations. Scaring your readers will take the shine off your document at this crucial moment.

MOVE 4: LOOK TO THE FUTURE   Looking to the future is a good way to end any document (Figure 8.14). A sentence or paragraph that looks to the future will leave your readers with a positive image.

AT A GLANCE

Five Moves in a Conclusion

- Move 1: Make an obvious transition.
- Move 2: Restate your main point.
- Move 3: Restress the importance of the subject.
- Move 4: Look to the future.
- Move 5: Say thank you and offer contact information.

When we have effectively managed the Curlew River, Darbey will likely see steady growth in population and industry. Once its reputation for flooding has been removed, people and businesses will likely move to this area for its riverside charm and outdoor activities. The town will experience a true revival.

MOVE 5: SAY THANK YOU AND OFFER CONTACT INFORMATION   You might end your document by saying thank you and offering contact information.

We appreciate your time and consideration. If you have any questions or would like to meet with us about this report, please contact the task force leader, Mary Subbock, at 555–0912 or e-mail her at msubbock@cdarbey.gov.

This kind of thank you statement leaves readers with a positive feeling and invites them to contact you if they need more information.

Organizing Cross-Cultural Documents

When communicating with international readers, you may need to alter the organization of a document to suit their different expectations. Most international readers know that North American documents tend to be direct and to the point. Even with

Link

For more information on writing cross-cultural texts, see Chapter 3, page 53.

Sample Conclusion

An obvious transition

The main point

Stresses the importance of the subject

A look to the future

A thank you with contact information

Conclusion

In conclusion, if Darbey is to survive and thrive, we need to take action now to address its increasing flooding problem. By restoring wetlands, developing greenways, and building levees, we can begin preparing for the flooding problems that are almost certainly a risk in the future.

The benefits clearly outweigh the costs. If we can reduce or eliminate flooding in Darbey, we will save our citizens millions of dollars in lost revenues and reconstruction. Moreover, Darbey will be viewed as a place with a future, because flooding will not continually undo all our hard work. When we have effectively managed the Curlew River, Darbey will likely see steady growth in population and industry. Once its reputation for flooding has been removed, people and businesses will likely move to this area for its riverside charm and outdoor activities. The town will experience a true revival.

We appreciate your time and consideration. If you have any questions or would like to meet with us about this report, please contact the task force leader, Mary Subbock, at 555-0912 or e-mail her at msubbock@cdarbey.gov.

Figure 8.14: This conclusion demonstrates all five closing moves that might be found in a document.

For more help with writing conclusions, go to
www.pearsonhighered.com/johnsonweb3/8.15

this awareness, though, the bluntness might still distract or even irritate them. So, your message may be more effective if you consider your readers' cultural preferences.

When organizing a cross-cultural document or presentation, perhaps the greatest concern is whether you are communicating with people from a high-context culture like many in Asia, the Middle East, and Africa or a low-context culture like those in North America or Europe. In a high-context culture, people expect a writer or speaker to provide a significant amount of contextual information before expressing the point or stating the purpose of the document. To most North Americans and Europeans, this up-front contextual information may make the document seem unfocused and its message sound distracted or vague.

To someone from a high-context culture, however, putting most or all of the contextual information up front signals that the writer is being careful, polite, and deliberate (Figure 8.15). This "indirect approach" signals sensitivity to the importance of the message and the refinement of the writer. In some cultures, this indirectness is intended to allow the reader to save face when problems are mentioned or criticisms are made later in the document.

Indirect Approach Introductions

The *indirect approach* primarily affects the introductions and conclusions of documents. In Asian cultures, including China, Japan, Korea, and India, writers and speakers often work from the most general information (placed in the introduction) toward the main point (found in the conclusion).

Usually, the first aim in an indirect approach document is to establish a relationship or call attention to the existing relationship. A Japanese introduction may discuss the weather or the changing of the seasons. A Chinese or Korean introduction may refer to any prior relationships between the writer's and reader's companies.

The example letter in Figure 8.15 plays it safe by mentioning the weather, health, and safety. Notice how the letter starts with general information and works toward more specific information. A letter like this one would be appropriate when the reader has met the writer but the relationship is still new. If the relationship were more established, the second paragraph describing the company would be removed.

Link

For more information on writing introductions in letters, see Chapter 17, page 457.

For many Asian readers, the preservation of harmony is important, so introductions will likely be subtle, striving to draw attention to common values and experiences. The individual is de-emphasized, while the community and situation are emphasized.

In Arab nations and many Islamic cultures, introductions may include calls for Allah's favor on the reader, family, or business. It is common to find highly ornate statements of appreciation, almost to the point of exaggeration. Arab writers may mention their own accomplishments and the high standing of their

An Indirect Approach Letter

10 February, 2008

Mr. Wu Choa
Northern Construction Company, Ltd.
2550 Hongqiao Road,
Shanghai, 200335, China

Dear Mr. Wu Choa,

Weather, health, and safety are common opening topics.

We hope you are in good health and that you returned home safely from our meeting in Shanghai. Our representatives in Shanghai tell us the weather has been mild and rainy. In our state of Missouri, the weather continues to be cold with more snow than usual, so we are expecting a late spring. We hope spring will bring less moisture.

Hammond Industries greatly appreciates your visit with our company's representatives in Shanghai. Our company was started 25 years ago by Charles Hammond in Missouri. Hammond Industries has been supplying solar panels to Asia for 15 years. We have relationships with 8 construction companies in China. Our products are recognized for their reliability and affordability. Our sales and service team in Shanghai is dependable and happy to respond to customers' needs.

Reinforces relationships with historical information

China is a growing market for our products. Our goal is to strengthen and expand our connections in the Shanghai province. We hope our products will continue to meet the needs of large construction companies in your province.

Becomes more specific by explaining corporate goals

Main Point: Products are available for purchase.

We have included a catalog and brochures that describe our products. Our representatives in Shanghai are also available to provide more information. Calls can be made to Mr. Sun, our sales representative. Any orders can be made through our website.

Sincerely,

Sally Gualandi

Sally Gualandi
Special Representative for Asia
Hammond Industries

Figure 8.15:
This example letter demonstrates the indirect approach common in many Asian cultures. This kind of letter would be written to a reader who had met the writer, but the relationship is still new.

acquaintances, while complimenting the reader's achievements and high status. This exaltation of self and acquaintances may seem unnecessary to the North American reader, but the intention is to build a relationship between writer and reader.

Letters in Mexico and some South American countries might start with polite inquiries about family members, especially spouses. They then usually move to the main point much like North American or European documents.

Indirect Approach Conclusions

An indirect approach conclusion usually states the main point of the document but not in an overly direct way. The most effective conclusion is one in which writer and reader reach a common understanding, often without the main point being said directly.

For example, in the letter in Figure 8.15, the writer is trying to persuade the reader to buy her company's products. But, she does not write, "We would like to sell you our products." Instead, she only makes the catalog and brochures available and explains where orders can be made. The reader in China will recognize that she is trying to sell something.

Link

For more information on writing conclusions in letters, see Chapter 17, page 460.

CHAPTER REVIEW

- The beginning of a document (introduction) builds a context. The middle (body) provides the content. And the end (conclusion) rebuilds the context.

- A genre is a predictable pattern for organizing information to achieve specific purposes.

- Outlining may seem old-fashioned, but it is a very effective way to sketch out the organization of a document.

- Presentation software can help you organize complicated information.

- Introductions usually include up to six opening moves: (1) define your subject, (2) state your purpose, (3) state your main point, (4) stress the importance of the subject, (5) provide background information, and (6) forecast the content.

- The body of larger documents is usually carved into sections. Each section has an opening, a body, and perhaps a closing.

- Sections usually follow patterns of arrangement to organize information.

- Conclusions usually include up to five closing moves: (1) make an obvious transition, (2) restate your main point, (3) restress the importance of the subject, (4) look to the future, and (5) say thank you and offer contact information.

- Some international readers, especially those from Asia, Africa, and the Middle East, will feel more comfortable with an "indirect approach" document that moves from the general to the specific.

Individual and Team Projects

1. Find a document that interests you at www.pearsonhighered.com/
johnsonweb3/8.16 (or you can find one on your own). Write a memo to your
instructor in which you critique the organization of the document. Does it have
a clear beginning, middle, and end? Does the introduction make some or all of the
six opening moves? Is the body divided into sections? Does the conclusion make
some or all of the five closing moves? Explain to your instructor how the document's organization might be improved.

2. Find an introduction that you think is ineffective. Then, rewrite the
introduction so that it includes a clear subject, purpose, and main point.
Also, stress the importance of the document's subject, provide some
background information, and forecast the structure of the document's
body. Sample introductions are available at www.pearsonhighered.com/
johnsonweb3/8.16.

3. On the Internet, find a home page that makes some or all of the six opening moves
discussed in this chapter. Write a memo to your instructor in which you compare
and contrast how home pages use the opening moves similarly and differently than
introductions in paper-based documents.

4. On the Internet or in a print document, find a section that uses one of the following
organizational patterns:

   - Cause and effect
   - Comparison and contrast
   - Better and worse
   - Costs and benefits
   - If…then
   - Either…or
   - Problem/needs/solution
   - Example

   Prepare a brief presentation to your class in which you show how the text you
found follows the pattern.
   You can find sample documents at www.pearsonhighered.com/
johnsonweb3/8.16 that will use some or all of these organizational patterns.

Collaborative Project

Around campus or at www.pearsonhighered.com/johnsonweb3/8.16, find a large document, perhaps a report or proposal. With presentation software, outline the document
by creating a slide for each section or subsection. Use the headings of the sections as
titles on your slides. Then, on each slide use bulleted lists to highlight the important
points in each section.

When you have finished outlining, look over the presentation. Are there any places where too much information is offered? Are there places where more information is needed? Where could you rearrange information to make it more effective?

Present your findings to your class. Using the presentation software, show the class how the document you studied is organized. Then, discuss some improvements you might make to the organization to highlight important information.

For additional technical writing resources, including interactive sample documents, document design tutorials and guidelines, and more, go to **www.mytechcommlab.com.**

The Bad News

In most ways, the project had been a failure. Lisa Franklin was on a chemical engineering team developing a polymer that would protect lightweight tents against desert elements like extreme sun, sandstorms, and chewing insects. Hikers and campers were interested in tents with this kind of protection. But the company Lisa worked for, Outdoor Solutions, wanted to begin selling tents to the military. Bulk sales to the U.S. Army and Marines would improve the company's bottom line, as well as open new opportunities to sell other products.

Despite a promising start, Lisa and the other researchers had not yet developed a stable polymer that would provide the desired protection. The best polymers they had developed were highly flammable. Tent materials covered with these polymers went up like an inferno when exposed to flame. The nonflammable polymers, meanwhile, seemed to break down within three months of use in desert conditions. Lisa's boss, Jim Franklin, was convinced they were close to a breakthrough, but they just couldn't find the right combination to create a nonflammable polymer that would last.

Unfortunately, Lisa's boss had been giving the company president the impression that the polymer was already a success. So, the president had secured a million-dollar loan to retool a factory to produce new lightweight tents coated with the polymer. Renovation was due to begin in a month.

Before starting the factory retooling, the president of the company asked for a final update on the polymer. So Jim wrote a progress report. Then he gave the report to Lisa for final revisions, because he was going out of town on a business trip. He said, "Do whatever you want to revise the report. I'm so frustrated with this project, I don't want to look at this report anymore. When you're done, send the report to the company president. I don't need to see it again."

In the report, Lisa noticed, Jim's facts were all true, but the organization of the report hid the fact that they had not developed a workable polymer. For example, when Jim mentioned that their best polymers were highly flammable, he did so at the end of a long paragraph in the middle of the report. It was highly unlikely that the president would be reading closely enough to see that important fact.

Lisa knew Jim was trying to hide the research team's failure to develop a workable polymer. She too didn't want to admit that they had failed. But she also thought it was important that the company president understand that they had not developed a successful polymer. After all, there was a lot of money on the line.

If you were in Lisa's place, how would you handle this situation?

In this chapter, you will learn:

- The importance of style in technical communication.

- The differences between plain and persuasive style.

- Strategies for writing plain sentences and paragraphs.

- How to tell when it's appropriate to use persuasive style.

- How to use techniques of persuasive style, including tone, similes,
  analogies, metaphors, and pace.

How you say something is often what you say. Your document's style expresses your attitude toward the subject. It reflects your character by embodying the values, beliefs, and relationships you want to share with the readers. In a word, style is about *quality*. Style is about your and your company's commitment to excellence.

One of the great advantages of writing with computers is the ability to move text around to improve style. By learning some rather simple techniques, you can make your writing clearer and more persuasive. By putting words and sentences in the proper places, you can make your documents more interesting and easier to understand. In this chapter, you will learn how to use these easy-to-learn stylistic techniques.

What Is Style?

In technical communication, style is not embellishment or ornamentation. Style is not artificial flavoring added to a document to make the content more "interesting." A few added adjectives or exclamation marks won't do much to improve your style.

Good style goes beyond these kinds of superficial cosmetic changes. It involves

- choosing the right words and phrases.
- structuring sentences and paragraphs for clarity.
- using an appropriate tone.
- adding a visual sense to the text.

Historically, rhetoricians have classified style into three levels: plain style, persuasive style, and grand style.

Two Common Styles in Technical Communication

AT A GLANCE

- Plain style—stresses clear wording and simple prose.
- Persuasive style—is used to influence people to accept your ideas and take action.

Plain style—Plain style stresses clear wording and simple prose. It is most often used to instruct, teach, or present information. Plain style works best in documents like technical descriptions, instructions, and activity reports.

Persuasive style—There are times when you will need to influence people to accept your ideas and take action. In these situations, persuasive style allows you to add energy and vision to your writing and speaking. This style works best with proposals, letters, articles, public presentations, and some reports.

Grand style—Grand style stresses eloquence. For example, Martin Luther King, Jr., and President John F. Kennedy often used the grand style to move their listeners to do what was right, even if people were reluctant to do it.

In this chapter, we will concentrate solely on plain and persuasive style, because they are most common in technical documents. Grand style is rarely used in technical communication because it often sounds too ornate or formal for the workplace.

Writing Plain Sentences

In the past, you have probably been told to "write clearly" or "write in concrete language" as though making up your mind to do so was all it took. In reality, writing plainly is a skill that requires practice and concentration. Fortunately, once a few basic guidelines have been mastered, plain style will become a natural strength in your writing.

GO TO THE NET

To see further discussions of style by classical rhetoricians, go to
www.pearsonhighered.com/johnsonweb3/9.1

Basic Parts of a Sentence

To start, let us consider the parts of a basic sentence. A sentence in English typically has three main parts: a subject, a verb, and a comment.

Subject—What the sentence is about

Verb—What the subject is doing

Comment—What is being said about the subject

English is a flexible language, allowing sentences to be organized in a variety of different ways. For example, consider these three variations of the same sentence:

Subject	Verb	Comment
The Institute	provided	the government with accurate crime statistics.
The government	was provided	with accurate crime statistics by the Institute.
Crime statistics	were provided	to the government by the Institute.

Notice that the *content* in these sentences has not changed; only the order of the words has changed. However, the focus of each sentence changes when the subject is changed. The first sentence is *about* the "Institute." The second sentence is *about* the "government." The last sentence is *about* "crime statistics." By changing the subject of the sentence, you essentially shift its focus, drawing your readers' attention to different issues.

Eight Guidelines for Plain Sentences

This understanding of the different parts of a sentence is the basis for eight guidelines that can be used to write plainer sentences in technical documents.

GUIDELINE 1: THE SUBJECT OF THE SENTENCE SHOULD BE WHAT THE SENTENCE IS ABOUT.   Confusion often creeps into texts when readers cannot easily identify the subjects of the sentences. For example, what is the subject of the following sentence?

1. Ten months after the Hartford Project began in which a team of our experts conducted close observations of management actions, our final conclusion is that the scarcity of monetary funds is at the basis of the inability of Hartford Industries to appropriate resources to essential projects that have the greatest necessity.

This sentence is difficult to read for a variety of reasons, but the most significant problem is the lack of a clear subject. What is this sentence about? The word "conclusion" is currently in the subject position, but the sentence might also be about "our experts," "the Hartford Project," or "the scarcity of monetary funds."

When you run into a sentence like this one, first decide what the sentence is about. Then, cut and paste to move that subject into the subject slot of the sentence. For example, when this sentence is restructured around "our experts," readers will find it easier to understand:

The subject (underlined) is what this sentence is "about." ────→ 1a. Ten months after the Hartford Project began, <u>our experts</u> have concluded through close observations of management actions that the scarcity of monetary funds is at the basis of the inability of Hartford Industries to appropriate resources to essential projects that have the greatest necessity.

This sentence is still rather difficult to read. Nevertheless, it is easier to understand because the noun in the subject slot ("our experts") is what the sentence is about. We will return to this sentence later in this chapter.

GUIDELINE 2: THE SUBJECT SHOULD BE THE "DOER" IN THE SENTENCE.   Readers tend to focus on who or what is doing something in a sentence. For example, which of the following sentences is easier to read?

2a. On Saturday morning, <u>the paperwork</u> was completed in a timely fashion by Jim.

2b. On Saturday morning, <u>Jim</u> completed the paperwork in a timely fashion.

Most readers would say that sentence 2b is easier to read because Jim, the subject of the sentence, is actually doing something. In the first sentence, the paperwork is merely sitting there. People make especially good subjects of sentences, because they are usually active.

GUIDELINE 3: THE VERB SHOULD STATE THE ACTION, OR WHAT THE DOER IS DOING.   Once you have determined who or what is doing something, ask yourself what that person or thing is actually doing. Find the *action* and turn it into the verb of the sentence. For example, consider these sentences:

In these sentences, it becomes harder and harder to figure out the action.

3a. The detective investigated the loss of the payroll.

3b. The detective conducted an investigation into the loss of the payroll.

3c. The detective is the person who is conducting an investigation of the loss of the payroll.

Sentence 3a is easy to understand because the action of the sentence is expressed in the verb. Sentences 3b and 3c are increasingly more difficult to understand, because the action, "investigate," is further removed from the verb of the sentence.

GUIDELINE 4: THE SUBJECT OF THE SENTENCE SHOULD COME EARLY IN THE SENTENCE.   Subconsciously, your readers start every sentence looking for the subject. The subject anchors the sentence, because it tells readers what the sentence is about. Consider these two sentences:

It's easier to locate the subject of this sentence because it is up front.

4a. If deciduous and evergreen trees experience yet another year of drought like the one observed in 1997, the entire <u>Sandia Mountain ecosystem</u> will be heavily damaged.

4b. <u>The entire Sandia Mountain ecosystem</u> will be heavily damaged if deciduous and evergreen trees experience yet another year of drought like the one observed in 1997.

For some interesting discussions of plain style, go to
www.pearsonhighered.com/johnsonweb3/9.3

The problem with sentence 4a is that it forces readers to hold all those details about trees, drought, and 1997 in short-term memory before it identifies the subject of the sentence. Readers almost feel a sense of relief when they come to the subject, because until that point they cannot figure out what the sentence is about.

Quite differently, sentence 4b tells readers what the sentence is about up front. With the subject early in the sentence, readers immediately know how to connect the comment with the subject.

The first sentence is harder to read because the action is in a nominalization.

GUIDELINE 5: ELIMINATE NOMINALIZATIONS.   Nominalizations are perfectly good verbs and adjectives that have been turned into awkward nouns. For example, look at these sentences:

5a. Management has <u>an expectation</u> that the project will meet the deadline.

5b. Management <u>expects</u> the project to meet the deadline.

In sentence 5a, "expectation" is a nominalization. Here, a perfectly good verb is being used as a noun. Sentence 5b is not only shorter than sentence 5a, it also has more energy because the verb "expects" is now an action verb.

Here are a whole bunch of nominalizations.

Consider these two sentences:

6a. <u>Our discussion</u> about the matter allowed us to make <u>a decision</u> on <u>the acquisition</u> of the new x-ray machine.

See how reducing them improves the readability?

6b. We <u>discussed</u> the matter and <u>decided</u> to <u>acquire</u> the new x-ray machine.

Sentence 6a includes three nominalizations—"discussion," "decision," and "acquisition"—making the sentence hard to understand. Sentence 6b is clearer, because the nominalizations "discussion" and "decision" have been turned into action verbs.

Why do writers use nominalizations in the first place? They use nominalizations for two reasons:

- First, humans generally think in terms of people, places, and things (nouns), so our first drafts are often filled with nominalizations, which are nouns. While revising, an effective writer will turn those first-draft nominalizations into action verbs.
- Second, some people mistakenly believe that using nominalizations makes their writing sound more formal or important. In reality, though, nominalizations only make sentences harder to read. The best way to sound important is to write sentences that readers can understand.

GUIDELINE 6: AVOID EXCESSIVE PREPOSITIONAL PHRASES.   Prepositional phrases are necessary in writing, but they are often overused in ways that make text too long and too tedious. Prepositional phrases follow prepositions like *in, of, by, about, over,* and *under.* These phrases are used to modify nouns.

Prepositional phrases become problematic when used in excess. For example, sentence 7a is difficult to read because it chains together too many prepositional phrases (the prepositional phrases are underlined and the prepositions are italicized). Sentence 7b shows the same sentence with fewer prepositional phrases:

7a. The decline *in* the number *of* businesses owned *by* locals *in* the town *of* Artesia is a demonstration *of* the increasing hardship faced *in* rural communities *in* the southwest.

The prepositional phrases have been reduced, clarifying the sentence's meaning. ──────▶ 7b. Artesia's declining number of locally owned businesses demonstrates the increased hardship faced by southwestern rural communities.

You should never feel obligated to eliminate all the prepositional phrases in a sentence. Rather, look for places where prepositional phrases are chained together into long sequences. Then, try to condense the sentence by turning some of the prepositional phrases into adjectives.

For example, the phrase "in the town of Artesia" from sentence 7a was reduced in sentence 7b to the adjective "Artesia's." The phrase "in rural communities in the southwest" was reduced to "southwestern rural communities." As a result, sentence 7b is much shorter and easier to read.

Figure 9.1 shows a website in which the authors write plainly. Notice how they have used minimal prepositional phrases.

GUIDELINE 7: ELIMINATE REDUNDANCY IN SENTENCES.   In our efforts to get our point across, we sometimes use redundant phrasing. For example, we might write "unruly mob" as though some kinds of mobs might be orderly. Or, we might talk about "active participants" as though someone can participate without doing anything.

Sometimes buzzwords and jargon lead to redundancies like, "We should collaborate together as a team" or "Empirical observations will provide a new understanding of the subject." In some cases, we might use a synonym to modify a synonym by saying something like, "We are demanding important, significant changes."

Redundancies should be eliminated because they use two or more words to do the work of one word. As a result, readers need to work twice as hard to understand one basic idea.

GUIDELINE 8: WRITE SENTENCES THAT ARE "BREATHING LENGTH."   You should be able to read a sentence out loud in one breath. At the end of the sentence, the period (.) signals, "Take a breath." Of course, when reading silently, readers do not actually breathe when they see a period, but they do take a mental pause at the end of each sentence. If a sentence runs on and on, it forces the readers to mentally hold their breath. By the end of an especially long sentence, they are more concerned about when the sentence is going to end than what the sentence is saying.

Need more examples of revised sentences? Go to
www.pearsonhighered.com/johnsonweb3/9.5

Writing Plainly

Actions are expressed in the verb phrases.

Sentences are breathing length.

There are almost no nominalizations or prepositional phrase chains.

Subjects are placed up front in sentences.

Source: Geothermal Energy Association, http://www.geo-energy.org.

Figure 9.1:
In this website, the Geothermal Energy Association uses plain language to tell its facts about geothermal energy.

The best way to think about sentence length is to imagine how long it takes to comfortably say a sentence out loud.

- If the written sentence is too long to say comfortably, it probably needs to be shortened or cut into two sentences. Avoid asphyxiating the readers with sentences that go on forever.
- If the sentence is very short, perhaps it needs to be combined with one of its neighbors to give it a more comfortable breathing length. You want to avoid making readers hyperventilate over a string of short sentences.

Creating Plain Sentences with a Computer

Computers give us an amazing ability to manipulate sentences. So, take full advantage of your machine's capabilities. First, write out your draft as you normally would, not

Link

For examples of different breathing-length sentences, see the discussion of pace in this chapter on page 226.

Eight Guidelines for Plain Sentences

- Guideline 1: The subject of the sentence should be what the sentence is about.
- Guideline 2: The subject should be the "doer" in the sentence.
- Guideline 3: The verb should state the action, or what the doer is doing.
- Guideline 4: The subject of the sentence should come early in the sentence.
- Guideline 5: Eliminate nominalizations.
- Guideline 6: Avoid excessive prepositional phrases.
- Guideline 7: Eliminate redundancy in sentences.
- Guideline 8: Write sentences that are "breathing length."

AT A GLANCE

paying too much attention to the style. Then, as you revise, identify difficult sentences and follow these seven steps:

1. Identify who or what is doing something in the sentence.

2. Turn that who or what into the subject of the sentence.

3. Move the subject to an early place in the sentence.

4. Identify what the subject is doing, and move that action into the verb slot.

5. Eliminate prepositional phrases, where appropriate, by turning them into adjectives.

6. Eliminate unnecessary nominalizations and redundancies.

7. Shorten, lengthen, combine, or divide sentences to make them breathing length.

With these seven steps in mind, let us revisit sentence 1, the example of weak style earlier in this chapter:

Original

1. Ten months after the Hartford Project began in which a team of our experts conducted close observations of management actions, our final conclusion is that the scarcity of monetary funds is at the basis of the inability of Hartford Industries to appropriate resources to essential projects that have the greatest necessity.

Now, let's apply the seven-step method for revising sentences into the plain style. First, identify who or what is doing something in the sentence, make it the subject, and move it to an early place in the sentence.

The experts are doing something.

1a. After a ten-month study, our experts have concluded that the scarcity of monetary funds is at the basis of the inability of Hartford Industries to appropriate resources to essential projects that have the greatest necessity.

The action is now in the verb.

Now eliminate the prepositional phrases, nominalizations, and redundancies.

1b. After a ten-month study, our experts concluded that Hartford Industries' budget shortfalls limit its support for priority projects.

In the revision, the "doers" ("our experts") were moved into the subject position and then moved to an early place in the sentence. Then, the action of the sentence ("concluded") was moved into the verb slot. Prepositional phrases like "after the Hartford Project" and "to appropriate resources to essential projects" were turned into adjectives. Nominalizations ("conclusion," "necessity") were turned into verbs or adjectives. And finally, the sentence was shortened to breathing length.

The resulting sentence still sends the same message—just more plainly.

Intercultural Style and Translation Software

In the *Hitchhiker's Guide to the Galaxy* by Douglas Adams, aliens are able to communicate with each other by using a Babel fish, which is described as a "yellow and leech-like" creature that can be slipped into the ear.

Unfortunately, these Babel fish are fiction, so we need to translate our own materials or have them translated for us. Professional translators can be expensive, especially if you are just trying to translate an e-mail from one of your overseas clients or figure out a website written in another language.

Several translation software programs are available for purchase, such as Power Translator, Systran, Translution, and @promt. The accuracy of these programs has improved greatly over the last decade. A free online translation tool is the Babel Fish website (babelfish.yahoo.com), available through Yahoo! (Figure A). Another free online translator is available through Promt (www.online-translator.com).

Translation software is far from perfect. Certainly, you should not rely exclusively on these tools to translate an important proposal or report into another language. To handle these kinds of projects, you will likely need to hire a translation service. However, for smaller texts, some helpful tips can minimize problems when using translation software:

HELP

Use basic sentences—Translation software typically tries to translate whole sentences rather than translate word-by-word. So, the software will struggle with complex sentences. Cut longer, more complex sentences into smaller, simpler sentences.

Use standard punctuation—Periods, commas, and question marks are fine. Less common punctuation marks like ellipses, dashes, colons, and semicolons often create difficulties for translation software. You can usually replace these uncommon marks with basic marks.

Use consistent words—In texts that will be translated, use the same word to mean the same thing. For example, if you called something an "animal" earlier in the document, you should avoid calling it a "mammal," "living being," "critter," or "quadruped" later in the document.

Avoid metaphors, sayings, and clichés—These kinds of phrases require knowledge of the culture of the source language. So, if you write something like "Our product slaughters the competition" or "She couldn't see the forest for the trees," the translation software will come back with some very strange results.

Remove any cultural, historical, or sports-related references—The translation software will not know what you mean if you say something like "That project was our battle of Gettysburg" or "He hit a home run with his presentation."

Back translate all text—After you translate something, have the software translate it back into English. Each time through the translator will magnify the differences in meaning, so the results are often funny. However, the backtranslation can signal places where the translation software completely missed the mark.

(continued)

Want to learn other methods for improving style at the sentence level? Go to
www.pearsonhighered.com/johnsonweb3/9.8

Writing Plain Sentences   213

Free Online Translation Tool

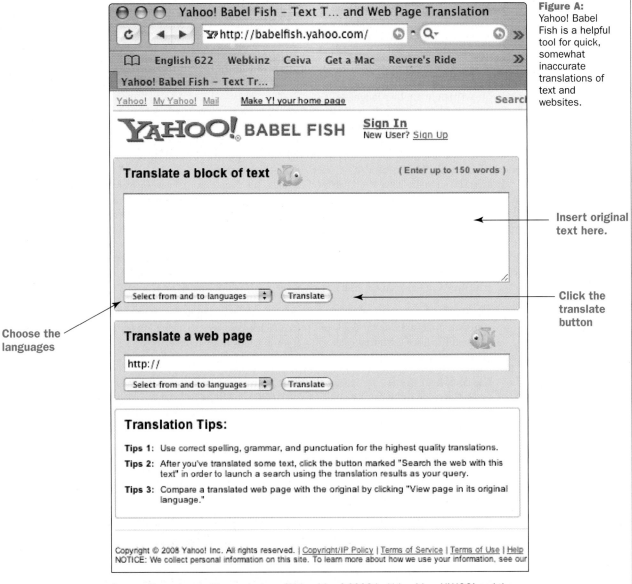

Figure A:
Yahoo! Babel Fish is a helpful tool for quick, somewhat inaccurate translations of text and websites.

Insert original text here.

Click the translate button

Choose the languages

Use the spellchecker on the original *and* translated texts—Since translation software usually tries to translate the whole sentence, any spelling errors or typos will cause it to make unnecessary guesses about what a sentence means.

To learn more about paragraphing, go to
www.pearsonhighered.com/johnsonweb3/9.9

So spellcheck the source text carefully. Then, after the text is translated, run it through the spellchecker again, using the target language. Your word processor likely has dictionaries from other languages preloaded, which will allow you to locate misspelling or other errors in the translation.

Avoid words that have double meanings—Try to use words that only have one meaning. For example, the word "compound" can have several meanings: "Working in the office compound, he compounded his problems when he used the compound to connect the compound joint." Here is another: "I was not content with the content of the proposal."

Minimize jargon and acronyms—The software will often make inaccurate guesses about the meanings of jargon and acronyms, unless you are using translation software that is specific to your field. So, simplify jargon words, and check how any acronyms were translated.

Avoid puns or other plays on words—The translation software won't understand the double meanings, and they will seem very odd in the translation.

Your translation software is not going to get everything right. So, you might warn the receiver of your text that you are using translation software. That way, he or she will forgive the errors and inconsistencies. For any critical documents, you should hire a human translator.

Writing Plain Paragraphs

Some rather simple methods are available to help you write plainer paragraphs. As with writing plain sentences, a computer really helps, because you can quickly move sentences around in paragraphs to clarify your meaning.

The Elements of a Paragraph

Paragraphs tend to include four kinds of sentences: a *transition* sentence, a *topic* sentence, *support* sentences, and a *point* sentence (Figure 9.2). Each of these sentences plays a different role in the paragraph.

TRANSITION SENTENCE   The purpose of a transition sentence is to make a smooth bridge from the previous paragraph to the present paragraph. For example, a transition sentence might state the following:

> Keeping these facts about West Nile virus in mind, let us consider the actions we should take.

> What should we do to help these learning disabled children in our classrooms?

Transition sentences are typically used when the new paragraph handles a significantly different topic than the previous paragraph. They help close the gap between the two paragraphs or redirect the discussion.

Types of Sentences in a Typical Paragraph

Figure 9.2: Paragraphs usually contain up to four types of sentences: transition, topic, support, and point sentences.

> **Transition Sentence (optional)**
>
> **Topic Sentence (needed)**
>
> **Support Sentences (needed)**
>
> > examples
> >
> > reasoning
> >
> > facts
> >
> > data
> >
> > definitions
> >
> > descriptions
>
> **Point Sentence (optional)**

TOPIC SENTENCE   The topic sentence is the claim or statement that the rest of the paragraph is going to prove or support:

> To combat the spread of the West Nile virus in Pennsylvania, we recommend three immediate steps: vaccinate all horses, create a public relations campaign to raise public awareness, and spray strategically.

> Children with learning disabilities struggle to cope in the normal classroom, so teachers need to be trained to recognize their special needs.

In technical documents, topic sentences typically appear in the first or second sentence of each paragraph. They are placed up front in each paragraph for two reasons.

- The topic sentence sets a goal for the paragraph to reach. It tells readers the claim the writer is trying to prove. Then, the remainder of the paragraph proves that claim.
- The topic sentence is the most important sentence in any given paragraph. Since readers, especially scanning readers, tend to pay the most attention to the beginning of a paragraph, placing the topic sentence up front ensures they will read it.

AT A GLANCE

Four Kinds of Sentences in a Paragraph

- Transition sentence (optional)
- Topic sentence
- Support sentence
- Point sentence (optional)

SUPPORT SENTENCES   The bulk of any paragraph is typically made up of support sentences. These sentences contain examples, reasoning, facts, data, anecdotes, definitions, and descriptions.

> First, we recommend that all horses be vaccinated against West Nile virus, because they are often the most significant victims of the virus. Last year, nearly 2000 horses died from the virus. Second, we believe a public relations campaign . . .

> Learning disabled children often struggle to understand their teacher, especially if the teacher lectures for periods longer than ten minutes.

Support sentences are used to prove the claim made in the paragraph's topic sentence.

GO TO THE NET

More examples of these kinds of sentences are available at
www.pearsonhighered.com/johnsonweb3/9.10

POINT SENTENCES   Point sentences restate the paragraph's main point toward the end of the paragraph. They are used to reinforce the topic sentence by restating the paragraph's original claim in new words. Point sentences are especially useful in longer paragraphs where readers may not fully remember the claim stated at the beginning of the paragraph. They often start with transitional devices like "Therefore," "Consequently," or "In sum" to signal that the point of the paragraph is being restated.

> These three recommendations represent only the first steps we need to take. Combating West Nile virus in the long term will require a more comprehensive plan.

> Again, learning disabled children often struggle undetected in the classroom, so teachers need to be trained to recognize the symptoms.

Point sentences are optional in paragraphs, and they should be used only occasionally when a particular claim needs to be reinforced. Too many point sentences will cause your text to sound too repetitious and even condescending.

Using the Four Types of Sentences in a Paragraph

Of these four kinds of sentences, only the topic and support sentences are needed to construct a good paragraph. Transition sentences and point sentences are useful in situations where bridges need to be made between paragraphs or specific points need to be reinforced.

Here are the four kinds of sentences used in a paragraph:

> 8a. How can we accomplish these five goals (transition sentence)? Universities need to study their core mission to determine whether distance education is a viable alternative to the traditional classroom (topic sentence). If universities can maintain their current standards when moving their courses online, then distance education may provide a new medium through which nontraditional students can take classes and perhaps earn a degree (support sentence). Utah State, for example, is reporting that students enrolled in its online courses have met or exceeded the expectations of their professors (support sentence). If, however, standards cannot be maintained, we may find ourselves returning to the traditional on-campus model of education (support sentence). In sum, the ability to meet a university's core mission is the litmus test to measure whether distance education will work (point sentence).

Here is the same paragraph with the transition sentence and point sentence removed:

> 8b. Universities need to study their core mission to determine whether distance education is a viable alternative to the traditional classroom (topic sentence). If universities can maintain their current standards when moving their courses online, then distance education may provide a new medium through which nontraditional students can take classes and perhaps earn a degree (support sentence). Utah State, for example, is reporting that students enrolled in its online courses have met or exceeded the expectations of their professors (support sentence). If, however, standards cannot be maintained, we may find ourselves returning to the traditional on-campus model of education (support sentence).

As you can see in paragraph 8b, a paragraph works fine without transition and point sentences. Nevertheless, transition and point sentences can make texts easier to read while amplifying important points.

GO TO THE NET
Want some more advice about organizing paragraphs? Go to
www.pearsonhighered.com/johnsonweb3/9.11

Aligning Sentence Subjects in a Paragraph

Now let's discuss how you can make paragraphs flow by weaving sentences together effectively. Have you ever read a paragraph in which each sentence seems to go off in a new direction? Have you ever run into a paragraph that actually feels "bumpy" as you read it? More than likely, the problem was a lack of alignment of the paragraph's sentence subjects. For example, consider this paragraph:

Notice how the subjects of these sentences (underlined) are not in alignment, making the paragraph seem rough to readers.

9. The lack of technical knowledge about the electronic components in automobiles often leads car owners to be suspicious about the honesty of car mechanics. Although they might be fairly knowledgeable about the mechanical workings of their automobiles, car owners rarely understand the nature and scope of the electronic repairs needed in modern automobiles. For instance, the function and importance of a transmission in a car are generally well known to all car owners, but the wire harnesses and printed circuit boards that regulate the fuel consumption and performance of their car are rarely familiar. Repairs for these electronic components can often run over $400—a large amount to a customer who cannot even visualize what a wire harness or printed circuit board looks like. In contrast, a $400 charge for the transmission on the family car, though distressing, is more readily understood and accepted.

There is nothing really wrong with this paragraph—it's just hard to read. Why? The paragraph is difficult to read because the subjects of the sentences change with each new sentence. Look at the underlined subjects of the sentences in this paragraph. They are all different, causing each sentence to strike off in a new direction. As a result, each sentence forces readers to shift focus to concentrate on a new subject.

With your word processor, you can easily revise paragraphs to avoid this bumpy, unfocused feeling. The secret is lining up the subjects in the paragraph.

To line up subjects, first ask yourself what the paragraph is about. Then, cut and paste words to restructure the sentences to line up on that subject. Here is a revision of paragraph 9 that focuses on the "car owners" as subjects:

Here, the paragraph has been revised to focus on people.

9a. Due to their lack of knowledge about electronics, some car owners are skeptical about the honesty of car mechanics when repairs involve electronic components. Most of our customers are fairly knowledgeable about the mechanical features of their automobiles, but they rarely understand the nature and scope of the electronic repairs needed in modern automobiles. For example, most people recognize the function and importance of a transmission in an automobile, but the average person knows very little about the wire harnesses and printed circuit boards that regulate the fuel consumption and performance of his or her car. So, for most of our customers, a $400 repair for these electronic components seems like a large amount, especially when these folks cannot even visualize what a wire harness or printed circuit board looks like. In contrast, most car owners think a $400 charge to fix the transmission on the family car, though distressing, is more acceptable.

In this revised paragraph, you should notice two things:

- First, the words "car owners" are not always the exact two words used in the subject slot. Synonyms and pronouns should be used to add variety to the sentences.
- Second, not all the subjects need to be "car owners." In the middle of the paragraph, for example, "$400 repair" is the subject of a sentence. This deviation from "car owners" is fine as long as the majority of the subjects in the paragraph

are similar to each other. In other words, the paragraph will still sound focused, even though an occasional subject is not in alignment with the others.

Of course, the subjects of the paragraph could be aligned differently to stress something else in the paragraph. Here is another revision of paragraph 9 in which the focus of the paragraph is "repairs."

This paragraph focuses on repairs.

9b. <u>Repairs</u> to electronic components often lead car owners, who lack knowledge about electronics, to doubt the honesty of car mechanics. The <u>nature and scope of these repairs</u> are usually beyond the understanding of most nonmechanics, unlike the typical <u>mechanical repairs</u> with which customers are more familiar. For instance, the <u>importance of fixing the transmission</u> in a car is readily apparent to most car owners, but <u>adjustments</u> to electronic components like wire harnesses and printed circuit boards are foreign to most customers—even though these electronic parts are crucial in regulating their car's fuel consumption and performance. So, <u>a repair</u> to these electronic components that costs $400 seems excessive, especially when the <u>repair</u> can't even be visualized by the customer. In contrast, a <u>$400 replacement</u> of the family car's transmission, though distressing, is more readily accepted.

You should notice that paragraph 9a is easier for most people to read than paragraph 9b. Paragraph 9a is more readable because sentences that have "doers" in the subject slots are easier for people to read. In paragraph 9a, the car owners are active subjects, while in paragraph 9b the car repairs are inactive subjects.

The Given/New Method

Another way to write plain paragraphs is to use the "given/new" method to weave sentences together. Described by Susan Haviland and Herbert Clark in 1974, the given/new method is based on the assumption that readers try to process new information by comparing it to information they already know. Therefore, every sentence in a paragraph should contain something the readers already know (the given) and something that the readers do not know (the new).

Consider these two paragraphs:

10a. Santa Fe has many beautiful places. Artists choose to strike off into the mountains to work, while others enjoy working in local studios. The landscapes are wonderful in the area.

10b. Santa Fe offers many beautiful places for artists to work. <u>Some artists</u> choose to strike off into <u>the mountains</u> to work, while others enjoy working in <u>local studios</u>. Both <u>the mountains</u> and <u>the studios</u> offer places to savor the wonderful landscapes in the area.

Both of these paragraphs are readable, but paragraph 10b is easier to read because words are being repeated in a given/new pattern. Paragraph 10a is a little harder to read, because there is nothing "given" that carries over from each sentence to its following sentence.

Using the given/new method is not difficult. In most cases, the given information should appear early in the sentence and the new information should appear later in the sentence. The given information will provide a familiar anchor or

Want to practice revising a few paragraphs? Go to
www.pearsonhighered.com/johnsonweb3/9.12

Writing Plain Paragraphs 219

context, while the new information will build on that familiar ground. Consider this larger paragraph:

Notice how the chaining of words in this paragraph makes it smoother to read.

11. Recently, an art gallery exhibited the mysterious paintings of <u>Irwin Fleminger</u>, a modernist artist whose vast Mars-like landscapes contain cryptic human artifacts. One of <u>Fleminger's paintings</u> attracted the attention of some young <u>schoolchildren</u> who happened to be walking by. At first, the <u>children</u> laughed, pointing out some of the <u>strange artifacts</u> in the painting. Soon, though, <u>the strange artifacts</u> in the painting drew the students into a critical awareness of the painting, and they began to ask their bewildered teacher what the artifacts meant. Mysterious and beautiful, <u>Fleminger's paintings</u> have this effect on many people, not just schoolchildren.

In this paragraph, the beginning of each sentence provides something given—typically an idea, word, or phrase drawn from the previous sentence. Then, the remainder of each sentence adds something new to that given information. By chaining together given and new information, the paragraph builds readers' understanding gradually, adding a little more information with each sentence.

In some cases, however, the previous sentence in a paragraph does not offer a suitable subject for the sentence that follows it. In these cases, transitional phrases can be used to provide readers given information in the beginning of the sentence. To illustrate,

12. This public relations effort will strengthen <u>Gentec's relationship</u> with leaders of the community. <u>With this new relationship in place</u>, the details of the project can be negotiated in terms that are fair to both parties.

In this sentence, the given information in the second sentence appears in the transitional phrase, not in the subject slot. A transitional phrase is a good place to include given information when the subject cannot be drawn from the previous sentence.

When Is It Appropriate to Use Passive Voice?

Before discussing the elements of persuasive style, we should expose a writing boogeyman as a fraud. Since childhood, you have probably been warned against using *passive voice*. You have been told to write in *active voice*.

One problem with this prohibition on passive voice is that passive voice is very common in technical communication. In some scientific fields, passive voice is the standard way of writing. So, when is it appropriate to use passive voice?

Passive voice occurs when the subject of the sentence is *being acted upon,* so the verb is in passive voice. Active voice occurs when the subject of the sentence is *doing the acting,* so the verb is an action verb. Here is an example of a sentence written in passive voice and a sentence written in active voice:

13a. The alloy was heated to a temperature of 300°C. (passive)

13b. Andy James heated the alloy to a temperature of 300°C. (active)

The passive voice sentence (sentence 13a) lacks a doer because the subject of the sentence, the alloy, is being acted upon. As a result, sentence 13b might be a bit easier to understand because a doer is the subject of the sentence. But, as you can see, sentence 13a is easy to read and understand, too.

For more information on the given/new method, go to
www.pearsonhighered.com/johnsonweb3/9.13

Passive voice does have a place in technical documents. A passive sentence is fine if:

- the readers do not need to know who or what is doing something in the sentence.
- the subject of the sentence is what the sentence is about.

For example, in sentence 13a, the person who heated the alloy might be unknown or irrelevant to the readers. Is it really important that we know that Andy James heated the alloy? Or, do we simply need to know that the alloy was heated? If the alloy is what the sentence is about and who heated the alloy is unimportant, then the passive voice is fine.

If you are wondering whether the passive voice should be used, consider your readers. Do they expect you to use the passive voice? Do they need to know who did something in a sentence? Go ahead and use the passive voice if your readers expect you to use passive voice or they don't need to know who did what.

Consider these other examples of passive and active sentences:

14a. The shuttle bus will be driven to local care facilities to provide seniors with shopping opportunities. (passive)

14b. Jane Chavez will drive the shuttle bus to local care facilities to provide seniors with shopping opportunities. (active)

15a. The telescope was moved to the Orion system to observe the newly discovered nebula. (passive)

15b. Our graduate assistant, Mary Stewart, moved the telescope to the Orion system to observe the newly discovered nebula. (active)

In both these sets of sentences, the passive voice may be more appropriate, unless there is a reason Jane Chavez or Mary Stewart needs to be singled out for special consideration.

When developing a focused paragraph, passive sentences can often help you align the subjects and use given/new strategies (i.e., make the paragraph more cohesive). For example, compare the following paragraphs, one written in passive voice and the other in active voice:

Paragraph written in passive voice

16a. We were shocked by the tornado damage. The downtown of Wheaton had been completely flattened. Brick houses were torn apart. A school bus was tossed into a farm field, as though some fairy-tale giant had crumpled it like a soda can. The town's fallen water tower had been dragged a hundred yards to the east. Amazingly enough, no one was killed by the storm. Clearly, lives were saved by the early warning that the weather bureau had sent out.

Same paragraph written in active voice

16b. The tornado damage shocked us. The tornado completely flattened the downtown of Wheaton. It had torn apart brick houses. It had tossed a school bus into a farm field, as though some fairy-tale giant had crumpled the bus like a soda can. The tornado had knocked over the town's water tower and dragged it a hundred yards to the east. Amazingly enough, the tornado did not kill anyone. Clearly, the early warning that the weather bureau sent out saved lives.

Most people would find paragraph 16a more interesting and readable because it uses passive voice to put the emphasis on the *damage* caused by the tornado. Paragraph 16b is not as interesting, because the emphasis is repeatedly placed on the *tornado*.

Do you want more information on using passive voice? See www.pearsonhighered.com/johnsonweb3/9.14

Used properly, passive voice can be a helpful tool in your efforts to write plain sentences and paragraphs. Passive voice is misused only when the readers are left wondering who or what is doing the action in the sentence. In these cases, the sentences should be restructured to put the doers in the subject slots.

Persuasive Style

Link

For more information on reader analysis, go to Chapter 3, page 40.

There are times when you will need to do more than simply present information clearly. You will need to influence your readers to take action or make a decision. In these situations, you should shift to *persuasive style*. When used properly, the persuasive style can add emphasis, energy, color, and emotion to your writing.

The following four persuasion techniques will help give your writing more impact. A combination of these, properly used, will make your writing more influential and vivid.

Elevate the Tone

Tone is the resonance or pitch that the readers will "hear" as they look over your document. Of course, most people read silently to themselves, but all readers have an inner voice that sounds out the words and sentences. By paying attention to tone, you can influence the readers' inner voice in ways that persuade them to read the document with a specific emotion or attitude.

One easy way to elevate tone in written texts is to first decide what emotion or attitude you want the readers to adopt. Then, use logical mapping to find words and phrases that evoke that emotion or attitude (Figure 9.3).

Mapping an Emotional Tone

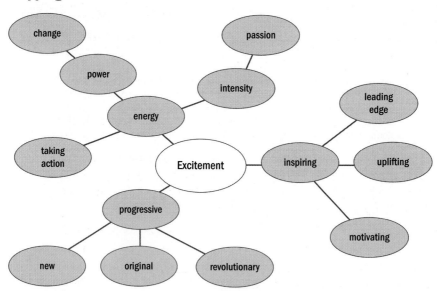

Figure 9.3: You can set an emotional tone by mapping it out and "seeding" the text with words associated with that tone.

Mapping an Authoritative Tone

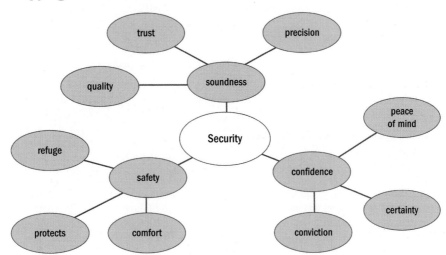

Figure 9.4:
If you use these words associated with "security" in your text at strategic points, the readers will perceive the sense of security that you are trying to convey.

For example, let us say you want your readers to feel excited as they are looking over your document. You would first put the word "excitement" in the middle of your screen or sheet of paper. Then, as shown in Figure 9.3, you can map out the feelings associated with that emotion.

You can use these words at strategic moments in your document. Subconsciously, the readers will detect this tone in your work. Their inner voice will begin reinforcing the sense of excitement you are trying to convey.

Similarly, if you want to create a specific attitude in your text, map out the words associated with it. For instance, let us say you want your document to convey a feeling of "security." A map around "security" might look like the diagram in Figure 9.4.

If you use these words associated with "security" at strategic points in your text, readers will sense the feeling you are trying to convey.

One warning: Use these words sparingly throughout the text. If you overuse them, the emotion or attitude you are setting will be too obvious. You need to use only a few words at strategic moments to set the tone you are seeking.

Use Similes and Analogies

Similes and analogies are rhetorical devices that help writers define difficult concepts by comparing them to familiar things. For example, let us say you are writing a brochure that describes an integrated circuit to people who know almost nothing about these devices. Using a simile, you might describe it this way:

> An integrated circuit is like a miniature Manhattan Island crammed into a space less than 1 centimeter long.

In this case, the simile ("X is like Y") creates an image that helps readers visualize the complexity of the integrated circuit.

Mandy Aslesen

BRAND SUPPORT SUPERVISOR, JAVA CITY COFFEE

Java City Coffee is a brewer and seller of eco-friendly coffees.

What are some strategies for being more persuasive?

Persuasive writing should always be a two-way street. As a representative for the Java City brand, I am often in a position of persuading store managers to make operational changes. But talk of "brand standards" and "vision statements" is not often tangible enough to convince store managers to change their daily routines.

The key to successful persuasion is to think of my needs as consistent and in step with the needs of the managers. Persuasion is more likely to succeed when all parties can sense the benefits of an action plan. Here are some tips to keep in mind:

- **Show sympathy.** You need to listen to people and *hear* what they are telling you. The quickest way *not* to persuade people is to make them feel like their perspective is being ignored or glossed over. By acknowledging their struggles, you show them that you are not out of touch with their daily experiences, and it leaves them more open to new suggestions.
- **Speak their language.** Although persuasion is a two-way street, the parties don't always meet in the middle. As the person attempting to persuade, it may be necessary for you to go that extra mile and write in the language that most affects others. Use phrases and jargon that are familiar to them and stay away from business-speak that they won't understand.
- **Show the benefit.** Once an audience senses that you understand and are on their side, it is your duty as the persuader to ensure they understand how your plans will benefit them. Change can be frightening or difficult, especially when familiar patterns have been settled into. Effective persuasion will attempt to alleviate these fears by being as clear as possible. If your audience has a firm sense of what the desired outcome will look and feel like, they will be motivated to believe that the benefits are worth the effort.

Here are a couple of other examples of similes:

A cell is like a tiny cluttered room with a nucleus inside and walls at the edges. The nucleus is like the cell's brain. It's an enclosed compartment that contains all the information that cells need to form an organism. This information comes in the form of DNA. It's the differences in our DNA that make each of us unique.

(*Source: Genetic Science Learning Center, http://gslc.genetics.utah.edu/units/cloning/ whatiscloning.*)

Everything that you see is made of things called atoms. An atom is like a very, very small solar system. As you know, the sun is at the center of the solar system. Just like the solar system, there is something at the center of an atom. It is called the nucleus. So the nucleus of an atom is like the sun in our solar system. Around the sun there are planets. They orbit the sun. In an atom, there are things called electrons that orbit the nucleus.

(*Source: Chickscope, http://chickscope.beckman.uiuc.edu/about/overview/mrims.html.*)

Analogies are also a good way to help readers visualize difficult concepts. An analogy follows the structure "A is to B as X is to Y." For example, a medical analogy might be

> Like police keeping order in a city, white blood cells seek to control viruses and bacteria in the body.

In this case, both parts of the analogy are working in parallel. "Police" is equivalent to "white blood cells," and "keeping order in a city" is equivalent to "control viruses and bacteria in the body."

Use Metaphors

Though comparable to similes and analogies, metaphors work at a deeper level in a document. Specifically, metaphors are used to create or reinforce a particular perspective that you want readers to adopt toward your subject or ideas. For example, a popular metaphor in Western medicine is the "war on cancer."

> **Off Target in the War on Cancer**
>
> We've been fighting the war on cancer for almost four decades now, since President Richard M. Nixon officially launched it in 1971. It's time to admit that our efforts have often targeted the wrong enemies and used the wrong weapons.
>
> (*Source: Devra Davis, Washington Post, November 4, 2007, http://www.washingtonpost. com/wp-dyn/content/article/2007/11/02/AR2007110201648.html.*)

> **Are We Retreating in the War on Cancer?**
>
> They are America's foot-soldiers in the war on cancer—young scientists whose research may someday lead to better treatments and even cures. But experts worry this small elite army is leaving the field in droves because government funding, which once allowed cancer research to flourish, is now drying up,
>
> (*Source: CBS News, May 20, 2008, http://www.cbsnews.com/stories/2008/05/20/ eveningnews/main4111776.shtml?source=related_story.*)

As shown in these examples, by employing this metaphor, the writer can reinforce a particular perspective about cancer research. A metaphor like "war on cancer" would add a sense of urgency, because it suggests that cancer is an enemy that must be defeated, at any cost. Metaphors are very powerful tools in technical writing because they tend to work at a subconscious level.

But what if a commonly used metaphor like "war on cancer" is not appropriate for your document? Perhaps you are writing about cancer in a hospice situation, where the patient is no longer trying to "defeat" cancer. In these cases, you might develop a new metaphor and use it to invent a new perspective. For example, perhaps we want our readers to view cancer as something to be managed, not fought.

> To manage their cancer, patients become supervisors who set performance goals and lead planning meetings for teams of doctors and nurses. Our doctors become consultants who help patients become better managers of their own care.

In a hospice situation, a "management" metaphor would be much more appropriate than the usual "war" metaphor. It would invite readers to see their cancer from a less confrontational perspective.

If you look for them, you will find that metaphors are commonly used in technical documents. Once you are aware of these metaphors, you can use the ones that already exist in your field, or you might create new metaphors to shift your readers' perspective about your subject. In Figure 9.5, for example, the writer uses metaphors to recast the debate over the use of nuclear energy.

Change the Pace

You can also control the readers' pace as they read through your document. Longer sentences tend to slow down the reading pace, while shorter sentences tend to speed it up. By paying attention to the lengths of sentences, you can increase or decrease the intensity of your text.

Long Sentences (low intensity)

These long sentences slow the reading down.

The effect of food insecurity on behavior problem indices suggests that children who experience food insecurity suffer more psychological and emotional difficulties than other children. Children who have experienced food insecurity exhibit more aggressive and destructive behaviors, and more withdrawn and distressed behaviors than other children. It appears that children are experiencing a great deal of psychological and emotional distress in response to issues of food insecurity within the family, and that children respond to this distress with a range of behavioral responses. The results clearly indicate that food insecurity negatively impacts children's psychological and emotional well-being, which is another important indicator of the well-being of children that may also have implications for other child development outcomes. Children who experience higher levels of psychological and emotional distress may also experience more difficulties in other areas such as school achievement, and they may interfere with a number of other activities in which children may be involved.

Short Sentences (high intensity)

Shorter sentences raise the "heartbeat" of the text.

The effect of children's food insecurity can be crucial. Food-insecure children have more psychological and emotional difficulties. They exhibit more aggressive and destructive behaviors. They are more withdrawn and distressed. As a result, they respond to this distress with a range of behavioral responses. The results show that food insecurity negatively affects children's well-being. This insecurity may have implications for other child development outcomes. These children may have more trouble at school. They may also struggle at a number of other activities.

If a situation is urgent, using short sentences is the best way to show that something needs to be done right away. Your readers will naturally feel compelled to take action. On the other hand, if you want your readers to be cautious and deliberate, longer sentences will decrease the intensity of the text, giving readers the sense that there is no need to rush.

AT A GLANCE

Persuasive Style Techniques

- Elevate the tone.
- Use similes and analogies.
- Use metaphors.
- Change the pace.

Balancing Plain and Persuasive Style

When you are drafting and revising a document for style, look for appropriate places to use plain and persuasive style. Minimally, a document should use plain style. Sentences should be clear and easy to read. Your readers should not have to struggle to figure out what a sentence or paragraph is about.

226   Chapter 9
Using Plain and Persuasive Style

GO TO THE NET

Want to learn more about metaphors? Go to
www.pearsonhighered.com/johnsonweb3/9.17

A Persuasive Argument

Patrick Moore: Going nuclear over global warming
By Patrick Moore

For years the Intergovernment Panel on Climate Change (IPCC) of the United Nations has warned us that greenhouse gas emissions from our fossil fuel consumption threaten the world's climate in ways we will regret. This year IPCC won the Nobel Peace Prize for its efforts.

It is the IPCC that former colleagues in Greenpeace, and most of the mainstream environmental movement, look to for expert advice on climate change. Environmental activists take the rather grim but measured language of the IPCC reports and add words like "catastrophe" and "chaos," along with much speculation concerning famine, pestilence, mass extinction and the end of civilization as we know it.

Until the past couple of years, the activists, with their zero-tolerance policy on nuclear energy, have succeeded in squelching any mention by the IPCC of using nuclear power to replace fossil fuels for electricity production. Burning fossil fuels for electricity accounts for 9.5 billion tons of global carbon dioxide emissions while nuclear emits next to nothing. It has been apparent to many scientists and policymakers for years that this would be a logical path to follow. The IPCC has now joined these growing ranks advocating for nuclear energy as a solution.

In its recently issued final report for 2007, the IPCC makes a number of unambiguous references to the fact that nuclear energy is an important tool to help bring about a reduction in fossil fuel consumption. Greenpeace has already made it clear that it disagrees. How credible is it for activists to use the IPCC scientists' recommendations to fuel apocalyptic fundraising campaigns on climate change and then to dismiss the recommendations from the same scientists on what we should do to solve it?

Greenpeace is deliberately misleading the public into thinking that wind and solar, both of which are inherently intermittent and unreliable, can replace baseload power that is continuous and reliable. Only three technologies can produce large amounts of baseload power: fossil fuels, hydroelectric, and nuclear. Given that we want to reduce fossil fuels and that potential hydroelectric sites are becoming scarce, nuclear is the main option. But Greenpeace and its allies remain in denial despite the fact that many independent environmentalists and now the IPCC see the situation clearly.

I have long realized that in retrospect we made a big mistake in the early years of Greenpeace when we lumped nuclear energy together with nuclear weapons as if they were all part of the same holocaust. We were totally fixated, and rightly so, on the threat of all-out nuclear war between the Soviet Union and the United States, and we thought everything nuclear was evil. We failed to distinguish the beneficial and peaceful uses of nuclear technology from its destructive and even evil uses.

The approach would be akin to including nuclear medicine with nuclear weapons just because nuclear medicine uses radioactive materials, most of which are produced in nuclear reactors. Nuclear medicine successfully diagnoses and treats millions of people every year, and it would be ludicrous to ban its use.

Greenpeace and company are basically stuck in the 1970s when it comes to their energy policy as it relates to climate change. They should accept the wisdom of the scientists at the IPCC and recognize that nuclear energy is a big part of the climate change solution. And they should stop misleading the public into thinking that wind and solar can do the job on their own. I will be the first to commend them for their courage.

Source: Sacramento Bee, December 12, 2007, http://www.sacbee.com/110/story/ 560569.html. This article is protected by copyright and should not be printed or distributed for anything except personal use. The Sacramento Bee, 2100 Q St., P. O. Box 15779, Sacramento, CA 95852 Phone: (916) 321-1000.

Figure 9.5:
In this article, the author uses numerous persuasive style techniques to add energy to his writing. Is the author persuasive?

The author uses metaphors to add energy and imagery.

The "argument is war metaphor" is used throughout the article.

An even pace is created with consistent sentence length.

The author uses an analogy to make an important comparison.

Persuasive style should be used to add energy and color. It should also be used in places in the document where readers are expected to make a decision or take action. The use of tone, similes, analogies, and metaphors in strategic places should encourage readers to do what you want. You can use short or long sentences to adjust the intensity of your prose.

In the end, developing good style takes practice. At first, revising a document to make it plain and persuasive might seem difficult. Before long, though, you will start writing better sentences while drafting. You will have internalized the style guidelines presented in this chapter.

CHAPTER REVIEW

- The style of your document can convey your message as strongly as the content; *how you say something is often what you say.*

- Style can be classified as plain, persuasive, or grand. Technical communication is most often concerned with plain or persuasive style.

- To write plain sentences, you can follow eight guidelines, which include (1) identifying your subject, (2) making the "doer" the subject of the sentence, (3) putting the doer's action in the verb, (4) moving the subject of the sentence close to the beginning of the sentence, (5) reducing nominalizations, (6) removing excess prepositional phrases, (7) eliminating redundancy, and (8) making sure that sentences are "breathing length."

- Writing plain paragraphs involves the use of four types of sentences: transitional and point sentences, which are optional, and topic and support sentences, which are necessary.

- Check sentences in paragraphs for subject alignment, and use the given/new method to weave sentences together.

- Techniques for writing persuasively include elevating the tone; using similes, analogies, and metaphors; and changing the pace.

Individual or Team Projects

EXERCISES AND PROJECTS

1. In your workplace or on campus, find three sentences that seem particularly difficult to read. Use the eight "plain style" guidelines discussed in this chapter to revise them to improve their readability. Make a presentation to your class in which you show how using the guidelines helped make the sentences easier to read.

2. In your workplace or on campus, find two paragraphs that seem difficult to read. First, align the subject in each paragraph to see if this technique improves the readability of the paragraph. Then, use given/new strategies to revise the paragraphs. Which method works better? Would a combination of subject alignment and given/new strategies be more appropriate? Present your revisions to your class, showing how you used the subject alignment and given/new techniques to improve the readability of the paragraphs.

For these exercises, sample sentences, paragraphs, and documents are available at
www.pearsonhighered.com/johnsonweb3/9.18

3. Find a document and analyze its style. Underline key words and phrases in the document. Based on the words you underlined, what is the tone used in the document? Does the document use any similes or analogies to explain difficult concepts? Can you locate any metaphors woven into the text? How might you use techniques of persuasive style to improve this document? Write a memo to your instructor in which you analyze the style of the document. Do you believe the style is effective? If so, why? If not, what stylistic strategies might improve the readability and persuasiveness of the document?

4. While revising some of your own writing, try to create a specific tone by mapping out an emotion or attitude that you would like the text to reflect. Weave a few concepts from your map into your text. At what point is the tone too strong? At what point is the tone just right?

5. Look over the Case Study at the end of this chapter. Imagine that you are Henry and write a memo to the managers of NewGenSport expressing your concerns about the safety of the product and its advertising campaign. Make sure you do some research on ephedra so you can add some technical support to your arguments.

Collaborative Project

Metaphors involving "war" are common in American society. We have had wars on poverty, cancer, drugs, and even inflation. More recently, "war on terrorism" has been a commonly used metaphor. With a group of classmates, choose one of these war metaphors and find examples of how it is used. Identify places where the metaphor seems to fit the situation. Locate examples where the metaphor seems to be misused.

With your group, discuss the ramifications of the war metaphor. For example, if we accept the metaphor "war on drugs," who is the enemy? What weapons can we use? What level of force is necessary to win this war? Where might we violate civil rights if we follow this metaphor to its logical end?

Then, try to create a new metaphor that invites people to see the situation from a different perspective. For example, what happens when we use "managing drug abuse" or "healing drug abuse" as new metaphors? How do these new metaphors, for better or worse, change how we think about illegal drugs and react to them?

In a presentation to your class, use examples to show how the war metaphor is used. Then, show how the metaphor could be changed to consider the situation from a different perspective.

For additional technical writing resources, including interactive sample documents, document design tutorials and guidelines, and more, go to **www.mytechcommlab.com.**

Going Over the Top

Henry Wilkins is a nutritionist who works for NewGenSport, a company that makes sports drinks. His company's best seller, Overthetop Sports Drink, is basically a fruit-flavored drink with added carbohydrates and salts. As a nutritionist, Henry knows that sports drinks like Overthetop do not really do much for people, but they don't harm them either. Overwhelmingly, the primary benefit of Overthetop is the water, which people need when they exercise. The added carbohydrates and salts are somewhat beneficial, because people deplete them as they exercise.

The advertisements for Overthetop paint a different picture, however. They show overly muscular athletes drinking Overthetop and then dominating opposing players. Henry thinks the advertisements are a bit misleading, but harmless. The product won't lead to the high performance promised by the advertising. But, Henry reasons, that's advertising.

One day, Henry was asked to evaluate a new version of Overthetop, called Overthetop Extreme. The new sports drink would include a small amount of an ephedra-like herbal stimulant formulated to enhance performance and cause slight weight loss. Henry also knew that ephedra had been linked to a few deaths, especially when people took too much. Ephedra was banned by the FDA in 2003, but this new "ephedralike" herbal stimulant was different enough to be legal. Henry suspected this stimulant would have many of the same problems as its sister drug, ephedra.

Henry's preliminary research indicated that 36 ounces (three bottles) a day of Overthetop Extreme would have no negative effect. But, more than a few bottles a day might lead to complications, perhaps even death in rare cases.

But before Henry could report his findings, the company's marketing team pitched its new advertising campaign for the debut of the new product. In the campaign, as usual, overly muscular athletes were shown guzzling large amounts of Overthetop Extreme and then going on to victory. It was clear that the advertisements were overpromising—as usual. But, they were also suggesting the product be used in a way that would put people at risk. The advertisements were technically accurate, but their style gave an inflated sense of the product's capabilities. In some cases, Henry realized, the style of the advertisements might lead to overconsumption, which could lead to problems.

The company's top executives were very excited about the new product. The advertisements made them only more enthusiastic. They began pressuring Henry to finish his evaluation of Overthetop right away so that they could start advertising and putting the product on store shelves.

Henry, unfortunately, was already concerned about the product's safety. But he was especially concerned that the advertisements were clearly misleading consumers in a dangerous way.

If you were Henry, how would you handle this situation? In your report to the company's management, how would you express your concerns? What else might you do?

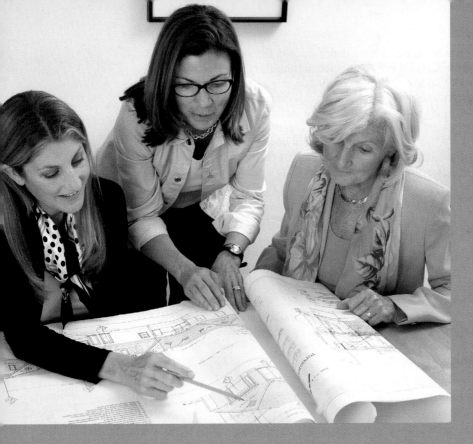

In this chapter, you will learn:

- How to design technical documents for electronic and paper-based media.

- To use the five principles of design: balance, alignment, grouping, consistency, and contrast.

- To use balancing techniques that enhance readability in a document.

- To use alignment and grouping strategies to add visual structure to text.

- To use consistency and contrast to balance uniformity and difference in the design of a document or interface.

- To anticipate the design expectations of international readers.

- How to use a process to design documents.

- About options for binding documents, as well as the sizes, colors, and weights of paper available.

Document design has become ever more important with the use and availability of computers. Today, readers expect paper-based documents to be attractive and easy to read, and include images and color. Meanwhile, screen-based documents like websites and multimedia texts are highly visual, so readers expect the interface to be well designed. Readers don't just *prefer* well-designed documents—they *expect* the design to highlight important ideas and concepts.

Imagine your own reaction to a large document with no headings, graphics, or lists. If you're like most people, you wouldn't even want to start reading. Your readers feel the same way. Like you, they want your document's design to make information easy to find and understand.

Five Principles of Design

People rarely read technical documents word for word, sentence by sentence. Instead, they tend to look over technical documents at various levels, skimming some parts and paying closer attention to others.

Good design creates a sense of order and gives readers obvious "access points" to begin reading and locating the information they need. Good design is not something to be learned in a day. However, you can master some basic principles that will help you make better decisions about how your document should look. Here are five principles to consider as you design documents:

Balance—The document looks balanced from left to right and top to bottom.

Alignment—Images and words on the page are aligned to show the document's structure, or hierarchy.

Grouping—Related images and words are placed near each other on the page.

Consistency—Design features in the document are used consistently, so the document looks uniform.

Contrast—Items in the document that are different look significantly different.

These principles are based on theories of Gestalt psychology, a study of how the mind recognizes patterns (Arnheim, 1969; Koffka, 1935). Designers of all kinds, including architects, fashion designers, and artists, have used Gestalt principles in a variety of ways (Bernhardt, 1986). You will find these five principles helpful as you learn about designing documents.

Design Principle 1: Balance

Balance is perhaps the most prominent feature of design in technical documents. On a balanced page or screen, the design features should offset each other to create a feeling of stability.

To learn more about Gestalt psychology and design, go to
www.pearsonhighered.com/johnsonweb3/10.1

Balanced and Unbalanced Page Layouts

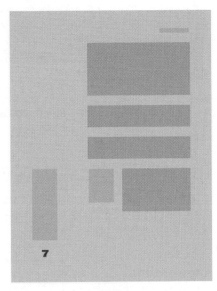

Report 7

7

Figure 10.1:
The balanced page on the left feels more stable and comfortable. The unbalanced page on the right creates more tension.

To balance a text, pretend your page or screen is balanced on a point. Each time you add something to the left side of the page, you need to add something to the right side to maintain balance. Similarly, when you add something to the top of the page, you need to add something to the bottom. Figure 10.1, for example, shows an example of a balanced page and an unbalanced page.

In Figure 10.1, the page on the left is balanced because the design features offset each other. The page on the right is unbalanced because the items on the right side of the page are not offset by items on the left. Meanwhile, the right page is top-heavy because the design features are bunched at the top of the page.

Balanced page layouts can take on many different forms. Figures 10.2 and 10.3 show examples of balanced layouts. The idea is not to create symmetrical pages (i.e., where left and right, top and bottom mirror each other). Instead, you want to balance the pages by putting text and images on all sides.

Balance is also important in screen-based documents. In Figure 10.4 (on page 236), the screen interface is balanced because the items on the left offset the items on the right.

Weighting a Page or Screen

When balancing a page or screen, graphic designers will talk about the "weight" of the items on the page. What they mean is that some items on a page or screen attract readers' eyes more than others—these features have more weight. A picture, for example, has more weight than printed words because readers' eyes tend to be drawn toward

A Balanced Design with Energy

The significantly larger title brings your eyes to the beginning of the text.

The elements on the page offset each other to create balance.

Notice how your eyes are drawn to the hawk. Animals or people in a picture usually attract the most attention from readers.

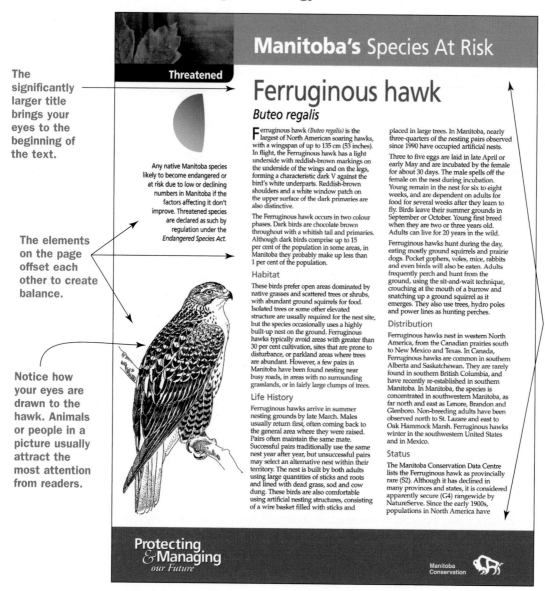

Manitoba's Species At Risk

Threatened

Any native Manitoba species likely to become endangered or at risk due to low or declining numbers in Manitoba if the factors affecting it don't improve. Threatened species are declared as such by regulation under the *Endangered Species Act.*

Ferruginous hawk

Buteo regalis

Ferruginous hawk (*Buteo regalis*) is the largest of North American soaring hawks, with a wingspan of up to 135 cm (53 inches). In flight, the Ferruginous hawk has a light underside with reddish-brown markings on the underside of the wings and on the legs, forming a characteristic dark V against the bird's white underparts. Reddish-brown shoulders and a white window patch on the upper surface of the dark primaries are also distinctive.

The Ferruginous hawk occurs in two colour phases. Dark birds are chocolate brown throughout with a whitish tail and primaries. Although dark birds comprise up to 15 per cent of the population in some areas, in Manitoba they probably make up less than 1 per cent of the population.

Habitat

These birds prefer open areas dominated by native grasses and scattered trees or shrubs, with abundant ground squirrels for food. Isolated trees or some other elevated structure are usually required for the nest site, but the species occasionally uses a highly built-up nest on the ground. Ferruginous hawks typically avoid areas with greater than 30 per cent cultivation, sites that are prone to disturbance, or parkland areas where trees are abundant. However, a few pairs in Manitoba have been found nesting near busy roads, in areas with no surrounding grasslands, or in fairly large clumps of trees.

Life History

Ferruginous hawks arrive in summer nesting grounds by late March. Males usually return first, often coming back to the general area where they were raised. Pairs often maintain the same mate. Successful pairs traditionally use the same nest year after year, but unsuccessful pairs may select an alternative nest within their territory. The nest is built by both adults using large quantities of sticks and roots and lined with dead grass, sod and cow dung. These birds are also comfortable using artificial nesting structures, consisting of a wire basket filled with sticks and

placed in large trees. In Manitoba, nearly three-quarters of the nesting pairs observed since 1990 have occupied artificial nests.

Three to five eggs are laid in late April or early May and are incubated by the female for about 30 days. The male spells off the female on the nest during incubation. Young remain in the nest for six to eight weeks, and are dependent on adults for food for several weeks after they learn to fly. Birds leave their summer grounds in September or October. Young first breed when they are two or three years old. Adults can live for 20 years in the wild.

Ferruginous hawks hunt during the day, eating mostly ground squirrels and prairie dogs. Pocket gophers, voles, mice, rabbits and even birds will also be eaten. Adults frequently perch and hunt from the ground, using the sit-and-wait technique, crouching at the mouth of a burrow and snatching up a ground squirrel as it emerges. They also use trees, hydro poles and power lines as hunting perches.

Distribution

Ferruginous hawks nest in western North America, from the Canadian prairies south to New Mexico and Texas. In Canada, Ferruginous hawks are common in southern Alberta and Saskatchewan. They are rarely found in southern British Columbia, and have recently re-established in southern Manitoba. In Manitoba, the species is concentrated in southwestern Manitoba, as far north and east as Lenore, Brandon and Glenboro. Non-breeding adults have been observed north to St. Lazare and east to Oak Hammock Marsh. Ferruginous hawks winter in the southwestern United States and in Mexico.

Status

The Manitoba Conservation Data Centre lists the Ferruginous hawk as provincially rare (S2). Although it has declined in many provinces and states, it is considered apparently secure (G4) rangewide by NatureServe. Since the early 1900s, populations in North America have

Protecting & Managing *our Future*

Manitoba Conservation

Source: Manitoba Conservation Wildlife and Ecosystem Protection Branch.

Figure 10.2: Graphic designers are especially careful about balance. This page layout uses the image of the hawk to balance two columns of written text. Meanwhile, the bold header and footer anchor the text at the top and bottom.

The green header and footer at the top and bottom of the page make it feel very balanced and stable.

A Simpler Design That Is Elegant

This image draws your eyes to it.

The two-column format balances the text.

Your eyes naturally flow down the page.

Birdbrain Breakthrough

STARTLING EVIDENCE THAT THE HUMAN BRAIN CAN GROW
NEW NERVES BEGAN WITH UNLIKELY STUDIES OF BIRDSONG

BY EDWIN KIESTER, JR., & WILLIAM KIESTER

THE BARN WHERE HE WORKS IS IN the horse country of Millbrook, New York, but it echoes with trills, tweets and obbligatos—the raucous music of more than a thousand caged canaries and finches. "Hear that one singing his heart out?" Fernando Nottebohm asks. "He has more than a dozen songs. He's telling the males, 'This is my territory.' He's telling the females, 'Hey, look at me.'"

Deconstructing birdsong may seem an unlikely way to shake up biology. But Nottebohm's research has shattered the belief that a brain gets its quota of nerve cells shortly after birth and stands by helplessly as one by one they die—a "fact" drummed into every schoolkid's skull. On the contrary, the often-rumpled Argentina-born biologist demonstrated two decades ago that the brain of a male songbird grows fresh nerve cells in the fall to replace those that die off in summer.

The findings were shocking, and scientists voiced skepticism that the adult human brain had the same knack for regeneration. "Read my lips: no new neurons," quipped Pasko Rakic, a Yale University neuroscientist doubtful that a person, like a bird, could grow new neurons just to learn a song.

Yet, inspired by Nottebohm's work, researchers went on to find that other adult animals—including human beings—are indeed capable of producing new brain cells. And in February, scientists reported for the first time that brand-new nerves in adult mouse brains appeared to conduct impulses—a finding that addressed lingering concerns that newly formed adult neurons might not function. Though such evidence is preliminary, scientists believe that this growing body of research will yield insights into how people learn and remember. Also, studying neurogenesis,

In a study of adult male canaries by Fernando Nottebohm (top), new brain cells appear green, while older ones are red.

or nerve growth, may lead them to better understand, and perhaps treat, devastating diseases such as Parkinson's and Alzheimer's, caused by wasted nerves in the brain.

Few would have predicted that canary courtship would lead to such a breakthrough. Nottebohm's bird studies "opened our eyes that the adult brain does change and develops new cells throughout life," says neurobiologist Fred Gage of the Salk Institute in La Jolla, California, whose lab recently found evidence of nerve cell regrowth in the human brain.

Nottebohm's research has achieved renown in biology and beyond. A scientist who advances an unconventional view and is later vindicated makes for compelling drama, presenting a hero who appeals to the rebel in us and a cautionary lesson to stay open-minded. Yet Nottebohm prefers being a revolutionary to a statesman. "Once I was in the 5 or 10 percent of scientists who believed in neurogenesis," he says. "Now 95 percent accept that position. I rather liked it better being in the minority."

He has been a bird lover since his boyhood, in Buenos Aires. "Listening to birds was sort of my hobby," he says. "Other boys had cars, I had birds. I liked to try to identify them by their songs." He obtained a doctorate at the University of California at Berkeley—yes, studying birds—before moving to Rockefeller University.

A key moment came in 1981 when he showed that the volume of the part of a male canary's brain that controls song-making changes seasonally. It peaks in the spring, when the need to mate demands the most of a suitor's musical ability, and shrinks in the summer. It then starts expanding again in the fall—a time to learn and rehearse new

SMITHSONIAN   JUNE 2002

36

MARTIN SCHOELLER/CORBIS OUTLINE; FERNANDO NOTTEBOHM

Source: Smithsonian, 2002, p. 36.

Figure 10.3:
This page from a magazine is simpler in design than the document shown in Figure 10.2. Notice how the elements on the page offset each other to create a balanced, stable look.

A Balanced Interface

The banner at the top of the screen offsets text and images lower on the page.

Figure 10.4: This screen is balanced, even though it is not symmetrical. The items on the left offset the items on the right.

Left and right, images and text are used to balance the screen.

Source: Fermilab.

pictures. Similarly, an animated figure moving on the screen will capture much more attention than static items.

Here are some basic weighting guidelines for a page or screen:

- Items on the right side of the page weigh more than items on the left.
- Items on the top of the page weigh more than items on the bottom.
- Big items weigh more than small items.
- Pictures weigh more than written text.
- Graphics weigh more than written text.
- Colored items weigh more than black-and-white items.
- Items with borders around them weigh more than items without borders.
- Irregular shapes weigh more than regular shapes.
- Items in motion weigh more than static items.

As you are balancing a page, use these weight guidelines to help you offset items. For example, if an image appears on the right side of the page, make sure that there is something on the left to create balance.

Using Grids to Balance a Page Layout

When designing a page or screen, your challenge is to create a layout that is balanced but not boring. A time-tested way to devise a balanced page design is to use a *page grid* to evenly place the written text and graphics on the page. Grids divide the page vertically into two or more columns. Figure 10.5 shows some standard grids and how they might be used.

Figure 10.6 (on page 240) shows the use of a three-column grid in a report. Notice how the graphics and text offset each other in the page layout.

Grids are also used to lay out screen-based texts. Even though screen-based texts tend to be wider than they are long, readers still expect the material to be balanced on the interface. Figure 10.7 (on page 241) shows possible designs using three- and four-grid templates. Figure 10.9 demonstrates the use of a six-column grid. Notice how the features of the interface are spaced consistently to create an orderly feel to the text.

Link

To learn more about designing screen interfaces, go to Chapter 14, page 364.

Grids for Designing Page Layouts

One-column grid

One-column grids offer simplicity, but not much flexibility.

Figure 10.5: Grids can help you place items on a page in a way that makes it look balanced.

(continued)

Two-column grid

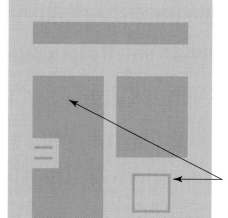

With more columns, you have more flexibility for design.

Three-column grid

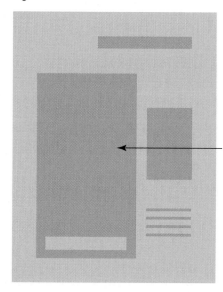

Notice how the text can go over two columns, leaving a large margin on one side.

Four-column grid

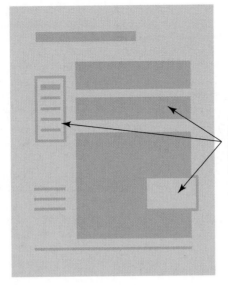

A four-column grid offers plenty of opportunities for creativity.

In many cases, the columns on a grid do not translate directly into columns of written text. Columns of text and pictures can often overlap one or more columns in the grid.

Using Other Balance Techniques

Word-processing programs and desktop publishing software also give you the ability to use professional design features like *margin comments, sidebars,* and *pullouts* (Figure 10.8). In screen-based documents, you can also use *navigation bars* and *banners* to balance the design. These features provide "access points" where people can begin reading the text.

MARGIN COMMENTS   Margin comments summarize key points or highlight quotations in the margin of the document. When a grid is used to design the page, one of the margins often leaves enough room to include an additional list, offer a special quote, or provide a simple illustration (Figure 10.8). Also, the page in Figure 10.6 shows how a whole column can be used to offer supplementary information to the main text.

NAVIGATION BARS   In screen interfaces, navigation bars are used to highlight links, usually as buttons or text (Figure 10.9). Navigation bars are typically placed along the left side of the screen or the top, but they can appear anywhere on the page.

Using Grids to Lay Out a Page

A three-column layout structures the whole page.

Images on left and bottom create balance with text.

An image can cross two columns, as this one does.

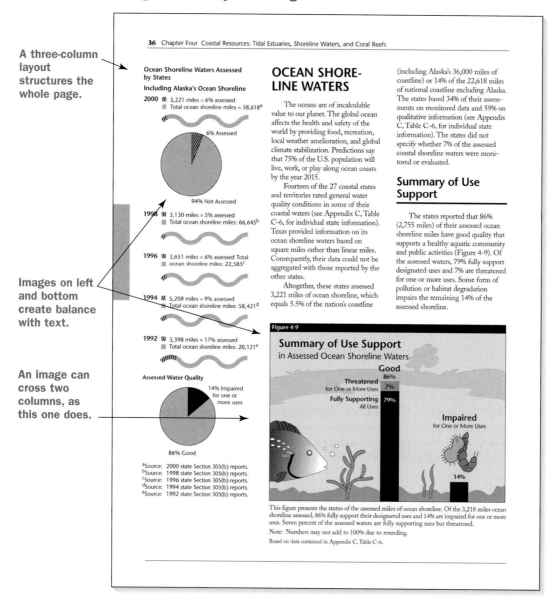

Figure 10.6:
A three-column grid was used to lay out this text in a balanced manner.

Source: U.S. Environmental Protection Agency, National Water Quality Inventory, 2000 Report, 2000, p. 36.

Grids for Interfaces

Figure 10.7: Items on screen-based pages should also be evenly placed. This approach creates a sense of stability.

Using Other Balancing Techniques for Print Documents

Margin comment

Pullout

Sidebar

Figure 10.8: Margin comments, pullouts, and sidebars can make a text more accessible.

Other Balancing Techniques for Interfaces

This banner anchors the interface at the top of the screen.

This navigation bar balances the text on the left side of the screen.

Source: Federal Bureau of Investigation, http://www.fbi.gov/fbihistory.htm.

Figure 10.9: Usually, what works on paper can work on screen. You can use pullouts, margin comments, and sidebars on a screen interface. Navigation bars and banners can also be added to screen-based documents to make them look balanced.

SIDEBARS Sidebars are used to provide examples and facts that support the main text. Sidebars should never contain essential information that readers must have in order to understand the subject or make a decision. Rather, they should offer supplemental information that enhances the readers' understanding.

PULLOUTS Pullouts are quotes or paraphrases drawn from the body text and placed in a special text box to catch the readers' attention. A pullout should draw its text from the page on which it appears. Often the pullout is framed with rules (lines) or a box, and the text wraps around it (Figure 10.8).

Link

To learn more advanced strategies for designing interfaces, turn to Chapter 14, page 365.

BANNERS Banners run along the top of the screen interface. In websites, they may include the name of the organization or an advertiser. In multimedia documents, they may identify the name of the document. Banners often include links, but they don't need to.

Want to learn about other balancing techniques? Go to
www.pearsonhighered.com/johnsonweb3/10.6

Jennifer Martin

INFORMATION ARCHITECT, IBM CORPORATION, CHICAGO

The Centers for IBM e-business Innovation help companies use design and technology to solve problems and deliver solutions.

How can design make an interface more accessible for people with disabilities?

Accessibility is one of the most important issues when developing a website. The term "accessibility" refers to improving access for everyone regardless of his or her abilities. These improvements can also benefit users of limited-function devices, older technologies, or slower connection speeds.

There are many simple techniques that you can use to make your website more accessible for people with disabilities:

- Make the page layout easily scannable by breaking content into manageable chunks, while including sufficient white space to help users with special needs.
- Verify that there is sufficient contrast between foreground and background elements to improve legibility for users with color vision deficiencies.
- Ensure that links and buttons are large enough so that users with limited fine motor skills are able to easily target them.
- Include heading tags and intra-page links such as "skip to main content." These tags and links are made invisible to those using traditional browsers and greatly improve navigation when using a screen reader.
- Set the Alt attribute to null (" ") for images that provide no value to visually impaired users, so that they will be ignored by screen readers.
- Use an accessibility checker to validate that your website complies with the World Wide Web Consortiums Content Accessibility Guidelines and any applicable laws.

Designing an accessible website should be thought of as something that is integral to the design process, rather than a separate task or skill.

Design Principle 2: Alignment

Items on a page or screen can be aligned vertically and horizontally. By aligning items *vertically* on the page, you can help readers identify different levels of information in a document. By aligning items *horizontally,* you can connect them visually so readers view them as a unit.

In Figure 10.10, for example, the page on the left gives no hint about the hierarchy of information, making it difficult for a reader to scan the text. The page on the right, meanwhile, uses alignment to clearly signal the hierarchy of the text.

Alignment takes advantage of readers' natural tendency to search out visual relationships among items on a page. If a picture, for example, is aligned with a block of text on a page, readers will naturally assume that they go together.

Using Vertical Alignment

 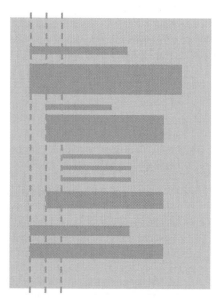

Figure 10.10: Alignment allows readers to see the hierarchy of information in a text.

In paper-based documents, look for ways you can use margins, indentation, lists, headings, and graphics to create two or three levels in the text. If you use a consistent alignment strategy throughout the text, you will design a highly readable and accessible document.

In technical documents, items are usually aligned on the left side. In rare cases, you might try aligning titles and headings on the right side. But, you should use centering only for titles, because it causes alignment problems in the text. Figure 10.11 shows how centering can create unpredictable vertical lines in the text.

Alignment is also very important in on-screen documents. To create a sense of stability, pay attention to the horizontal and vertical alignments of features on the interface. For example, the screen in Figure 10.12 shows how you can align text and graphics to make an interface look stable.

To see sample texts that use alignment well and poorly, go to
www.pearsonhighered.com/johnsonweb3/10.7

Proposal for an
Improved
Wireless Network on Campus

Written to Sally Johnson, Vice President for Electronic Advancement

Abstract: In this proposal, we argue that Western State Community College needs to improve its wireless network on campus. Currently, the network only covers specific "hot spots" like the Student Union, the Biotechnology Center, and the Computer Science Center. As a result, a majority of students cannot take advantage of it. We recommend creating a wireless cloud over the campus by establishing a grid of Cisco 350 Series Wireless Access Points (WAPs) with a maximum transmission rate of 11 Mbps. The network could be made secure with IPSec and a Cisco 3030 VPN Concentrator. The cost for this network would likely be about $120,000. The benefits, however, would greatly outstrip the costs, because the university could minimize its hardwired computer classrooms.

The Innovation Team

Introduction to Technical Communication, Humanities 381

Figure 10.11:
Centering is fine for headings, but too much centered material can make the text look chaotic.

Design Principle 3: Grouping

The principle of grouping means that items on a page that are near each other will be seen as one unit. Grouping allows you to break up the information on a page by dividing the text into scannable blocks.

Humans naturally see items that are placed near each other as a whole unit. So, if two items are placed near each other, like a picture and a caption, readers will assume that they belong together. In Figure 10.12, notice how pictures are put near paragraphs so that they are seen as units. The banner at the top of the page is supposed to be seen as a block unto itself.

Grouping is also referred to as "using white space" to frame items on the page. White spaces are places where no text or images appear on the page and include:

- the margins of the document.
- the space around a list.
- the area between an image and the body text.
- the space between two paragraphs.

An Interface That Uses Alignment and Grouping Well

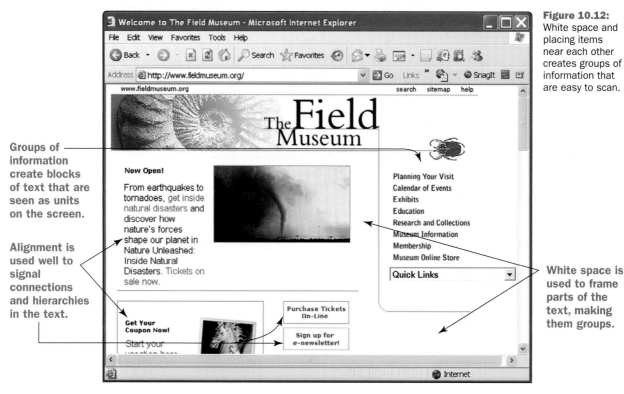

Groups of information create blocks of text that are seen as units on the screen.

Alignment is used well to signal connections and hierarchies in the text.

Figure 10.12: White space and placing items near each other creates groups of information that are easy to scan.

White space is used to frame parts of the text, making them groups.

Source: The Field Museum, Chicago, Illinois, http://www.fieldmuseum.org.

These spaces create frames around the items on the page so readers can view them as groups. For example, the white space around a vertical list (like the one in Figure 10.12) helps readers see that list as one unit.

Using Headings

Link

For more advice on using headings to organize a document, see Chapter 8, page 178.

One way to group information is to use headings. When used properly, headings will help your readers quickly understand the structure of your document and how to use it.

Your computer's word-processing program makes it easy for you to use headings by changing fonts and font sizes. Or, you can use your word processor's Style feature to create standard headings for your documents (see the Help Box on page 255).

Different types of headings should signal the various levels of information in the text.

- **First-level headings** should be sized significantly larger than second-level headings. In some cases, first-level headings might use all capital letters ("all caps") or small capital letters ("small caps") to distinguish them from the font used in the body text.
- **Second-level headings** should be significantly smaller and different from the first-level headings. Whereas the first-level headings might be in all caps, the second-level headings might use bold lettering.
- **Third-level headings** might be italicized and a little larger than the body text.
- **Fourth-level headings** are about as small as you should go. They are usually bolded or italicized and placed on the same line as the body text.

Figure 10.13 shows various levels of headings and how they might be used.

Levels of Headings

DOCUMENT TITLE

FIRST-LEVEL HEADING

This first-level heading is 18-point Avant Garde, bolded with small caps. Notice that it is significantly different from the second-level heading, even though both levels are in the same typeface.

Second-Level Heading

This second-level heading is 14-point Avant Garde with bolding.

Third-Level Heading

This third-level heading is 12-point Avant Garde italics. Often no space appears between a third-level heading and the body text, as shown here.

Fourth-Level Heading. This heading appears on the same line as body text. It is designed to signal a new layer of information without standing out too much.

Figure 10.13: The headings you choose for a document should be clearly distinguishable from the body text and from each other so that the readers can see the levels in the text.

In most technical documents, the headings should use the same typeface throughout (e.g., Avante Garde, Times, Helvetica). In other words, the first-level heading should use the same typeface as the second-level and third-level headings. Only the font size and style (bold, italics, small caps) should be changed.

Headings should also follow consistent wording patterns. A consistent wording pattern might use gerunds (-*ing* words) to lead off headings. Or, to be consistent, questions might be used as headings.

Inconsistent Headings

Global Warming

What Can We Do about Global Warming?

Alternative Energy Sources

Consistent Headings Using Gerunds

Defining Global Warming

Doing Something about Global Warming

Finding Alternative Energy Sources

Consistent Headings Using Questions

What Is Global Warming?

How Can We Do Something about Global Warming?

Are Alternative Energy Sources Available?

Headings should also be specific, clearly signaling the content of the sections that follow them:

Unspecific Headings

Kinkajous

Habitat

Food

Specific Headings

Kinkajous: Not Rare, But Hard to Find

The Habitat of Kinkajous

The Food and Eating Habits of Kinkajous

Headings serve as *access points* for readers, breaking a large text into smaller groups. If headings are used consistently, readers will be able to quickly access your document to find the information they need. If headings are used inconsistently, readers may have difficulty understanding the structure of the document.

Want more information on using headings? Go to
www.pearsonhighered.com/johnsonweb3/10.8

Using Borders and Rules

In document design, *borders* and straight lines called *rules* can be used to carve a page into smaller groups of information. They can also help break the text into more manageable sections for the readers.

Borders completely frame parts of the document (Figure 10.14). Whatever appears within a border should be able to stand alone. For example, a bordered warning statement should include all the information readers need to avoid a dangerous situation. Similarly, a border around several paragraphs (like a sidebar) suggests that they should be read separately from the main text.

Rules are often used to highlight a banner or carve a document into sections. They are helpful for signaling places to pause in the document. But, when they are overused, they can make the text look and feel too fragmented.

Using Rules and Borders to Group Information

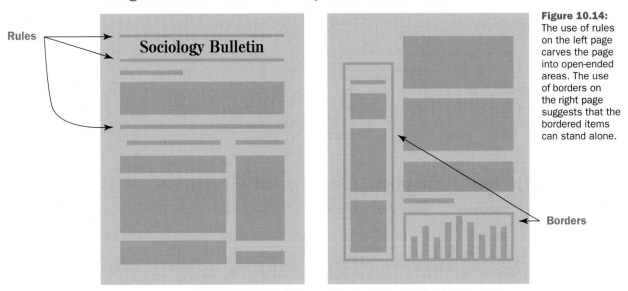

Rules

Sociology Bulletin

Figure 10.14: The use of rules on the left page carves the page into open-ended areas. The use of borders on the right page suggests that the bordered items can stand alone.

Borders

Borders and rules are usually easy to create with any word processor. To put a border around something, highlight that item and find the Borders command in your word processor (Figure 10.15). In the window that appears, you can specify what kind of border you want to add to the text.

Rules can be a bit more difficult to use, so you might want to use a desktop layout program, like Adobe InDesign or QuarkXPress. However, if your document is small or

Making Borders

Select the kind of border you want to use.

Select the color and width of the border's lines.

Select where you want the borders.

Figure 10.15: The Borders command in your word processor will allow you to put boxes around items or draw lines.

simple, you can use the Draw function of your word processor to draw horizontal or vertical rules in your text.

Design Principle 4: Consistency

The principle of consistency suggests that design features should be used consistently throughout a document or website:

- Headings should be predictable.
- Pages should follow the same grid.
- Lists should use consistent bulleting or numbering schemes.
- Page numbers should appear in the same place on each page.

Consistency is important because it creates a sense of order in a document while limiting the amount of clutter. A consistent page design will help your readers access information quickly, because each page is similar and predictable. When design features are used inconsistently, readers will find the document erratic and hard to interpret.

Figure 10.16 shows a consistent layout for a Bluetooth® headset user's manual. In this sample text, notice how the consistent use of design features creates a predictable, useful document.

GO TO THE NET

More advice about using rules and borders is available at **www.pearsonhighered.com/johnsonweb3/10.9**

Consistent Layout

Consistent use of icons and headings

Consistent fonts

Consistent use of images

Consistent use of page numbers

Figure 10.16:
These pages from the same user's manual are consistent in many ways.

Source: Jabra BT160 Bluetooth User Manual, pp. 2–5. GN Netcom, Inc.

Choosing Typefaces

Consistency should be an important consideration when you choose typefaces for your document. As a rule of thumb, a document should not use more than two typefaces. Most page designers will choose two typefaces that are very different from each other, usually a *serif* typeface and a *sans serif* typeface.

A serif typeface like Times Roman, New York, or Bookman has small tips (serifs) at the ends of the main strokes in each letter (Figure 10.17). Sans serif typefaces like Arial or Helvetica do not have these small tips.

Serif and Sans Serif Typefaces

Serifs

No Serifs

Figure 10.17: A serif typeface like Bookman (left) includes the small tips at the ends of letters. A sans serif typeface like Helvetica (right) does not include these tips.

Serif fonts, like Times, New York, or Bookman, are usually perceived to be more formal and traditional. Sans serif typefaces like Helvetica seem more informal and progressive. So, designers often use serif fonts for the body of their text and sans serif in the headings, footers, captions, and titles (Figure 10.18). Using both kinds of typefaces gives a design a progressive and traditional feel at the same time.

Your choice of typefaces, of course, depends on the needs and expectations of your readers. Use typefaces that fit their needs and values, striking the appropriate balance between progressive and traditional design.

Using Different Typefaces for Different Purposes

Helvetica (sans serif font)

Bookman (serif font)

Which Is Better? Serif or Sans Serif?

Serif typefaces are often used in traditional-looking texts, especially for the body text. Studies have suggested inconclusively that serif typefaces like Bookman are more legible, but there is some debate as to why. Some researchers claim that the serifs create horizontal lines in the text that make serif typefaces easier to follow. These studies, however, were mostly conducted in the United States, where serif typefaces are common. In other countries, like Britain, where sans serif typefaces are often used for body text, the results of these studies might be quite different.

Figure 10.18: Often designers will choose a sans serif typeface for headings and a serif typeface for the main text.

For websites that discuss the history of typefaces, go to **www.pearsonhighered.com/johnsonweb3/10.10**

Labeling Graphics

Graphics such as tables, charts, pictures, and graphs should be labeled consistently in your document. In most cases, the label for a graphic will include a number.

Graph 5: Growth of Sales in the Third Quarter of 2004

Table C: Data Set Gathered from Beta Radiation Tests

Figure 10: Diagram of Wastewater Flow in Hinsdale's Treatment Plant

The label can be placed above the graphic or below it. Use the same style for labeling every graphic in the document. Then locate the labels consistently on each graphic.

Link

For more information on labeling graphics, see Chapter 11, page 279.

Creating Sequential and Nonsequential Lists

Early in the design process, you should decide how lists will be used and how they will look. Lists are very useful for showing a sequence of tasks or setting off a group of items. But, if they are not used consistently, they can create confusion for the readers.

When deciding how lists will look, first make decisions about the design of sequential and nonsequential lists (Figure 10.19).

Sequential (numbered) lists are used to present items in a specific order. In these lists, you can use numbers or letters to show a sequence, chronology, or ranking of items.

Nonsequential (bulleted) lists include items that are essentially equal in value or have no sequence. You can use bullets, dashes, or check boxes to identify each item in the list.

Lists make information more readable and accessible. So, you should look for opportunities to use them in your documents. If, for example, you are listing steps to describe how to do something, you have an opportunity to use a sequential list. Or, if you are creating a list of four items or more, ask yourself whether a nonsequential list would present the information in a more accessible way.

AT A GLANCE

Checking for Consistency

The following items should be used consistently in your document:

• Typefaces (serif and sans serif)
• Labeling of graphics
• Lists (sequential and nonsequential)
• Headers and footers

Just make sure you use lists consistently. Sequential lists should follow the same numbering scheme throughout the document. For example, you might choose a numbering scheme like (1), (2), (3). If so, do not number the next list 1., 2., 3. and others A., B., C., unless you have a good reason for changing the numbering scheme. Similarly, in nonsequential lists, use the same symbols when setting off lists. Do not use bullets (•) with one list, check marks (✓) with another, and boxes (■) with a third. These inconsistencies only confuse the readers. Of course, there are situations that call for using different kinds of nonsequential lists. If you need lists to serve completely different purposes, then different symbols will work—as long as they are used consistently.

Steps for Taking a Digital Picture

Follow these simple steps:

1. Turn on the Digital Camera
2. Adjust Settings
3. Focus the Subject
4. Shoot the Picture
5. Check the Results
6. Turn Off the Camera

This sequential list orders the information into steps.

These nonsequential lists signal that all the items in the lists are equal in value.

Items to Bring on This Sierra Club Outing

- Comfortable Hiking Boots
- Wide Brim Hat
- Sunscreen
- Water (at least one liter)
- Healthy Snacks (trailmix, granola bars)
- Map and Compass
- Rain Gear
- Pocket Knife
- Money (10 dollars will be enough for most emergencies)

Do Not Bring the Following Items:

- Guns or Other Weapons
- Electronic Devices (radios, televisions)
- Pets (the mountain lions might eat them)
- Bad Attitudes

Figure 10.19: The sequential list on the left shows an ordering of the information, so it requires numbers. The nonsequential list on the right uses bullets because there is no particular ordering of these items.

Inserting Headers and Footers

Even the simplest word-processing software can put a header or footer consistently on every page. As their names suggest, a header is text that runs across the top margin of each page in the document, and a footer is text that runs along the bottom of each page (Figure 10.20, p. 256).

Headers and footers usually include the company's name or the title of the document. In documents of more than a couple of pages, the header or footer (not both) should include the page number. Headers and footers often also include design features like a horizontal rule or a company logo. If these items appear at the top or bottom of each page, the document will tend to look like it is following a consistent design.

For advice about using lists, go to
www.pearsonhighered.com/johnsonweb3/10.11

Using Styles and Templates

Word-processing programs like Word or WordPerfect include Styles functions that will help you manage the formatting of your documents. When you apply a style to, say, a heading, paragraph, caption, or list, the computer will automatically change the text into a predetermined format.

For example, let us say you want all your second-level headings to be set in 12-point Helvetica Italics. You can define that style with the Styles function. From then on, whenever you use a second-level heading, you can simply select that style from the Style menu (Figure A). The computer will do the formatting for you.

That's handy, but here is the real advantage of using styles: If you ever need to change the styles in your document (say you decide to use the Arial typeface instead of Helvetica for headings), you simply need to redefine that style. The computer will go through your document and change all the headings to that new style.

Styles are useful when you are designing a larger document, especially when you are working with a team. If you and your team members decide on the styles of the document up front, each person can follow the same formatting. Then, when you put the document together, you will need to do only some minimal editing to make the format consistent.

Word processors also include style *templates* that you can follow. These templates will help you format letters, memos, résumés, reports, and presentations.

Using the Styles Function

The Styles menu drops down.

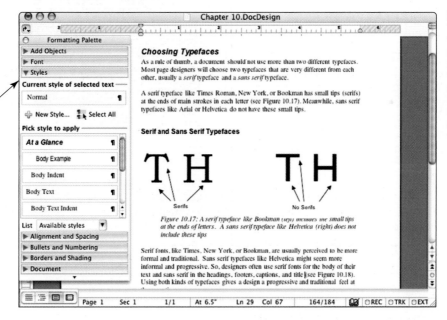

Figure A: The Styles function on your computer can help keep your formatting consistent. It is especially helpful on larger documents and team projects.

(continued)

For websites that will help you use the "Styles" menu, go to
www.pearsonhighered.com/johnsonweb3/10.12

Design Principle 4: Consistency

The easiest way to use a template is to start out with one. When you create a document with the New function, your word processor will usually offer you a selection of templates you can use. Choose the template you need. Then, you can modify the template to handle the individual needs of the document you are writing.

If you want to apply a template to a document after you have started writing, find the Templates function in your word processor. You can then apply the template to your document. After you apply the template, you may need to make some adjustments to words or images so the template fits the text you have written.

In some cases, you may want to create your own template. Perhaps, for example, you want to make your own letterhead. To make your own template, design a document on a blank page. Then, select "Save As." In the Save As box, you can choose "Save Document as a Template." From then on, whenever you create a new file you can select this template from the collection of other templates.

Link

For more information about making templates, go to Chapter 17, page 479.

Headers and Footers

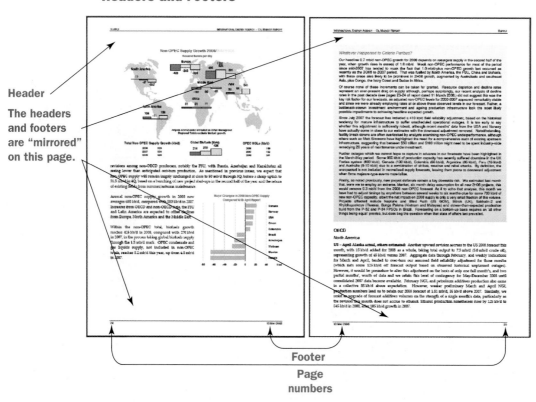

Header

The headers and footers are "mirrored" on this page.

Footer

Page numbers

Figure 10.20: A header appears consistently across the top of each page, except the first. A footer appears consistently at the bottom. A page number usually appears in either the header or the footer.

Source: Oil Market Report, May 13, 2008.

GO TO THE NET

Need a template? Go to
www.pearsonhighered.com/johnsonweb3/10.13

Five Principles of Document Design

AT A GLANCE

- Balance
- Alignment
- Grouping
- Consistency
- Contrast

Design Principle 5: Contrast

Contrast makes items look distinct and different, adding energy and sharpening boundaries among the features on the page or screen.

A good guideline is to "make different things on the page look very different." Contrast, as shown in Figure 10.21, makes design elements lively.

Your uses of contrast when designing a page should be considered carefully. Word processors offer many tools for adding contrast in ways that capture readers' attention. Sometimes, though, you can accidentally create contrast problems with different colors or shading in the background—or just too much clutter on the page. In this section are some helpful tips for using contrast in documents and interfaces.

Adding Shading and Background Color

When used properly, shading and background color can help highlight important text in a document. However, these design features can also make texts hard to

Contrast in a Webpage

Figure 10.21: In this webpage, contrast is used to catch the reader's eye and make the text easier to read.

Large headings grab the reader.

Color adds contrast to the text.

Reverse type (white on dark purple) draws the eye.

Images contrast with text.

Source: Medline Plus, http://www.medlineplus.gov.

read. For example, Figure 10.22 shows how a lack of contrast can make text hard to read.

Background color or images can also cause text on computer screens to be difficult to read. So, use them carefully and be sure to check how backgrounds work (or don't work) with the text on the screen.

It is fine to use shading, background color, and background images. However, make sure the words on the page contrast significantly with their background.

Shading and Contrast

This text is set against a gray background. It is difficult to read because there is not enough contrast between the words and the background against which they are set.	This text is set against a white background. It is easier to read, because the contrast between the words and the white background is sharp and definite.

Figure 10.22: Shaded versus unshaded text: The shaded text on the left is more difficult to read because the words do not contrast significantly with the background shading.

Highlighting Text

There are several different ways to use highlighting, a form of contrast, to make words stand out in a text. With a computer, you can highlight text with ALL CAPS, SMALL CAPS, **bolding**, *italics*, <u>underlining</u>, and color.

ALL CAPS SHOULD BE USED IN MODERATION, USUALLY ONLY FOR HEADINGS OR SHORT WARNING STATEMENTS. AS YOU CAN SEE, THESE SENTENCES ARE HARD TO READ. ALSO, ALL CAPITALS SUGGESTS THAT THE DOCUMENT IS SHOUTING.

SMALL CAPS ARE A LITTLE EASIER TO READ THAN ALL CAPS BUT STILL SHOULD BE USED SPARINGLY. THEY ARE ESPECIALLY USEFUL FOR HEADINGS.

Bolding is useful for highlighting individual words or whole sentences. When used in moderation, bolding helps words or sentences stand out. When used too often, however, the highlighting effect is reduced because readers grow used to the bolded text.

Italics, similarly, can be used to add emphasis to words or sentences. A whole paragraph in italics, however, is hard to read, so use italics sparingly—only when you need it.

To view webpages with contrast problems, go to
www.pearsonhighered.com/johnsonweb3/10.14

Color is perhaps the most prominent way to highlight words or sentences. It is especially helpful for making headings more prominent. However, color can be distracting when it is overused. So, use it selectively and consistently.

Underlining is rarely used now that we have other ways to highlight text. Underlining was effective when typewriters were widely used, because it was the only way (except for all caps) to highlight information. Computers, however, now give us better options for highlighting like italics, bolding, and color.

Highlighting is an easy way to create contrast in a document. When used selectively, it can make important words or sentences pop off the page. If it is overused, though, readers become immune to it.

Using Font Size and Line Length

Your decisions about font size, line length, and line spacing should depend on the readers of your document.

APPROPRIATE FONT SIZE   The font size you choose for your text depends on how your readers will be using the text.

For most readers, text in 11-point or 12-point font sizes is easy to read.

Fonts smaller than 10 points can be difficult to read, especially when used for a longer stretch of text. When a page length is required for a document, it is often tempting to use a smaller font, so more words can be put on a page. But most people quickly grow tired of straining their eyes to read the text. You are better off using a normal-sized font with fewer words. That way, at least the readers will actually read the text.

Large font sizes (above 14 points) should be used only for special purposes, such as documents written for older readers or situations where the text needs to be read at a distance. When used at length, these larger sizes can make a text look childish. Extra-large font sizes usually suggest that the writer is simply filling a page to hide a lack of content.

PROPER LINE LENGTH   Line length is also an important choice in a document. Readers who scan usually prefer a shorter line length, making columns especially useful for documents that will be read quickly.

However, when lines are
short, they force
readers to quickly look
back and forth.
Eventually, these short
lines will frustrate
readers because they
require such rapid
movement of the eyes.
They also seem to suggest
fragmented thinking on
the part of the writer,
because the sentences
look fragmented.

Longer line lengths can have the opposite effect. Readers quickly grow tired of following the same line for several inches across the page. The lines seem endless, giving readers the impression that the information is hard to process. More important, though, readers will find it difficult to locate the next line on the left side. A line should never be wider than 6 inches across.

As you consider font size and line length, you should anticipate how your readers will use your document. If they will be scanning, shorter line lengths will help them read quickly. If they will not be scanning, longer lines are fine. Then, choose the font size or line length that suits their needs.

Cross-Cultural Design

As the global economy grows, designing documents for cross-cultural readers may be one of the greatest challenges facing technical communication. Today, most international readers are adjusting to Western design practices. But, with the global reach of the Internet and the growth of economies around the world, international readers are beginning to expect documents and interfaces to reflect their own cultural design conventions.

When designing cross-culturally, your first consideration is whether your document or interface needs a "culturally deep" or a "culturally shallow" design.

- **Culturally deep** documents and interfaces use the language, symbols, and conventions of the target culture to reflect readers' design preferences and expectations. To develop a culturally deep design, you probably need help from designers or consultants who are familiar with the target culture and understand its design expectations.
- **Culturally shallow** documents and interfaces usually follow Western design conventions, but they adjust to reflect some of the design preferences of the cultures in which they will be used. They also avoid any cultural taboos of the people who are likely to use the text. Culturally shallow designs tend to be used in documents or interfaces that need to work across a variety of cultures.

Unless your company is targeting its products or services to a specific culture (e.g., a nation like Korea or Zimbabwe), most of your documents or interfaces will need to be culturally shallow so that they can work across a variety of cultures.

Culturally shallow designs usually consider four design issues: use of color, use of people, use of symbols, and direction of reading.

Use of color—Choice of colors in a cross-cultural document can influence how readers interpret the message, because colors can have different meanings across cultures. For instance, the use of red in Japan signals anger, while in China red signals happiness. The use of red in Egypt symbolizes death. Meanwhile, the color green in France symbolizes criminality, while in the United States green symbolizes moving forward or environmental consciousness. Figure 10.23 shows how some common colors are perceived across cultures. When designing your document or interface, you should use colors that reflect the expectations of the likely readers (or at least avoid colors that have negative associations).

Images of people—Cross-cultural texts should use images of people carefully. Avoid big smiles, highly emotional expressions, suggestive behavior, and flashy clothing. In pictures, interactions between women and men should avoid sending mixed signals. In some cultures, especially Islamic cultures, images of people are used only when "needed." The definition of "need" varies among Islamic subcultures, but images tend to be used only for purposes of identification.

Colors in Other Cultures

Color	Japan	France	China	Egypt	United States
Red	Anger, danger	Aristocracy	Happiness	Death	Danger, stop
Blue	Villainy	Freedom, peace	Heavens, clouds	Virtue, faith, truth	Masculine, conservative
Green	Future, youth, energy	Criminality	Ming Dynasty, heavens	Fertility, strength	Safe, go, natural
Yellow	Grace, nobility	Temporary	Birth, wealth, power	Happiness, prosperity	Cowardice, temporary
White	Death	Neutrality	Death, purity	Joy	Purity, peace, marriage

Figure 10.23: Colors can have very different meanings in different cultures. In some cases the meanings of colors may even be contradictory among cultures.

Source: From Patricia Russo and Stephen Boor, "How Fluent Is Your Interface? Designing for International Users." In S. Ashlund, K. Mullet, A. Henderson, E. Hollnagel, and T. White, eds., Proceedings of the INTERACT '93 and CHI '93 Conference on Human Factors in Computing Systems, pp. 342–347, Table 1. © 1993 ACM, Inc. Reprinted by permission. http://doi.acm.org/10.1145/169059.169274.

Link

For more
information on
international and
cross-cultural
symbols, go to
Chapter 11,
page 296.

Use of symbols—Common symbols can have very different meanings in different cultures. For example, in many cultures, the "OK" hand signal is highly offensive. Uses of crescent symbols (i.e., moons) or crosses can have a variety of religious meanings. White flowers or a white dress can signify death in many Asian cultures. To avoid offending readers with symbols, a good approach is to use only simple shapes (e.g., circles, squares, triangles) in cross-cultural documents.

Direction of reading—Many cultures in the Middle East and Asia read right to left instead of left to right. As a result, some of the guidelines for balancing a page design discussed earlier in this chapter should be reversed. For example, a document or interface that reads right to left tends to be anchored on the right side. Otherwise, the text will look unbalanced to a right-to-left reader. Figure 10.24, for example, shows a website that is designed right to left for Middle Eastern readers.

Link

For more help
on working with
cross-cultural
readers, go to
Chapter 3,
page 53.

Cross-cultural design can be very challenging. The secret is to consult with people from the target culture and/or use consultants to help you design your documents and interfaces. Then, be ready to learn from your mistakes.

A Right-to-Left Interface Design

Figure 10.24:
These interfaces demonstrate a right-to-left versus left-to-right design of the same webpage. Note how the Middle Eastern webpage (top) is anchored on the right, mirroring the design in the Western webpage (bottom).

Source: Microsoft, http://www.microsoft.com.

Using the Principles of Design

Balance, alignment, grouping, consistency, contrast: These five basic design principles should help you create easy-to-read page layouts and screen interfaces that highlight important information and attract your readers.

In many ways, designing documents is like drafting the written text in your documents. You should go through a process—in this case, a *design process.*

Analyze readers and the document's context of use.

Use thumbnails to sketch out the design.

Design the document.

Revise and edit the design.

As with drafting, each stage in the process takes you closer to a finished document.

Analyze Readers and the Document's Context of Use

You should begin designing a document by first looking back at your analyses of your readers and the contexts in which they will use your document. Different kinds of readers will have different expectations and preferences for the design of your document. A more traditional reader, for example, might prefer a conservative layout with simple headings and a one-column or two-column layout. A more progressive reader might prefer a bolder layout with more color and contrast.

As you consider your readers, try to match your design to the tone you are setting in the written part of the document. Positive, exciting themes in the written text should be reflected in the color and boldness of the design. Serious or somber themes should be reflected with subdued color schemes and a plainer design.

The context in which your document will be used should also play an important role in helping you decide the appropriate design. Pay special attention to the *physical context* in which the readers will be looking over your document. Will they be in a meeting? At a worksite? In their living rooms?

Each of these contexts might shape how you will design the document. For example, if you know your readers will be discussing your document in a meeting, you might use margin notes and lists to highlight information that will be discussed at the meeting. That way, your readers will be able to quickly find "talking points" in your document that they can share with the others in the meeting.

Use Thumbnails to Sketch Out the Design

Next, start sketching out how you think the design will look. Professional graphic designers often use *thumbnails* to sketch out possible layouts for their documents or screen interfaces. Thumbnails can be sketched by hand, usually with pencil, or with a computer drawing program (Figure 10.25). They are miniature versions of the pages.

Thumbnailing is especially helpful when you are working with a team. With little effort, you can quickly sketch out sample pages or screen interfaces. When the group agrees, you can then use these sketches to help you lay out the page.

Sample Page and Interface Thumbnails

Figure 10.25: Thumbnails are miniature sketches of possible page or screen layouts. They help you find a good design with a minimum amount of effort.

The simplest way to thumbnail pages is to fold a regular sheet of paper into four parts (i.e., fold it once lengthwise and once horizontally). The folds will then give you four separate "pages" to practice on. For computer interfaces, simply turn the sheet sideways. You will have four rectangles that are each about the shape of a screen.

One nice thing about designing with thumbnails is how easy they are to use. It takes only seconds to sketch out a few layouts. If you or other people on your team don't like the sketches, you can toss them in the recycling bin. Then, start again. Thumbnails require much less effort than an attempt to develop a full layout of a page or document. As a result, you will have more freedom to be creative and to try out a few different designs for the document.

GO TO THE NET

For more advice on thumbnailing, go to
www.pearsonhighered.com/johnsonweb3/10.15

Design the Document

Usually, the design is developed as the document is drafted. You don't need to wait until the entire document is written before you start laying out some of its pages. In fact, it is better to begin designing *before* the whole document is drafted. That way, the design might help you (1) identify gaps in the content; (2) find places where visuals, like graphs and charts, are needed; and (3) locate places in the document where supplemental text, like sidebars or margin notes, might be added to enhance the reading.

When you or your team have drafted a few pages of written text, you should try turning your thumbnails into page designs. Of course, the text will change as the drafting continues. But, the actual written words will help you see how the document is going to look on paper or on the screen. Then, you can start using templates and/or grids to lay out information on the page. Make sure you create a flexible enough design for your document, leaving plenty of room for graphics and other visual elements.

As the drafting of the document nears completion, use your sample pages to guide the design of the rest of the pages.

Revise and Edit the Design

Revising and editing a document's design is similar to revising and editing the written text. Start looking over the document globally, paying attention to issues of balance, alignment, grouping, consistency, and contrast. Then, examine the document page by page to see if the visuals are consistent and appropriate.

Finally, proofread the design. While proofreading, you are going to find places where smaller alterations and corrections are needed. You should take the time to iron out these little wrinkles. After all, small inconsistencies and errors in the design will draw readers' attention away from the written text.

A Primer on Binding and Paper

In the technical workplace, most documents are printed on standard-sized paper—the kind you use in your copier or printer. There will be times, though, when you will need to make decisions about how your document will be bound, as well as decisions about the size, weight, and color of the paper. As copiers become more sophisticated, you may find yourself making these decisions with more frequency.

Here is a quick primer to help you choose binding and paper.

Binding

In most cases, technical documents are bound with a paper clip, a staple, or perhaps a three-ring binder. But if you are interested in a more permanent type of binding, here are a few of your options.

PERFECT BINDING   In this method, glue is applied to the back edge of the pages, and the pages are then inserted into a cover (Figure 10.26). Perfect binding is a relatively

Different Kinds of Bindings

Perfect Binding

Comb Binding

Spiral Binding

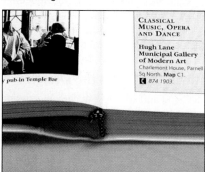

Stitched Binding

Figure 10.26: The four most common kinds of bindings include perfect binding, comb binding, spiral binding, and sewn or stitched binding. All four have their advantages and disadvantages.

inexpensive option for binding larger documents. The downside is that, with time, the binding will eventually come loose. When the binding begins to fail, pages will fall out of the document.

COMB BINDING   For comb binding, slots are punched along the back edge of the pages. Then, a machine is used to insert a plastic comb binder. This kind of binding is usually the least expensive, especially for smaller documents. However, comb binding often comes loose with heavy usage. Also, over time, the plastic combs become brittle and break.

SPIRAL BINDING   This method is similar to comb binding, except holes are drilled or punched into the pages and a wire spiral is threaded through the holes. Compared to comb binding, spiral binding is much more secure but is also more expensive. This binding is great for documents that need to lay flat when open.

SEWN OR STITCHED BINDING   Here, sheets of paper are sewn or stapled together down the middle fold. Most hardback books are made with one of these two methods.

To learn more about binding documents, go to
www.pearsonhighered.com/johnsonweb3/10.16

Expensive hardbacks that are meant to last will be sewn with thread. Less expensive hardbacks will have two or three staples down the middle fold. In large books, sections of the book (called *signatures*) are sewn or stitched together. Then, these signatures are placed side by side and bound into the cover.

Selecting the Paper

Your choice of paper depends on the needs of your readers and the contexts in which the document will be used.

SIZE   A standard sheet of paper in North America is 8.5 × 11 inches, while most other countries use a metric sheet of paper called A4, which is slightly larger at 21.0 centimeters × 29.7 centimeters. You can fold these sheets in half lengthwise or widthwise to make a nice booklet. You can fold a sheet into three panels to make a brochure. Meanwhile, a double sheet (11 × 17 inches or 42 centimeters × 29.7 centimeters) is available in many copiers. You can fold one of these sheets in half to make two standard sheets.

Other sizes can be used. You should talk to a printer or copy shop about what kinds of sizes are available. If you want an irregular size of paper, the cost of printing will go up significantly.

WEIGHT   Paper comes in a variety of weights. The paper that you use in your printer or a copier is probably *20-pound bond,* which is a relatively lightweight paper. You can choose heavier paper if you want to add firmness to the document. For a cover, you might even choose a 60-pound or heavier card stock to make the document stiffer.

COLOR   Paper comes in a variety of different colors. When choosing a color, you should keep two things in mind. First, you want the words and images to *contrast* significantly with the page. Black words printed on a gray-colored sheet of paper, for instance, will often be difficult to read. Second, you do not want the color to interfere with any duplicating that the readers might need to do. A mauve-colored document, for example, is not going to duplicate well on a copy machine. To avoid problems with contrast and duplication, in most cases your best choices for paper colors are white or off-white.

GLOSSY OR MATTE   Increasingly, people are able to choose whether the paper is glossy or matte. Glossy paper is smooth and shiny and is often used in magazines or annual reports. Matte paper has a rougher texture and is not shiny, like the paper in your printer or copier. Most professional printers can offer you glossy paper for important documents (but at a higher expense).

Computers are giving us more options for binding and paper choices. As copiers grow more sophisticated, you might find yourself making these kinds of choices about the binding and paper used in your documents.

Interested in learning more about paper? Go to
www.pearsonhighered.com/johnsonweb3/10.17

A Primer on Binding and Paper   267

- Good design allows readers to (1) easily scan a document, (2) access the document to find the information they need, (3) understand the content more readily, and (4) appreciate the appearance of the document.

- The five principles of document design are balance, alignment, grouping, consistency, and contrast.

- On a balanced page, elements offset each other to create a stable feeling in the text. Unbalanced pages add tension to the reading process.

- Alignment creates relationships among items in a text and helps readers determine the hierarchical levels in the text.

- Grouping divides text into "scannable" blocks by using headings, rules, and borders to make words and images easier to comprehend.

- Consistency makes documents more accessible by making features predictable; inconsistent documents are harder to read and interpret.

- Contrast can cause text features or elements to stand out, but contrast should be used with restraint so that text elements work together rather than compete against each other.

- Documents and interfaces that need to work cross-culturally can use a culturally shallow design or a culturally deep design.

- Use a design process that includes (1) analyzing your readers and your document's context of use, (2) using thumbnails to rough out the design, (3) designing the document, and (4) revising and editing the design.

- You can choose from a variety of bindings and paper sizes, weights, and colors.

Individual or Team Projects

1. On campus or at your workplace, find a poorly or minimally designed document. If you look on any bulletin board, you will find several good documents you can use. In a memo to your instructor, critique this document using the five design principles discussed in this chapter. Explain how the document fails to follow the principles.

2. Read the Case Study at the end of this chapter. Using the design principles discussed in this chapter, sketch out some thumbnails of a better design for this document. Then, using your word processor, develop an improved design. In a memo to your instructor, compare and contrast the old design with the new one, showing why the new design is superior.

3. On the Internet, find examples of badly designed websites and well-designed websites. In a memo to your instructor, compare two websites, using the design principles in this chapter to point out the websites' strengths and weaknesses. In your memo, make some suggestions for improvements.

4. Find a document that illustrates good design. Then, change some aspect of its rhetorical situation (purpose, readers, context). Redesign the document to fit that altered rhetorical situation. For example, you might redesign a user's manual to accommodate the needs of 8-year-old children. You might turn a normal-sized document into a poster that can be read from a distance. In a memo to your instructor, discuss the changes you made to the document. Show how the changes in the rhetorical situation led to alterations in the design of the document. Explain why you think the changes you made were effective.

5. On the Internet, find an international company's website that is intended to work cross-culturally. In a presentation to your class, explain why you believe the site is culturally shallow or culturally deep. How have the designers of the website made adjustments to suit the expectations of people from a different culture? How might they improve the design to make it more effective for the target readers?

Collaborative Project

With other members of your class, choose a provider of a common product or service (for example, a car manufacturer; mobile phone, computer, clothing, or music store; museum; or theater). Then, find the websites of three or four competitors for this product or service.

Using the design principles you learned in this chapter, critique these websites by comparing and contrasting their design. Considering its target audience, which website design seems the most effective? Which is the least effective? Explain your positive and negative criticisms in some depth.

Then, using thumbnails, redesign the weakest site so that it better appeals to its target audience. How can you use balance, alignment, grouping, consistency, and contrast to improve the design of the site?

In a presentation to your class, discuss why the design of one site is stronger and others are weaker from a design perspective. Then, using an overhead projector, use your thumbnails to show how you might improve the design of the weakest site you found.

For additional technical writing resources, including interactive sample documents, document design tutorials and guidelines, and more, go to **www.mytechcommlab.com.**

Sample documents and websites for critiquing and redesigning are available at
www.pearsonhighered.com/johnsonweb3/10.18

Scorpions Invade

Lois Alder is a U.S. Environmental Protection Agency biologist in western Texas who specializes in integrated pest management (IPM). The aim of IPM is to use the least toxic methods possible for controlling insects, rodents, and other pests. Lois's job is to help companies, schools, and government organizations develop IPM strategies to minimize their use of pesticides, especially where people live, work, and play.

Recently, a rural school district outside El Paso, Texas, was experiencing an infestation of scorpions. Scorpions were finding their way into classrooms, and they were regularly found in the schoolyards around the buildings. Teachers and children were scared, though no one had been stung yet. Parents were demanding that the school district do something immediately.

In a conference call, Lois talked to the principals and custodians of the schools. The custodians said they just wanted to "nuke" the scorpions by spraying large amounts of pesticides in and around the buildings. But this solution would expose the children, teachers, and staff at the schools to high levels of toxins. Lois convinced them that they should first try an IPM approach to the problem. She said she would quickly set up a training program to help them. Lois knew, though, that she needed to send these school officials some information right away.

From the EPA website, Lois downloaded a well-written chapter, "IPM for Scorpions in Schools" (Figure A) (http://www.epa.gov/pesticides/ipm/schoolipm/). The design of the document, however, was a bit bland, and Lois was concerned that school officials, especially the custodians, would not read it carefully. Moreover, she knew that some of the custodians had poor reading skills.

So, she decided she would need to redesign the document to make it more accessible. She wanted to highlight the need-to-know information while stripping out any information that the school officials would not need. If you were Lois, how would you use the principles of design discussed in this chapter to redesign this document? How could you make this document more accessible to the people who need to use it? How could you make it more attractive and interesting to its likely readers?

A Document Design That Could Be Improved

Figure A:
This document's design is rather minimal and perhaps less than effective for the readers who might use it. How could the design be improved?

Chapter 13
IPM for Scorpions in Schools

Introduction

Scorpions live in a wide variety of habitats from tropical to temperate climates and from deserts to rain forests. In the United States they are most common in the southern states from the Atlantic to the Pacific. All scorpions are beneficial because they are predators of insects.

The sting of most scorpions is less painful than a bee sting. There is only one scorpion of medical importance in the United States: the sculptured or bark scorpion, Centruroides exilicauda (=sculpturatus). The danger from its sting has been exaggerated, and its venom is probably not life-threatening. This species occurs in Texas, western New Mexico, Arizona, northern Mexico, and sometimes along the west bank of the Colorado River in California.

Identification and Biology

Scorpions range from 3/8 to 8 1/2 inches in length, but all scorpions are similar in general appearance.

Scorpions do not lay eggs; they are viviparous, which means they give birth to live young. As embryos, the young receive nourishment through a kind of "placental" connection to the mother's body. When the young are born, they climb onto the mother's back where they remain from two days to two weeks until they molt (shed their skin) for the first time. After the first molt, the young disperse to lead independent lives. Some scorpions mature in as little as six months while others take almost seven years.

All scorpions are predators, feeding on a variety of insects and spiders. Large scorpions also feed on small animals including snakes, lizards, and rodents. Some scorpions sit and wait for their meal to come to them while others actively hunt their prey. Scorpions have a very low metabolism and some can exist for 6 to 12 months without food. Most are active at night. They are shy creatures, aggressive only toward their prey. Scorpions will not sting humans unless handled, stepped on, or otherwise disturbed.

It is rare for scorpions to enter a building since there is little food and temperatures are too cool for their comfort. There are some exceptions to this rule. Buildings

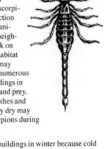

in new developments (less than three years old) can experience an influx of scorpions because the construction work has destroyed the animals' habitat. In older neighborhoods, the heavy bark on old trees provide good habitat for scorpions, and they may enter through the more numerous cracks and holes in buildings in search of water, mates, and prey. Also, buildings near washes and arroyos that are normally dry may become refuges for scorpions during summer rains.

Scorpions do not enter buildings in winter because cold weather makes them sluggish or immobile. They are not active until nighttime low temperatures exceed 70°F. Buildings heated to 65½ or 70°F provide enough warmth to allow scorpions to move about. Scorpions found inside buildings in cold weather are probably summer visitors that never left. Although scorpions prefer to live outdoors, they can remain in buildings without food for long periods of time.

Stings

A scorpion sting produces considerable pain around the site of the sting, but little swelling. For four to six hours, sensations of numbness and tingling develop in the region of the sting, then symptoms start to go away. In the vast majority of cases, the symptoms will subside within a few days without any treatment.

If the sting is from a bark scorpion, symptoms can sometimes travel along nerves, and tingling from a sting on a finger may be felt up to the elbow, or even the shoulder. Severe symptoms can include roving eyes, blurry vision, excessive salivation, tingling around the mouth and nose, and the feeling of having a lump in the throat. Respiratory distress may occur. Tapping the sting can produce extreme pain. Symptoms in children also include extreme restlessness, excessive muscle activity, rubbing at the face, and sometimes vomiting. Most vulnerable to the sting of the bark scorpion are children under five years and elderly persons who have

IPM for Schools   103   **Chapter 13 · Scorpions**

Source: U.S. Environmental Protection Agency, http://www.epa.gov/pesticides/ipm/schoolipm.

(continued)

Avoiding Scorpion Stings

Schools in areas where encounters with scorpions are likely should teach children and adults how to recognize scorpions and to understand their habits. Focus on scorpion biology, behavior, likely places to find them, and how to avoid disturbing them.

At home, children and parents should be taught to take the following precautions to reduce the likelihood of being stung:

- Wear shoes when walking outside at night. If scorpions are suspected indoors, wear shoes inside at night as well.

- Wear leather gloves when moving rocks, boards, and other debris.

- Shake out shoes or slippers before they are worn, and check beds before they are used.

- Shake out damp towels, washcloths, and dishrags before use.

- When camping, shake out sleeping bags, clothes, and anything else that has been in contact with the ground before use.

- Protect infants and children from scorpions at night by placing the legs of their cribs or beds into clean wide-mouthed glass jars and moving the crib or bed away from the wall. Scorpions cannot climb clean glass.

an underlying heart condition or respiratory illness. The greatest danger to a child is the possibility of choking on saliva and vomit.

Antivenin for the bark scorpion is produced at Arizona State University in Tempe and is available in Arizona but not in other states. The therapeutic use of antivenin is still experimental. People have been treated without antivenin for many years, and in areas where antivenin is unavailable, people are monitored closely by medical professionals until the symptoms subside.

DETECTION AND MONITORING

To determine where scorpions are entering, inspect both the inside and outside of the building at night (when scorpions are active) using a battery-operated camp light fitted with a UV (black) fluorescent bulb. Scorpions glow brightly in black light and can be spotted several yards away.

Always wear leather gloves when hunting scorpions. Places to check inside the building include under towels, washcloths, and sponges in bathrooms and kitchen; under all tables and desks, since the bark scorpion may climb and take refuge on a table leg or under the lip of a table; and inside storage areas. Outside, check piles of rocks and wood, under loose boards, and in piles of debris. After the following treatment strategies have been implemented, monitoring with the black light should continue to verify population reduction.

MANAGEMENT OPTIONS

Physical Controls

In most cases, physical controls will be adequate to manage a scorpion problem.

Removal of Scorpions

Any scorpions found during monitoring can be picked up using gloves or a pair of kitchen tongs, and transferred to a clean, wide-mouthed glass jar. Scorpions cannot climb clean glass. You can also invert a jar over a live scorpion and then slide a thin piece of cardboard under the mouth of the jar to trap the scorpion inside. Once a scorpion is captured, drop it into a jar of alcohol or soapy water (water without soap will not work) to kill it.

Habitat Modification

If you discover areas near school buildings that harbor a number of scorpions, you can try to alter the habitat to discourage them. Wood piles, rocks, loose boards, and other debris should be removed from the immediate vicinity of the building.

If there are slopes on the school grounds that are faced with rip rap (large rocks placed on a slope or stream bank to help stabilize it), or other similar areas highly attractive to scorpions, place a barrier of aluminum flashing between the rip rap and the school to prevent scorpion access. The flashing must be bent in an "L" shape away from the building. The other edge of the flashing should be buried a short distance from the rocks, deep enough in the soil so that the L shape will not fall over and lean on the rip rap. Make the height of the barrier before the bend greater than the height of the rip rap to prevent scorpions from standing on the rocks and jumping over the barrier.

Carry a caulking gun during nighttime inspections inside and outside the building to seal any cracks and crevices found. If scorpions are entering through weep holes in windows or sliding doors, cover the holes with fine-mesh aluminum screening, available from hardware or lumber stores. The ends of pipes that are designed as gray water drains should be fitted with loose filter bags, or makeshift end-pieces made from window screen. The screened end will prevent scorpion access to the drainpipe, sink, and other parts of the building.

It is important to continue nighttime patrols and caulking until all entryways have been located and sealed, and all the scorpions in the building have been captured and killed. Once the access routes are sealed, and all indoor scorpions have been removed, only doorways provide access, unless the scorpions ride in on logs and other materials. Glazed tiles can be placed around the perimeter of the buildings, and under or around doors and windows as part of the decor and as practical scorpion barriers. Scorpions have difficulty crossing smooth tiles unless the grout line is wide. Wood storage should be elevated above the ground since scorpions like contact with moist soil. Before bringing materials such as logs inside, bang them on a stone to dislodge any scorpions.

Traps

A simple trap made of damp gunny sacks laid down near the building in the evenings may be useful for monitoring and trapping. Scorpions may seek out the moist environment under the sacks where they can be collected in the morning. This trap is most effective when used before summer rains.

Chemical Controls

In general, preventing scorpion problems is better than trying to kill these creatures with pesticide. Spraying the perimeters of buildings is not only unnecessary but also ineffective. Scorpions can tolerate a great deal of pesticide in their environment. The pesticide will only be harmful to humans, especially children, and to other wildlife. Using physical controls along with education to reduce the fear of scorpions will help prevent unnecessary treatments.

First Aid for Scorpion Stings

Most scorpion stings are similar to a bee or wasp sting, and like bee or wasp stings, the majority of scorpion stings can be treated at school or the victim's home.

First aid for a scorpion sting includes the following:

- Calm the victim.
- Do not use a tourniquet.
- Wash the area with soap and water.
- Apply a cool compress (an ice cube wrapped in a wet washcloth), but do not apply ice directly to the skin or submerge the affected limb in ice water.
- To reduce pain, over-the-counter pain relievers such as aspirin or acetaminophen can help.
- Elevate or immobilize the affected limb if that feels more comfortable.
- Do not administer sedatives such as alcohol.

BIBLIOGRAPHY

Bio-Integral Resource Center (BIRC). 1996, 1997 directory of least-toxic pest control products. IPM Practitioner 18(11/12):1-39.

Cloudsley-Thompson, J.L. 1968. Spiders, Scorpions, Centipedes and Mites. Pergamon Press, New York. 278 pp.

Keegan, H.L. 1980. Scorpions of Medical Importance. Jackson University Press, Jackson, MS. 140 pp.

Mallis, A. 1982. Handbook of Pest Control. Franzak and Foster, Cleveland, OH. 1101 pp.

Olkowski, W., S. Daar, and H. Olkowski. 1991. Common-Sense Pest Control: Least-toxic solutions for your home, garden, pets and community. Taunton Press, Newtown, CT. 715 pp.

Polis, G.A., ed. 1990. The Biology of Scorpions. Stanford University Press, Stanford, CA. 587 pp.

Smith, R.L. 1982. Venomous animals of Arizona. Cooperative Extension Service, College of Agriculture, University of Arizona, Tucson, AZ. Bulletin 8245. 134 pp.

Smith, R.L. 1992. Personal communication. Associate Professor, Entomology Dept., University of Arizona at Tucson.

This document was produced for USEPA (Document #909-B-97-001) by the Bio-Integral Resource Center, P.O. Box 7414, Berkeley, CA 94707, March 1997.

CHAPTER

11

Creating and Using Graphics

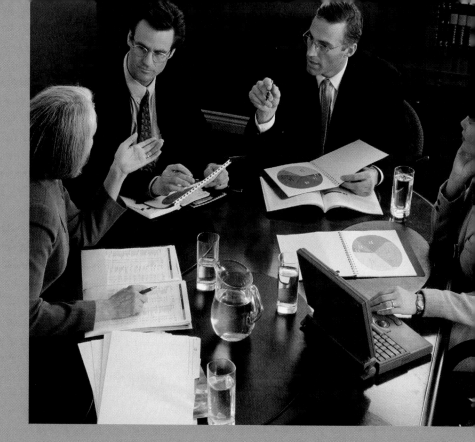

In this chapter, you will learn:

- The importance of visuals in documents and presentations.

- Four guidelines for using visuals effectively.

- How to use tables, charts, and graphs.

- Strategies for taking photographs and using them in documents and presentations.

- How drawings, icons, and clip art can be used effectively to enhance understanding.

- How to use video and audio in multimedia and Internet-based documents.

G raphics are an essential part of any technical document or presentation. Your readers will often pay more attention to the visuals in your document than to the written text. For example, think about how you began reading this chapter. More than likely, you did not begin reading at the top of this page. Instead, you probably took a quick glance at the graphics in the chapter to figure out what it is about. Then, you started reading the written text. Your readers will approach your documents the same way.

Guidelines for Using Graphics

As you draft your document, you should look for places where graphics could be used to support the text. Graphics are especially helpful in places where you want to reinforce important ideas or help your readers understand complex concepts or trends.

Graphics should be used to enhance and clarify your message—to slice through the details and numbers to make the information easier to process. In Figure 11.1,

Written Text vs. Table

In 2007, the state of Michigan reported 16 new cases of West Nile Virus (WNV) among humans. Four of those cases resulted in death. The most cases were in Wayne County, which saw 7 total cases with 5 males and 2 females contracting the virus. Wayne County had two deaths, half of the state's total for 2007. Kent, Macomb, and Oakland Counties each had two cases apiece. Kent County reported two females (ages 19–65) with WNV, Macomb County had two females (one 19–65 and one over 65), and Oakland County counted two males (19–65). The Macomb County victim over 65 died. Kalamazoo, Lapeer, and St. Clair Counties each had one reported case. Kalamazoo County's victim, which resulted in death, was a male over 65. Lapeer County reported one male (19–65), and St. Clair County reported one female (19–65).

Figure 11.1: Putting figures into a table makes them much easier to access. The table cannot completely replace the written text, but it can reinforce it by organizing the information more effectively.

2007 HUMAN WNV CASES

COUNTY	AGE (YEARS)	MALE	FEMALE	TOTAL CASES	DEATHS
Kalamazoo	0–18				
	19–65				
	>65	1		1	1
Kent	0–18				
	19–65		2	2	
	>65				
Lapeer	0–18				
	19–65	1		1	
	>65				
Macomb	0–18				
	19–65		1	1	
	>65		1	1	1
Oakland	0–18				
	19–65	2		2	
	>65				
St. Clair	0–18				
	19–65		1	1	
	>65				
Wayne	0–18	1		1	
	19–65	3	1	4	
	>65	1	1	2	2
TOTALS (includes probable and confirmed cases, all clinical syndromes)		9	7	16	4

Source: State of Michigan, http://www.michigan.gov/emerging diseases.

GO TO THE NET

To see more examples of text-to-table conversions, go to
www.pearsonhighered.com/johnsonweb3/11.1

for example, the written text and the table provide essentially the same information. And yet, the information in the table is much easier to access.

To help you create and use graphics effectively and properly, there are four guidelines you should commit to memory.

Guideline One: A Graphic Should Tell a Simple Story

A graphic should tell the "story" about your data in a concise way. In other words, your readers should be able to figure out in a quick glance what the graphic says. If your readers need to pause longer than a moment, there is a good chance they will not understand what the graphic means.

Figure 11.2, for example, shows how a graph can tell a simple story. Almost immediately, a reader will recognize that Southern Asia has by far the highest proportion of underweight children. Meanwhile, South-Eastern Asia and Eastern Asia have seen significant improvements.

This first guideline—tell a simple story—also applies to photographs in a document (Plotnik, 1982). At a glance, your readers should be able to figure out what story a photograph is telling. The photograph in Figure 11.3, for example, is not complex, but it tells a clear story about the markings on a Desert Checkerspot butterfly.

A Graph That Tells a Simple Story

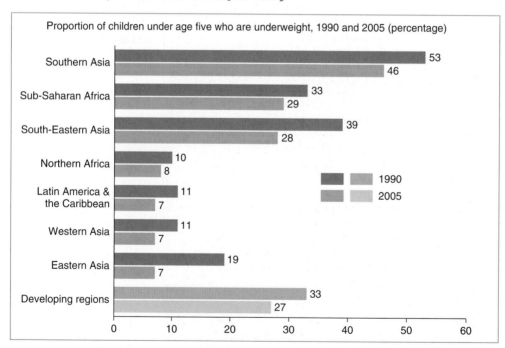

Figure 11.2: This graph tells a simple story about hunger that readers can grasp at a glance.

Source: United Nations, The Millennium Development Goals Report, 2007.

For websites that offer other guidelines for using graphics, go to
www.pearsonhighered.com/johnsonweb3/11.2

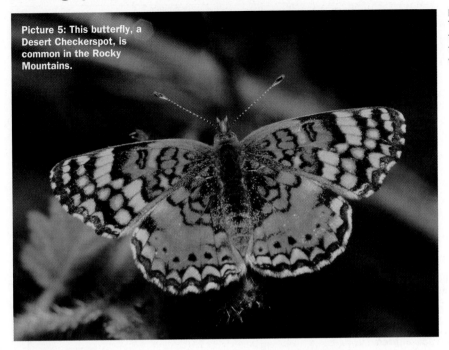

Picture 5: This butterfly, a Desert Checkerspot, is common in the Rocky Mountains.

Figure 11.3:
This photograph tells a simple story that reinforces the written text.

Source: Corel.

Guideline Two: A Graphic Should Reinforce the Written Text, Not Replace It

Graphics should be used to support the written text, but they cannot replace it altogether. Since technical documents often discuss complex ideas or relationships, it is tempting to simply refer the readers to a graphic (e.g., "See Chart 9 for an explanation of the data"). Chances are, though, that if you cannot explain something in writing, you won't be able to explain it in a graphic either.

Instead, your written text and visual text should work with each other. The written text should refer readers to the graphics, and the graphics should support the written information. For example, the written text might say, "As shown in Graph 2.3, the U.S. Federal Deficit has deepened dramatically in this decade." The graph would then support this written statement by illustrating this trend (Figure 11.4).

The written text should tell readers the story that the graphic is trying to illustrate. That way, readers are almost certain to understand what the graphic is showing them.

Guideline Three: A Graphic Should Be Ethical

Graphs, charts, tables, illustrations, and photographs should not be used to hide information, distort facts, or exaggerate trends. In a bar chart, for example, the scales can be altered to suggest that more growth has occurred than is actually the case (Figure 11.5).

A Graph That Reinforces the Written Text

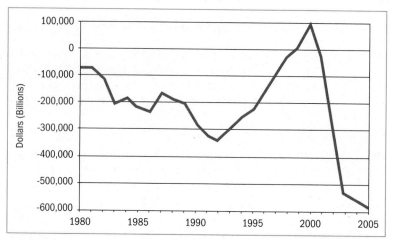

Source for data: U.S. Treasury Department, Financial Management Service, http://www.fms.treas. gov/mts.index.html.

Figure 11.4:
A line graph typically shows a trend over time. This graph shows the U.S. government's deficit trends.

Unethical and Ethical Bar Charts

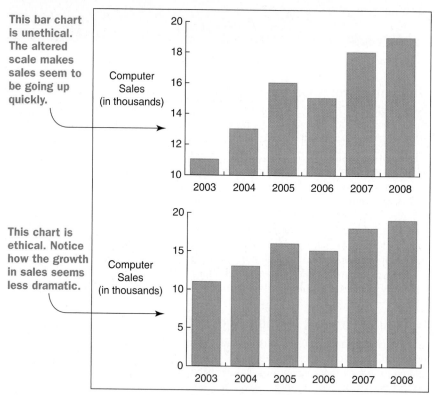

This bar chart is unethical. The altered scale makes sales seem to be going up quickly.

This chart is ethical. Notice how the growth in sales seems less dramatic.

Figure 11.5:
The top bar chart is unethical because the y-axis has been altered to exaggerate the growth in sales of computers. The second bar chart presents the data ethically.

In a line graph, it is tempting to leave out data points that won't allow a smooth line to be drawn. Likewise, with computers, photographs can be distorted or doctored.

A good rule of thumb with graphics—and a safe principle to follow in technical communication altogether—is to always be absolutely honest with the readers. Your readers are not fools, so attempts to use graphics to distort or stretch the truth will eventually be detected. Once detected, unethical graphics can erode the credibility of an entire document or presentation (Kostelnick & Roberts, 1998). Even if your readers only *suspect* deception in your graphics, they will begin to doubt the honesty of the whole text.

Link

For more information on the ethical use of data, see Chapter 5, page 108.

Guideline Four: A Graphic Should Be Labeled and Placed Properly

Proper labeling and placement of graphics help readers move back and forth between the print and visual features. Each graphic should be labeled with an informative title (Figure 11.6). Other parts of the graphic should also be carefully labeled:

- The x- and y-axes of graphs and charts should display standard units of measurement.
- Columns and rows in tables should be labeled so the readers can easily locate specific data points.
- Important features of drawings or illustrations should be identified with arrows or lines and some explanatory text.
- The source of the data used to make the graphic should be clearly identified underneath.

Labeling of a Graphic

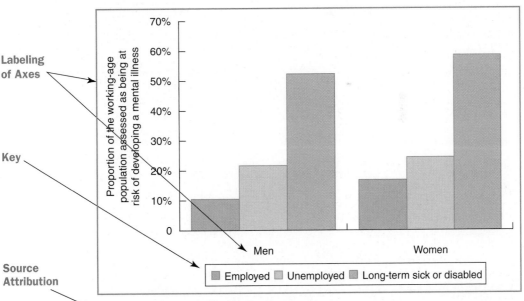

Labeling of Axes

Key

Source Attribution

Figure 11.6: Good labeling of a graphic is important so that readers can understand it.

Source: British Household Panel Survey, University of Essex, Institute for Social and Economic Research; the data is the average for the five years to 2005/06; updated June 2007. New Policy Institute, http://www.poverty.org.uk.

Want to see other unethical graphics? Go to **www.pearsonhighered.com/johnsonweb3/11.4**

Guidelines for Using Graphics 279

AT A GLANCE

Guidelines for Using Graphics

- A graphic should tell a simple story.
- A graphic should reinforce the written text, not replace it.
- A graphic should be ethical.
- A graphic should be labeled and placed properly.

Assuming you include a title with the graph, an explanatory caption is not needed. Nevertheless, a sentence or two of explanation in a caption can often help reinforce or clarify the story the graphic is trying to tell.

When placing a graphic, put it on the page where it is referenced or, at the farthest, put it on the following page. Your readers will rarely flip more than one page to look for a graphic. Even if they *do* make the effort to hunt down a graphic that is pages away, the effort will take them out of the flow of the document, inviting them to start skimming.

Readers should be able to locate a graphic with a quick glance. Then, they should be able to quickly return to the written text to continue reading. When labeled and placed properly, graphics work seamlessly into the flow of the whole text.

Link

For more information on designing page layouts, see Chapter 10, pages 232–262.

Displaying Data with Graphs, Tables, and Charts

To decide which graphic is best for the data you want to display, first decide what story you want to tell. Then, choose the type of graphic that best fits that story. The chart in Figure 11.7 will help you decide which one works best.

Choosing the Appropriate Graphic

The Story to Be Told	Best Graphic	How Data Are Displayed
"I want to show a trend."	Line graph	Shows how a quantity rises and falls, usually over time
"I want to compare two or more quantities."	Bar chart	Shows comparisons among different items or the same items over time
"I need to present data or facts for analysis and comparison."	Table	Displays data in an organized, easy-access way
"I need to show how a whole is divided into parts."	Pie chart	Shows data as a pie carved into slices
"I need to show how things, people, or steps are linked together."	Flowchart	Illustrates the connections among people, parts, or steps
"I need to show how a project will meet its goals over time."	Gantt chart	Displays a project schedule, highlighting the phases of the work

Figure 11.7: Different kinds of graphics tell different stories. Think about what story you want to tell. Then, locate the appropriate graph, table, or chart for that story.

For more information on labeling and placing graphs, go to
www.pearsonhighered.com/johnsonweb3/11.5

Line Graphs

Line graphs are perhaps the most familiar way to display data. They are best used to show measurements over time. Some of their more common applications include the following:

> **Showing trends**—Line graphs are especially good at showing how quantities rise and fall over time (Figure 11.8). Whether you are illustrating trends in the stock market or charting the changes in temperature during a chemical reaction, a line graph can show how the quantity gradually increases or decreases. When two or more lines are charted on a line graph, you can show how quantities rise and fall in tandem (or don't).

A Line Graph Showing a Trend

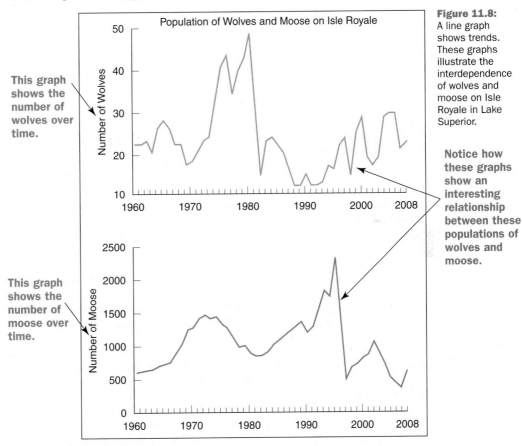

This graph shows the number of wolves over time.

This graph shows the number of moose over time.

Figure 11.8: A line graph shows trends. These graphs illustrate the interdependence of wolves and moose on Isle Royale in Lake Superior.

Notice how these graphs show an interesting relationship between these populations of wolves and moose.

Data Source: http://www.isleroyalewolf.org.

Showing relationships between variables—Line graphs are also helpful when charting the interaction of two different variables. Figure 11.9, for example, shows a line graph that illustrates how a rise in the temperature of a gas is accompanied by a rise in the volume of gas.

A Line Graph Showing a Relationship Between Variables

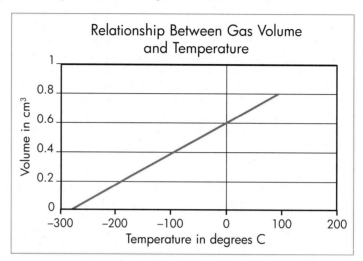

Figure 11.9: Here, the volume of a gas is plotted against the temperature. In this case, an extrapolation of the line allows us to estimate "absolute zero," the temperature at which all molecular activity stops.

Source: The Safetyline Institute, http://www.safetyline.wa.gov.au/institute/level2/course16/lecture47/147_02.asp.

In a line graph, the vertical axis (y-axis) displays a measured quantity such as sales, temperature, production, growth, etc. The horizontal axis (x-axis) is usually divided into time increments such as years, months, days, or hours. As shown in Figure 11.8, in a line graph, the x- and y-axes do not need to start at zero. Often, by starting one or both axes at a nonzero number, you can better illustrate the trends you are trying to show.

The x-axis in a line graph usually represents the "independent variable," which has a consistently measurable value. For example, in most cases, time marches forward steadily, independent of other variables. So, time is often measured on the x-axis. The y-axis often represents the "dependent variable." The value of this variable fluctuates over time.

You can use more than one line to illustrate trends in a line graph. Depending on your printer, computers also give you the ability to use colors to distinguish the lines. Or, you can use dashes, dots, and solid lines to help your readers distinguish one line from the others.

The drawback of line graphs is their inability to present data in exact numbers. For example, in Figure 11.8, can you tell exactly how many wolves were counted in 2001? Line graphs are most effective when the trend you are showing is more significant than the exact figures.

GO TO THE NET

For more examples of line graphs, go to
www.pearsonhighered.com/johnsonweb3/11.7

Teresa Lynch, M.D.

RESIDENT PHYSICIAN OF INTERNAL MEDICINE AND PEDIATRICS,
RUSH UNIVERSITY MEDICAL CENTER, CHICAGO, ILLINOIS

Rush University Medical Center is an important research hospital that specializes in urban needs.

AT WORK

How do you use graphics in your everyday work?

Over the course of my training as a resident physician and even in undergraduate and medical school, I have used graphics to supplement and amplify my written communication.

Probably the most important area in which I use graphics is in the presentation of research data. Most data collected in the lab or in clinical studies get written up as a document and submitted for publication in a scientific or medical journal. Because these journals have very strict criteria regarding document length and content, a well-placed graphic and short explanatory paragraph can substitute for monotonous lists of research data that, in addition to occupying precious space, can be both boring to read and difficult to interpret.

Furthermore, graphics often determine who will actually read your article. Most people will scan the graphics first to determine whether or not they have any interest in the subject matter. If there are no graphics, there is a good chance they won't read the text. Well-placed graphics, on the other hand, will gather an audience for your research.

I also use graphics in my day-to-day work as a physician. On one occasion, I may create an algorithm, or diagram, to explain the proper treatment of a disease process to the medical students I work with. Later that same day, I may create a line graph to show the trend in a person's disease state as I contemplate the next step in his/her treatment. In all of these instances, graphics provide a useful visual summary of the information I need to convey to the various people with whom I work.

Bar Charts

Bar charts are used to show quantities, allowing readers to make visual comparisons among measurements. The width of the bars is kept the same, while the length of the bars varies to represent the quantity measured. Like line graphs, bar charts can be used to show trends over time. They are also useful for showing increases or decreases in volume (Figure 11.10).

There are a wide variety of ways to use bar charts. A bar chart can be turned on its side, making the bars run horizontally (see Figure 11.2). Multiple bars can also be grouped together, allowing you to show how different measurements compare (Figure 11.10).

Computers can be used to enhance bar charts even further. Coloring and shading the bars will enhance the readers' ability to interpret the data and identify trends.

Tables

Tables provide the most efficient way to display data or facts in a small amount of space. In a table, information is placed in horizontal rows and vertical

A Bar Chart

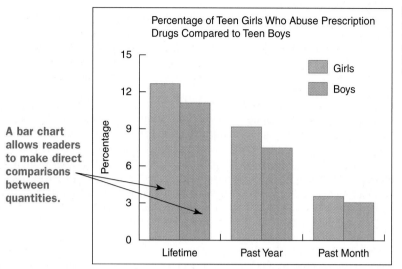

Percentage of Teen Girls Who Abuse Prescription Drugs Compared to Teen Boys

A bar chart allows readers to make direct comparisons between quantities.

Figure 11.10: A bar chart is especially effective for showing volumes.

Source: Substance Abuse and Mental Health Services Administration, 2005 National Survey on Drug Use and Health.

Inserting a Table

Identify how many columns and rows are needed.

In some word processors, the computer can format the table automatically for you.

Figure 11.11: The Table function in your word processor allows you to make a simple table.

To see more sample bar charts, go to **www.pearsonhighered.com/johnsonweb3/11.8**

columns, allowing readers to quickly find specific numbers or words that address their needs.

Creating a table takes careful planning, but computers can do much of the hard work for you. For simpler tables, you can use the Table function on your word-processing software (Figure 11.11). It will allow you to specify how many rows and columns you need (make sure you include enough columns and rows for headings in the table). Then, you can start typing your data or information into the cells.

If the Table function in your word processor is not enough for your needs, spreadsheet programs like Microsoft Excel and Corel Quattro Pro also allow you to make quick tables (see the Help box in this chapter).

After creating the basic table, you should properly label it. In most cases, the table's number and title should appear above it (Figure 11.12). Down the left column, the *row headings* should list the items being measured. Along the top row, the *column headings* should list the qualities of the items being measured. Beneath the table, if needed, a citation should identify the source of the information.

Parts of a Table

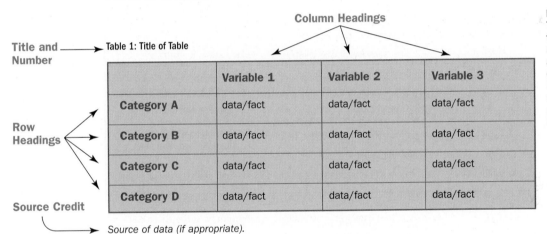

Figure 11.12: The parts of a table are rather standard. Rows and columns cross each other in ways that allow readers to locate specific pieces of information.

In some cases, tables can be used to present verbal information rather than numerical data. In Figure 11.13, for example, the table is being used to verbally provide health information. With this table, readers can quickly locate their age and find the cancer test they need.

When adding a table to your document, think about what your readers need to know. It is often tempting to include large tables that hold all your data. However, these large tables might clog up your document, and readers will find it hard to locate specific information. You are better off creating small tables that focus on the specific information you want to present. Move larger tables to an appendix, especially if they present data not directly referenced in the document.

A Table That Presents Verbal Information

CANCER DETECTION			
TEST OR PROCEDURE			
Age	**Frequency**	**Females**	**Males**
18–20	One time	Complete Health Exam	Complete Health Exam
	Yearly	Pap smear	
	Monthly	Skin self-exam, Breast self-exam	Skin self-exam
20–40	Every 3 years	Complete Health Exam, Clinical breast exam, Pelvic exam	Complete Health Exam
	Yearly	Pap smear	
	Monthly	Skin self-exam, Breast self-exam	Skin self-exam, Testis self-exam
40–50	Every 3 years	Complete Health Exam	Complete Health Exam, Prostate specific antigen (PSA) blood test
	Yearly	Cinical breast exam, Mammogram, Endometrial biopsy, Pap smear, Pelvic exam, Digital rectal exam, Stool blood test	Digital rectal exam, Stool blood test
	Monthly	Skin self-exam, Breast self-exam	Skin self-exam, Testis self-exam
50–65	Every 5–10 years	Colonoscopy, Procto, Double-contrast barium enema (DCBE)	Colonoscopy, Procto, Double-contrast barium enema (DCBE)
	Yearly	Complete Health Exam, Cinical breast exam, Endometrial biopsy, Mammogram, Pap smear, Pelvic exam, Digital rectal exam, Stool blood test	Complete Health Exam, Prostate specific antigen (PSA) blood test, Digital rectal exam, Stool blood test
	Monthly	Skin self-exam, Breast self-exam	Skin self-exam, Testis self-exam
65+	Every 5–10 years	Colonoscopy, Procto, Double-contrast barium enema (DCBE)	Colonoscopy, Procto, Double-contrast barium enema (DCBE)
	Yearly	Complete Health Exam, Cinical breast exam, Mammogram, Endometrial biopsy, Pap smear, Pelvic exam, Digital rectal exam, Stool blood test	Complete Health Exam, Prostate specific antigen (PSA) blood test, Digital rectal exam, Stool blood test
	Monthly	Skin self-exam, Breast self-exam	Skin self-exam, Testis self-exam

Source: National Foundation for Cancer Research.

Figure 11.13: Tables can also present verbal information concisely. In this table, a great amount of information is offered in an easy-to-access format.

GO TO THE NET

Need more help making a table? Go to
www.pearsonhighered.com/johnsonweb3/11.9

Pie Charts

Pie charts are useful for showing how a whole divides into parts (Figure 11.14). Pie charts are popular, but they should be used sparingly. They take up a great amount of space in a document while usually presenting only a small amount of data. The pie chart in Figure 11.14, for instance, uses a third of a page to plot a mere eleven data points.

A Pie Chart

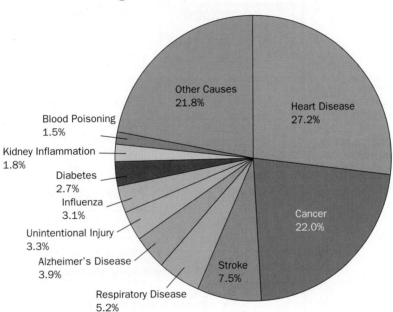

Ten Leading Causes of Death Among Women

Figure 11.14: A pie chart is best for showing how a whole can be divided into parts.

Source: U.S. Department of Health and Human Services, Women's Health *2007.*

Pie charts are difficult to construct by hand, but your computer's spreadsheet program (Excel or Quattro Pro) can help you create a basic pie chart of your data. When labeling a pie chart, you should try to place titles and specific numbers in or near the graphic. For instance, in Figure 11.14, each slice of the pie chart is labeled and includes a measurement to show how the pie was divided. These labels and measurements help readers compare the data points plotted in the chart.

The key to a good pie chart is a clear story. For example, what story is the pie chart in Figure 11.14 trying to tell? Heart disease and cancer are the most significant causes of death among women.

Flowcharts

Flowcharts are used to visually guide readers through a series of decisions, actions, or steps. They typically illustrate a process described in the written text.

Do you want to see other pie charts? Go to
www.pearsonhighered.com/johnsonweb3/11.10

Displaying Data with Graphs, Tables, and Charts 287

A Flowchart

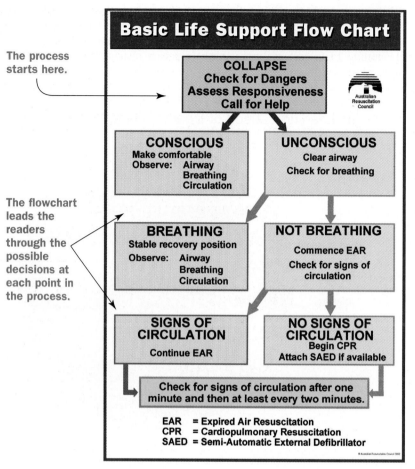

The process starts here.

The flowchart leads the readers through the possible decisions at each point in the process.

Basic Life Support Flow Chart

COLLAPSE
Check for Dangers
Assess Responsiveness
Call for Help

Australian Resuscitation Council

CONSCIOUS
Make comfortable
Observe: Airway
Breathing
Circulation

UNCONSCIOUS
Clear airway
Check for breathing

BREATHING
Stable recovery position
Observe: Airway
Breathing
Circulation

NOT BREATHING
Commence EAR
Check for signs of circulation

SIGNS OF CIRCULATION
Continue EAR

NO SIGNS OF CIRCULATION
Begin CPR
Attach SAED if available

Check for signs of circulation after one minute and then at least every two minutes.

EAR = Expired Air Resuscitation
CPR = Cardiopulmonary Resuscitation
SAED = Semi-Automatic External Defibrillator

Source: Australian Resuscitation Council, http://www.resus.org.au/public/bls_flow_chart.pdf.

Figure 11.15: A flowchart is often useful for illustrating a process.

Link

For more information on writing instructions, go to Chapter 20, pages 569–574.

Arrows are used to connect parts of the flowchart, showing the direction of the process.

As shown in Figure 11.15, flowcharts are helpful for illustrating instructions, especially when judgment calls need to be made by the user of the instructions. A flowchart typically cannot replace written instructions, especially if the steps are complex. But, it can illustrate the steps in the process to help readers understand the written text.

Flowcharts can be found in a variety of other forms, such as organization charts or circuit diagrams. An organization chart illustrates the hierarchy of decision making in an organization. In a circuit diagram, meanwhile, a flowchart is used to chart the path of electricity.

GO TO THE NET

Other flowcharts can be viewed at
www.pearsonhighered.com/johnsonweb3/11.11

Gantt Charts

Gantt charts have become quite popular in technical documents, especially proposals and progress reports. Gantt charts, like the one in Figure 11.16, are used to illustrate a project schedule, showing when the phases of the project should begin and end.

Visually, these charts illustrate the interrelations among different aspects of a large project. That way, people who are working on one part of the project will know what other teams are doing. Another benefit to a Gantt chart is that it gives readers an overall sense of how the project will proceed from beginning to end.

Gantt charts are becoming increasingly simple to create, because project-planning software like ArrantSoft, Artemis Project Management, and Microsoft Project can easily generate them for use in technical documents.

Link

For more help with planning a project timeline, go to Chapter 4, page 70.

A Gantt Chart

These lines show the progress of the project.

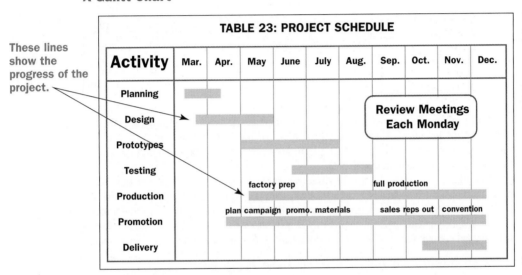

Figure 11.16: A Gantt chart is often used to show how the stages in a project interrelate. In this chart, you can easily see how the promotion of the product is started, even before the product is fully tested.

Using Pictures, Drawings, and Screen Shots

Increasingly, computers give you the ability to include pictures, drawings, and video in documents. Even if you are not artistic, you can quickly use a digital camera, a drawing program, a scanner, or a video camera to add life to your documents.

The purpose of a picture, drawing, or video is to show what something looks like. These kinds of visuals are especially helpful when your readers may not be familiar with something, like an animal or a piece of equipment. They are also helpful for showing the condition of something, like a building under construction or damage to a car.

Photographs

Digital cameras and scanners are making the placement of photographs in technical documents easier than ever. A good first step is to ask what *story* you want the photograph to tell. Then, set up a shot that tells that story.

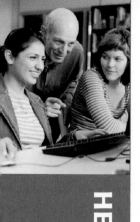

Making Visuals with a Spreadsheet Program

Most word-processing programs, like Word or WordPerfect, come bundled with a companion spreadsheet program, like Excel or Quattro Pro. These spreadsheet programs allow you to make quick graphs and charts from a data set.

You can then insert these charts and graphs directly into your document. Overall, making and inserting graphics is not too difficult, because your word-processing program and spreadsheet program are designed to be compatible. With a few clicks of the mouse, you can copy and paste the graph you made with the spreadsheet program directly into your document.

To make a chart or graph, open a spreadsheet and enter the data you would like graphed (see Figure A).

Data in a Spreadsheet

HELP

Data points are entered here.

Labels are added in the spreadsheet.

	A	B	C	D	E	F	G	H
		1940	1950	1960	1970	1980	1990	2000
1								
2	La Plata County	17.9	20.5	22.8	24.5	15.2	13.3	12.8
3	Montezuma County	18.8	19.2	21.6	23.4	16.1	13.6	11.7
4	Archuleta County	19.9	20.3	23.6	22.1	14	12.7	9.2
5								
6								

Average Rainfall

Figure A:
To make a table, chart, or graph in a spreadsheet, begin by entering the data into the "worksheet."

In the toolbar, which is usually at the top of the page, the spreadsheet program will have a button that allows you to look at the variety of graphs and charts available. You might even be able to preview different graphs and charts to see which one best illustrates the story you want the data to tell.

Once you find the type of graph or chart that works for your data, let the spreadsheet program graph the data (see Figure B). At this point, you will be able to make adjustments to the labeling on the graph. Don't forget to title the graph and label the x-axis and y-axis.

The graphs you can make in a spreadsheet program are usually rather plain, but they will be good enough for most workplace documents. They are also quick and easy to make.

A Graph Generated with a Spreadsheet Program

Here, the graphic is inserted into the spreadsheet.

Figure B:
The spreadsheet will generate the visual for you. This visual can then be pasted into a document.

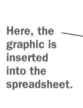

Need help using your spreadsheet program to make graphs? Go to
www.pearsonhighered.com/johnsonweb3/11.13

PHOTOGRAPHING PEOPLE   If you need to include a picture of a person or a group of people standing still, take them outside and photograph them against a simple but scenic background. Photographs taken in the office tend to look dark, depressing, and dreary. Photographs taken outdoors, on the other hand, imply a sense of openness and freethinking. When photographing people working, a good strategy is to show people doing what they *actually* do (Figure 11.17).

A Photograph of a Person in Action

Source: Corel.

If you need to photograph people inside, put as much light as possible on the subjects. If your subjects will allow it, use facial powder to reduce the glare off their cheeks, noses, and foreheads. Then, take their picture against a simple backdrop to reduce background clutter.

If you are photographing an individual, take a picture of his or her head and shoulders. People tend to look uncomfortable in full-body pictures.

One general photography guideline that works well in most situations, especially when photographing people, is the "Rule of Thirds." The Rule of Thirds means the focal point of a picture (e.g. a subject's eyes, the key feature of an object) will appear where the top third of the picture begins. For example, in Figure 11.17, the welder's goggles, which are the focal point of this picture, appear where the top third of the picture meets the middle third. Similarly, in Figure 11.18, notice how the focal point of the pottery (the design and bulge) is where the top third of the picture starts.

PHOTOGRAPHING OBJECTS   When taking pictures of objects, try to capture a close-up shot while minimizing any clutter in the background (Figure 11.18). It is often a good idea to put a white drop cloth behind the object to block out the other items and people in the background. Make sure you put as much lighting as possible on the object so it will show up clearly in your document.

A Photograph of an Object

Note the plain background behind the subject of the photograph.

Figure 11.18: When photographing objects, try to reduce the amount of clutter around your subject.

Source: The Internet Public Library, http://www.ipl.si.umich.edu/div/pottery/image15.htm.

When photographing machines or equipment, try to capture them close-up and in action. After all, a picture of equipment sitting idle on the factory floor is rather boring. But if you show the machine being used or focus on the moving parts, you will have a much more dynamic picture.

PHOTOGRAPHING PLACES   Places are especially difficult to photograph. When you are at the place itself, snapping a picture seems simple enough. But, the pictures often come out flat and uninteresting. Moreover, unless people are in the picture, it is often difficult to tell the scale of the place being photographed.

When photographing places, focus on people doing something in that place. For example, if you need to photograph a factory floor, you should show people doing their jobs. If you are photographing an archaeological site, include someone working on the site. The addition of people will add a sense of action and scale to your photograph.

Inserting Photographs and Other Images

A digital camera will usually allow you to save your photographs in a variety of memory sizes. High-resolution photographs (lots of pixels) require a lot of memory in the camera and in your computer. They are usually saved in a format called a .tiff file. Lower-resolution photographs (fewer pixels) are saved as .gif or .jpg files. Usually, .gif and .jpg files are fine for print and online documents. However, if the photograph needs to be of high quality, a .tiff file might be the best choice.

Once you have downloaded an image to your computer, you can work with it using software programs like Microsoft Paint or Adobe Photoshop (Figure 11.19). These programs will allow you to touch up the photographs or, if you want, completely alter them.

GO TO THE NET

To learn more tips about photography, go to
www.pearsonhighered.com/johnsonweb3/11.14

When you have finished touching up or altering the image, you can then insert it into your document or presentation. Most word-processing programs have an Insert Picture command. To insert the picture, put your cursor where you want the image to appear in the document. Then, select "Insert Picture." A box will open that allows you to locate the image on your computer's hard drive. Find and select the image you want to insert.

At this point, your computer will insert the image into your document. Usually, you can then do a few simple alterations to the file, like cropping, with the Picture toolbar in your word processor.

Working with Images

Here are tools for altering the images.

Colors can be added or altered with this tool.

Figure 11.19: Software such as Adobe Photoshop allows you to work with photographs and other kinds of images.

Illustrations

Illustrations are often better than photographs at depicting buildings, equipment, maps, and schematic designs. Whereas photographs usually include more detail than needed, a good illustration highlights only the most important features of the subject.

LINE DRAWINGS AND DIAGRAMS   A line drawing or diagram is a semirealistic illustration of the subject being described. You can create simple drawings and diagrams with the Draw function of most word-processing programs. As the drawings grow more complex, however, most writers will hire professional artists to transform rough sketches into finished artwork.

Line drawings offer several advantages. They can provide a close-up view of important features or parts. They can also be easily labeled, allowing you to point out important features to the readers.

In some ways, however, drawings and diagrams are less than realistic. For example, the diagram of the rabies virus in Figure 11.20 does not look exactly like the actual virus. Instead, it shows only how the larger parts of the virus are interconnected and work together.

A Diagram

Labels are added to identify features.

Explanatory text can be added to clarify the meaning of the diagram.

Envelope (Membrane)   Matrix Protein   Glycoprotein

Ribonucleoprotein

Rabies virions are bullet-shaped with 10-nm spikelike glycoprotein peplomers covering the surface. The ribonucleoprotein is composed of RNA encased in nucleoprotein.

Figure 11.20: A drawing is only partially realistic. It concentrates on relationships instead of showing exactly what the subject looks like.

Source: Centers for Disease Control, http://www.cdc.gov/rabies/virus.htm.

MAPS   Maps offer a view of the subject from above. You can use them to show geographic features like streets, buildings, or rivers. They can be used to portray the rooms in a building or illustrate a research site (Figure 11.21). In some cases, they can be used to show the readers where a particular event occurred, allowing them to see where a particular place fits into an overall geographic area. When including a map, zoom in on the area that is being discussed in the text.

Maps may be easier to generate than you think. Today, Internet sites like Mapquest (www.mapquest.com), Google Maps (maps.google.com), and Topozone

A Map

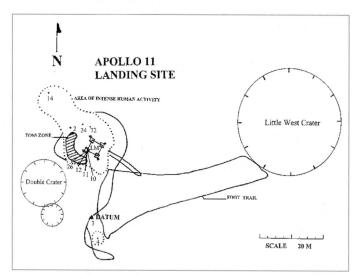

N

APOLLO 11 LANDING SITE

14   AREA OF INTENSE HUMAN ACTIVITY

Little West Crater

TOSS ZONE

2 24 72

LM

26 12 11
10

Double Crater

FOOT TRAIL

DATUM
3

SCALE   20 M

Figure 11.21: A map is useful for showing a place from above.

Source: New Mexico State University, http://www.spacegrant.nmsu.edu/lunarlegacies/images/scan2.gif.

To see examples of line drawings, go to
www.pearsonhighered.com/johnsonweb3/11.16

(www.topozone.com) allow you to create very detailed maps. Before trying to draw your own map, you might let these Internet sites do it for you.

Link

For more information on copyright law, go to Chapter 5, page 111.

ICONS AND CLIP ART   Icons play an important role in technical documentation. In some documents, they are used as warning symbols. They can also serve as signposts in a text to help readers quickly locate important information (Figure 11.22). If you need to use an icon, standard sets of symbols are available on the Internet for purchase or for free.

Clip art drawings are commercially produced illustrations that can be purchased or used for free. Usually, when you purchase a collection of clip art, you are also purchasing the rights to use that clip art in your own documents.

It is tempting to advise you not to use clip art at all. When desktop publishing first came into the workplace, clip art was an original way to enhance the message and tone of a document. But now, most readers are tired of those little pictures of people shaking hands, pointing at whiteboards, and climbing ladders. In some cases, clip art becomes decorative fluff that takes readers' attention away from the document's message. Use it sparingly and only when it *truly* contributes to your message.

Common Icons

Figure 11.22: Icons are widely available on the Internet. The person in the middle is supposed to be sneezing, but, as with many icons, it could convey an unintentional meaning.

Sources: Centers for Disease Control, http://www.cdc.gov/diabetes/pubs/images/balance.gif, and International Association for Food Protection, http://www.foodprotection.org.

Screen Shots

With a combination of keystrokes, most computers will allow you to make a *screen shot* of your computer screen. Essentially, a screen shot is a picture of whatever is on the screen (Figure 11.23). You can use screen shots to insert a variety of different images into your document or presentation.

You can make screen shots quickly with a PC or a Mac. With a PC, press the Print Screen button on your keyboard. The computer will put an image of the screen on your "clipboard." Look for it there. With a Mac, press the three keys Apple-Shift-3 at the same time. You should hear a camera clicking sound. The picture will then appear on your screen's desktop.

You can insert the screen shot into your document with the Insert Picture command in your word processor. Using your Drawing tools, you can then "crop" (trim) the image to remove any extra details.

A Screen Shot

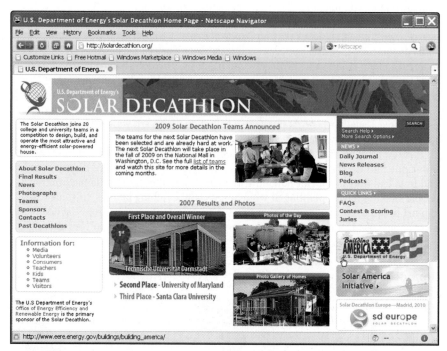

Figure 11.23: Screen shots are a simple way to add images to your document or presentation.

Source: Office of Energy Efficiency and Renewable Energy, http://www.eere.energy.gov/ solar_decathlon.

Screen shots are useful in a variety of situations. For example, you can use a screen shot to put a picture of a webpage in a printed document. Or, after drawing an illustration, you can make a screen shot of it and add it to a webpage.

Using Cross-Cultural Symbols

Symbols often translate among cultures better than words. They can also be more memorable and enhance comprehension for second-language readers (Horton, 683).

Symbols, however, don't always translate exactly across cultures, so you need to check your use of symbols in documents and websites with readers from other cultures. Otherwise, your symbols might lead to unintended consequences. For example, international dockworkers have been known to roughly toss boxes labeled with the broken wine glass symbol (meaning "fragile"), because they assumed the boxes contained broken glass.

In another case, the "Mr. Yuk" poison symbol has had mixed results in its bid to replace the traditional skull-and-crossbones symbol (Figure 11.24). One problem is that "Yuk" is a common name in Asia, especially Korea. Meanwhile, the use of "Mr." suggests elder status to many Asian children, implying the face deserves added respect.

In a research study, a majority of international children did not understand the Mr. Yuk image or see it as negative, and a few thought the symbol meant the product was good to

Mr. Yuk Versus the Skull-and-Crossbones Symbol

Figure 11.24: The poison symbol, Mr. Yuk, was intended to avoid problems with the traditional skull-and-crossbones. But, the symbol has its own problems when it crosses cultures. The skull-and-crossbones shown here is from the European Union's standardized set of symbols.

eat (Jackson-Smith). In North America, some health organizations now prohibit the use of Mr. Yuk because of these kinds of problems. The symbol was so heavily promoted that it now has a friendly undertone for children, attracting them to dangerous products.

To avoid misunderstandings, designers have developed symbols that are intended to cross cultures. The American Institute of Graphic Arts (AIGA) created the symbol system that is familiar to North Americans and used globally (Figure 11.25). The European Union and International Standards Organization (ISO) have also created sets of international symbols that are widely used.

International Symbols

Figure 11.25: The AIGA, European Union, and International Standards Organization (ISO) have created a set of symbols that are able to work internationally.

Here are a few helpful guidelines for using symbols cross-culturally:

Keep human icons simple—Icons of humans should not be more than simple pictographs (Figure 11.26). Distinctive clothing or facial features could lead to unintended interpretations or confusion. Smiles, frowns, winks, or smirks can have very different meanings across cultures, so symbols that use faces are particularly problematic.

For websites that offer free icons and clip art, go to
www.pearsonhighered.com/johnsonweb3/11.18

People in Symbols

Figure 11.26: Simple pictographs are often used for human icons that need to cross cultures. These are examples of icons used at the Olympics.

Use hand signals carefully—Just about any hand signal is considered offensive in some culture, including the thumbs-up signal, "OK" sign, V-symbol, a pointing finger, and even the palm out "halt" signal. If you can imagine an entire user's manual that uses an extended middle finger to point to things, you will get the idea about why hand signals can be problematic.

Avoid culture-specific icons—Mailboxes, phonebooths, and eating utensils, among other items, can look very different in other cultures, so symbols representing them might not translate. The typical North American mailbox on a street corner, for example, looks nothing like the canister mailboxes in England, while some cultures don't have public mailboxes at all. In another case, much of the world uses chopsticks for eating, so a fork would not properly symbolize "eat" or "food" to many readers.

Avoid religious symbols—Using crosses, crescents, stars, wings, candles, yin and yang, and other religious symbols can be interpreted very differently in other cultures. The symbol for the Red Cross, for example, is the Red Crescent in Islamic cultures, and the Red Crystal is used in Israel.

Avoid animal symbols and mascots—Animals can mean very different things in other cultures. In Western societies, the owl symbolizes wisdom, but in Southeast Asia, owls are considered unintelligent and vicious. Rats are considered clever and intelligent in many Asian countries, while in Western countries they are thought to be diseased and threatening. In some Islamic cultures, dogs are considered "unclean," making them particularly bad cartoon mascots for products. Meanwhile, the word *mouse* is not associated with computers in some cultures, so using a mouse symbol to represent a computer's pointing device would be confusing.

Link

For more information on cross-cultural readers, go to Chapter 3, page 53.

Symbols can be very helpful in technical documents because they enhance translation and comprehension. Your best approach is to use internationally accepted symbols whenever they are available and to always check your use of symbols with likely cross-cultural readers.

Using Video and Audio

One of the options available with computers is the use of video and audio in multimedia or Internet-based documents. With a digital video or audio recorder, you can add movies, music, and sound to any on-screen document.

Video

The ability to add video to a text is a mixed blessing. On the one hand, nothing will grab readers' attention like a video presentation. A video, for example, might show an important experiment or a current event. It might show an expert talking about a product or service (Figure 11.27). On the other hand, poorly made videos seem amateurish and quickly become tedious to watch. Another problem is that videos can take a long time to download.

If you want to add a video to your document, first decide exactly what you want the video to do. Then, write a script that achieves that goal in the least amount of time, because long videos take up lots of memory (and they can get a little boring).

As you create a video, think of it as a moving photograph in your document. Many of the same guidelines for shooting photographs will work with videos. For example, try to reduce background clutter so the viewers can focus on the subject. Also, put as much light on the subject as possible. And, if you are filming people, you might want to put some powder on their foreheads, noses, and cheeks to reduce glare.

After you shoot the video, you can use video editing software like Apple iMovie, Adobe Premiere, and Microsoft Movie Maker to edit the final piece into a concise format.

Using a Video

Keep the background in a video simple.

Videos should be used to make short, simple points.

Figure 11.27: Video editing software can help you make videos to be inserted into multimedia documents, websites, or presentations.

Source: The Gallup Organization, http://www.gallup.com.

When you are finished filming and editing the video, you can place it into your document, usually with the Insert Movie command. Even simple word processors like Microsoft Word and Corel WordPerfect will allow you to put a video into the text. Videos can also be inserted into presentations made with Microsoft PowerPoint, Corel Presentation, and other presentation software packages.

Audio, Podcasting, and Music

You can also add audio clips or background music to your electronic documents, including websites. Most word-processing programs, presentation software, and website design software will allow you to "insert" music or audio. The best way to insert music is from a compact disc (CD) or by downloading a song in MP3 format off the Internet.

Several software applications, like Apple iTunes or Windows Media Player, will allow you to download music and podcasts legally off the Internet, usually at minimal cost. Music and podcasts can be downloaded legally at apple.com, emusic.com, pandora.com, and rhapsody.com.

When you select Insert, the computer will usually ask you to select a file. Choose the audio track or music file to be inserted. Then, if you want the music to play continuously, tell the computer to "Loop" the music when you are in the "Audio" text box.

Podcasting is becoming a popular way to insert audio files into documents, especially websites. Podcasts are radiolike audio files that can be downloaded to computers and MP3 players (not only Apple iPods). If you have a microphone and a computer with Internet access, you can start creating podcasts right away. Software packages for making podcasts include Podcast Station, Propaganda, Audacity, and Adobe Audition. Many companies offer a "podcasting kit" that includes a microphone and software (Figure 11.28).

Podcasting

Figure 11.28: A podcasting mic, like this one from Alesis, will allow you to add professional-quality audio to websites or multimedia documents.

Source: Alesis, http://www.alesis.com.

For links to legal music download sites, go to **www.pearsonhighered.com/johnsonweb3/11.20**

- Computers have made including graphics in technical documents easier. So, readers have come to expect them.

- Graphics should: (1) tell a simple story, (2) reinforce the text, not replace it, (3) be ethical, and (4) be properly labeled and placed on the page.

- Various kinds of graphs, tables, and charts allow you to tell different stories with data or facts.

- Digital cameras and scanners are making the placement of photographs in documents easier than ever.

- Use icons and clip art only when they enhance the readability and comprehension of the document. Clip art, especially, can simply clutter a document.

- Graphics need to be carefully considered when documents need to work cross-culturally. Images and symbols can have very different meanings in other cultures.

- Increasingly, computers allow us to insert video and audio into documents. Treat these items as you would treat photographs.

Individual or Team Projects

1. On the Internet, find a chart or graph that you can analyze. Using the four guidelines for graphics discussed in this chapter, critique the chart or graph by discussing its strengths and places where it might be improved. Present your findings to your class.

2. Find a set of data. Then, use different kinds of charts and graphs to illustrate trends in the data. For example, you might use a bar chart, line graph, and pie chart to illustrate the same data set. How does each type of graphic allow you to tell a different "story" with the data? What are the strengths and limitations of each kind of graphic? Which kind of chart or graph would probably be most effective for illustrating your data set?

3. On the Internet, find some written text that could be turned into a graph, table, or chart. Then, create a visual that would complement this text. In a memo to your instructor, show how the text and visual would complement each other by referring readers back and forth between text and visuals.

4. Using a digital camera, practice taking pictures and inserting those pictures into documents. Take pictures of people, objects, and places. When taking pictures of people, compare pictures taken inside and outside. Take full-body pictures and head shots. When taking pictures of objects, first leave the background behind the object cluttered. Then, use a backdrop to unclutter the picture. When photographing places, try to make images that tell a story about the place.

When you are finished, compare and contrast your photographs. Which types of photographs seem to work best in a document? What kinds of photographs tend not to work?

Collaborative Project

With a group of classmates, locate a large document that has few or no visuals. Then, do a "design makeover" in which you find ways to use visuals to support and clarify the written text. Try to include at least one visual for every two pages in the document. Use graphs, photographs, and drawings to illustrate important points in the document. Then, add icons and clip art to reinforce important points or themes in the document.

When you are finished, write a brief report to your instructor in which your group discusses how you made over the document. Critique the original draft of the document, showing how the lack of adequate visuals made the information in the document hard to access. Then, discuss the ways in which your revised version improves on the original. Finally, discuss some of the following issues about the amount and types of visuals used in this kind of document:

- At what point are there too many graphics?
- Do some graphics work better than others?
- How can you balance the written text with visuals to avoid making the document too text-heavy or visual-heavy?
- How do the needs and characteristics of the expected readers of the document shape the kinds of visuals that are used?

Your report might offer some additional guidelines, beyond the ones discussed in this chapter, for using visuals more effectively.

For additional technical writing resources, including interactive sample documents, document design tutorials and guidelines, and more, go to **www.mytechcommlab.com.**

You can find sample documents for the collaborative project assignment at
www.pearsonhighered.com/johnsonweb3/11.22

Lying with the Truth

The prior quarter had been tough at Ventrec Industries. Sales were declining as the economy turned downward. Management was looking just about everywhere in the budget to cut costs. So, the CEO asked each division manager to "defend" his or her budget and find places to cut. Everyone knew that the company was preparing contingency plans to lay off people, if necessary, to bring the budget under control.

Needless to say, this directive was not popular with the middle managers and engineers at the company. All of the managers, including Brad Fillmore's boss, Gail Zin, believed they had already slashed their budgets to the bone.

At a division meeting, Gail showed up with an eight-page proposal she had already sent to management that identified some strategic but minor cuts. In the proposal, she recommended her division forego upgrading computers for another year. She recommended they do better at going "paperless" to cut down on costs. She suggested that onsite visits to clients be slashed in half, with face-to-face meetings replaced by teleconferencing whenever possible.

These cuts seemed minor but reasonable to Brad. However, as he flipped through the report, the final bar chart caught his eye. Gail's overall argument in the proposal was that her division had already slashed its budget dramatically, so she shouldn't be asked to cut even more. Her bar chart did seem to show a significant cut in expenditures over the past six years (Figure A).

West Central Division Annual Expenditures

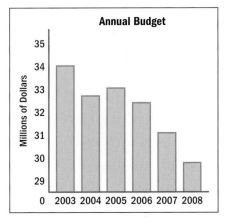

Figure A:
Gail's Bar Chart

(continued)

However, Brad noticed that Gail had not set the y-axis to 0, thus exaggerating the trend. She was telling the "truth," but it was clear she was trying to use a graph to spin the information to her advantage. Technically, what she was doing was unethical.

"Surely," Brad thought, "upper management isn't going to fall for this old trick. They're not stupid up there."

After the meeting, Brad dropped by Gail's office and pointed out the problem with the graph. Gail immediately grew defensive. "Listen Brad," she said sharply, "I'm doing everything I can to keep this division together. If that means I need to distort a graph, I'll do it. Every data point in that graph is true. Now, I want you to keep quiet. If management hears that I'm playing loose with the numbers—which I'm not—we might see some bigger problems around here. So, if I hear about this graph again, I'm going to know who said something, and I'm not going to be happy."

If you were Brad, how might you handle this situation? Clearly, Gail and Brad both want the proposal to be successful, but he is almost certain management will catch the distorted graph. If so, Gail might be in trouble, and he would probably get the blame. Should he approach Gail again? Should he tell someone else?

For examples of unethical graphs, go to
www.pearsonhighered.com/johnsonweb3/11.23

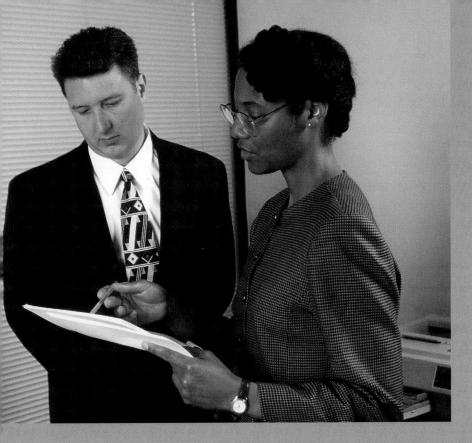

In this chapter, you will learn:

- How to use revising and editing as a form of quality control.

- How to apply the four levels of editing to your text.

- How to use copyediting symbols.

- Strategies for revising your draft by reviewing its rhetorical situation (subject, purpose, readers, and context of use).

- Strategies for reviewing the content, organization, and design of your document.

- How to revise your sentences, paragraphs, headings, and graphics.

- How to proofread for grammar, punctuation, and word usage.

- Strategies for document cycling and usability testing.

Your computer's ability to make rough drafts look like finished documents can be both an advantage and a disadvantage. An advantage is the ability to quickly produce a document that looks finished. The disadvantage, unfortunately, is that this finished-looking document might fool you into skipping the revision and editing phase of the writing process.

Your readers, however, won't be fooled. They will spot the weaknesses in content, organization, style, and design. Meanwhile, your unedited document will probably include mistakes in grammar, spelling, and usage.

Link

For more information on quality control, turn to Chapter 4, page 86.

Revising and editing are important because they are forms of *quality control* in a document. Your documents should reflect the quality standards that are held by your company (or maintain an even higher quality standard). The revising and editing phase is where your documents will go from "adequate" to "excellent."

Levels of Edit

Revising and editing requires more than checking a document for errors and perhaps changing a few sentences. In most cases, this kind of "revising" is not sufficient for technical documents.

Instead, when you are revising and editing, you need to approach the document from several different points of view. To help them better approach documents, professional editors use a tool called the "levels of edit" to assess how much editing a document needs before the deadline:

> **Level 1: Revision**—revises the document as a whole, which is why this level of edit is often called "global editing." Revision pays attention to the document's subject, purpose, readers, and context of use.
>
> **Level 2: Substantive editing**—pays special attention to the content, organization, and design of the document.
>
> **Level 3: Copyediting**—concentrates on revising the style for clarity, persuasion, and consistency, especially at the sentence and paragraph levels.
>
> **Level 4: Proofreading**—catches only the grammar mistakes, misspellings, and usage problems.

Which level of edit is appropriate for your document? The answer to this question depends on how much time you have and on the quality needed in the document. Given enough time, a writer will ideally go through all four levels, beginning with revision and ending with proofreading (Figure 12.1). But, sometimes the deadline is looming. So, as you begin the revising and editing phase, start out by determining what level of editing is possible and/or needed to produce the desired quality of document. Then, begin reworking the document at that level.

Revising: Level 1 Editing

While you were drafting the document, you probably revised it as you wrote, sharpening your ideas on the subject, reconsidering your purpose, and adjusting the design. Now that the document is completely drafted, you can start revising it as a whole.

To see other variations of the levels of edit, go to
www.pearsonhighered.com/johnsonweb3/12.1

The Levels of Editing

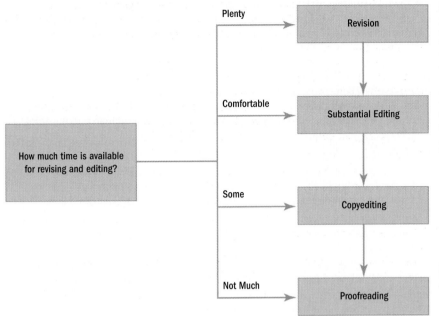

Figure 12.1: Ideally, writers would always go through all four levels of edit, from revision to proofreading. Sometimes, though, limited time determines what level is appropriate to finish a document.

Revision is a process of "re-visioning" the document. In other words, you are trying to gain a new perspective on your text so you can ensure that its subject and purpose are appropriate for your intended readers. You also want to revise the document so it will work in its context of use.

To revise (re-vision) your document, look back at your initial decisions about the rhetorical situation at the beginning of your writing process:

SUBJECT   Check whether your subject needs to be narrowed or broadened.

- Has your subject changed or evolved?
- Did you limit or expand the scope of your subject?
- Has your document strayed from the subject anywhere?

PURPOSE   Make sure the document is achieving its purpose.

- What do you want the document to achieve?
- Is your document's purpose still the same?
- Has your purpose become more specific or has it broadened?

READERS   Looking back at your original profile of your readers, think about the characteristics of the primary readers and other possible readers.

Judy Prono, Ph.D.

TECHNICAL EDITOR AND TEAM LEADER, LOS ALAMOS NATIONAL LABORATORY, LOS ALAMOS, NEW MEXICO

Los Alamos National Laboratory is a U.S. government research facility.

AT WORK

How are the levels of edit used at your workplace?

Since the mid-1980s, we have used levels of edit at Los Alamos National Laboratory to define editing expectations for both authors and editors. In 1986, a panel of authors helped us condense our four levels into three: proofreading, grammar, and full edit.

- A proofreading edit corrects glaring errors that could prove embarrassing to author and laboratory alike.
- The grammar edit ensures document clarity for audiences of technical peers.
- The full edit polishes technical documents and is the recommended edit for documents intended for broader audiences than those of technical peers.

Beyond simply improving the writing, our levels also encompass issues such as the sequencing and clarity of tables, figures, and equations; acronym and number usage; and lab publication policies.

Our levels of edit help authors control the cost of editing: The more comprehensive an edit, the more time consuming and hence more costly it is. However, we encourage authors to choose their edits in terms of a document's purpose and audience. An archival report intended simply to document an experiment requires less editing than a progress report intended to update key research sponsors. Similarly, technical audiences, given their expertise, will infer the significance and relationships of data that less technical audiences need stated explicitly. Documents intended for the latter thus need more extensive editing to ensure clarity.

Although, occasionally, editing comes down to the money or time available, in general, our levels of edit have proved helpful in defining editing needs and priorities.

- Do you now know more about your primary readers' needs?
- Have you fully anticipated your readers' values and attitudes?
- Do you need to more thoroughly address the secondary, tertiary, and gatekeeper readers in your document?

Link

For more information on defining the rhetorical situation, go to Chapter 2, page 20.

CONTEXT OF USE   Consider the contexts in which your document might be read or used.

- Do you better understand the physical places where your readers will read or use the document?
- Do you better understand the economic, political, and ethical issues that will influence how your readers interpret your document?
- Have you anticipated the personal, corporate, and industry-related issues that will also shape your readers' interpretation?

Certainly, these are difficult questions to ask as you finish your document. But, they are worth asking. You need to make adjustments to the document

Revising the Document

When revising, look back at your original notes about the rhetorical situation.

Do you need to sharpen your purpose statement?

Do you have a better understanding of the readers?

Has the context of use changed?

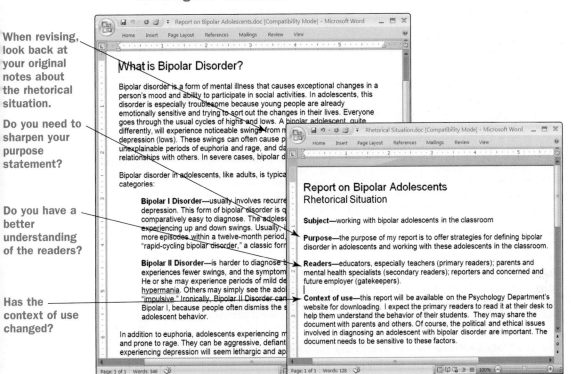

What is Bipolar Disorder?

Bipolar disorder is a form of mental illness that causes exceptional changes in a person's mood and ability to participate in social activities. In adolescents, this disorder is especially troublesome because young people are already emotionally sensitive and trying to sort out the changes in their lives. Everyone goes through the usual cycles of highs and lows. A bipolar adolescent, quite differently, will experience noticeable swings from depression (lows). These swings can often cause unexplainable periods of euphoria and rage, and da relationships with others. In severe cases, bipolar d

Bipolar disorder in adolescents, like adults, is typica categories:

Bipolar I Disorder—usually involves recurre depression. This form of bipolar disorder is q comparatively easy to diagnose. The adoles experiencing up and down swings. Usually, more episodes within a twelve-month period, "rapid-cycling bipolar disorder," a classic form

Bipolar II Disorder—is harder to diagnose experiences fewer swings, and the symptom He or she may experience periods of mild de hypermania. Others may simply see the ado "impulsive." Ironically, Bipolar II Disorder can Bipolar I, because people often dismiss the s adolescent behavior.

In addition to euphoria, adolescents experiencing m and prone to rage. They can be aggressive, defiant experiencing depression will seem lethargic and ap

Page: 1 of 1   Words: 346

Report on Bipolar Adolescents
Rhetorical Situation

Subject—working with bipolar adolescents in the classroom

Purpose—the purpose of my report is to offer strategies for defining bipolar disorder in adolescents and working with these adolescents in the classroom.

Readers—educators, especially teachers (primary readers); parents and mental health specialists (secondary readers); reporters and concerned and future employer (gatekeepers).

Context of use—this report will be available on the Psychology Department's website for downloading. I expect the primary readers to read it at their desk to help them understand the behavior of their students. They may share the document with parents and others. Of course, the political and ethical issues involved in diagnosing an adolescent with bipolar disorder are important. The document needs to be sensitive to these factors.

Page: 1 of 1   Words: 128

Guidelines for Revising (Level 1)

AT A GLANCE

- Subject—Is the subject too narrow or too broad?
- Purpose—Does the document achieve its stated purpose?
- Readers—Is the document appropriate for the readers?
- Context of use—Is the document appropriate for its context of use?

if its purpose has evolved or you have gained a better understanding of your readers.

Once you have reconsidered the document's rhetorical situation, you can work through the text to see if it stays focused on the subject and achieves its purpose (Figure 12.2). Look for places where you can revise the text to better suit your readers and the contexts in which they will use the information you are providing them.

Revision at this global level requires some courage. You may discover that parts of your document need to be completely rewritten. In some cases, the whole document may need to be reconceived. But, it is better to be honest with yourself at this point and make those changes. After all, a document that fails to achieve its purpose is a waste of all the time you've spent.

Substantive Editing: Level 2 Editing

While doing substantive editing, you should concentrate on the content, organization, and design of the document (Figure 12.3). A good approach to substantive editing is to consider the document from three different perspectives:

Need more help with revision? Go to
www.pearsonhighered.com/johnsonweb3/12.2

Substantive Editing: Level 2 Editing

309

Substantive Editing

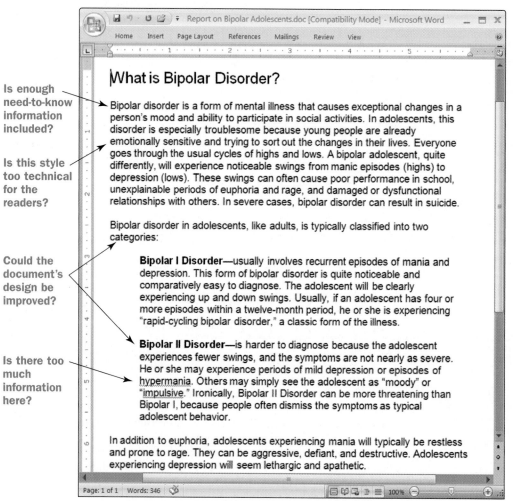

Is enough need-to-know information included?

Is this style too technical for the readers?

Could the document's design be improved?

Is there too much information here?

Figure 12.3: Substantive editing urges you to ask questions about the appropriateness of the content, style, and design of your document. When answering these questions, think about the needs of your readers.

CONTENT   Look for any gaps or digressions in the content.

- Are there any places (gaps) where you are lacking proof or support for your claims?
- Do you need to do more research to support your points?
- Are there any places where you have included information that the readers do not need to know to make a decision or take action?

ORGANIZATION   A document should conform to a recognizable genre, and it should have an identifiable introduction, body, and conclusion.

Guidelines for Substantive Editing (Level 2)

AT A GLANCE

- Content—Are there any digressions or gaps in content?
- Organization—Does the document conform to a recognizable genre or pattern?
- Design—Do the page layout and graphics enhance the readability of the document?

- Are there any places where you have deviated from the organizational pattern of the genre you are following? Are these deviations helpful toward achieving your purpose? Or, should you reorganize the document to suit the genre?
- Does the introduction clearly identify the subject, while stating your purpose and your main point? Should the introduction include more background information or stress the importance of the subject?
- Does the conclusion restate your main point, restress the importance of the subject, and look to the future?

DESIGN The document should be designed for the readers and the contexts in which it will be used.

Link

For more information about organization, see Chapter 8, pages 175–201.

- Is the text readable (scannable) in the situations and places where people will use it?
- Does the design reflect your readers' values and attitudes? Is it straightforward for conservative readers, or is it more innovative for progressive readers?
- Does the design properly use principles of balance, alignment, grouping, consistency, and contrast?
- Does the design clarify the structure of the text with titles and subheads?
- Do the graphics support the text, and do they clarify difficult points?

Link

For more document design strategies, see Chapter 10, page 232.

Copyediting: Level 3 Editing

When you are copyediting, you should concentrate on improving style and consistency, especially at the sentence and paragraph levels. You should also look over the headings and graphics to make sure they are appropriate and accurate.

While copyediting, you might find the Track Changes feature of your word processor especially helpful (Figure 12.4). It will show places where changes were made to the text. Later, you can decide if you want to keep those changes or not. This feature is especially helpful if you are working with a team on a document. That way, any changes are recorded for the others in the group to approve.

When copyediting, pay close attention to the sentences, paragraphs, headings, and graphics.

Link

For help on improving style at the sentence level, go to Chapter 9, page 210.

SENTENCES Look over the sentences to make sure they are clear and concise.

- Are the subjects of the sentences easy to locate?
- Do the verbs of the sentences express the actions of the sentences?
- Can you eliminate any unnecessary prepositional phrases?
- Are the sentences breathing length?

Tracking Changes While Copyediting and Proofreading

Color text highlights changes made to text.

These lines in the margin help locate changes.

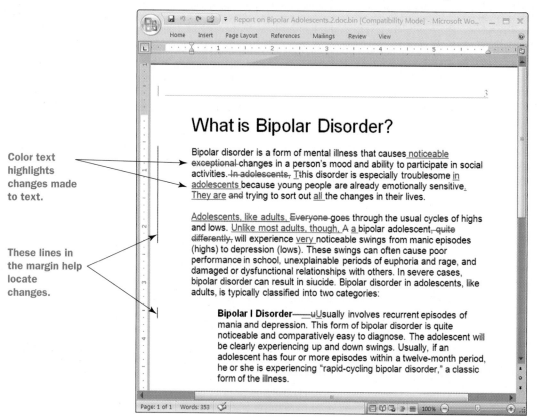

Figure 12.4:
With the Track Changes function, your word processor can keep track of the changes you (or someone else) make to the document.

PARAGRAPHS   Make sure the paragraphs support specific claims. Rework the sentences in the paragraphs to improve the flow of the text.

Link

For more information on using headings, see Chapter 10, page 246.

- Does each paragraph have a clear topic sentence (a claim) and enough support to back it up?
- Would any paragraphs be stronger if you included a transition sentence at the beginning or a point sentence at the end?
- Are the subjects in the paragraph aligned, or could you use given/new strategies to smooth out the text?
- Would transitions or transitional phrases help bridge any gaps between sentences?

HEADINGS   The headings should be easy to understand and consistently used.

- Do the headings in the document properly reflect the information that follows them?
- Do the headings make the document scannable, highlighting places where important information can be found?
- Are there clear levels of headings that help readers identify the structure of the document and the importance of each part of the document?

To learn more about copyediting, go to
www.pearsonhighered.com/johnsonweb3/12.4

Guidelines for Copyediting (Level 3)

AT A GLANCE

- Sentences—Are the sentences clear and concise?
- Paragraphs—Do the paragraphs have a clear topic sentence and support?
- Headings—Do the headings help the readers scan for important information?
- Graphics—Do the graphics support the written text?

GRAPHICS   Look over the graphics in the document to make sure they support the written text. Check the graphics for accuracy.

- Does each graphic tell a simple story?
- Does each graphic support the written text without replacing it?
- Are the graphics clearly titled and referred to by number in the written text?

Link

For more help using graphics, turn to Chapter 11, page 275.

Proofreading: Level 4 Editing

Proofreading assumes that the document is complete in almost every way. Now you need to focus only on the mechanical details of the document, like the grammar, spelling, punctuation, and word usage. While proofreading you should focus on marking "errors" and making only minor stylistic changes to the text.

Grammar

In technical documents, correct grammar and punctuation are expected. You should always remember that your readers *will* notice the grammatical mistakes. So, these mistakes can often sabotage an otherwise solid document.

Most word-processing programs have a grammar checker, but these checkers are notoriously unreliable. So follow any advice from the grammar checker cautiously, because your computer will often flag grammatically correct sentences. There is no substitute for mastering grammar rules.

As you proofread the document, pay attention to your own reactions. When you stumble or pause, chances are good that you have found a grammatical mistake. Mark these places, and then identify what grammatical mistake has been made. Figure 12.5 describes some of the more common grammatical errors.

Link

For more on grammar rules, go to Appendix A, page A-2.

In the Grammar and Punctuation Guide (Appendix A in the back of this book), you will find examples and remedies for these common errors. If you are not familiar with one or more of these errors, you should spend a little time learning how to avoid them.

In the end, most readers can usually figure out the meaning of a document even when it has errors in grammar. But these errors make the document look sloppy or seem unprofessional. After stumbling over a few errors, readers will begin to doubt the quality or soundness of the document and the information it contains. Even worse, they may doubt your abilities or commitment to quality. Readers expect correct grammar in technical documents.

Punctuation

Punctuation reflects the way we speak. For example, a *period* is supposed to reflect the amount of time (a period) that it takes to say one sentence. If you were to read a document out loud, the periods would signal places to breathe. Similarly, commas are used to signal pauses. When you come across a comma in a sentence, you pause slightly.

By understanding these physical characteristics of punctuation, you can learn how to use the marks properly (Figure 12.6).

Link

For a more detailed discussion of punctuation, go to Appendix A, page A-2.

Want to improve your grammar? Go to **www.pearsonhighered.com/johnsonweb3/12.5**

Proofreading: Level 4 Editing   313

Common Grammatical Errors

Error	Explanation
comma splice	Two or more distinct sentences are joined only by a comma.
run-on sentence	The sentence is composed of two or more distinct sentences.
fragment	The sentence is incomplete, usually missing a subject or verb.
dangling modifier	A modifier (usually an introductory phrase) implies a different subject than the one in the sentence's subject slot.
subject-verb disagreement	A singular or plural subject does not agree with the verb form.
misused apostrophe	An apostrophe is used where it doesn't belong (usually confusing *it's* and *its*).
misused comma	A comma signals an unnecessary pause in a sentence.
pronoun-antecedent disagreement	A pronoun does not agree with a noun earlier in the sentence.
faulty parallelism	A list of items in a sentence is not parallel in structure.
pronoun case error	The case of a pronoun is incorrect (usually due to confusion about when to use *I* or *me*).
shifted tense	Sentences inconsistently use past, present, and future tenses.
vague pronoun	It is unclear what the pronoun refers to.

Figure 12.5:
Here are the usual grammatical culprits. If you avoid these simple errors, your document will have almost no grammatical problems.

Punctuation is intended to help readers understand your text. But when the marks are misused, they can create a great amount of confusion. The Grammar and Punctuation Guide (Appendix A) includes a more detailed explanation of punctuation usage.

Physical Characteristics of Punctuation

Punctuation Mark	Physical Characteristic
capitalization	signals a raised voice to indicate the beginning of a sentence or a proper name
period [.]	signals a complete stop after a statement
question mark [?]	signals a complete stop after a question
exclamation mark [!]	signals a complete stop after an outcry or objection
comma [,]	signals a pause in a sentence
semicolon [;]	signals a longer pause in a sentence and connects two related, complete statements
colon [:]	signals a complete stop but joins two equal statements; or, it indicates the beginning of a list
hyphen [-]	connects two or more words into a compound word
dash [—]	sets off a comment by the author that is an aside or interjection
apostrophe [']	signals possession, or the contraction of two words
quotation marks [" "]	signal a quotation, or when a word or phrase is being used in a "unique way"
parentheses [( )]	enclose supplemental information like an example or definition

Figure 12.6:
Punctuation mirrors the physical characteristics of speech.

For quick answers to grammar issues, go to
www.pearsonhighered.com/johnsonweb3/12.6
For short essays on grammar issues, go to
www.pearsonhighered.com/johnsonweb3/12.7

Spelling and Typos

Spelling errors and typos can be jarring for your readers. One or two spelling errors or typos in a document may be forgiven, but several errors will cause your readers to seriously question your commitment to quality. Here are some ways to avoid those errors.

Guidelines for Proofreading (Level 4)

- Grammar—Are all the sentences grammatically correct?
- Punctuation—Are the sentences properly punctuated?
- Spelling and typos—Are there any spelling errors or typos?
- Word usage—Are all the words used properly?

AT A GLANCE

Use the spell check feature on your computer—Most word-processing programs come with a spell checker that is rather reliable. Even if you are a good speller, the spell checker will often catch those annoying typos that inevitably find their way into texts (Figure 12.7).

However, a spell checker is not perfect, so you will still need to pay careful attention to spelling in your documents. After all, words may be spelled correctly, but they may not be the words you intended. Here is a sentence, for example, that has no errors according to a spell checker:

> Eye sad, they're our many places four us to sea friends and by good she's stakes in Philadelphia.

To avoid these embarrassing errors, don't rely exclusively on the spell checker.

Running the Spell Checker

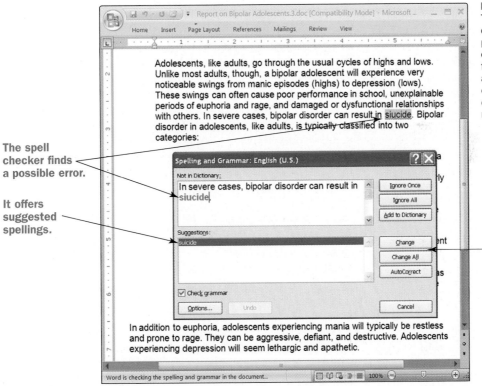

The spell checker finds a possible error.

It offers suggested spellings.

You can choose to make changes, add the word to your online dictionary, or ignore the "error."

Figure 12.7: The spell checker on your word processor is especially helpful for locating typos and other errors. Grammar checkers are not nearly as reliable.

For websites that discuss the history of punctuation marks, go to
www.pearsonhighered.com/johnsonweb3/12.8
For help using your spell checker, go to
www.pearsonhighered.com/johnsonweb3/12.9

Proofreading: Level 4 Editing 315

Keep a dictionary close by—Technical documents often use terms and jargon that are not in your computer's spell checker. A dictionary is helpful in these cases for checking spelling and usage.

On the Internet, you may be able to find dictionaries that define specialized words in your field. There are, for example, numerous online dictionaries for engineering or medicine.

Overall, though, you should be quick to grab the dictionary or look up a word online when there are any doubts about the spelling of a word. The best editors are people who do not hesitate to look up a word, even when they are almost certain that they know how the word is spelled.

Word Usage

In English, many words seem the same, but they have subtle differences in usage that you should know (Figure 12.8). There are several good *usage guides* available in print that explain the subtle differences among words. More usage guides are becoming available over the Internet.

Common Usage Problems in Technical Documents

Confused Words	Explanation of Usage
accept, except	*accept* means to receive or agree to; *except* means to leave out
affect, effect	*affect* is usually used as a verb; *effect* is usually used as a noun
anyone, any one	*anyone* is a pronoun that refers to a person; *any one* means any one of a set
between, among	*between* is used for two entities; *among* is used for more than two entities
capitol, capital	*capitol* is the seat of a government; *capital* is money or goods
criterion, criteria	*criterion* is singular; *criteria* is plural
ensure, insure	*ensure* means to make certain; *insure* means to protect with insurance
complement, compliment	to *complement* is to complete something else or make it whole; a *compliment* is a kind word or encouragement
discreet, discrete	*discreet* means showing good judgment; *discrete* means separate
farther, further	*farther* refers to physical distance; *further* refers to time or degree
imply, infer	*imply* means to suggest indirectly; *infer* means to interpret or draw a conclusion
its, it's	*its* is always possessive; *it's* is always a contraction that means "it is"; *its'* is not a word
less, fewer	*less* refers to quantity; *fewer* refers to number
personal, personnel	*personal* refers to an individual characteristic; *personnel* refers to employees
phenomenon, phenomena	*phenomenon* is singular; *phenomena* is plural
precede, proceed	*precede* means to come before; *proceed* means to move forward
principle, principal	a *principle* is a firmly held belief or law; a *principal* is someone who runs a school
their, there, they're	*their* is a possessive pronoun; *there* is a place; *they're* is a contraction that means "they are"
whose, who's	*whose* is a possessive pronoun; *who's* is the contraction of "who is"
your, you're	*your* is a possessive pronoun; *you're* is a contraction that means "you are"
who, whom	*who* is a subject of a sentence; *whom* is used as the object of the sentence

Figure 12.8: You can avoid some of the most common usage problems in technical documents by consulting this list.

Want to try out some great online dictionaries? Go to **www.pearsonhighered.com/johnsonweb3/12.10**

If you are unsure about how a word should be used, look it up. Dictionaries are usually sufficient for answering usage questions. When a dictionary cannot answer your question, though, look up the word in an online or print usage guide.

Using Copyediting Symbols

While editing, you might find it helpful to use the same editing symbols that professional editors use. To mark any stylistic changes and inconsistencies, editors have developed a somewhat universal set of copyediting marks (Figure 12.9). They are easy to use and widely understood. See Figure 12.10 for examples of how they are used.

Editing Symbols and Their Uses

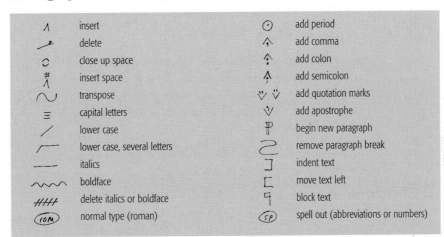

∧	insert	⊙	add period
ℓ	delete	⋀	add comma
⌒	close up space	⋀	add colon
#	insert space	⋀	add semicolon
∼	transpose	⌄ ⌄	add quotation marks
≡	capital letters	⌄	add apostrophe
/	lower case	¶	begin new paragraph
⌐	lower case, several letters	⌒	remove paragraph break
—	italics	⌐	indent text
∼∼∼	boldface	⌐	move text left
####	delete italics or boldface	⌐	block text
(rom)	normal type (roman)	(sp)	spell out (abbreviations or numbers)

Figure 12.9: Copyediting symbols offer a standardized method for editors and writers to work together on a document.

Lost in Translation: Cross-Cultural Editing

Today, translation software from Language Weaver, free services from World Lingo (www.worldlingo.com), and Altavista's Babel Fish (babelfish.altavista.com) make it possible to translate documents from one language to another with up to 80 percent accuracy. But that means 20 percent of any "machine-translated" document might be flawed or confusing. The limited success of these programs is often due to cultural subtleties that can cause confusing results.

When working with cross-cultural documents, extra effort needs to be put into the revising and editing process. Several companies have experienced some classic gaffes when they have not taken the extra time to edit their texts from their readers' cultural point of view. For example, the baby food brand name Gerber means "vomit" in French. Meanwhile, the Germanic word for travel is "Fahrt," which causes some classic problems with travel-related products that are sold in both Central Europe and English-speaking countries. For example, the Swedish furniture maker IKEA once sold a mobile workbench called the "Fartfull" in the United States and Europe.

Moreover, slogans often don't translate well. In 1996, Japanese designers at Panasonic created a web browser that used Woody Woodpecker as a mascot. Japanese executives

Link

For other humorous translation problems, go to Chapter 3, page 54.

Need to check the usage of a word? Go to
www.pearsonhighered.com/johnsonweb3/12.11

Lost in Translation:
Cross-Cultural
Editing

317

Figure 12.10: Copyediting marks can be used to identify changes in the text.

Eruption History of Kilauea

Can you be more specific here? Are more accurate estimates available?

When Kilauea began to form is not known, but various estimates are 300,000-600,000 years ago. The volcano had been active ever since, with no prolonged periods of quiescence known. geologic studies of surface exposures and examination of drillhole samples show that Kilauea is made mostly of lava flows, locally interbedded with deposites of explosive eruptions. Probably what we have seen hapen in the past 200 years is a good guide to what has happened ever since Kilauea emerged from the sea as an island perhaps 50,000-100,000 *years* ago.

Lava Erupts from Kilauea's Summit and Rift Zones

Define the jargon in this paragraph.

Throughout its history Kilauea has erupted from three main areas: its summit and two rift zones. Geologists debate whether Kilauea has Always had a caldera at the summit or whether it is a relatively recent feature of *the* past few thousand years. It seems most likely that the caldera has come and gone throughout the life of Kilauea.

The summit of the volcano is high because eruptions are more frequent there than at any other single location on the volcano.

It's hard to figure out what you're describing here. A diagram would help.

However, more eruptions actually occur on the long rift zones than in the summit area, but they are not localized instead *eruptions* construct ridges of lower elevation than the summit. Eruptions along the east and southwest rift zones have build ridges reaching outward from the summit some 125 KM and 35 KM respectively.

Most eruption*s* are relatively gentle, sending lava flows downslope from fountains a few meters to a few hundred meters high. OVER and over again these eruptions occur, gradually building up the volcano and giving it a gentle, shield-like form. Every few decades to centuries, however, powerful explosions spread ejecta across the landscape. Such explosions can be lethal, as the one in 1790 that killed scores of people in a war party near the summitting of Kilauea. Such explosions can take place from either the bridging summit or the upper rift zones.

Source: U.S. Geological Survey, http://hvo.wr.usgs.gov/kilauea/history/main.html. Errors in the text were added and did not appear in the original.

Want to learn more about using copyediting symbols? Go to **www.pearsonhighered.com/johnsonweb3/12.12**

were horrified to learn that the English meaning of their slogan "Touch Woody—The Internet Pecker" was not exactly what they had intended (Yoshida, *EE Times,* 1996). Fortunately, most cross-cultural gaffes cause only confusion, not embarrassment. Nevertheless, you need to carefully check your documents when editing.

International business specialists Carol Leininger and Rue Yuan (1997) offer the following advice for cross-cultural documents:

Use short, direct sentences that follow subject, verb, object order—Second-language readers and translation software will be more successful if they can easily locate the subjects and verbs of sentences. Longer sentences should be cut into shorter sentences.

Use positive sentences and minimize negative sentences—Negative sentences sometimes translate more harshly than originally intended. A negative sentence that offers a simple caution to the reader can translate into one that makes dire predictions of harm or death.

Use a limited set of words—Most international companies, such as Caterpillar and IBM, have developed standard language guides of English words to be used in international documents. Documents that use these words are easier for people and translation software to translate.

Avoid humor or jokes—Jokes are highly cultural and situational, so they rarely translate well into other cultures and languages. Usually, they are just confusing, but sometimes they can be embarrassing for the writer or insulting to the reader.

Minimize jargon and slang—Jargon words and slang phrases are also culturally dependent and difficult to translate. These terms should be translated into their common meanings even though they might lose some of their original flair.

Check any sayings, clichés, or idioms—These turns of phrase often do not translate well. For example, in North America people "cross their fingers" for luck, but in Germany, people "hold their thumbs." Machines "run" in English, but they "walk" in Spanish.

Avoid obvious metaphors—Metaphors cannot be completely avoided, but obvious ones should be removed. In German, calling a company a "shooting star" would suggest that it will be successful long into the future. In North America, "shooting star" means the company will rise quickly and fail spectacularly. Meanwhile, sports metaphors like, "She hit a home run" or "He just punted" will be confusing to most international readers. Metaphors that use body parts (e.g., "I'll keep an eye on the project") or animals (e.g., "He's a workhorse") can have very different and disturbing meanings when translated.

Check slogans—Slogans usually rely on a cultural twist of words, so they are particularly risky when translated. In Taiwan, Pepsi's slogan "Come alive with the Pepsi Generation" translated into, "Pepsi will bring your ancestors back from the dead" (Pendergrast, 1994).

Check product names—Names of products can also translate in embarrassing ways. Products like the Pinto, Puffs, Waterpik, and latte, among others, have sexually suggestive meanings in other languages. The Chevy Nova didn't sell well in Mexico and Latin America because "no va" means "It doesn't go" in Spanish.

To ensure that your documents will work cross-culturally, your best strategy is to user-test your documents with readers from likely target cultures because, unfortunately, translation software will rarely catch the subtleties of languages. So, you need to do it. Meanwhile, your translation dictionary probably won't include insulting phrases or sexually suggestive slang.

HELP

Online Copyediting

Professional editors still usually prefer to edit text on paper. On paper, it is easy to suggest changes and keep track of the version being edited. Increasingly, though, editing is being handled online, because it is easier to collaborate with other writers and cycle documents in electronic form. Also, people are becoming more comfortable with electronic documents, so they prefer to work online.

Here are some online editing techniques you can use.

Creating and using styles—As you edit the document, use the Style feature of your word-processing program to define the formatting of headings, indented text, numbered and bulleted lists, examples, and so forth. At first, setting styles might slow your editing down, but as you move through the document, you will find that the preset styles will save you time in the long run.

Version control—Most word-processing programs will allow you to keep track of multiple versions of your document. Each version is automatically saved under a slightly different name. Then, you can use the Versions feature of your program to see past versions and the dates they were created. You can also use the Save As feature to save different versions. For example, with each major revision, you might change the name slightly (e.g., ProReport.1, ProReport.2, and so on).

Track changes—You can also track any changes automatically with the Track Changes feature. Your word processor will highlight various changes in a different color, allowing the original author or your co-workers to see the changes you made.

Copyediting symbols—With the Font feature of your computer, you can simulate some of the copyediting symbols used on paper (Figure A).

Hidden text—This feature allows you to make small insertions into your text (Figure B). All word processors will allow you to "hide" text. Hidden text appears on the screen, but it does not appear in the printed version of the document. Usually the Hidden Text feature is located in the same place as the "Bold" or "Italics" option. You simply highlight the text you want hidden and select "Hidden."

Online Copyediting Symbols

Copyedit Action	Online Mark	How to Do It
delete	delete this text	strike through
italicize	italicize this text	underline
bold	bold this text	underline wave
capitalize	capitalize this text	double underline
insert text	insert word	hidden text
insert text	insert more than one word	"Comments" feature

Figure A:
Computers can simulate many of the copyediting symbols.

Comments—To signal that you want to insert more than a word or a phrase, use the Comments feature in your word-processing program. This feature lets you add a marginal balloon or hidden note that will not appear in the printed text (Figure B). When readers put their cursors over that highlighted text, the note pops up on the screen.

Link

For more help on using Outline View, see Chapter 8, page 180.

Endnotes and footnotes—Some online editors like to signal queries by using endnotes or footnotes. Using the Insert Footnote feature of your word processor, you can type a note in the footnote window at the bottom of your screen.

Outlining view—The Outline View feature is helpful for gaining an overall sense of the structure of the text. As discussed in Chapter 8, you can use electronic outlining to help you organize the document into a logical structure.

Online Editing

Hidden text signals an insertion.

Figure B:
Online editing is not as flexible as copyediting on paper, but it does have advantages for sending or cycling a document electronically.

Marginal balloons signal questions for the author.

For more online editing tips, go to
www.pearsonhighered.com/johnsonweb3/12.13

Lost in Translation:
Cross-Cultural
Editing

321

Document Cycling and Usability Testing

When you are completing your document, it is important that you gain an outside perspective. Often, while drafting, we become too close to our documents. Consequently, we can no longer edit or assess our own work objectively. Two ways to gain that outside perspective are *document cycling* and *usability testing*.

Document Cycling

Document cycling is a method for letting others at your company look over your draft. When you *cycle* a document, you pass it among your co-workers and supervisors to obtain feedback (Figure 12.11).

Computers give you the ability to quickly send your document around to others for their suggestions for improvement. You can send it as an attachment or make multiple paper copies. Then, your supervisors, colleagues, and even your primary readers can look it over and offer ideas for improvement.

When revising and editing your documents, it is important to let others look over your work. Document cycling is an important part of a *quality feedback loop*, a central principle of quality management. If you rely on yourself alone to edit your work, the quality of your document might be low.

Usability Testing

Usability testing means trying out your document on real readers. This kind of testing on real readers can be informal or formal, depending on the importance of your document and the time you have to test it.

Informal usability testing—Minimally, you should ask other people to look over the document and mark places where they stumble or find it difficult to understand its meaning.

Formal usability testing—More formally, you can run experiments with sample readers to measure how well they can understand and use your document. Instructions and user's manuals are commonly user-tested in this way. But, increasingly, focus groups of readers are being used to test persuasive documents like proposals. In some cases, companies will create *usability laboratories* to observe the reactions of readers as they use documents.

Document Cycling in the Workplace

Figure 12.11: Document cycling gathers feedback by letting others look over the document.

To learn more about usability testing, go to
www.pearsonhighered.com/johnsonweb3/12.14

Types of Usability Testing

	Usability Test	How It Is Conducted
informal testing	document markup	Readers are asked to read through a document, marking places where they stumble or fail to understand.
	read and locate test	Readers are asked to locate specific kinds of information in a document. They are timed and videotaped.
	summary test	Readers are asked to summarize the important information in a document.
	protocols	Readers are asked to talk out loud as they are using the text. Their comments are taped and transcribed.
	journal or tape recording	Readers are asked to keep a written or taped journal at their workplace to record their experiences with the document.
	surveying	Readers are given a questionnaire after they use the document, asking them about their experience.
	interviewing	Readers are interviewed about their experiences using a document.
	focus groups	Groups of readers look over a document and discuss their reactions to the work.
formal testing	laboratory testing	Through cameras and a one-way mirror, readers are carefully observed using a text.

Figure 12.12: A variety of different methods are available to user-test a document.

The more similar your test subjects are to your target readers, the better the results of your testing will be. In other words, the folks around the office are fine for an informal test. But, if you want a more accurate assessment of how your document will be used, you should try to locate a group of people who are most like your readers. Figure 12.12 lists some of the more common methods for testing usability, ranking them from informal methods to formal methods.

Most usability testing is designed to answer four questions (Figure 12.13):

Can they find it? — *Read-and-locate tests* are used to determine whether users can locate important parts of the document and how quickly they can do so. Often, the users are videotaped and timed while using the document.

Can they understand it? — *Understandability tests* are used to determine if the users retain important concepts and remember key terms. Users are often asked to summarize parts of the document or define concepts.

Can they do it? — *Performance tests* are used to determine whether users can perform the actions the document describes. These tests are often used with instructions and procedures.

Is it safe? — *Safety tests* are used to study whether the activities described in the document, especially in instructions or user's manuals, are safe. These tests carefully watch for possible safety problems by having sample readers use the product documentation.

Usability Questions

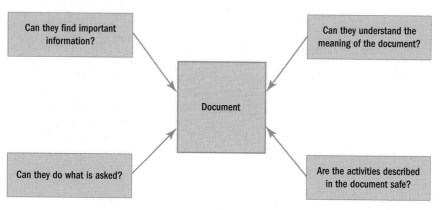

As you devise a usability test or series of usability tests, you should set quantifiable objectives that will allow you to measure *normal* and *minimal* user performance with the document. In other words, first define what you would expect *typical* users to be able to accomplish with your document. Then, define what you would *minimally* expect them to be able to do.

READ-AND-LOCATE TESTS: CAN THEY FIND IT?   To run a read-and-locate test, list five to seven important pieces of information that you want readers to locate in the document. Then, while timing and/or videotaping them, see how long it takes them to find that information.

Videotaping your subjects is especially helpful, because you can observe how readers go about accessing the information in your document. Do they go right to the beginning or the middle? Do they flip through the text looking at the headings or graphics? Do they look at the table of contents or index (if these features exist)?

After your subjects locate the major pieces of information you asked them to look for, have them tell you about these major points orally or in writing. Then, check their answers against your original list. If they successfully found four or five items from your list of most important pieces of information, your document is likely well written and well designed. If, however, they struggled to find even a few of your major points, you probably need to revise your document to ensure that the important information is easy to locate.

UNDERSTANDABILITY TESTS: CAN THEY UNDERSTAND IT?   When running an understandability test, you want to determine how well the users of your document grasped its meaning. Before running the test, write down your document's purpose and main point. Then, write down three important concepts or points that anyone should retain after reading the document.

Give your readers a limited amount of time to read through the document or use it to perform a task. Then have them put the document away so they cannot use it. Verbally or in writing, ask them:

- What is the purpose of this document?
- What is the document's main point?
- Can you tell me three major points that are made in the document?

If their answers to these questions are similar to the ones you wrote down, your document is likely understandable. If, however, your readers struggle to answer these questions, or get them wrong, you should think seriously about revising the document to highlight the information you intended your readers to retain.

Usability Testing a Document

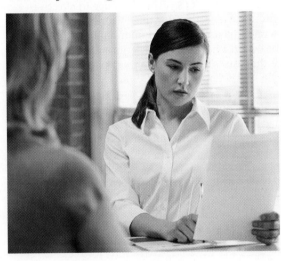

To test the usability of a document, you can run experiments with people who represent real readers. Videotaping is an especially good way to see how people actually use a document.

PERFORMANCE TESTS: CAN THEY DO IT?   Almost all technical documents are written to help readers take some kind of action. A set of instructions, obviously, asks readers to follow a procedure. A report might make some recommendations for change.

To do a performance test, have the users perform the procedure the document describes. Or, ask them to react to your recommendations. Here again, videotaping the users is a good way to keep a record of what happened. Did they seem to find the document easy to use? Where did they stumble or show frustration? When did they react positively or negatively to the tasks or ideas described in the document?

Ultimately, performance tests are designed to find out whether the users can do what the document asks of them. But it is also important to determine their attitude toward performing these tasks. You want to ensure not only that they *can* do it, but also that they *will* do it.

SAFETY TESTS: IS IT SAFE? Above all, you want your documentation and products to be safe. It is impossible to reduce all risk of injury, but you should try to reduce the risk as much as possible. Today, it is common for companies to be sued when their documentation or products are shown to be inadequate. Often, in product liability lawsuits, documents like instructions and user's manuals are used to prove or deny a company's negligence for an injury.

Without putting test subjects at risk themselves, safety tests are usually designed to locate places where users may make potentially injurious mistakes. They also ask readers about the warnings and cautions in the document to determine whether the reader noticed and understood these features.

SETTING OBJECTIVES AND MEASURING RESULTS The challenge to effective usability testing is to first identify some objectives for the document. These objectives could refer to (1) how well the users can find information, (2) how well they understand important ideas, and (3) how well they perform tasks described in the document. Then, measure the results of your usability testing against these objectives.

It's often quite sobering to watch people fumble around with your document, misunderstand its meaning, and not follow its directions. But, the results of your tests should help you revise the document to improve its usability.

No form of usability testing will ensure that your document is a success. However, feedback from users is usually the best way to gain new insights into your document and to solicit suggestions for improvement.

- Revising and editing are forms of quality control that should be a regular part of your writing process.

- Documents and presentations can be edited at four different levels: revision, substantive editing, copyediting, and proofreading.

- Editorial tools such as copyediting marks are helpful even for noneditors.

- Document cycling is a process of circulating your text among colleagues and your supervisor. You can use e-mail attachments to send your work out for review by others.

- Usability testing can involve informal or formal methods to test the effectiveness of your document. You might use sample readers to test documents that will be used by a broad readership.

Individual or Team Projects

1. Find a document on campus or at your workplace that needs editing. Edit the document by working backward from a level 4 edit (proofreading) to a level 1 edit (revision). As you apply each level of edit to the document, pay attention to the different kinds of actions and decisions you make at each level. How is the document evolving as you edit it?

2. Exchange a text with a member of your class. Then, do a level 2 edit (substantive editing) of the draft, using copyediting symbols to reflect the changes you think should be made to the document. Write a cover letter to the author in which you explain the changes you want made. Hand in a copy of this documentation to your instructor.

3. Find a text on the Internet that needs to be edited. Do a level 3 edit (copyediting) of the text, using the copyediting marks shown in this chapter. Write a memo to your instructor in which you discuss how you edited the text. Discuss some of the places where you struggled to mark the changes you wanted to make.

4. Find a text that needs editing and use the online editing strategies discussed in this chapter to edit it. In an e-mail to your instructor (with the edited version attached), discuss some of the differences and difficulties between paper-based and online editing. Discuss which form of editing you prefer and why.

Collaborative Project

With your group, locate a longer document on the Internet that needs editing. The document might be a report or proposal, or perhaps even a website. Then, do a level 2 edit (substantive editing) on the document, including the two levels of editing below it (copyediting and proofreading).

As you are editing, set up a document cycling routine that allows you to distribute your document among your group members. You can cycle the document in a paper version or use online editorial strategies to keep track of versions and changes in the document.

When your group has finished editing the document, revise it into a final form. Then, conduct usability tests on the document, using other members of your class as subjects.

In a memo to your instructor, discuss the evolution of the document at each stage of the editorial process. Tell your instructor (1) how you used substantive editing to improve the document, (2) how you cycled the document among your team members, and (3) how you used usability testing to identify places where the document might be improved.

For additional technical writing resources, including interactive sample documents, document design tutorials and guidelines, and more, go to **www.mytechcommlab.com.**

Just My (Bad) Style

Gerald Williams was a manager for an electrical engineering company. The company was growing rather quickly. Only a couple years out of college, Gerald was named the project manager in charge of a new microelectronics design team.

The first thing he did was hire some new members for his team. He was fortunate to find some of the best and brightest electrical engineers available in his area.

He had one problem, though. One of the employees he hired, Darren Yander, was an intense guy who did not handle criticism well. He was one of those people who believed he did everything right.

Unfortunately, he also was a mediocre writer. His documents were hard to understand, and they were loaded with grammatical errors and typos. Gerald spoke with Darren about his writing skills on a couple occasions. He told Darren that he needed to spend more time revising and editing his drafts.

Darren would usually dismiss Gerald's concerns by saying he had a unique "style" that worked well for him: "All you want is for me to write like you. Well, I have a style of my own. I don't think my creativity should be smothered with some boring bureaucratic style."

Gerald decided that arguing with Darren about revising and editing his documents wasn't worth it. Almost all of Darren's documents were used exclusively within their team, so Darren's writing style didn't seem to matter.

Unfortunately, one of Darren's documents did go beyond the design team. One of his reports found its way into the hands of Margie Washington, one of the vice presidents of the company. Margie came to Gerald's office and asked him about an "odd" report that was written by someone in his group. Gerald immediately recognized that the report was written by Darren in his "style." The report was difficult to read and was very convoluted. It was also littered with grammatical errors and typos.

Gerald explained to Margie that he and Darren were working on improving his writing skills. Margie said, "You better do something about this problem—quick. I don't care if the guy is bright and hardworking. If another report like this one is sent to upper management, they're going to start looking for someone to fire. You're his boss. Do something."

After Margie left, Gerald called Darren to his office and explained the problem with the report. Darren grew very defensive. "It's not my fault upper management is too stupid to know good writing when they see it." He refused to make revisions to the report. He also refused to work on improving his writing skills.

If you were Gerald, what would you do in this situation? How would you handle it in a way that would allow all sides to be satisfied?

CHAPTER

13

Using E-Mail and Instant Messaging

In this chapter, you will learn:

- The importance of e-mail and instant messaging in the workplace.

- About e-mail as a workplace communication tool and how it is used in work settings.

- The basic features and functions of e-mail.

- Strategies for managing large amounts of e-mail in the workplace.

- The use of netiquette to avoid problems with e-mail.

- How to use instant messaging and understand its basic functions.

If you are like most people, e-mail is already an important part of your life, and instant messaging (IMing) or texting is becoming just as important. E-mail and instant messaging are now essential communication tools in technical workplaces. E-mail is used in situations that once called for memos, letters, or phone calls. Meanwhile, instant messaging gives us the ability to chat with others in "real time" through a variety of technologies, including computers, mobile phones, and personal digital assistants (PDAs).

More than likely, you already know how to use e-mail and are familiar with instant messaging. So, we won't spend too much time on the basics in this chapter. Instead, you will learn how to use these communication tools more effectively and more efficiently in the technical workplace.

E-Mail Basics

Today, you probably don't think twice about using e-mail to communicate with your friends, classmates, and instructors. Similarly, in the technical workplace, e-mail is a common way to interact with others, including your supervisors, co-workers, and clients.

What Is E-Mail?

But what exactly is e-mail? E-mail is an electronic system for sending and receiving messages and files over a computer network. E-mail is still relatively new in the workplace compared to the phone or paper documents. As a result, its purpose and uses are still being defined. Should it completely replace memos and letters in the workplace? Can e-mail be used to convey crucial information or confidential details? When is e-mail better than face-to-face communication? When is it worse?

Different expectations for e-mail usage can lead to misunderstandings because people often have very different ideas about how e-mail should be used. For example, as shown in Figure 13.1, a formal request for information might be overlooked by someone who believes e-mail is only for informal purposes. Or worse, it might receive a casual or flippant response that would seem very unprofessional. (Susan, in Figure 13.1, probably has a big problem.)

E-Mail Guidelines to Keep in Mind

When you are using e-mail, here are a few guidelines to keep in mind:

E-mail is increasingly used for professional purposes—Not long ago, e-mail was primarily used for quick comments or nonessential information. Today, e-mail is the principal form of communication in most workplaces, so people expect e-mail messages to be professional and nonfrivolous.

E-mail is a form of public communication—Your readers can purposely or mistakenly send your e-mail messages to countless others. So, you should not say things in an e-mail that you would not say openly to your supervisors, co-workers, or clients.

E-mail is increasingly formal—In the past, readers forgave typos, misspellings, and bloopers in e-mail messages, especially when e-mail was new and difficult to use. Today, readers expect e-mails to be more formal and reflect the quality expected in other forms of communication.

E-mail standards and conventions are still being formed—The use of e-mail in the workplace is still being worked out. So, you need to pay close attention to how e-mail is used in your company and in your readers' companies.

Different Expectations for E-Mail Usage

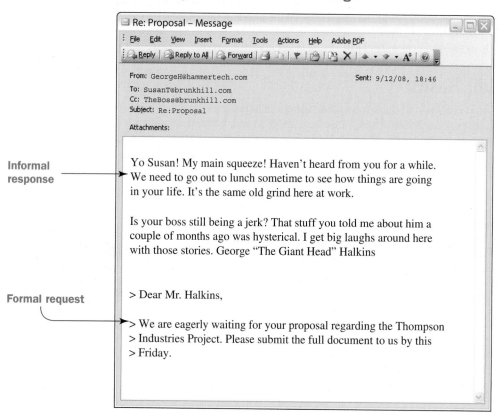

Informal response →

Formal request →

Legal Issues with E-Mail

You should also keep in mind that legal constraints shape how e-mail should be used in the workplace. Here are four things to remember:

Your e-mails can be saved and used against you—Courts of law treat e-mail as written communication, equivalent to a memo or letter. Your messages can be saved and used against you or your company in court cases.

Your workplace e-mails are owned by your employer—Employers are within their rights to read your e-mail without your knowledge or permission. Also, deleted e-mails can be retrieved from the company's servers, and they can be used in a legal case.

For websites that offer free e-mail service, go to
www.pearsonhighered.com/johnsonweb3/13.2

Copyright laws protect e-mail—E-mail, like any other written document, is protected by copyright law. So, you need to be careful not to use e-mails in any way that might violate copyright law. For example, if you receive an e-mail from a client, you cannot immediately post it to your company's website without that client's permission.

Harassment and discrimination cases often hinge on evidence found in e-mails—Careless e-mails about personal relationships or appearances can be saved and used against the sender in a court case. Indiscreet comments about gender, race, or sexual orientation can also have unexpected consequences. Your "harmless" dirty jokes sent to your co-workers might end up being used by a lawyer to prove that you are creating a "hostile workplace environment."

Link

For more information on copyright law, go to Chapter 5, page 111.

Link

For more information on ethics, see Chapter 5, page 92.

Features of E-Mail

Typical e-mail messages will have a *header* and *body* (Figure 13.2). They also have additional features like *attachments* and *signatures*.

Basic Features of an E-Mail

For attaching files

To, cc, bcc, and subject lines

Message area (Fred's response)

Previous e-mail from Cathryn (to which Fred is responding)

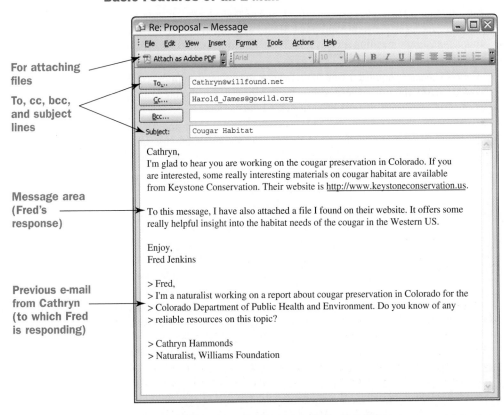

Figure 13.2:
A basic message has several different parts. In this message, the sender, Fred, is replying to Cathryn and copying (cc'ing) the message to Harold.

For examples of corporate e-mail policies, go to
www.pearsonhighered.com/johnsonweb3/13.3
To read some classic e-mail blunders, go to
www.pearsonhighered.com/johnsonweb3/13.4

Header

The header has lines for the following items:

To line—This line contains the e-mail address of the person or people to whom you are sending the e-mail.

cc and bcc lines—The cc and bcc lines are used to copy the message to people who are not the primary readers, like your supervisors or others who might be interested in your conversation. The cc line shows your message's recipient that others are receiving copies of the message too. The bcc line ("blind cc") allows you to copy your messages to others without anyone else knowing.

Subject line—The subject line signals the topic of the e-mail with a concise phrase.

Attachments line—This line signals whether there are any additional files, pictures, or programs "attached" to the e-mail message.

Message Area

The *message area* is where you will type your message to your readers. Like other written documents, your message should have an introduction, a body, and a conclusion. You should also be as brief as possible, because most people won't read long e-mail messages.

Introduction—The *introduction* should (1) define the subject, (2) state your purpose, and (3) state your main point. If you want the reader to do something, you should mention it right here in the introduction, because most readers will not read your whole message.

Body—The *body* should provide the information needed to support your e-mail's main point or achieve your e-mail's purpose. You should strip your comments down to only need-to-know information.

Conclusion—The *conclusion* should (1) restate the main point and (2) look to the future. Here is not the place to tell your reader something important that you didn't mention earlier in the e-mail or that you need him or her to do something, because most readers will not read your whole e-mail.

The message area might also include these other kinds of text:

Reply text—When you reply to a message, most e-mail programs allow you to copy parts of the original message into your message. These parts are often identified with > arrows running down the left margin (Figure 13.2).

Links—You can also include direct links to websites (Figure 13.2). Most e-mail software will automatically recognize a webpage address like http://www.predatorconservation.org and make it a live link in the e-mail's message area.

Attachments—Attachments are files, pictures, or programs that your readers can download to their own computer. If you would like to add an attachment to your e-mail message, click on the button that says "Attach Document" or "Attachment" in your e-mail software program. Most programs will then open a window that allows you to find and select the file you want to attach. If you attach a file to your e-mail message, you should notify the readers in the message area that a file is attached. Otherwise, they may not notice it.

GO TO THE NET

To learn about how e-mail has evolved, go to
www.pearsonhighered.com/johnsonweb3/13.5
For helpful advice about using e-mail, go to
www.pearsonhighered.com/johnsonweb3/13.6

Blake Murnan

ASSISTANT MARKETING MANAGER, MARKETING SERVICES, AGRILIANCE

Agriliance is a multinational company that offers agricultural products and services.

How do you avoid "career-limiting moves" when using e-mail?

It's amazing how many times people make career-limiting moves over e-mail. They say something over e-mail that ends up hurting their career.

What makes e-mail valuable as a communication tool is what makes it so perilous. Once released from your computer, e-mails become difficult to control or recall. In the end, you do not know who is going to end up with your e-mail. Therefore, as a professional in the workplace, you should take great care in how you construct an e-mail.

Here are some items to be aware of when constructing or responding to an e-mail:

- **Recipient**—Know who you are sending your e-mail to. It sounds obvious, but do you want to be the one who sends a sensitive e-mail intended to go to an individual but accidentally goes to the whole company? It happens—be careful.
- **Carbon copy (cc)**—Do not develop the reputation as a person who carbon copies (ccs) everyone, because people will begin to overlook your important e-mails. Direct your e-mails to the people who need to take action and copy the people who really need to be in the loop.
- **Blind copy (bcc)**—Always remember that there may be more people observing your actions than you think.
- **Spelling**—Whether misspellings are unintentional or intentional, the recipient may think you are incompetent or demonstrating a lack of respect by not taking the time to correctly communicate. Most e-mail programs have automatic spellcheck—use it.
- **E-mail police**—Your company may monitor your e-mails for items that are not work-related or may create a liability for the company. Therefore, try to limit non-work e-mails to a minimum.

Always remember that your e-mails are a reflection of you, as a professional, and therefore should be constructed and managed with great care.

You might write something like, "To this e-mail, I have attached a file that…."

Signature—E-mail programs usually let you create a *signature file* that automatically puts a *signature* at the end of your messages.

> Frank Randall, Marketing Assistant
> Genflex Microsystems
> 612–555–9876
> frandall@genflexmicro.net

Signature files allow you to personalize your message and add contact information. By creating a signature file, you can avoid typing your name, title, phone number, etc., at the end of each message you write.

For a list of emoticons, go to
www.pearsonhighered.com/johnsonweb3/13.7
Want to find a cool signature file? Go to
www.pearsonhighered.com/johnsonweb3/13.8

Managing E-Mail

Before long, e-mail messages begin to pile up in your Inbox, and you start to feel like you will never be able to read them all. This is called an *e-mail glut*. To avoid an e-mail glut, you need to develop a strategy for managing your e-mail messages. A good e-mail management strategy follows four steps.

Step 1: Prioritize

When you open your e-mail Inbox, begin by mentally prioritizing your messages. Some messages, like those from your supervisor or clients, probably need immediate attention. Other messages, like personal messages and requests for information, can be handled after the most urgent e-mails.

Step 2: Read and Reply

E-mail messages tend to fall into three basic categories: action-item messages, informational messages (FYI), and junk. E-mails requiring action should receive the highest priority. E-mails that are informational messages can be read and deleted. Junk can be deleted immediately.

Now that you have prioritized your e-mails, start working with the e-mails that require you to take action. Try using the "3-D method":

Delegate—Can someone else or should someone else do this work for you? If someone else can, hit the Forward button and send him or her the message.

Do now—Can you answer this request right now? If you can do what is being requested in two minutes or less, you should do it immediately.

Do later—Some actions require more time. Jot down these tasks on your "to do" list. Also, write down a day and time when you will do these tasks.

> **AT A GLANCE**
>
> ### Reading and Replying to E-Mail
>
> - Delegate—Should someone else do this work?
> - Do now—Can I handle this request right now?
> - Do later—Should I handle this request later?

Link

For more information on using "to do" lists with online calendars, see Chapter 6, page 120.

Step 3: Sort and Archive

When you are finished reading and responding, you should then sort the remaining e-mails into folders. Even the simplest e-mail programs will allow you to create folders for sorting your messages (Figure 13.3).

In most e-mail programs, you can also create *filters* or *rules* for your e-mail that will automatically sort your messages into folders. In your e-mail program, look for a function called "Create Filter" or "Create Rule" or something similar. When this command is selected, a box will usually appear that allows you to create filters that sort e-mails by *sender, subject line,* or *keywords.* Filters are a handy way to sort e-mail, especially if you receive many messages each day.

You might also decide to *archive* more important e-mail messages as a way to keep a record of your decisions and actions. E-mails can be archived in

GO TO THE NET

Tired of unwanted e-mail? For help, go to
www.pearsonhighered.com/johnsonweb3/13.9

Mailboxes and Folders

E-mails are sorted into these folders.

Figure 13.3: Your e-mail program will allow you to sort your e-mails into folders. Some programs allow you to sort files automatically by keywords.

Managing E-Mail	

AT A GLANCE

Managing E-Mail

- Step 1: Prioritize—Which messages are the most important?
- Step 2: Read and reply—Which messages need action right now?
- Step 3: Sort and archive—Which messages should be saved?
- Step 4: Purge—Which messages can be deleted?

electronic form (e.g., on your hard drive) or print form (printed out and filed).

Step 4: Purge

No matter how hard you try, e-mail messages will still pile up. So, once a week, you should set aside time to *purge* your e-mail account. Start the purging process by doing something about all the messages piling up in your Inbox and your e-mail account's other folders.

Be realistic about the messages you save. Will you ever need these messages again? If these old messages are important, ask yourself whether they should then be archived on your hard drive or printed out and filed. If they don't need to be kept, delete them.

GO TO THE NET

To learn more about creating filters, go to
www.pearsonhighered.com/johnsonweb3/13.10
To find more tips on managing e-mail, go to
www.pearsonhighered.com/johnsonweb3/13.11

Managing E-Mail 337

E-Mail Netiquette

E-mail has its own etiquette, or "netiquette." Here are a dozen netiquette guidelines that you can follow in the workplace:

Guideline 1: Be concise—Keep the length of your messages under a screen and a half. Longer messages probably won't be read carefully.

Guideline 2: Provide only need-to-know information—Decide who needs to know what you have to say, and send them only information they need—nothing more.

Guideline 3: Treat the security of the message about the same as a message on a postcard—Recognize that anyone can pass your message along or use it against you. If the message is confidential or proprietary, e-mail is not an appropriate way to send it.

Guideline 4: Don't say anything in an e-mail that you would not say in a meeting or to legal authorities—Hey, accidents happen. You might send something embarrassing to the entire organization. Or, someone else might do it for you.

Guideline 5: Never immediately respond to a message that made you angry or upset—Give yourself time to cool off. You should never write e-mail when you are angry, because these little critters have a way of returning to bite you.

Guideline 6: Avoid using too much humor, especially irony or sarcasm—In an e-mail, your witty quips rarely come off exactly as you intended. Remember that your e-mails can be easily misinterpreted or taken out of context. The American sense of humor, meanwhile, is often disturbing to international audiences; so if your message is going overseas, keep the humor to a minimum.

Guideline 7: Be extra careful about excerpting or forwarding the e-mail of others—If there is the remotest chance that someone else's message could be misunderstood, paraphrase the meaning in your own message instead of forwarding his or her message or copying parts of it.

Guideline 8: Don't be unpleasant in an e-mail if you would not be unpleasant face to face—E-mail provides a false sense of security, much like driving in a car. Remember that those are real people on that information superhighway, and they have feelings. Plus, they can save your abusive or harassing messages and use them as evidence against you.

Guideline 9: Never send anything that could be proprietary—There have been some classic cases where classified information or trade secrets have been accidentally sent to competitors or the media via e-mail. If the information is proprietary, use traditional routes like the regular mail or a courier.

Guideline 10: Don't flame, spam, or chain-letter people at work—If you must, do these things from your personal computer at home. At work, assume these activities will ultimately be used against you. You might be fired.

Guideline 11: Think twice before you send "urgent" messages that are making their way across the net—There are lots of hoaxes about computer viruses "erasing your hard drive." These hoaxes gain new life when some gullible person starts sending them to all his or her co-workers. Some virus alerts are real, but most are not. Wait a day or two to see if the threat is real before you warn everyone you know.

Guideline 12: Be forgiving of the grammatical mistakes of others—but don't make them yourself—Some people treat e-mail informally, so they rarely revise (though they should). Grammar mistakes and misspellings happen—so you should forgive the sender. On the other hand, these misteaks and typos make the sender look stoopid. Don't send out sloppy e-mail messages.

Using E-Mail Internationally

The speed of e-mail makes it an ideal way to communicate and build relationships with international clients and co-workers. In many cases, e-mail has replaced both phone calls and letters, because it has the immediacy of the phone while giving readers the time to translate and consider the message being sent. When working internationally, you will discover that many clients and co-workers prefer to conduct business via e-mail.

A challenge with using e-mail internationally is that North Americans tend to view e-mail as an "informal" or even "intimate" medium for communication. As a result, they regularly stumble over social norms and conventions of other cultures. Too quickly, Americans try to become too friendly and too informal. Also, in e-mail, Americans can be sloppy with grammar, spelling, and word usage, causing significant problems for non-English speakers who are trying to translate.

Here are some tips for using e-mail internationally:

Allow time to form a relationship—Introduce yourself by name and title, and provide some background information about your company and yourself. Tell the readers where you are writing from and where you are in relation to a major city. Don't rush into making a request, because doing so will often come off as pushy or rude.

Use titles and last names—Titles are often much more important in other cultures than in the United States. Minimally, you should use titles like Mr., Ms., or Dr. (Mrs. can be risky). If you know the proper titles from the readers' culture, such as Madame, Herr, Signora, then you should use them. Eventually, your international clients or co-workers may want to move to a first-name relationship, but it's often a good idea to let them make that first move.

Focus on the facts—In your message, concentrate on factual issues regarding the who, what, where, when, how, and why. Cut out other nonessential information because it will cloud your overall message.

(continued)

Want to learn more about e-mail netiquette? Go to
www.pearsonhighered.com/johnsonweb3/13.12
Want to know if a virus is a hoax? Go to
www.pearsonhighered.com/johnsonweb3/13.13

E-Mail Netiquette   339

Talk about the weather—If you want to personalize your message, talking about the weather is a safe topic. People are often curious about the weather in other parts of the world. It's a safe, universal topic that will allow you to get beyond just factual information. (Hint: Convert all temperatures to Celsius and measurements into metric.)

Use attachments only when needed—In some parts of the world, e-mail systems cannot handle large attachments. Inbox quotas may be too small or baud rates may be too slow. A good approach is to send an initial e-mail that asks if an attachment would be welcome. Then, if the reader tells you it will work, you can send it in a follow-up e-mail.

Use plain text—You should assume that your readers can only receive plain text. So, turn off any special characters (smart quotes, dashes, emoticons, etc.) because they will often come out as gibberish at the other end. Also, assume that any embedded website addresses and e-mail addresses won't be shown as links. You should spell these addresses out in full, so the readers can cut and paste them for their own use.

Limit or avoid photographs and graphics—Photographs, background images, and other graphics don't always transfer properly when they are sent internationally, and they can require a great amount of memory. Plus, photographs can mean something unexpected to readers from other cultures.

Avoid clichés at the closing—Commonly used closings like "Do not hesitate to contact me," "If there's a problem, just holler" or "Don't be afraid to call," do not translate well into other languages and may be confusing to the readers.

Avoid humor—Attempts to be funny or tell jokes can backfire. Humor usually relies on cultural knowledge, so its meaning can be misinterpreted by the readers. In some cases, the humor might be seen as insulting.

Create a simple signature file with your contact information—International e-mails are often printed out, which can cause the sender's e-mail address and other contact information to be separated from the message. A concise signature file that appears at the bottom of each e-mail should include your name, title, e-mail address, postal address, phone number, and corporate website.

Use simple grammar and proofread carefully—Simple sentences are easier for human and machine translators to interpret. Complex grammar or grammar mistakes greatly increase problems with translation.

Using Instant Messaging at Work

Instant messaging has become an important tool in the technical workplace. Texting, meanwhile, is also becoming more popular as a way to keep in touch.

Unlike e-mail, instant messaging happens in real time. When you hit "Send," your message will immediately appear on the receivers' screens (Figure 13.4). They can then respond, sending you an instant message.

To find a listserv on a subject that interests you, go to
www.pearsonhighered.com/johnsonweb3/13.14

An Instant Messaging Interface

These buttons add or remove participants.

These buttons allow you to page or e-mail a participant.

Add — Remove ⊘ | Page 📱 | Mail ✉

Instant Messenger

Tom: Hey Folks, let's get to work on this report.

Jose: Yeah, let's go.

Discussion in progress

Sara: I've been looking over the draft Tom sent out, and I think it looks pretty good so far. I think the introduction needs a little work. Right now, it doesn't really explain clearly why waste treatment is a problem here at the Toledo plant.

Tom: Isn't that obvious? I mean, doesn't everybody see waste as a problem?

Sara: I'm not sure. It's obvious to us, but the people at the main office don't always see things the same way.

Jose: I'm with Sara. I think we need to be very clear about the importance of the problem up front (the introduction).

Contact Groups

▼ Project 1
  ○ Jose
  ○ Sara
  ○ Tom
▶ Project 2
▶ Project 3
▶ Private

Audio ———○———— 🔊

Video ▶ ⏹ 💬

Message **B** *I* U̲ 💬

These buttons add audio and video.

Sara's next message

All we need to do is add a couple sentences that state some of the more obvious issues. You know, show them the cost of doing nothing.

Figure 13.4: Instant messaging allows you to speak synchronously with others. Here Sara is about to send a message to Tom and Jose.

Some instant messaging programs will allow you to send live audio or video. With a microphone and/or digital camera (webcam) mounted on your computer, you can let other people see and hear you talking.

Several companies, such as Yahoo!, AOL, Microsoft, and Google, will give you free access to instant messaging. To use their systems, you will need to download their software to your computer from their websites.

Most instant messaging programs ask you to sign in with a password. Then, you can create a list of other people with whom you want to converse. To begin a conversation, you can check your IM program to see who is on the network. If they are on the network, you can send them a message.

Want to find websites that offer free instant messaging? Go to **www.pearsonhighered.com/johnsonweb3/13.15**

Using Instant Messaging at Work 341

Here are a few guidelines for using instant messaging at work:

Schedule instant messaging meetings—If you are going to use instant messaging to meet with people, you should schedule a time for the virtual meeting. Without a scheduled meeting, you may not be able to pull people away from their other work.

Write concise messages—Instant messaging conversations usually move along quickly, especially if you are talking with more than one person. So, keep your messages short and easy to read. You might learn to use some of the common instant messaging abbreviations to keep the messages short (Figure 13.5). The common emoticons for e-mail also work well.

Cut the chitchat—Instant messaging can be fun, but time is valuable. When collaborating with others, keep the personal comments to a minimum.

Instant messaging is a valuable new tool that is becoming as common as e-mail in the workplace. As companies look to cut costs and improve efficiency, instant messaging is a good way to replace in-person meetings, while allowing people to collaborate through their computers.

Instant Messaging Abbreviations

Abbreviation	Meaning
AFAIK	as far as I know
B4N	bye for now
BBL	be back later
BRB	be right back
BTDT	been there done that
BTW	by the way
CUL	see you later
F2F	face to face
G	grin
GDR	grinning, ducking, and running
HTH	hope this helps
IDK	I don't know
IMO	in my opinion
JTLYK	just to let you know
LOL	laughing out loud
LTL	lets talk later
TTFN	ta ta for now
TX	thanks
WFM	works for me

Figure 13.5: These abbreviations can help you communicate more efficiently and add emotion to your messages.

342   Chapter 13
Using E-Mail and Instant Messaging

GO TO THE NET

For websites that discuss using IM, go to
www.pearsonhighered.com/johnsonweb3/13.16
More IM abbreviations can be found at
www.pearsonhighered.com/johnsonweb3/13.17

Using Text Messaging at Work

There are some important differences between instant messaging and text messaging, so let's focus on texting, or Short Message Service (SMS), for a moment. One difference is that text messaging is usually done with a mobile phone or PDA, while IMing tends to be tied to a computer. Second, most phone services will only allow you a maximum of 160 characters in a text message, forcing you to be more concise. And, third, while texting, you might need to use the number pad to create words, unless you have a phone or PDA with a full keyboard. As a result, there are some limitations to texting that you will need to work around, especially if you are going to use texting at work.

If you haven't used texting before, here's how to start:

1. Go to the menu of your mobile phone or PDA.

2. Select the option called "Messages" or something similar.

3. Enter the phone number of the person who will receive the text message.

4. Type in your message.

If you are using the number pad, you will need to press the number that corresponds to the letter you are typing (Figure 13.6). Pressing 2 means 'a.' Pressing 2 twice is 'b.' Pressing 5 is 'j' and pressing 5 twice is 'k.' It takes practice to learn how to text quickly.

Texting at Work

Figure 13.6: Texting on a mobile phone usually requires some fast fingerwork on the number pad. Each number corresponds to three or four letters, so you need to press a number from one to four times to bring up the letter you want.

Due to the layout of the number pad and the limited characters in a message, texting uses an even more cryptic language than instant messaging. Vowels are commonly removed from any word and *text message shorthand* relies more on numbers and symbols than IMing does. For example, "b4n" means "bye for now" and "ne1" means "anyone"

It's impossible to remember all the shorthand abbreviations and acronyms. So, if you're not sure what an acronym or abbreviation means, try sounding it out. When you're typing a text message, make up your own abbreviations and acronyms when needed.

So, what about your boss or your professor? Is he or she going to be able to understand your shorthand text messages? Maybe, but probably not. So, you will need to spell words out, except when you are using an obvious IM acronym.

A good guideline for using texting at work is to send shorthand messages only to people who are your age and younger or at your rank in the company or below. People who are older or higher up than you should receive spelled-out text messages.

Of course, there are exceptions. Some bosses and professors can text with anyone. But if they can't, your shorthand message might just look like gibberish to them. That can lead to misunderstandings and frustration, which will probably be blamed on you.

Texting is with us to stay, and it will become more and more important in the technical workplace. As full keyboards become more common and restrictions on character length loosen, the cryptic abbreviations and acronyms will probably fade away—maB N.

CHAPTER REVIEW

- E-mail is an essential tool in the workplace, and instant messaging is quickly growing in popularity. These tools allow you to communicate, collaborate, and connect to others.

- E-mail is a hybrid technology that combines the functions of the phone, memos, and letters.

- In the workplace, e-mail should be used for professional purposes only because it is a public form of communication and should always convey professionalism.

- An e-mail has the following features: header, message area, signature, and attachments.

- Managing e-mail requires some discipline and diligence. To avoid e-mail glut, you need to (1) prioritize, (2) read and reply, (3) sort and archive, and (4) purge your messages.

- As you are reading and responding to your e-mails, remember to delegate, do now, or do later.

- Netiquette involves guidelines for politeness and ethical behavior with e-mail.

- Instant messaging and texting are synchronous ways to communicate and hold "virtual meetings" with other people.

Individual and Team Projects

1. Use a search engine to find a special-interest group on the Internet that you might like to join. Send the webmaster a polite e-mail that requests information about the group. Pay special attention to the content and organization of your e-mail. Also, pay attention to netiquette, trying to avoid any statements that might offend the recipient of your e-mail.

2. One of your co-workers likes to send out jokes to your whole work team. The jokes are tasteful, but they clog up your Inbox. Write an e-mail to your co-worker in which you tactfully ask him to reduce the number of jokes he sends out.

3. Look at your own e-mail account. Use the e-mail management scheme described in this chapter (prioritizing, reading and replying, sorting and archiving, and purging) to clean up your Inbox and purge old e-mails. Try to manage your e-mail for three weeks with this system, spending some time once a week cleaning up your e-mail account.

4. Find an e-mail that could be misread to sound angry or nasty. Then, rewrite the e-mail in a way that avoids the potential nastiness. If you do not have a sample e-mail, ask one of your classmates or co-workers to send you an e-mail that uses an angry tone.

5. Read the Case Study at the end of this chapter. Pretend you are Jane. Write an e-mail to your co-workers in which you explain your complaint to your former supervisor. Trade your e-mail with one of the other people in your class to see whether you would handle Jane's problem similarly.

Collaborative Project

1. With a group of classmates, develop your own set of netiquette guidelines for your class or university. What are some guidelines for proper use of e-mail in the classroom? How can these guidelines be expanded to suit the entire university?

2. Using instant messaging, collaborate with a group of classmates to compose a memo to your college or company. In the memo, discuss a specific problem on campus (e.g., the need for better food, safety, faster computers) and ask what can be done about the problem. Pay attention to how instant messaging helps or impedes your ability to compose the memo. When you are finished with the memo, discuss how the writing process changes with instant messaging.

3. With a group, use e-mail or instant messaging to have a practice "flame" about a specific topic (e.g., television program, politician, musician, college team). Without being offensive, argue emphatically about the topic, trying to score points on the other members of the flame. As you flame with your group, notice your reactions to their e-mails. Does each e-mail add to the stress and tension, rarely resolving the conflict? What are some ways you might stop the flaming? How can you and your group restore good relations using e-mail?

Exposed

Jane Fredrickson was asked to join an existing work team at her office. The group had been together for a long time, and they all got along really well. Jane was excited about working with some new people.

Unfortunately, the closeness of her new work team soon became a problem. The others on the team had their own language and code of conduct. They joked around with each other with instant messaging. They sent each other pictures of their children in the bathtub. They sent each other dirty jokes. Frankly, Jane was becoming fed up with their behavior.

So, one day she decided to vent her annoyance in an e-mail to Harold Grimes, her former supervisor. After listing her complaints, using words like "annoying," "idiotic," and "unprofessional," she wrote, "I wish someone would do something about these people and their juvenile antics."

A week later, Jane's supervisor, Alisha Jones, sent an e-mail to the work team, asking them to put an end to the fun and games (Figure A). To her e-mail, she attached a couple of files that spelled out the company's rules for using e-mail and instant messaging. She also mentioned that there had been a complaint.

To Jane's horror, part of her original message to Harold was copied below Alisha's message. Harold had apparently forwarded part of Jane's original message to Alisha. And, Alisha had simply forwarded it to the group with her own message added on.

Needless to say, the other members of Jane's work team were going to be upset. Even though her name was not on the message, Jane knew they would figure out who had complained about them. What should she do now?

Problems with Using E-Mail

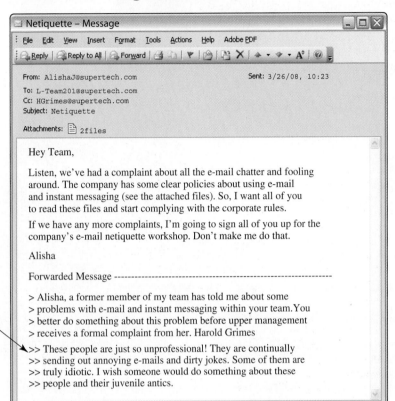

Figure A:
Part of Jane's message is copied below Harold's.

Netiquette – Message

File Edit View Insert Format Tools Actions Help Adobe PDF

Reply | Reply to All | Forward

From: AlishaJ@supertech.com   Sent: 3/26/08, 10:23
To: L-Team201@supertech.com
Cc: HGrimes@supertech.com
Subject: Netiquette

Attachments: 2files

Hey Team,

Listen, we've had a complaint about all the e-mail chatter and fooling around. The company has some clear policies about using e-mail and instant messaging (see the attached files). So, I want all of you to read these files and start complying with the corporate rules.

If we have any more complaints, I'm going to sign all of you up for the company's e-mail netiquette workshop. Don't make me do that.

Alisha

Forwarded Message ---

> Alisha, a former member of my team has told me about some
> problems with e-mail and instant messaging within your team. You
> better do something about this problem before upper management
> receives a formal complaint from her. Harold Grimes
>> These people are just so unprofessional! They are continually
>> sending out annoying e-mails and dirty jokes. Some of them are
>> truly idiotic. I wish someone would do something about these
>> people and their juvenile antics.

Oops. Here is part of Jane's original message.

For additional technical writing resources, including interactive sample documents, document design tutorials and guidelines, and more, go to **www.mytechcommlab.com.**

CHAPTER
14
Designing Websites

In this chapter, you will learn:

- How websites are both similar to and different from print documents.

- The basic features of websites.

- How to plan and research a website.

- How to organize and draft a website.

- How to use web-authoring software.

- How to use style to make a website more readable.

- How design principles are used to create an effective website interface.

- The importance of good revision, editing, and proofreading in creating websites.

In a short time, websites have become an essential form of communication in technical workplaces. Today, websites are regularly written and designed by engineers, scientists, and other technical personnel. In some high-tech fields it is expected that new employees will know how to write for the web.

The first thing you should remember about websites is that *they are documents.* They may look different from paper-based documents and they may be used differently, but they are still *written texts* with words and images. As a result, you can apply many of the communication strategies in this book to developing these screen-based documents.

An important difference, though, is that people tend to read websites visually and spatially, scanning from block of information to block of information. They *navigate* within the website, among these blocks of information. Because websites are visual-spatial texts, they are composed and designed differently than paper-based documents. This chapter will show you how to take advantage of the visual-spatial qualities of on-screen documents.

Basic Features of a Website

A website is a group of related webpages that are linked together into a whole document. Websites have a few different kinds of pages.

>   **Home page**—The home page is usually the first page of the site. It identifies the subject and purpose of the website, while forecasting its overall structure. In many ways, the home page is like an introduction and a table of contents for the website (Figure 14.1).

Similarities Between Websites and Other Documents

Website	Paper-Based Document
Home page	Introduction, or table of contents
Node page	Chapter or section
Page	Page
Linking	Turning a page
gifs, jpegs, tiffs	Graphics (pictures, drawings)
Navigating	Reading and scanning
Surfing the web	Reading and scanning
Site map	Index
Search engine	Index
Navigation bar	Table of contents or tabs

Want to learn more about how the Internet works? Go to
www.pearsonhighered.com/johnsonweb3/14.1

Basic Features of a Website   349

A Home Page

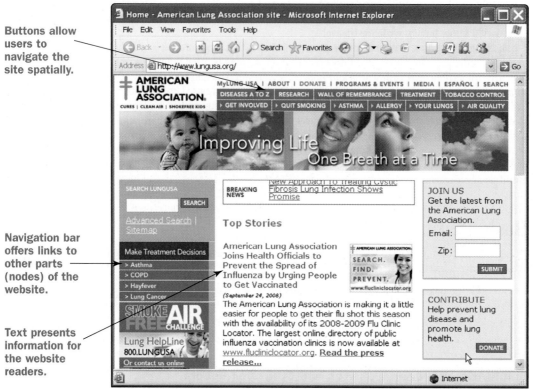

Buttons allow users to navigate the site spatially.

Navigation bar offers links to other parts (nodes) of the website.

Text presents information for the website readers.

Source: American Lung Association, http://www.lungusa.org.

Figure 14.1:
A website uses visual-spatial strategies to organize information. In this website from the American Lung Association, notice how the text and images are presented visually for easy access to the information on the page and site.

Node pages—These pages tend to be linked to the home page. They subdivide the website's content into smaller topics. For example, a university's home page will have links that go to node pages like *Colleges and Departments, Libraries, Students,* and *Faculty and Staff.* These are all nodes in the website.

Pages—Individual pages contain the facts, details, images, and other information that readers are seeking.

Navigation pages—The navigation pages help readers find the information they are seeking. These pages include site maps that show the contents of the site like an index in a book. They also include search engines that allow readers to search the site with keywords.

Splash page—This optional page comes before the home page and acts like a cover to a book. Splash pages are mostly decorative, using animation or images to set a specific tone before readers enter the website through the home page.

To see some interesting splash pages, go to
www.pearsonhighered.com/johnsonweb3/14.2

Mitch Curren
COPYWRITER, BEN & JERRY'S ICE CREAM

Mitch Curren writes copy for both printed materials and the web for Ben & Jerry's Ice Cream in South Burlington, Vermont.

How is writing for a website different from other kinds of writing?

In many ways, writing for the web is similar to other kinds of writing. As Ben & Jerry's Chief S'creamwriter & LexiConeHead, I'm responsible for writing and editing copy that communicates the company's *brand* voice, the voice that conveys Ben & Jerry's unique brand personality in ways our customers find recognizable and engaging.

Whether I'm creating copy for the back of an ice cream pint, for a printed brochure, or for our website and e-newsletter audience, ensuring brand voice and style are constant objectives. I also work closely with various project managers to meet project-specific goals and objectives.

Quite often that means I put on my "Composition 101" instructor's hat to make sure folks' copy requests contain basic copy-directional essentials:

- What's the message?
- Who's your audience?
- What do you want the message to achieve?

While the same writing basics and project dynamics apply to copy written for our website, the dynamics of the web itself are different. Most people don't "read" web content the same way they'd read a printed document; hypertext and hyper-links enable all kinds of browsing, surfing, and scanning opportunities no linear print materials can match.

Those and other opportunities are what I keep in mind when I write stuff for our website. Whatever topic I'm covering or message I'm conveying needs to be organized into informational "chunks"—often with links to additional chunks—that offer a quick gist for folks who like to scan as well as links for folks who want further information.

In the end, though, don't be too concerned about all the website jargon. If you think about it, all these features of websites are very similar to features in paper-based documents. For example, the home page is just like the introduction to any other document. Selecting a link is just like turning a page in a book or magazine.

One significant difference, however, involves the organization of the information. As shown in Figure 14.2, websites usually use spatial (or nonlinear) organizational patterns. In a spatial organizational pattern, readers can follow a variety of paths after reading the home page. The website lets readers move around freely in the document to find the information they need. Keep in mind that websites come in many shapes and sizes, and they are usually not as symmetrical as the diagram shown in Figure 14.2.

Basic Pattern for a Website

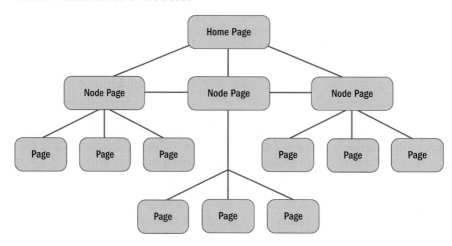

Figure 14.2: Websites are spatial rather than linear. They allow readers to move in a variety of different directions. In this example, readers can go directly to any of three pages from the node page without going through other pages.

Planning and Researching a Website

When planning a new website, you should start by making some strategic guesses about the kinds of people who might be interested in using it. Begin the planning phase by answering the Five-W and How Questions about your website:

Who will be visiting this website?

Why are they using this website?

What information or kind of information are they looking for?

Where will the information in the site be useful to them?

When will they use this information?

How might readers use this website?

With these questions answered, you can think more deeply about how the website should be developed. Consider the website's subject, purpose, readers, and context of use.

Subject

Clearly define the boundaries of your website by determining what is "inside" the subject area and what is "outside." Websites can hold almost limitless amounts of information. So, you should include only information that your readers might need. For example, the boundaries of your subject might be defined in the following way:

Link

For more information on defining the boundaries of a subject, go to Chapter 2, page 21.

My website will be about Nobel physicist Richard Feynman's career as a scientist and professor. It won't be about his personal life.

We want to create a website that explains the paleontology research being pursued at Dinosaur National Monument and allows paleontologists to network with each other.

To see other organizational structures for websites, go to
www.pearsonhighered.com/johnsonweb3/14.3

Purpose

Like any other document, you should define the purpose of your website. For the most part, websites are meant to be informative, meaning they are developed to provide information to readers, rather than to persuade them. Consequently, some keywords in a statement of purpose might include the following:

Link

For more tips on defining a document's purpose, see Chapter 2, page 21.

to inform	*to demonstrate*
to explain	*to illustrate*
to train	*to report*
to acquaint	*to summarize*
to teach	*to collect*
to instruct	*to familiarize*
to present	*to show*
to display	*to exhibit*

A purpose statement for a website might read something like the following sentences:

> This site is designed to present the life and works of Richard Feynman.

> Our intent is to familiarize paleontologists with prior research at Dinosaur National Monument so they can plan future explorations and digs.

Make sure you clearly define your purpose *before* you start creating the website. You should be able to state your website's purpose in one sentence. If you cannot boil it down into one sentence, your website will probably not be focused enough.

Readers

Even though just about anyone can use your website, you should target specific readers by paying attention to their needs. As discussed in Chapter 3, you should begin analyzing your readers with a Writer-Centered Chart (Figure 14.3).

PRIMARY READERS (ACTION TAKERS)   Your primary readers will use the information on your website to make a decision or take action. What are their needs, values, and attitudes? Are they researching your company's services or products? Are they collecting information for a report? Are they looking for help to solve a problem?

SECONDARY READERS (ADVISORS)   These readers will likely be advisors or supervisors to the primary readers. Are they reviewing your information for accuracy, or are they confirming facts? Are they advising the primary readers about a decision? What is their level of expertise with your website's subject?

TERTIARY READERS (EVALUATORS)   Other readers also may have various reasons for visiting your site. How might your company's competitors use the information on your website? How might a journalist or lawyer use the information? What kinds of proprietary or personal information would not be appropriate for public access?

GO TO THE NET

For worksheets that will help you analyze your website's rhetorical situation, go to **www.pearsonhighered.com/johnsonweb3/14.4**

Planning and Researching a Website

353

A Writer-Centered Chart for a Website

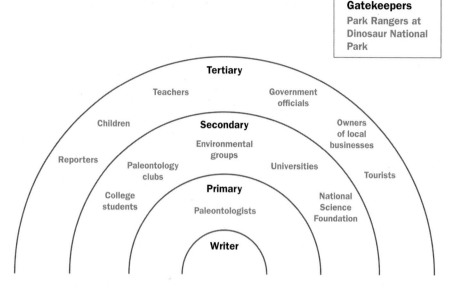

Gatekeepers
Park Rangers at Dinosaur National Park

Figure 14.3: The writer-centered chart is a helpful way to identify and sort out the readers who would most likely be interested in using your website.

GATEKEEPER READERS (SUPERVISORS)   Your gatekeepers are mostly interested in the accuracy of the site and whether it reflects the image and values the company wants to promote. Specifically, your supervisors will want to check your site to make sure it reflects the mission of the company.

Link

For more information on analyzing readers, see Chapter 3, page 40.

Context of Use

You should also consider the physical, economic, ethical, and political issues that will shape how your readers use your website.

PHYSICAL CONTEXT   Involves the physical places where people might use your website. Will readers be accessing the site from home, in their office, in a meeting room, or perhaps at an Internet cafe? Are they at a desk, on the factory floor, or in an airport? Are they accessing through a wireless network? If so, is the network Wi-Fi, WiMAX, or Bluetooth? How fast is their connection to the Internet?

ECONOMIC CONTEXT   Involves the financial issues that shape how readers will interpret your website. What is their likely economic situation or status? What are the trends in the industry or national economy? What kinds of economic decisions are you asking readers to make?

Link

For more information on ethics, see Chapter 5, page 92.

ETHICAL CONTEXT   Involves the personal, social, and environmental issues that your website touches. Where does your website involve issues of rights, laws, utility, or caring? How should these ethical issues be handled?

AT A GLANCE

Defining a Website's Rhetorical Situation

- Subject: What information is inside the scope of the website, and what isn't?
- Purpose: In one sentence, what is the purpose of the website?
- Readers: Who will be reading the site, and what kind of information are they looking for?
- Context: What physical, economic, ethical, and political factors will shape how the website is written and read?

POLITICAL CONTEXT   Involves the micropolitical and macropolitical issues that might influence how your readers interpret the website. For example, does your website step on some toes within the company? How does it fit in with your company's overall political stance?

As you think about the rhetorical situation for your website, put yourself in your readers' place. Try to anticipate the kinds of information they are looking for and why they need that information. Putting yourself in your readers' place will help you enormously as you are writing and designing the website.

Link

For more information on defining a document's context of use, go to Chapter 3, page 46.

Websites for International and Cross-Cultural Readers

It is becoming increasingly important to design websites for international and cross-cultural readers. Up to this point, English has been the unofficial language of the web, but increasingly, non-English-speaking users around the world are accessing the Internet.

Of course, it would be impossible to anticipate the needs of all potential readers around the world, but you can make your website more usable in a few important ways.

Translate the website—The global marketplace allows your company to attract potential clients and customers around the world. So, if your company regularly does business with people from another country or culture, it might be a good idea to make your website, or at least parts of your website, available in the readers' language. You can use Google Translate (translate.google.com) to translate your website for you, but it isn't as reliable as a human translator (Figure 14.4).

Use common words—Try to use words that are commonly defined in English. The meanings of slang and jargon words change quickly, sometimes leaving international readers confused.

Avoid clichés and colloquialisms—Informal American English includes phrases like "piece of cake" or "miss the boat" that might be meaningless to people from other cultures. Also, sports metaphors like "kickoff meeting" or "hit a home run" sound very odd to people who are not familiar with American football and baseball.

Avoid cultural icons—Symbols, especially religious symbols, should be avoided where possible and carefully used where necessary.

Minimize humor—American humor does not translate well into other cultures. So, attempts to be funny on your website might be offensive or just confusing.

Link

For more information on writing for international and cross-cultural readers, go to Chapter 3, page 53.

Chapter 3 discusses writing for international and cross-cultural audiences in more depth. In most cases, these guidelines for writing for cross-cultural readers are applicable to websites also.

GO TO THE NET Want to know more about ethical issues involving websites? Go to www.pearsonhighered.com/johnsonweb3/14.6

Planning and Researching a Website 355

Translating a Website with Google Translate

Figure 14.4:
Here is a Norwegian translation of the Professional Writing Program's website at Purdue University done with *Google Translate* (translate. google.com).

Source: Purdue University, Professional Writing Program, http://www.digitalparlor.org/pwenglish/.

Organizing and Drafting a Website

Organizing and drafting a website is not all that different from organizing and drafting a print document.

Organizing the Website

Professional website developers often prefer to start creating a new site by mapping out its contents on a whiteboard, a large piece of paper, or a computer screen. Logical mapping is a good way to sort out the content of the website and to develop an efficient organizational scheme for it.

To map out the contents of the site, start by writing your website's purpose in the middle of your screen or a sheet of paper. Then, use lines and circles to begin identifying the contents of the site (Figure 14.5).

You might also try a low-tech method that is popular with professional web designers—using sticky notes to map out your website on a blank wall. This

For more advice about writing websites for international and cross-cultural readers, go to
www.pearsonhighered.com/johnsonweb3/14.7

Mapping Out a Website

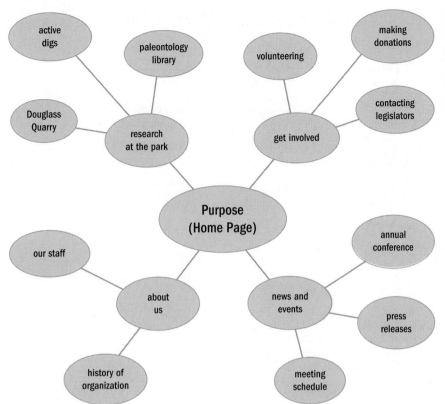

low-tech method allows you and your team to move the notes around to look at different ways to organize the site. Another advantage to sticky notes is that they can be added or crumpled up with ease.

Creating Levels in the Website

Your logical map and research will likely give you a sense of how many levels are needed in the website. The map shown in Figure 14.5, for example, might translate into a structure like the one shown in Figure 14.6.

How many levels should you create in a website? Professional website designers use the following guidelines to determine the number of levels needed in a website:

* a maximum of three links for the most important information.
* a maximum of five links for 80 percent of all information.
* a maximum of seven links for all information.

Levels in a Website

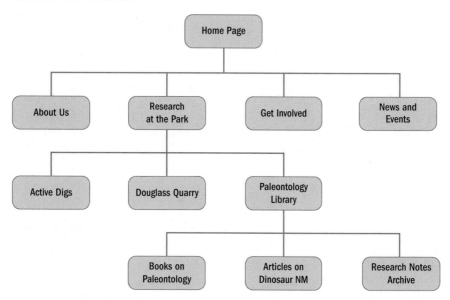

Figure 14.6: A logical map can be turned into levels of information. Each level offers information that is increasingly specific.

These guidelines are helpful, because if you force your readers, especially customers, to wade through too many pages, you risk losing them. If you make them work too hard, they will grow frustrated and give up.

Like any guidelines, though, these are not rules to be followed absolutely, because some situations resist your efforts to adhere to the guidelines. Nevertheless, if you notice that your website's structure goes against these guidelines, you might look for ways to reduce the number of levels between your readers and the information they want.

Drafting the Home Page

Your home page should be similar to the introduction in a paper-based document. The home page should set a context by clearly signaling the subject, purpose, and main point of the site. It might also stress the importance of the site's information to your readers.

SUBJECT   Your site's subject should be clear as soon as readers access the home page. You don't need to say something like, "The subject of this site is…," but you can use a title or a graphic to quickly indicate what the site is about (Figure 14.7).

PURPOSE   Likewise, the purpose of your site should also be clear on the home page.

> In this website, you will find helpful information about paleontology research and resources at Dinosaur National Monument. This information will support the planning of your own research at the site.

A Home Page

Forecast the structure of the site.

Define the subject.

State the purpose.

State the main point.

Stress the importance.

Offer background information.

Source: Ocean Alliance, http://www.whale.org.

Figure 14.7:
The home page is an introduction to the website. Notice how this home page makes all the common moves found in the introduction of a paper-based document. These introductory moves can be made in a few different places on the screen.

Often, websites do not directly state their purpose on the home page. In these situations, the purpose of the website should be obvious to the readers.

MAIN POINT   Since websites tend to be more informative than persuasive, they often do not state a main point on the home page. Nevertheless, you should think about the overall point you want to stress.

> Being informed about prior and current research at Dinosaur National Monument is essential to designing your own research project, especially if you are proposing a dig.

STATEMENT OF IMPORTANCE   The importance of your subject may be immediately obvious to you, but it might not be obvious to your website's readers. You should stress the importance of the subject on the home page to encourage readers to use the site.

> Resources at the monument are very limited, so projects are approved only if they are likely to succeed and do not duplicate prior projects. The information in this website

Link

For more information on writing introductions, see Chapter 8, page 182.

should help you avoid proposing research projects that will not be approved. This website might also help you locate other paleontologists with whom you can collaborate on large projects.

In some cases, a picture or graphic can also signal the importance of the site.

BACKGROUND INFORMATION   On most home pages, background information is limited, because only a limited amount of space is available on the screen. You might, however, give the readers some background information on your subject to help them gain a quick overview.

The Dinosaur National Monument Paleontologist Network was created in 1986 to foster communication among paleontologists who work in the park. Since then, our network has grown to include over 200 scientists.

FORECASTING   Usually home pages include a navigation bar that forecasts the structure of the site. The navigation bar can appear just about anywhere, but it typically appears along the top of the page or the left-hand side (Figure 14.7).

Drafting Node Pages

Node pages direct traffic on your website. These pages typically follow links from the home page, further dividing the subject of the site. Each node page introduces readers to one of the site's major topics. For example, the node page shown in Figure 14.8 was a link displayed on the home page in Figure 14.7.

A well-designed node page allows readers to select among other further topics. In essence, it serves as a miniature home page to this part of the site. In large websites, node pages are very much like home pages because they tell readers the *subject, purpose,* and *main point* of this part. They include mostly contextual information that helps direct readers toward more specific details further in the website.

Drafting Basic Pages

There really is no sharp distinction between node pages and basic pages. As readers move further into the site, the pages should become more detail oriented.

Basic pages increasingly provide the content (e.g., facts, data, examples, details, descriptions) that readers are looking for. For example, the basic page shown in Figure 14.9 includes mostly facts, examples, and reasoning. More than likely, this information is why readers came to the website in the first place. But they needed to navigate through the home page and node pages to arrive here.

In a basic page, you should keep the discussion limited to a topic that can be handled in about a screen and a half. Readers of websites rarely have enough patience to scroll down further. So, if you need more than two screens to cover a particular issue, you might need two or more basic pages.

To see other sample node pages, go to
www.pearsonhighered.com/johnsonweb3/14.9

A Node Page

Identify the subject of the node.

Offer links to basic pages.

Figure 14.8: A node page introduces a specific topic in the website. Here is the "Research" node page for this website. It allows readers to find more information on research about whales.

The navigation bar for the website is consistently present on the screen.

Source: Ocean Alliance, http://www.whale.org/research/index.html.

Drafting Navigational Pages

Websites also include other kinds of pages that can help readers navigate the site.

SITE MAPS   A site map is like an index in a book. The site map lists all the topics covered in a website, large or small. That way, if readers are having trouble finding what they want, they can go straight to the site map to find a link to the information they need.

HELP   Software programs like RoboHelp, DocToHelp, and Help & Manual will allow you to create Help areas for your website. Usually, only the largest websites include a Help area. These Help areas offer additional information and usually provide definitions or tutorials that readers may need.

SEARCH ENGINES   Larger websites may also include an internal search engine that helps readers locate information by typing in keywords. Search engines are available,

A Basic Page

Notice that the navigation bar is the same as before.

Specific information about an issue is included.

Source: Ocean Alliance, http://www.whale.org/wci/wci_rightwhale.html.

Figure 14.9:
A basic page contains the specific information that readers are looking for. This basic page (linked from the "Research" node page) discusses a specific issue: the right whale.

Types of Website Pages

AT A GLANCE

- Home page— introduction page for the website
- Node pages—opening pages for sections of the website
- Basic pages—pages containing the content readers are looking for
- Navigational pages— site maps, Help area, search engine

usually free, from the larger search engines like Excite (excite.com), Lycos (lycos.com), and Google (google.com). However, putting a search engine into your website takes a little more programming skill.

A Warning About Copyright and Plagiarism

Inventing and collecting the content for websites is not much different from creating the content for print documents. You still need to thoroughly research the subject in electronic, print, and empirical sources. The information you invent or collect will be the substance of the site.

To avoid any problems with copyright or plagiarism, remember that the same rules apply to websites as apply to print documents. If you want to use an image off someone else's website, you need to ask permission. Also, you cannot take passages of text from other sites and use them in your own unless you properly cite them.

Link

To learn more about copyright and plagiarism, go to Chapter 5, page 112.

For more information on copyright issues and the Internet, go to
www.pearsonhighered.com/johnsonweb3/14.11

Using Web-Authoring Software

Not long ago, website developers needed to know how to use HTML (hypertext markup language) to create a website. HTML is a coding language that tells the computer how something should look on the screen.

Fortunately, there is now a variety of "web-authoring" software packages available that make designing webpages much easier. Web-authoring software is similar to word-processing software, except that it's used for writing webpages. The web-authoring software converts what you type on the screen into HTML code (see Figure A). You can also lay out pages, insert images, choose colors, and include graphics. This kind of authoring software is called WYSIWYG, because what-you-see-is-what-you-get when you design the webpage.

Countless web-authoring software packages are available. The most popular commercial packages include Adobe GoLive, Macromedia Dreamweaver, and Microsoft Expression Web Designer. Figure A shows Dreamweaver in use.

Using Web-Authoring Software

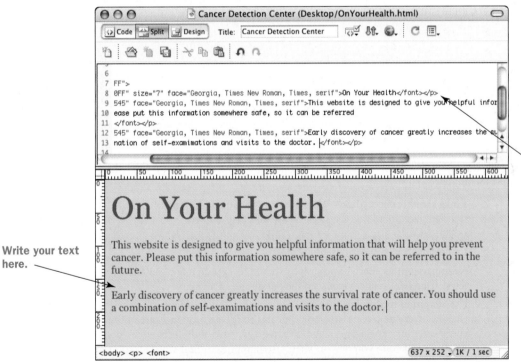

Write your text here.

Figure A: Web-authoring software makes developing a website much easier. Dreamweaver, the web-authoring tool shown here, works like a word processor.

HTML code is automatically written here by the authoring program.

(continued)

Creating a webpage is not difficult with web-authoring software.

1. Open a file by choosing "New Page."
2. Type in a title for the page.
3. Type in some text.
4. Highlight text for links and type in the address of the link.
5. Use the buttons on the design pallette for color, lines, and shapes to add some pizzazz.
6. Use the Insert command from the pull-down menu to add images.
7. Save the page.

Each software package uses different commands, but if you know how to use a word processor, you can figure out these commands.

When you have finished creating your webpage, you will need to "FTP" (file transfer protocol) the file to a server (a large computer that connects your computer to the Internet). Here's where things become a bit more complicated. Your university, company, or Internet service provider (ISP) should have instructions available for putting your webpage on the Internet.

The first time you create a website might be a little difficult, because there is much to learn. You will find, though, that creating websites is not difficult once you learn how to do it. Web-authoring software eliminates many of the complications that once existed with HTML coding.

Using Style in a Website

As in any other document, the style of a website is important. The difference between websites and paper-based documents, though, is that readers are even more likely to "raid" the website for the information they need.

When raiding for information, few readers will have the patience to actually read long paragraphs or sentences on a screen. So, here are some strategies for improving the readability of your website.

Links should reflect titles—When readers click on a link, the title of the selected page should be the same as the link they clicked.

Keep sentences short—On average, sentences in websites should be shorter than sentences in paper-based documents.

Keep paragraphs short—Paragraphs should be kept to a few sentences or fewer. That way, readers can scan the paragraph in a glance.

Use mapping to develop themes—You can use logical mapping to set themes or create a tone for your website. To create a theme, put in the center of your screen or a sheet of paper the word that best represents the theme or tone you desire. Then, use mapping to find words that are associated with that theme or tone (Figure 14.10). As with paper-based documents, use these words strategically throughout the site. When your readers encounter these words, they will subconsciously sense the theme or tone you are trying to create.

Link

For more information about using logical mapping to improve style, turn to Chapter 9, page 223.

Mapping a Theme for a Website

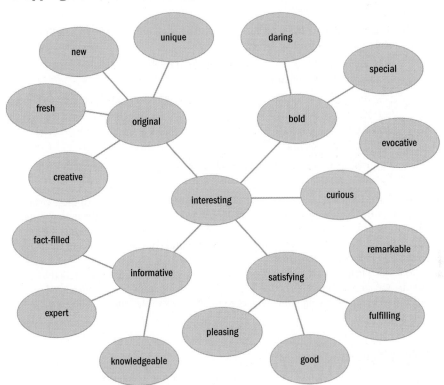

Designing the Website

The design of your website is important for several reasons.

Readers prefer attractive, well-designed webpages—They are more likely to spend time looking over webpages that please their eyes.

Well-designed websites are more navigable and more effective—If readers can quickly locate paths to the information or products they are seeking, they will be much more willing to do business with your company.

Well-designed websites give readers confidence—A professional-looking website will make your company and its products seem more trustworthy.

Here, we will go over a few basic principles of web design. As you improve you will want to learn how to use Cascading Style Sheets (CSS) to design your webpages and sites.

Designing the Interface

The *interface* of your website concerns how the pages look on the screen. The design of your website's interface determines in great part whether your website will be attractive to readers and whether it will be easy to use.

As discussed in Chapter 10, when designing an interface for a website, you should use the *Five Principles of Design:*

Balance—Items placed on the screen offset each other to create a feeling of stability on the page or screen.

Alignment—Aligned items on the page help readers identify different levels of information in the interface.

Grouping—Items that are near each other on the screen will be seen as being "grouped."

Consistency—Each page in the website should look similar to the others, minimizing chaos in the interface.

Contrast—Contrast sharpens the relationships between images and text, headings and text, and text and the background.

Link

For more information on designing interfaces, see Chapter 10, page 263.

The interface shown in Figure 14.11, for example, demonstrates all of these principles successfully. Notice how the text is balanced from side to side. Information is also aligned in clear vertical lines, and you can see where specific information has been grouped into larger blocks. Meanwhile, consistency and contrast are used to make the interface interesting but consistent.

These five principles should be used to develop a consistent interface, called a template, for your webpages. The template will give your website a standardized, predictable design, allowing your readers to quickly locate the information they are seeking.

Adding Images

Images for your website can be saved as separate files on your computer and on the server. These files can include any pictures, graphs, or drawings you might want to include on your webpage. They also include elements of the interface, such as banners on the screens, corporate logos, and icons. Images can be saved in a variety of *file formats*. The two formats most commonly used for websites are *jpeg* and *gif* formats (Figure 14.12).

jpeg (joint photographic experts group)—The jpeg file format is widely used for photographs and illustrations with many colors. Images in jpeg format can use millions of different colors, allowing them to better capture the subtleties of photographs. The main limitation of jpeg images is their higher memory requirements, causing them to download more slowly, especially on computers with slow connections to the Internet.

gif (graphic interchange format)—The gif format is primarily used for illustrations, logos, and simple graphics. Gif images can use a maximum of only 256 colors, making them less useful for photographs. Their advantage is that they use less memory than jpeg files, making them quicker to download.

You can tell the difference between jpeg and gif files by the extensions on their file names. A jpeg file will have the extension ".jpg" added to it (e.g., cougarpic.jpg), and a gif file will have the extension ".gif" added (e.g., cougarpic.gif).

A Well-Designed Interface

Good balance, using design features in ways that offset each other on the screen

Good contrast, making the text easy to read

Good grouping, putting related information together in blocks

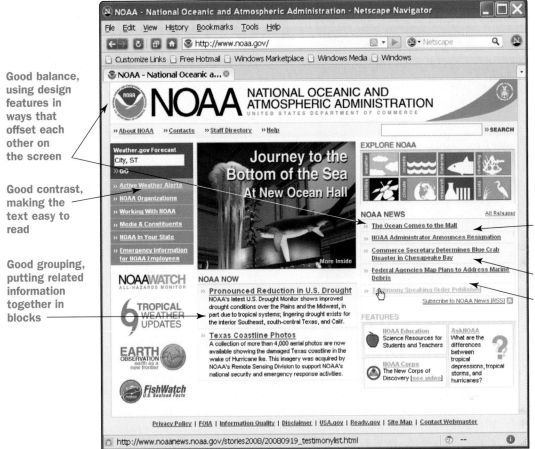

Source: National Oceanic and Atmospheric Administration (NOAA), http://www.noaa.gov.

Figure 14.11:
This home page uses all the design principles successfully. Notice how the interface uses balance, alignment, grouping, consistency, and contrast to create an attractive and functional page.

Good alignment, creating a vertical line of items

Good consistency, with similar items designed the same

To put images in your webpages, you need to do three things.

1. **Create or capture the image.** You can use a digital camera, scanner, or drawing software to create an image. Then, you can use a program like Adobe Photoshop to turn the file into a jpeg or gif. Or, you might find an image on a website or CD-ROM and download it to your hard drive.

   To capture an image off a website, put your cursor on it and hold down the button on your mouse. The browser will bring up a window that gives you a few choices. Choose "Save Image As." The browser will let you save the image to your computer's hard drive.

Want to know more about designing interfaces? Go to
www.pearsonhighered.com/johnsonweb3/14.14

Designing the Website

367

Two Major Types of Screen-Based Images

Figure 14.12: Most images on websites use either the jpeg or the gif format. The image of Hurricane Katrina on the left is a jpeg. The Purdue logo on the right is a gif.

Sources: NASA, http://rapidfire.sci.gsfc.nasa.gov/gallery/?search=katrina, and Purdue University.

If the image is large, you can "compress" it with Adobe Photoshop, iPhoto, or MS Office Picture Manager. A compressed image will load faster for your readers.

2. **Place the image.** Put the image file in the folder on your computer that holds the other files for your website. When possible, keep images with the pages that reference them. Otherwise, there is a chance the image will not be transferred with the page, causing the image to not be shown on the page (see "Uploading, Testing, and Maintaining Your Website" on the next page).

3. **Insert the image.** Using your web-authoring software, open the page in which the image should appear. Then, use the Insert Image command (or equivalent) to add the image to the page.

When you are finished inserting the image, it should appear on the webpage.

Revising, Editing, and Proofreading

Before you put your website on the Internet, you should revise the site just as you would revise a paper-based document. As discussed in Chapter 12, you should use the four levels of editing (revising, substantive editing, copyediting, and proofreading) to improve the quality of the site.

Revising and editing are not simply ways to polish the website. You need to look closely at the content, organization, style, and design to check if they are appropriate

GO TO THE NET

Want to learn more about using images? Go to
www.pearsonhighered.com/johnsonweb3/14.15

for your readers and the context in which the site will be used. When you are revising the website, ask yourself these questions:

Does the website highlight information my readers want?

Are the subject and purpose of the website obvious on the home page?

Is the site organized in a way that will be logical to my site's readers?

Is the style of the website appropriate? Does it set the right tone?

Is the interface easy to navigate?

Is the interface attractive and interesting?

Proofreading a website is especially difficult, because seeing errors on the screen is often harder than seeing them on paper. So, you might print out the screens of your website and proofread them on paper.

When you have finished revising and editing the website, do some usability testing to ensure quality. Minimally, you should ask your co-workers and friends to try using the site, most likely on your computer.

Depending on the importance of the site, you might also set up formal usability tests with sample readers. You can do the same kinds of read-and-locate tests, understandability tests, performance tests, and even safety tests that you might use to test the usability of a print document.

Link

For more information on usability testing, go to Chapter 12, page 322.

Uploading, Testing, and Maintaining Your Website

Once you are finished revising, editing, and testing the usability of your website, it is time to upload it to the Internet. Once it is uploaded, you will need to test the site yet again to see if it works the way you want it to. After correcting any problems, you will need to maintain the website to keep it up to date.

Uploading the Website

To make your website available to the public, you need to upload it to a server—a large computer that connects your computer to the Internet. Before uploading to the server, most writers of websites prefer to complete the website on their own computer's hard drive. Then, they use a file transfer protocol (FTP) software program to copy the whole website to the server at once (Figure 14.13).

When you copy your webpage files, don't forget to also copy your image files along with them. If the image files are not included with the webpage files, the pictures, illustrations, graphs, and other images you want placed on your website will not appear.

Testing the Website

Once your website is on the server, it is time to test it. Use at least a couple different browsers (e.g., Firefox, Opero, Explorer, Safari) to look over the pages on the site. Try out the links and resize the browser window a few different ways. Usually, you will find

Uploading the Website

Your Hard Drive

Home page file → homepage.html

File for a node page → learnautism.html

Files for basic pages → library.html, whatis.html, othersites.html

Graphics files → autisticboy.jpg, hcentre.jpg, pagebanner.gif

Drag and drop files with your mouse

Figure 14.13: An FTP program copies files from your computer's hard drive to a server. The server is connected to the Internet.

Files and folders in server computer are shown here.

Source: Whisper Technology, http://www.whispertech.com/surfer/index.htm.

that the website has some formatting problems that did not show up in your web-authoring program's version of the site. For example, perhaps an image failed to appear or the spaces between lines are uneven. Another common problem is that links to other pages don't work, which means that readers will see error messages when they choose links to these pages.

To fix these problems, reopen the files with your web-authoring program. Make the corrections and changes needed. Then, re-upload your revised pages to the website.

The testing phase can be a little frustrating at times, but it is important. You want to make certain that your website looks the way you intend it to. Those little problems won't fix themselves, and they will be annoying to your readers. Once you have fixed these small problems, though, your website will convey the sense of quality you want to project.

Maintaining the Website

Most websites cannot be left alone for long. You will find yourself wanting to add or change information on your website regularly. You will find new links to other websites, or you will discover that information on your site is out of date. You should plan to regularly update your website, usually weekly or monthly.

To update a page, go back to the file saved on your hard drive and open it in your web-authoring program. Make the changes you want and then use FTP to copy the updated file onto the server. The server will replace the old file with your new file.

Most companies hire or contract a webmaster to maintain their site. The webmaster regularly updates the site, incorporating any changes and eliminating any dated materials. When you are your own site's webmaster, though, you should revisit your site regularly, looking for items to update.

Want to learn more about FTP? Go to
www.pearsonhighered.com/johnsonweb3/14.17

- Websites are important for both corporate and personal communication.

- The basic features of a website are a home page, node pages, basic pages, and navigational pages.

- Prepare to develop a website by first defining its subject, purpose, readers, and context of use. Consider international and multicultural audiences as well.

- Mapping is a good way to plan the content of a website; an initial map can be followed by a more detailed plan of the levels that will be used in the website.

- The home page of a website is like the introduction to a document. Tell readers the subject, purpose, and main point of the site, while offering background information, stressing the importance of the subject, and forecasting the site's structure.

- Choose a consistent style to make your website easier to scan and interpret.

- The design of the website should follow the five principles of design: balance, alignment, grouping, consistency, and contrast.

- Revising and editing are especially important in websites, because you will change the text regularly.

- When your website is finished, you need to upload, test, and maintain it.

Individual or Team Projects

1. Write a critique of a website you find on the Internet. Look at its content, organization, style, and design. Does it achieve its purpose? Is it appropriate for its intended readers? Can you find information easily? Write your critique in memo or e-mail form for your instructor.

2. On paper or whiteboard, map out the contents of a website you find on the Internet. Identify its home page, node pages, basic pages, and navigational pages. Then, in a presentation to your class, discuss the organization of the website. Show its strengths and identify any places where the website's organization might be improved.

3. Imagine a website you would like to build. Identify its subject, purpose, readers, and context of use. Then, diagram the site on paper or a whiteboard, showing how it would be organized. Thumbnail some sample pages on paper, sketching out how a few pages would look on the site (home page, node pages, basic pages). Attach these drawings to a memo of transmittal to your instructor in which you discuss the content and organization of your website.

4. Find a website at your college or workplace that is ineffective. Write a report to a specific reader in which you discuss the shortcomings of the site and make

recommendations for improvement (you don't need to send the report). In your report, consider the content, organization, style, and design of the site. In your recommendations, show your sample reader pages and organizational schemes that would improve the site.

Collaborative Project

You and your group have been asked to develop a "virtual tour" of your university or workplace. The tour will allow visitors to your website to look around and familiarize themselves with important places on your campus or in your office. The tour should allow visitors to quickly navigate the site, finding the information they need to locate important places and people.

Start out by discussing and describing in depth the subject, purpose, readers, and context of use for the site. Then, using paper or a whiteboard, describe how the site would be organized. Make decisions about the style of the site and how the various pages should be designed.

Write a proposal to your university administrators or company managers, showing how this virtual tour would be a nice addition to the university's or company's website. Your proposal should offer a work plan for making the website a reality and describe its costs and benefits, especially the benefits to visitors.

Then, if the software and hardware are available, create part of the website. Of course, you probably cannot create the whole virtual tour, but you can make the home page, a couple of node pages, and a few sample basic pages that show important places on campus or your workplace.

For additional technical writing resources, including interactive sample documents, document design tutorials and guidelines, and more, go to **www.mytechcommlab.com.**

Slowed to a Crawl

Alice Franklin worked for Kent Medical's Outreach Division, which was part of a large medical services company. She worked her way up through the company as a biomedical engineer, but she jumped at the chance to work with the Outreach Division because it allowed her to work with people in other parts of the world.

Her job was to help Kent Medical deliver free medical services and supplies to people in poor countries that did not have advanced medical services. She went home every night knowing she had made a difference. Plus, her trips to South America and Africa were exhilarating and fulfilling.

A few months ago, Alice's boss, Kathryn Young, had hired Zigda Communications, a high-priced web design firm, to redesign the Outreach Division's website. The website was a critical part of the division's public relations efforts and fundraising. It was also the portal through which clients requested help and communicated with Kent Medical.

Zigda agreed to do the work for a fraction of their usual fee, and they did an incredible job redesigning the site. The Flash animation was amazing, and the embedded videos were very effective at telling the story of Kent Medical's efforts to help people in other parts of the world. A week after the site went online, it was featured as a "Best Website" on a major news website. That week, the website got over 700,000 hits, more than it ever had before. Plus, the Outreach Division had already raised over $1.2 million in donations through online contributions.

However, Alice noticed a problem. The Outreach Division's clients in South America and Africa were not accessing the site as much as they had before. Requests for services and supplies had dropped off dramatically.

Trying to figure out what was happening, Alice called up one of the missing clients in Brazil. A representative explained to her that they were not able to access the website because their computers were using dial-up connections. So, all those fancy animations, video, and forms were overwhelming their computers.

Even when they *could* get through to a request form for medical services and supplies, the new forms were far too advanced for their browsers. Upgrading the browsers, however, would clog up their computers' memory and cause their computers to crash.

Alice wasn't sure what to do. Hiring Zigda had been the decision of her boss. Bringing up this problem was going to be very embarrassing because the "successful" new website was undermining their core mission of providing medical services and supplies to their clients. And yet, the website was also generating millions of dollars in new donations. How might you handle this situation if you were in Alice's place?

In this chapter, you will learn:

- How to set goals and make a plan for finding a career position.

- Methods for using both web-based and traditional networking.

- How to design and prepare a résumé in both electronic (scannable) and traditional formats.

- Techniques for writing a persuasive application letter.

- How to create a targeted, professional portfolio that highlights your background and experience.

- Strategies for effective interviewing.

Finding a good job requires energy, dedication, and optimism—and readiness to hear the word "no." There will be exciting moments, such as receiving a call for an interview or being offered a job. There will be moments when you doubt yourself and your abilities. Emotional ups and downs are a normal part of the job-searching process.

Your computer gives you access to many job-searching tools—so use them. The least effective job seekers are those who wait for the Sunday paper, send out a few résumés, and wait by the phone for a response. The most effective job seekers are those who spend each day actively searching for a job along several paths.

Job searching is not a one-time event. You will probably look for a job at least a few times during your lifetime. U.S. Department of Labor statistics show that people tend to change careers four to six times in their lives, including twelve to fifteen job changes. In other words, being able to find a job is now becoming a necessary skill in a successful career.

Setting Goals, Making a Plan

From the beginning, you should adopt a professional attitude about looking for a job. To approach the task professionally, you should start your search by first setting some specific goals and developing a plan for reaching those goals.

Setting Goals

So, before you begin sending your résumé to every company you can think of, spend some time setting goals for your search. Begin by answering the Five-W and How Questions:

> *What are my needs and wants in a job/career?*
>
> *Who would I like to work for?*
>
> *Where would I like to live?*
>
> *When do I need to be employed?*
>
> *Why did I choose this career path in the first place?*
>
> *How much salary, vacation, and benefits do I need?*

Type your answers into a file, or write them on a piece of paper. Be specific about things like location, salary, vacation, and the amount of time you have to find a position. If you don't know the answers to specific questions, like salary, then use Internet search engines to help you find these answers. If you are still in school, you can ask your professors to give you a good estimate of salaries in your field.

Your first reaction to being unemployed might be to say, "I'll take any job available." That's a common response to an uncomfortable situation like job searching. Instead, spend a little time thinking about the quality of life you desire and how your career fits into that overall life picture. Once you know your goals, you can start looking for a job that suits you.

To see employment trends, go to
www.pearsonhighered.com/johnsonweb3/15.1

Using a Variety of Job-Seeking Paths

With your goals in mind, you can begin developing a plan for finding a good job. Fortunately, your computer offers a door into an amazing number of job-seeking pathways.

JOB SEARCH ENGINES   The numerous job search engines available on the Internet are good places to start looking for jobs. By entering a few keywords, you should be able to locate a variety of job opportunities. Most search engines will allow you to limit your search by region, job title, or industry. Some search engines will even allow you to post your résumé so employers can find you. For example, Monster.com, shown in Figure 15.1, has a link on its home page that takes you to its résumé-posting area.

Here are some of the more popular job search engines:

4Jobs.com   *Usajobs.opm.gov*

Indeed.com   *Hotjobs.yahoo.com*

SimplyHired.com   *Jobbankusa.com*

Careerbuilder.com   *Job-hunt.org*

Truecareers.com   *Monster.com*

Jobster.com   *Worktree.com*

Collegegrad.com

Also, popular Internet search engines such as Google.com include job search areas that you can use.

Job search engines are fast becoming the best way to find a job, especially in technical fields. They offer easy and inexpensive contacts between job seekers and employers. Employers are beginning to prefer search engines because these tools can be used to run a nationwide search for the best people. You should check these sites regularly and post your résumé on at least a few of them.

SOCIAL NETWORKING WEBSITES   More than ever, people are finding jobs through their connections on social networking sites like Facebook, MySpace, and Second Life. There are also business-centered social networking sites like LinkedIn and Ryze that you should join because they are designed to help people do career networking. You should turn your social networking site(s) into a job finding tool by including your résumé and any other information you would like potential employers to see. In some cases, employers may find you before you find them because they use social networking sites to search for potential new employees.

Before you go on the job market, however, you should spend some time cleaning up your current social networking sites. Interviewers regularly do background checks on their candidates through these sites. Even if you think your profile is safely locked, some companies have been known to use connections like alumni from your university to gain access to profiles. So, while you are on the job market, your site should not have anything on it that you wouldn't want to share with a potential employer.

PERSONAL NETWORKING   Someone you know is probably aware of a job available in your field. Or, they know someone who knows about a job. Make a list of your

For websites that will help you answer these goal-setting questions, go to
www.pearsonhighered.com/johnsonweb3/15.2
For links to job search sites, go to
www.pearsonhighered.com/johnsonweb3/15.3

An Internet Job Search Engine

Post your online résumé in this area.

Here, you can run keyword searches for jobs.

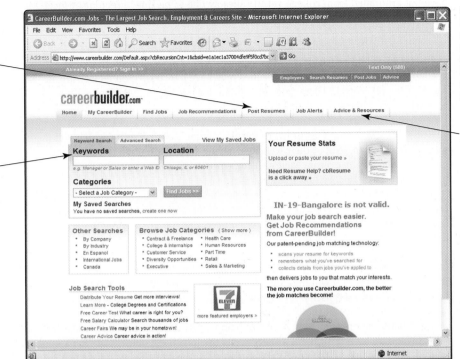

Figure 15.1: Monster.com is one of the more popular job search engines. It offers a variety of tools to aid your search.

Advice about interviewing can be found in this area.

Source: CareerBuilder.com homepage. Copyright 2008 CareerBuilder, LLC. Reprinted with permission.

friends, relatives, and professors who might be able to help you find a job. Then, send each of these people an e-mail that tells them you are "on the market" looking for a job. You might even attach a résumé to your e-mail so they can look it over and perhaps forward it to a potential employer.

More than likely, you will be pleasantly surprised by the response to these e-mails. Your friends and relatives know more people than you realize. Meanwhile, your professors are often aware of opportunities available in your area.

PROFESSIONAL NETWORKING   Most career tracks have professional groups associated with them. Engineers, for example, have the Institute of Electrical and Electronics Engineers (IEEE), while medical practitioners have the American Medical Association (AMA). Technical writers have the Society for Technical Communication (STC). These professional groups are especially helpful for networking with people who are already employed in your field.

Most large professional groups have a local chapter you can join. Chapter meetings offer great opportunities to contact people who have jobs similar to the one you want. Also, local chapters usually have websites that post job openings in your area.

Want to find a professional group in your major? Go to **www.pearsonhighered.com/johnsonweb3/15.4**

Setting Goals, Making a Plan

377

You should become involved with these groups as soon as possible, even if you have not graduated from college yet. It takes awhile to become a regular at meetings, but once people get to know you, they can be very helpful with finding job opportunities.

COLLEGE PLACEMENT OFFICE Most colleges have a placement office that is available to students. The placement office may have jobs posted on its website, or you can visit the office itself. There, you can sign up for interviews and speak with a counselor about improving your job-searching skills.

TARGETING Make up a list of ten to twenty "target" companies for which you might want to work. Then, look at their websites, paying special attention to each company's human resources office. From each website, write down notes about the company's mission, products, history, and market.

If one of your targeted companies does not have a job available, send its human resources department a copy of your résumé and an application letter. In your letter, tell the target company's human resources officer that you are sending materials for their files in case a position becomes available.

CLASSIFIED ADVERTISEMENTS In the classifieds section of a newspaper, especially the Sunday edition, you will find job advertisements. Keep in mind, though, that newspapers carry advertisements for only a few jobs in any given area. *Most jobs are not advertised in the paper.* But these ads are worth checking once a week.

Most major newspapers now have companion websites that list online classifieds. If you are looking for a position in a specific city or state, these online classifieds may be helpful. They include all the jobs that are printed in the classified advertisements in the newspaper.

The secret to effective job searching is to set clear goals and have a strategy for reaching those goals. With the variety of electronic tools available, you have many different paths to follow to find a position (Figure 15.2). The most successful job seekers use them all.

Preparing a Résumé

A résumé is a summary of your background, experience, and qualifications. Usually, résumés for entry-level jobs fit on one page. Résumé s for advanced positions might extend to two pages. You need to spend some serious time and effort on your résumé, because it may be one of the most important documents you will ever write.

Your résumé needs to make a good first impression because it is usually the first item that employers will look over. If your résumé reflects the qualifications and experience they are looking for, they may read your other materials and give you a call. If your résumé is poorly written or poorly designed, chances are slim that you will ever get your foot in the door.

Using your computer, you can revise and redesign your résumé to suit the needs of each job. You can make strategic decisions about how to organize the materials on your résumé so that you can highlight the strengths an employer is looking for.

For links to several major newspapers'
classified ads, go to
www.pearsonhighered.com/johnsonweb3/15.5

The Job-Searching Cycle

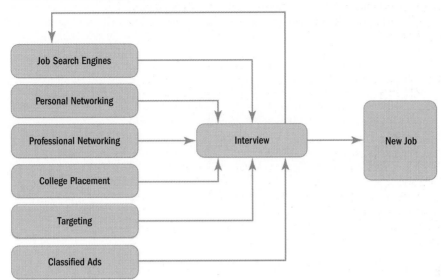

Figure 15.2: There are many tools available for finding a job and building a career. You should take advantage of all of them.

Types of Résumés

Résumés tend to follow one of two organizational approaches: the *chronological* approach and the *functional* approach.

> **Chronological approach**—Organizes the résumé according to education and work experience, highlighting a job seeker's qualifications in a few areas. A chronological résumé might be organized into sections such as Education, Work Experience, Skills, and Awards/Activities.

> **Functional approach**—Organizes the résumé according to your talents, abilities, skills, and accomplishments. A functional résumé would be organized into sections such as Leadership, Design Experience, Communication Skills, and Training Abilities.

By far, the chronological approach is the most common in technical fields, especially for entry-level jobs. It allows you to highlight the details of your experience. Meanwhile, the functional approach is not advantageous for new graduates because their experiences are limited and have not given them enough opportunities to demonstrate their abilities. For most jobs, you should submit a chronological résumé unless you are specifically asked for a functional résumé.

Chronological Résumé

Chronological résumés can be divided into the following sections, which can be organized a variety of different ways:

- Name and contact information
- Career objective/career summary

- Educational background
- Related work experience
- Other work experience
- Skills
- Awards and activities
- References

The headings in this list are the ones commonly used in chronological résumés, but other headings are available. Use headings that best highlight your qualifications and set the specific tone you desire (Figures 15.3, 15.4, and 15.5).

Link

For more information on using fonts effectively, go to Chapter 10, page 246.

NAME AND CONTACT INFORMATION   At the top of the page, your résumé should include a heading that states your name, address, phone number, e-mail address, and perhaps a website address. To catch the reader's eye, your name should use a larger font size. The heading can be left justified, centered, or right justified, depending on the design of your résumé. Some people like to add a *rule* (a line) to set off the heading from the rest of the information in the résumé.

CAREER OBJECTIVE OR CAREER SUMMARY (OPTIONAL)   A *career objective* is a phrase or sentence that describes the career you are seeking. Your career objective should specify the type of position you are applying for and the industry in which you want to work.

> Seeking a position as a Physician's Assistant in a research hospital.
>
> A computer programming career working with mainframes in the defense industry.
>
> Looking for a career as a school psychologist working with children who have behavioral problems.
>
> Mechanical Engineer developing nanotechnological solutions for biotech applications.

Avoid writing a career objective that is too narrow or too broad. A career objective that is too narrow might eliminate you from some potential jobs. For example, if you specify that you are looking for a job at a "large engineering firm," the human resources officer at a small engineering firm might assume you are not interested in her company's job. On the other hand, an objective that is too broad, like "a job as an electrical engineer," might give the impression that you are not sure about what kind of career would suit your talents.

A *career summary* is a brief sentence or paragraph that describes your career to this point. Career summaries are typically used by people with years of experience.

> I have been employed in the hospitality industry for 10 years, working my way up from desk clerk to assistant manager at a major downtown hotel. My specialty is coordinating catering services for large conferences.
>
> My experience as a webmaster includes designing and managing a variety of interactive sites that have been used by large companies to promote their products and solicit new business.

Your career objective or career summary should convey a sense of what makes you unique or interesting as an applicant. A career objective or career summary is not required in a résumé. In fact, some résumé experts suggest that these statements unnecessarily take up space on a résumé, because they rarely do more than describe

To find other headings for your résumé, go to
www.pearsonhighered.com/johnsonweb3/15.6

Chronological Résumé

Name and contact information are prominently displayed.

Career objective describes position sought.

Education is highlighted by placement high in the résumé.

Work experience lists duties with bullets.

Skills are listed separately for emphasis.

Awards and memberships are placed later in the résumé.

Anne Franklin

834 County Line Rd.
Hollings Point, Illinois 62905

Home: 618-555-2993
Mobile: 618-555-9167
e-mail: afranklin@unsb5.net

Career Objective

A position as a naturalist, specializing in agronomy, working for a distribution company that specializes in organic foods.

Educational Background

Bachelor of Science, Southern Illinois University, expected May 2009.
Major: Plant and Soil Science
Minor: Entomology
GPA: 3.2/4.0

Work Experience

Intern Agronomist, December 2007–August 2008
Brighter Days Organic Cooperative, Simmerton, Illinois
- Consulted with growers on organic pest control methods. Primary duty was sale of organic crop protection products, crop nutrients, seed, and consulting services.
- Prepared organic agronomic farm plans for growers.
- Provided crop-scouting services to identify weed and insect problems.

Field Technician, August 2007–December 2008
Entomology Department, Southern Illinois University
- Collected and identified insects.
- Developed insect management plans.
- Tested organic and nonorganic pesticides for effectiveness and residuals.

Skills

Computer Experience: Access, Excel, Outlook, PowerPoint, and Word. Global Positioning Systems (GPS). Database Management.
Machinery: Field Tractors, Combines, Straight Trucks, and Bobcats
Communication Skills: Proposal Writing and Review, Public Presentations, Negotiating, Training, Writing Agronomic and Financial Farm Plans.

Awards and Memberships

Awarded "Best Young Innovator" by the Organic Food Society of America, 2007
Member of Entomological Society of America

References Available Upon Request

Chronological Résumé

James L. Mondragon

576 First Avenue, Rolla, Missouri 65408
Phone: 573-555-4391, e-mail: bigmondy12@umr.edu

Career Summary

My interests in chemical engineering started with a childhood fascination with plastic products. I enrolled at the University of Missouri–Rolla because of their strong chemical engineering program, especially in the area of applied rheology and polymeric materials. I have also completed a co-op with Vertigo Plastics in St. Louis. My background includes strong computer modeling skills, especially involving polymeric materials.

Work Experience

Vertigo Plastics, Inc., St. Louis, Missouri, 5/07–1/08, 5/08–1/09
Co-op Chemical Engineer
Performed inspections of chemical equipment and plant equipment affected by chemical systems (such as boilers and condensers). Monitored the performance of chemical systems at various sites throughout the Vertigo Plastics system. Performed calculations and wrote reports summarizing the performance of those systems. Helped troubleshoot problems.

Other Work Experience

To Go Pizza, *Server,* Springfield, Missouri, 2/01–7/03

Educational Background

University of Missouri–Rolla, BSE, Expected May 2009
Major: Chemical Engineering
Advanced Coursework included Chemical Engineering Fluid Flow, Chemical Engineering Heat Transfer, Chemical Engineering Thermodynamics I & II, Process Dynamics and Control, Chemical Engineering Reactor Design, Chemical Engineering Economics, Chemical Process Safety, Chemical Process Design, Chemical Process Materials.

Activities and Awards

Awarded Stevenson Scholarship in Engineering, 2007
Treasurer, UMR Student Chapter, American Institute of Chemical Engineers (AIChE), 2007–Present
Member, Tau Beta Pi, 2007

**Dossier with References Available at UMR Placement Services
(573-555-2941)**

Marginal annotations (left):

Name and contact information are prominently displayed.

A career summary is used to summarize experience.

Work experience is featured because it is placed high in the résumé.

Other work experience is listed concisely.

A paragraph lists classes taken.

Résumé ends by mentioning where a dossier can be obtained.

Marginal annotation (right):

Figure 15.4: James Mondragon's résumé uses a paragraph style. It uses a sans serif font for headings (Myriad) and serif font (Times) for the body text. The headings are centered.

Figure 15.5:
This résumé from a student who returned to college features work experience ahead of educational experience.

Elizabeth Bryson Young

2382 Appian Way, Evanston, Illinois 60204
(312) 555-4930, eyoung2382@gmail.com

Name and contact information is placed up front.

Professional Experience

June 2002–May 2006 **Laboratory Scientist**, *Cathert Laboratories, Chicago, Illinois*
- Co-wrote $1.8 million in grants for research
- Researched cell and molecular biology of human leukemia
- Developed pharmacology applications of new anti-cancer drugs
- Updated laboratory specifications

Dates of work experience are prominent.

July 1999–June 2002 **Senior Biological Technician**, *Houston Biotech, Houston, Texas*
- Supervised technicians and staff in 50-person laboratory
- Ensured quality control and sanitary standards in laboratory
- Budgeted $350,000 per year in materials and supplies

Biological Technician
- Provided technical support and services to our scientists and engineers working in the laboratory
- Isolated, identified, and prepared specimens for examination
- Promoted to Senior Biological Technician in December 2002

Numbers help quantify experience.

May 1999–July 1999 **Biological Technician**, *Centers for Disease Control, Washington, DC*
- Surveyed cancer patients and families about cancer histories
- Collected data and developed medical databases
- Assisted oncologists with preparations of samples

Education

June 2009 **Bachelor of Arts, Professional Writing (English),** Virginia Tech University, Blacksburg, Virginia

Coursework listed here.

Courses Completed: Science and Medical Writing, Technical Editing, Publishing, Website Development, Advanced Scientific and Technical Communication.

Undergraduate Thesis: Medical Writing as a New Trend in Technical Communication

December 1999 **Bachelor of Science, Biology,** Michigan State University, East Lansing, Michigan

(continued)

Computer Skills

Word, InDesign, Photoshop, Illustrator, Excel, Dreamweaver, PowerPoint

Awards

AMWA Best Biotech Article in 2004 for "Turning a Corner on Leukemia," *Biotech News*, March 2004

Researcher of the Year, Cathert Laboratories, 2003

Professional Organizations

American Medical Writers Association (AMWA)

Society of Technical Communication (STC)

American Association for the Advancement of Science (AAAS)

Volunteer Work

Health Volunteers Overseas, 2001-2005

Habitat for Humanity, 1996-present

Northern Chicago Urban Ministries, 2008-present

References

Thomas VanArsdale, Chief Scientist
Cathert Laboratories
1208 Discovery Lane
Chicago, IL 60626
(312) 654-3864
tvanarsdale@cathertlabs.com

Franklin Charleston, Assistant Professor
Department of English
North Carolina State University
Tompkins Hall, Box 8105
Raleigh, NC 27695-8105
(919) 515-3866
franklcharleston@ncsu.edu

Christina Smith, Associate Professor
Department of English
Virginia Tech University
323 Shanks Hall (0112)
Blacksburg, VA 24061
(540) 231-6501
chrsmith34@vt.edu

Volunteer work adds depth to experience and shows maturity.

References listed on second sheet.

the position being applied for. If you are looking for room to put more information on your résumé, you might leave out the career objective or career summary.

EDUCATIONAL BACKGROUND   Your educational background should list your most recent college degree and other degrees in reverse chronological order—most recent to least recent. You can list the degree you are working on now as "expected" with the month and year when you should graduate.

Name each college and list your major, minor, and any distinctions you earned (e.g., scholarships or summa cum laude or distinguished-scholar honors) (Figure 15.3). You might also choose to mention any coursework you have completed that is related to your career (Figure 15.4 and 15.5).

Any specialized training, such as welding certification, experience with machinery, or military training, might also be listed here with dates of completion.

If you are new to the technical workplace, you should put your educational background high on your résumé. Your degree is probably your most prominent achievement, so you want to highlight it. If you have years of professional work experience, you may choose to put your educational background lower on your résumé, allowing you to highlight your years of experience.

RELATED WORK EXPERIENCE   Any career-related jobs, internships, or co-ops that you have held should be listed, starting with the most recent and working backward chronologically. For each job, include the title of the position, the company, and dates of employment (month and year).

Below each position, list your workplace responsibilities. As demonstrated in Figures 15.3, 15.4, and 15.5, you can describe these responsibilities in a bulleted list or in a brief paragraph. Use action verbs and brief phrases to add a sense of energy to your work experience. Also, where possible, add any numbers or details that reflect the importance of your responsibilities.

> Coordinated a team of 15 student archaeologists in the field.

> Participated in the development of UNIX software for a Cray supercomputer.

> Worked with over 100 clients each year on defining, updating, and restoring their water rights in the Wilkins Valley.

Verb-first phrases are preferable to full sentences, because they are more scannable and require less space. Some action verbs you might consider for your résumé include the following:

adapted	*devised*	*organized*
analyzed	*directed*	*oversaw*
assisted	*equipped*	*planned*
collaborated	*examined*	*performed*
collected	*exhibited*	*presented*
compiled	*implemented*	*proposed*
completed	*increased*	*recorded*
conducted	*improved*	*researched*
constructed	*instructed*	*studied*
coordinated	*introduced*	*supervised*
corresponded	*investigated*	*taught*
designed	*managed*	*trained*
developed	*observed*	*wrote*

You might be tempted to exaggerate your responsibilities at a job. For example, a cashier at a fast food restaurant might say he "conducted financial transactions." In

GO TO THE NET

Want more advice on writing résumés? Go to
www.pearsonhighered.com/johnsonweb3/15.7
For sample résumés, go to
www.pearsonhighered.com/johnsonweb3/15.8

Preparing a Résumé   385

reality, nobody is fooled by these kinds of puffed-up statements. They simply draw attention to a lack of experience and, frankly, a mild lack of honesty. There is nothing wrong with simply and honestly describing your experiences.

OTHER WORK EXPERIENCE   Almost everyone has worked at jobs that were not related to his or her desired career. If you have worked at a pizza place, waited tables, or painted houses in the summer, those jobs can be listed in your résumé. But, they should not be placed more prominently than your related work experience, nor should they receive a large amount of space. Instead, simply list these jobs in reverse chronological order, with names, places, and dates.

> Pizza Chef. Giovanni's. Lincoln, Nebraska. September 2007–August 2008
>
> Painter. Campus Painters. Omaha, Nebraska. May 2007–September 2007
>
> Server. Crane River Brewpub and Cafe. Omaha, Nebraska. March 2005–May 2007

Do not offer any additional description. After all, most people are well aware of the responsibilities of a pizza chef, painter, and server. Any more description will only take up valuable space on your résumé.

Don't underestimate the importance of non-career-related jobs on your résumé. If you do not have much professional work experience, any kind of job shows your ability to work and hold a job. As your career progresses, these items should be removed from your résumé.

SKILLS   Résumés will often include a section that lists career-related skills. In this area, you may choose to mention your abilities with computers, including any software or programming languages you know how to use. If you have been trained on any specialized machines or know how to do bookkeeping, these skills might be worth mentioning. If you have proven leadership abilities or communication skills (technical writing or public speaking) you might also list those.

> *Computer Skills:* Word processing (Word, WordPerfect), desktop publishing (InDesign, Quark), web design (Dreamweaver, FrontPage), and data processing (Excel, Access).
>
> *Leadership Abilities:* President of Wilkins Honor Society, 2007–2008. Treasurer of Lambda Kappa Kappa Sorority, 2005–2008. Volunteer Coordinator at the Storehouse Food Shelter, 2006 to present.

The skills section is a good place to list any training you have completed that does not fit under the "Educational Background" part of your résumé.

AWARDS AND ACTIVITIES   List any awards you have won and any organized activities in which you have participated. For example, if you won a scholarship for your academic performance, list it here. If you are an active member of a club or fraternity, show those activities in this part of your résumé. Meanwhile, volunteer work is certainly worth mentioning, because it shows your commitment to the community and your willingness to take the initiative.

REFERENCES   Your references are the three to five people who employers can call to gather more information about you. Your references could include current or

Helpful services for managing your letters of reference can be found at
www.pearsonhighered.com/johnsonweb3/15.9

Sections in a Chronological Résumé

AT A GLANCE

- Name and contact information
- Career objective/ career summary
- Educational background
- Related work experience
- Other work experience
- Skills
- Awards and activities
- References

former supervisors, professors, colleagues, and professionals who know you and your work. They should be people you trust to offer a positive account of your abilities. Each reference listing should include a name, title, address, phone number, and e-mail address.

References can take up a large amount of space on a résumé, so they are typically not listed on the résumé itself. Instead, a line at the bottom of the résumé states, "References available upon request." Then, the references—listed on a separate sheet of paper under the heading "References"—can be sent to any employer who requests them.

If the employer asks for references to appear on the résumé, you can add them at the bottom of your résumé, usually on a second page (Figure 15.5). To avoid causing your résumé to go over two pages, you may need to list them in two or three columns on the second page.

Functional Résumé

The functional résumé is less common than the chronological résumé, especially for new college graduates. This type of résumé is designed to highlight the job applicant's abilities and skills by placing them up front in the résumé. The advantage of this type of résumé is its ability to boil years of experience down to a few strengths that the job applicant would like to highlight.

Figure 15.6 shows an example of a functional résumé. In this example, note how the résumé places the applicant's strengths, including awards, high in the document. Then, the remainder of the résumé concisely lists details about the applicant's employment background, education background, and professional memberships.

Designing the Résumé

The design of your résumé should reflect your personality and the industry in which you want to work. A résumé for an engineering firm, for example, will probably be somewhat plain and straightforward. A résumé for a graphic artist position at a magazine, on the other hand, should demonstrate some of your skills as a designer. There are, of course, exceptions. Some progressive engineering firms, for example, might prefer a layout that reflects your innovative qualities.

Most word-processing programs include résumé templates that you can use to lay out your information. If you use one of these templates, alter the design in some way, because unfortunately, many thousands of people have access to the same templates. So, employers often receive several résumés that look identical. You want yours to stand out.

If you decide to design your own résumé, Chapter 10 in this book offers design principles that are helpful toward creating a design for your résumé. These principles are balance, alignment, grouping, consistency, and contrast. All of these principles should be used to design your résumé.

Balance—Pay attention to the vertical and horizontal balance of the page. Your résumé should not be weighted too heavily toward the left or right, top or bottom.

A Functional Résumé

Name and contact information are placed up front.

The objective describes the position sought.

Qualities are summarized here, including awards.

Employment and education histories are very concise, listing only the details.

Walter David Trimbal

818 Franklin Drive
Atlanta, Georgia 30361
404-555-2915

Objective: Senior architect position in a firm that specializes in urban revitalization projects.

Leadership Experience

- Managed design team for additions/alterations for small-scale commercial buildings in downtown Atlanta.
- Led planning charette for Bell Hill Neighborhood renovation.
- Awarded a 2006 "Archi" for design of Delarma Commerce Center.

Technical Expertise

- Experienced with the latest developments in computer-aided drafting hardware and software.
- Able to resolve conflicting building and zoning codes, while preparing site plans.

Community Involvement

- Founding Member of the Better Atlanta Commission in 1999 and served as board member until 2007.
- Served on the Architecture Public Involvement Committee for Atlanta Metropolitan Council of Governments from 2006–2007.

Employment Background

Vance & Lipton—Architects, Senior Architect, Atlanta, GA, 2003 to present.
Fulton County Planning Department, Planning Architect, Atlanta, GA, 1999–2003.
Ronald Alterman—Architect, Intern, Boston, MA, 1998–1999.

Education Background

B.A. in Architecture, Boston College, Boston, MA, 1999.
A.A. in Computer-Aided Drafting, Augusta Technical College, Augusta, GA, 1996.

Professional Memberships

National Council of Architectural Registration Boards
Greater Atlanta Architects Guild

References Available Upon Request

Figure 15.6:
A functional résumé puts the applicant's abilities and skills up front where an employer will see them. Other features, such as employment history and education, are minimized.

Kim Isaacs
EXECUTIVE DIRECTOR, ADVANCED CAREER SYSTEMS, INC.
Advanced Careers Systems is a résumé-writing and career development company.

How has the Internet changed the job search process?

The Internet has changed the job search dramatically, and while the web makes the search easier in many ways, it also requires the seekers to be savvier than ever. Because job seekers can search the Internet for positions, they can now apply for jobs that they might not have even heard about in the past. They can also apply for international jobs as easily as national ones. Seekers can go to major online career sites and set up "job agents" that enable them to enter information about their goals and receive notifications when matching jobs become available.

Because the candidate pool is so large, many employers report that they're constantly inundated with new résumés, and employers have the luxury of seeking a "perfect match." That's why it's so important, for example, to make sure that electronic résumés use keywords that will be found on a résumé-tracking system. Even if your résumé is excellent, it can be overlooked without the right wording and phrasing.

Despite the increase in using the Internet to search for positions, online recruiting still accounts for a small fraction of actual hires (the figure keeps changing, but it's around 5%). Of course job seekers should use the Internet to research companies, scour for opportunities, and take advantage of the career tools available, but they still need to focus most of their effort on old-fashioned job search methods such as networking.

Alignment—Different levels of information should be consistently indented to make the résumé easy to scan. Don't just align everything at the left margin; instead, use vertical alignment to create two or three levels in the text.

Grouping—Use white space to frame groups of information. For example, a job listed on your résumé with its responsibilities should be identifiable as a chunk of text. Sometimes using rules, especially horizontal lines, is a good way to carve a résumé into quickly identifiable sections (Figure 15.3).

Consistency—The design of your résumé should be internally consistent. Use bolding, italics, and font sizes consistently. Bullets or other symbols should also be used consistently throughout the résumé.

Contrast—Titles and headings should be noticeably different from the body text. To contrast with the body text, you might choose a different serif or sans serif font for your titles and headings. You can increase the font sizes and/or use bolding to make the résumé more scannable.

> **Link**
> For more information on page design, turn to Chapter 10, page 237.

A helpful strategy for designing your résumé is to collect résumés from other people. There are also numerous books and websites available that offer ideas about designing résumés. You can use these sample résumés as models for designing your own.

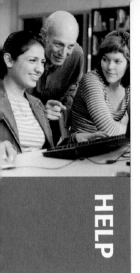

Designing a Scannable/Searchable Résumé

Companies are increasingly asking for *scannable* résumés (Figure A). These kinds of résumés are scannable by computers, which sort and rank the best candidates. Also, résumés posted on job search engines need to be searchable through keywords.

How are scannable/searchable résumés used by employers? Usually, after all the résumés are scanned by the computer, a human resources officer or recruiter will enter ten keywords that describe the position. Then, the computer returns a ranked list of the applicants who matched the most keywords. To survive the cut, you need to find a way to anticipate the keywords that will be entered. Here's a hint: The job advertisement probably contains many of the keywords the employer will be looking for.

Here are some ideas for developing a scannable/searchable résumé:

- Use well-known keywords to describe your skills and experience.
- Use terms in predictable ways. For example, you should write "Managed a team of technicians" rather than "Responsible for guiding a contingent of technical specialists."
- Include acronyms specific to your field (e.g., CAD, TQM, APA, IEEE).
- Use common headings found in résumés: Career Objective, Work Experience, Skills, Qualifications, Education, Honors, Publications, Certifications.
- At the end of your résumé, make a list of any additional traits or skills you possess: time management, dependability, efficiency, leadership, responsibility. These may be used as keywords.

If you are making a paper-based scannable résumé:

- Use white 8½-by-11-inch paper, printed on one side only.
- Do not fold or staple the paper.
- Place your name on its own line at the top of the page.
- Use a standard address format below your name.
- List each phone number on its own line.
- Use standard typefaces like Arial, Helvetica, Times, New York, or Garamond. The computer may have trouble reading other fonts.
- Don't use a font size smaller than 10 points for any of the text.
- Don't use italics, underlining, shadows, or reverse type (white type on a black background) because scanners have trouble reading them.
- Don't use vertical and horizontal lines, graphics, boxes, or shading.
- Don't use two- or three-column formats. One column is easier for the computer to scan.

Fortunately, scanning machines and computers do not care about the length or design of your résumé. Your scannable résumé should be plain in design, and it can be longer than your regular résumé, if needed.

If you suspect your résumé will be scanned, your best strategy is to make two résumés. One should be your regular résumé and the other should be a scannable résumé. Clearly identify the scannable résumé for the employer by placing a cover note on it. That way, the employer won't be confused by the submission of two résumés.

Need more help making a scannable résumé? Go to
www.pearsonhighered.com/johnsonweb3/15.11

A Scannable Résumé

Figure A:
A scannable
résumé removes
much of the
formatting and
design, allowing
the computer to
more easily
locate keywords.
Compare this
résumé with the
regular résumé
in Figure 15.3.

Name and contact information is plainly presented. →

Anne Franklin
834 County Line Rd.
Hollings Point, IL 62905
Home: 618-555-2993
Mobile: 618-555-9167
e-mail: afranklin@unsb5.net

Career Objective: A position as a naturalist, specializing in agronomy, working for a distribution company that specializes in organic foods.

Headings are predictable and easy to locate. →

Educational Background

Bachelor of Science, Southern Illinois University, expected May 2009.
Major: Plant and Soil Science
Minor: Entomology
GPA: 3.2/4.0

Work Experience

Intern Agronomist, December 2007–August 2008
Brighter Days Organic Cooperative, Simmerton, IL

Italics have been removed for easier scanning. →

Consulted with growers on organic pest control methods. Primary duty was sale of organic crop protection products, crop nutrients, seed, and consulting services.
Prepared organic agronomic farm plans for growers.
Provided crop-scouting services to identify weed and insect problems.

Field Technician, August 2007–December 2008
Entomology Department, Southern Illinois University

Bullets have been removed to simplify text. →

Collected and identified insects.
Developed insect management plans.
Tested organic and nonorganic pesticides for effectiveness and residuals.

Skills

Computer Experience: Access, Excel, Outlook, PowerPoint, and Word. Global Positioning Systems (GPS). Database Management.
Machinery: Field Tractors, Combines, Straight Trucks, and Bobcats
Communication Skills: Proposal Writing and Review, Public Presentations, Negotiating, Training, Writing Agronomic and Financial Farm Plans.

Awards and Memberships

Awarded "Best Young Innovator" by the Organic Food Society of America (OFSA), 2007
Member of Entomological Society of America (ESA)

Writing Effective Application Letters

Your résumé will provide the employer with facts and details about your education, work experience, and skills. It cannot, however, convey a full sense of your individuality or your interest in the company that has a job available. Your application letter, which will accompany your résumé, should offer a fuller picture of you as a person. It should also discuss how you can make a contribution to the company.

An effective application letter strives to prove that you are uniquely qualified for the available position. Two common mistakes are made in application letters. First, applicants simply restate the information available on the résumé, failing to demonstrate why they are the right person for the job. Second, they discuss why the position would be good *for them* (e.g., "A job at Gurson Industries would help me reach my goal to become an electrical engineer working with sensors"). To put it bluntly, employers don't really care whether their job is good for you.

Instead, your letter should prove to potential employers that your education, experience, and skills will allow you to make a valuable contribution to *their* company. Put the emphasis on *their* needs, not yours. Your letter should fit on one page for an entry-level position. Two pages would be a maximum for any job.

Content and Organization

Link

For more information on writing letters, go to Chapter 17, page 457.

Like any letter, an application letter will have an introduction, body, and conclusion (Figure 15.7). It will also include common features of a letter such as the header (your address and the employer's address), a greeting ("Dear"), and a closing salutation ("Sincerely") with your signature.

INTRODUCTION   You should begin your letter by making up to five moves commonly found in introductions of any document: Identify your subject, state your purpose, state your main point, stress the importance, and offer background information.

Subject and purpose

Dear Ms. Sims:

Background information

Main point and importance

I would like to apply for the Organic Agronomist position you advertised through HotJobs.com on March 19. My experience with organic innovations in plant and soil science as well as my minor in entomology would allow me to make an immediate contribution to your company.

This introduction makes all five introductory moves, but all five moves are not needed for a successful introduction. Minimally, an introduction should identify the subject (the position being applied for), state your purpose ("I would like to apply for your job"), and state your main point ("I am uniquely qualified for your position").

BODY   In the body of the letter, you should include two to three paragraphs that show how your educational background, work experience, and skills fit the employer's needs. You should organize the body of your letter to highlight your strengths. If your

The Basic Pattern of an Application Letter

The introduction identifies your subject, purpose, and main point, while offering background information and stressing the importance of the subject.

The body discusses your education, experience, and skills. The number of topics you cover will depend on your background.

The conclusion thanks your readers for their time and looks to the future.

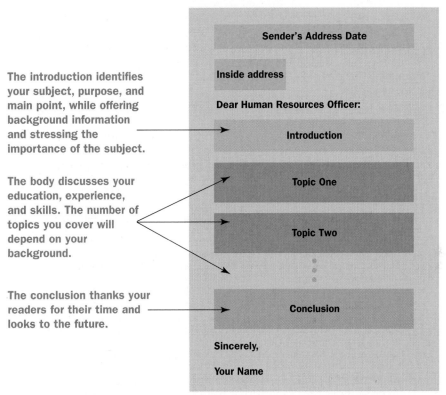

Figure 15.7: An application letter includes the common features of a letter, like the sender's address, date, inside address, introduction, body, and closing salutation.

Sender's Address Date

Inside address

Dear Human Resources Officer:

Introduction

Topic One

Topic Two

Conclusion

Sincerely,

Your Name

educational background is your best asset, put that paragraph right after the letter's introduction (Figure 15.8). If your work experience is stronger than your education, then put that information up front (Figure 15.9).

Remember that you are making an argument, so each paragraph should start out with a claim, and the rest of the paragraph should support that claim with examples, facts, and reasoning. Here is a sample paragraph that supports a claim about an agronomist's educational background.

A clear claim. ──────▶ My education and research as an organic agronomist would benefit your company significantly. As a Plant and Soil Science major at Southern Illinois University, I have been studying and researching environmentally safe alternatives to pesticides. Specifically, my mentor, Professor George Roberts, and I have been working on using benevolent insects, like ladybird beetles (*Coleomegilla maculata*), to control common pests on various vegetable plants. We have also developed several varieties of organic insecticidal soaps that handle the occasional insect infestation.

Support for ──────▶ that claim.

In the body, you need to back up your claims with facts, examples, details, and reasoning—you need proof. You also need to do more than simply restate information that can be found on your résumé. Instead, you should breathe life into your letter by telling stories about yourself.

GO TO THE NET
To see sample application letters, go to
www.pearsonhighered.com/johnsonweb3/15.12

Writing Effective Application Letters

393

Letter of Application Emphasizing Education

Figure 15.8: A letter of application should make an argument, not simply restate items that can be found on the résumé.

834 County Line Rd.
Hollings Point, Illinois 62905
April 2, 2009

Valerie Sims, Human Resources Manager
Sunny View Organic Products
1523 Cesar Chavez Lane
Sunny View, California 95982

Dear Ms. Sims:

Opening paragraph states the subject, purpose, and main point.

I would like to apply for the Organic Agronomist position you advertised through HotJobs.com on March 19. My experience with organic innovations in plant and soil science as well as my minor in entomology would allow me to make an immediate contribution to your company.

Education is discussed up front with examples.

My education and research as an organic agronomist would benefit your company significantly. As a Plant and Soil Science major at Southern Illinois University, I have been studying and researching environmentally safe alternatives to pesticides. Specifically, my mentor, Professor George Roberts, and I have been working on using benevolent insects, like ladybird beetles (*Coleomegilla maculata*), to control common pests on various vegetable plants. We have also developed several varieties of organic insecticidal soaps that handle the occasional insect infestation.

Work experience is used to show potential contributions to employer.

I also worked as an intern for Brighter Days Organic Cooperative, a group of organic farmers, who have an operation similar to Sunny View. From your website, I see that you are currently working toward certification as an organic farm. At Brighter Days, I wrote eleven agronomic plans for farmers who wanted to change to organic methods. My work experience in the organic certification process would be helpful toward earning certification for Sunny View in the shortest amount of time.

Other skills are highlighted to show unique abilities.

Finally, I would bring two other important skills to your company: a background in farming and experience with public speaking. I grew up on a farm near Hollings Point, Illinois. When my father died, my mother and I kept the farm going by learning how to operate machinery, plant the crops, and harvest. We decided to go organic in 1999, because we always suspected that my father's death was due to chemical exposure. Based on our experiences with going organic, I have given numerous public speeches and workshops to the Farm Bureau and Future Farmers of America on organic farming. My farming background and speaking skills would be an asset to your operation.

Conclusion ends on a positive note and offers contact information.

Thank you for this opportunity to apply for your opening. I look forward to hearing from you about this exciting position. I can be contacted at home (618-555-2993) or through e-mail (afranklin@unsb5.net).

Sincerely,

Anne Franklin

Anne Franklin

Letter of Application Emphasizing Work Experience

Opening paragraph uses background information to make personal connection.

Work experience with co-op is highlighted.

Paragraph on education makes connections to employer's needs.

Conclusion indirectly requests the interview.

576 First Avenue
Rolla, Missouri 65408
March 12, 2009

Mr. Harold Brown, Human Resources Director
Farnot Plastic Solutions
4819 Renaissance Lane
Rochester, New York 14608

Dear Mr. Brown:

Last week, you and I met at the University of Missouri–Rolla engineering job fair. You mentioned that Farnot Plastics might be interviewing entry-level chemical engineers this spring to work on applications of polymeric materials. If a position becomes available, I would like to apply for it. With my experience in applied rheology and polymeric materials, I would be a valuable addition to your company.

My work experience includes two summers as a co-op at Vertigo Plastics, a company similar in size, products, and services to Farnot. My responsibilities included inspecting and troubleshooting the plant's machinery. I analyzed the production process and reported on the performance of the plant's operations. While at Vertigo, I learned to work with other chemical engineers in a team-focused environment.

My education in chemical engineering at University of Missouri–Rolla would allow me to contribute a thorough understanding of plastics engineering to Farnot. In one of the best programs in the country, I have excelled at courses in thermodynamics, chemical process design, and chemical process materials. In addition, my work in the university's state-of-the-art chemical laboratories has prepared me to do the prototype building and vacuum forming that is a specialty of your company. I also have experience working with the CAD/CAM systems that your company uses.

The enclosed résumé highlights my other qualifications. I enjoyed speaking with you at the job fair, and I would appreciate an opportunity to talk to you again about opportunities at Farnot. If you would like more information or you would like to schedule an interview, please call me at 573-555-4391 or e-mail me at bigmondy12@umr.edu.

Sincerely,

J. L. Mondragon

James L. Mondragon
Enclosure: Résumé

Figure 15.9:
The organization of the application letter should highlight strengths. Here, work experience is being highlighted by appearing immediately after the introduction.

CONCLUSION   Your conclusion should make three moves: thank the reader, offer contact information, and look to the future. Your goal is to leave a positive impression at this point.

Thank you statement.

Look to the future.

Contact information.

> Thank you for this opportunity to apply for your opening. I look forward to hearing from you about this exciting position. I can be contacted at home (618–555–2993) or through e-mail (afranklin@unsb5.net).

Keep the conclusion concise, and avoid any pleading for the position. Employers will not look favorably on someone who is begging for the job.

Style

Another way an application letter differs from a résumé is in its style. You want to adopt a style that conveys a sense of your own personality and your interest in the position available.

"YOU" ATTITUDE   Put the emphasis of the letter on your readers by using the *"you" attitude.* By strategically using the words "you" and "your" in the letter, you can discuss your qualifications from your readers' point of view.

> Your company would benefit from my mechanical engineering training and hands-on experience as a certified welder.

> Your group is one of the fastest growing medical practices in Denver. As a member of your medical staff, I would be reliable and hardworking.

Link

For more information on using the "you" style in letters, go to Chapter 17, page 471.

To put stress on the readers rather than on yourself, you might also change "I" sentences to "my" sentences. The first sentence above, for example, uses the phrase "my mechanical engineering training" rather than "I have been trained in mechanical engineering." This subtle change lessens the overuse of "I" in the letter, allowing you to put more emphasis on your readers.

ACTIVE VOICE   In stressful situations, like writing application letters, you might be tempted to switch to passive voice. Passive voice in application letters will make you sound detached and even apathetic.

> *Passive:* The senior project on uses of nanotechnology in hospitals was completed in November 2008.
> *Active:* I completed my senior project on the uses of nanotechnology in hospitals in November 2008.

> *Passive:* The proposal describing the need for a new bridge over the Raccoon River was completed and presented to the Franklin City Council.
> *Active:* My team completed the proposal for a new bridge over the Raccoon River, and we presented it to the Franklin City Council.

Link

For more help on using active voice, see Chapter 9, page 220.

Why active voice? The active voice shows that you took action. You were in charge, not just a passive observer of the events around you.

NONBUREAUCRATIC TONE   Avoid using business clichés to adopt an artificially formal tone. Phrases such as "per your advertisement" or "in accordance with your needs" only make you sound stuffy and pretentious.

Bureaucratic: Pursuant to your advertisement in the *Chicago Sun-Times,* I am tendering my application toward your available position in pharmaceutical research.

Nonbureaucratic: I am applying for the pharmaceutical researcher position you advertised in the *Chicago Sun-Times.*

Employers are not interested in hiring people who write in such a pretentious way—unless they are looking for a butler.

THEMES   Think of one quality that sets you apart from others. Do you work well in teams? Are you a self-starter? Are you able to handle pressure? Are you quality minded? These are some themes that you might mention throughout your letter.

AT A GLANCE

Style in an Application Letter

• "You" attitude—Put the emphasis on the employer.

• Active voice—Put yourself, not events, in charge.

• Nonbureaucratic tone—No one wants to hire bureaucrats, so don't sound like one.

• Themes—Point out what makes you different or attractive.

While working on the Hampton County otter restoration project, I was able to communicate clearly and effectively, helping us win over a skeptical public that was concerned about the effects of introducing these animals to local streams.

Beyond simply telling readers that you have a special quality, use examples to show them how you have used this quality to solve problems in the past. Some qualities you might mention include the ability to

• make public presentations.
• be a leader.
• manage time effectively.
• motivate yourself and others.
• follow instructions.
• meet deadlines.
• work well in a team.
• write clearly and persuasively.
• handle stressful situations.

Weave one of these qualities into your letter or address it directly. One theme should be enough, because more than one theme will make you sound like you are boasting.

Revising and Proofreading the Résumé and Letter

When you are finished writing your résumé and application letter, you should spend a significant amount of time revising and proofreading your materials. These documents need to be revised just like any other documents.

Content—Did you provide enough information? Did you provide too much? Do the résumé and letter answer the interviewers' basic questions about your qualities and abilities? Do your materials match up with the job advertisement?

Organization—Are your materials organized in ways that highlight your strengths? Can readers easily locate information about your education, work experience, and skills?

Style—Do your materials set a consistent theme? Is the tone projected in the letter of application appropriate?

Design—Is the résumé highly scannable? Does it follow the design principles of balance, alignment, grouping, consistency, and contrast?

When you are finished revising, proofread everything carefully. Have as many people as possible look over your materials, especially your résumé. For most jobs, even the smallest typos can become good excuses to pitch an application into the recycle bin. Your materials need to be nearly flawless in grammar and spelling.

Creating a Professional Portfolio

Increasingly, it is becoming common for employers to ask applicants to bring a *professional portfolio* to their interview. Sometimes, interviewers will ask that a portfolio be sent ahead of the interview so they can familiarize themselves with the applicant and his or her abilities.

A portfolio is a collection of materials that you can use to demonstrate your qualifications and abilities. Portfolios include some or all of the following items:

- Résumé
- Samples of written work
- Examples of presentations
- Descriptions and evidence of projects
- Diplomas and certificates
- Awards
- Letters of reference

These materials can be placed in a three-ring binder with dividers and/or put on a website or CD-ROM.

Your personal portfolio should not be left with the interviewer, because it may be lost or not returned. Instead, you should create a "giveaway" portfolio with copies of your work that you can leave with the employer. If the interviewer asks to keep the portfolio, you can hand over your giveaway portfolio, explaining that you have original documents in your main portfolio.

Even if interviewers do not ask you to bring a portfolio, putting one together is still well worth your time. After all, if you walk into an interview with just your résumé and a smile, you will have little evidence to prove that you are the right person for the job. If you bring a portfolio, you will have a collection of materials to show your interviewers, allowing you to make the strongest case possible that you have the qualifications and experience they need.

Collecting Materials

If you have not made a portfolio before, you are probably wondering if you have enough materials to put one together. Here are some steps you can take toward finding materials for your portfolio:

Search your hard drive—Your personal computer probably holds documents and presentations that you can print out. The materials may be old, but a little revision and polishing will make these materials suitable for your portfolio.

Look in your current files—You will probably be surprised by the amount of materials you have stuffed away in file cabinets, boxes, and closets. Start

A Print Portfolio

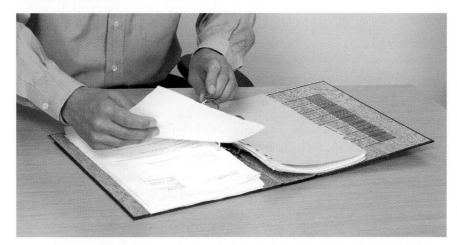

A portfolio is a useful tool to bring to an interview. It holds all your work in an accessible package so that you can support your claims during your interview.

pulling out old projects, presentations, reports, certificates, diplomas, and awards that you have saved over the years.

Start saving materials—From now on, any projects you do in your classes or at your workplace should be saved for your portfolio. Save copies of these items on your hard drive in a special folder called "Portfolio." For print documents, you should designate a desk drawer or file cabinet where you can store items for your portfolio.

Find opportunities to create materials—If you do not have much material for your portfolio, then you will need to create some. Internships and co-ops are especially good ways to gain experience and fill your portfolio. If these opportunities are not available, look for ways to volunteer with worthy organizations or causes. Volunteering is a good way to gain experience while adding materials to your portfolio.

Ask for letters of reference—When you have good experiences with professors or other people, ask them for letters of reference for your portfolio. Even if you are not actively looking for a job, ask for these letters when the relationship is still strong. Good letters are hard to obtain after your classes or other experiences fade into the past.

Something to keep in mind is that much of your written work in college is suitable for your portfolio. Your class-related materials do not need to be directly applicable to the job you are applying for. Instead, potential employers are interested in seeing evidence of your success and your everyday abilities. Your documents written for class will show evidence of success.

Even nontechnical materials—like your critical analysis of Beethoven's Ninth Symphony that you wrote for a course in classical music—show your ability to do research, adopt a critical perspective, and write at length. These items are appropriate if you are looking for materials to fill a portfolio.

Organizing Your Portfolio

Now that you have your materials and portfolio ready, you can organize the materials according to various schemes. For example, you might organize materials year by year or job by job. A particularly good way to organize a portfolio is to follow the categories in a typical résumé:

Cover sheet—A sheet that includes your name, address, phone number, and e-mail address.

Educational background—Diplomas you have received, workshops you have attended, a list of relevant courses, and college transcripts.

Related work experience—Printed materials from your previous jobs, internships, co-ops, and volunteer work. You might include performance reviews, news articles that mention you, or brochures about specific projects. You might even include photographs of places you have worked, projects you have worked on, and people you have worked with.

Specialized skills—Certificates of completion for training, including any coursework you have completed outside of your normal college curriculum.

Writing samples and publications—Examples of your written work, presentations, or websites.

Awards—Any award certificates or letters of congratulation. You might include letters that mention any scholarships you have received.

Other interests—Materials that reflect your other activities, such as volunteer work, sports, or hobbies. Preferably, these materials would be relevant to the kinds of jobs you are seeking.

References—Letters of reference from teachers, colleagues, co-workers, or employers. Also, it is helpful to keep a list of references with phone numbers and addresses where they can be contacted.

Assembling the Portfolio in a Binder

Your portfolio needs to be easy to use at an interview. If your portfolio is nothing more than a hodgepodge of paper stuffed in a three-ring binder, you will find it difficult to locate important information.

So, go to an office supply store and purchase the following items:

Three-ring binder—The binder should be at least 2 inches wide to hold your materials, but it should not be too large to be comfortably carried to an interview. Find one that is suitable for a formal situation like an interview.

Dividers with tabs—Tabbed dividers are helpful for separating your materials into larger categories. The tabs are helpful for finding materials quickly, especially in a stressful situation like an interview.

Pocketed folders or clear plastic sleeves—Put your materials in pockets or clear plastic sleeves. That way, you can easily insert and remove your materials when you want.

As you assemble the portfolio, keep professionalism and ease of use in mind:

- Copies of your résumé can usually be placed in a pocket on the inside cover of the three-ring binder. You may need them if interviewers do not have a copy of your résumé in front of them.
- Labels can be used to provide background information on each item in the portfolio. These labels are helpful in jogging your memory during a stressful interview.

Creating an Electronic Portfolio

If you know how to create a basic website, you can create an electronic portfolio for yourself (Figure 15.10). An electronic portfolio has several advantages. It can

- be accessed from anywhere there is a networked computer, including an interviewer's office.
- include multimedia texts such as movies, presentations, and links to websites you have created.
- include materials and links to information that would not typically be found in a nonelectronic portfolio. For example, you might put links to your university and academic department to help interviewers learn about your educational background.

Electronic Portfolio

Links take the reader to materials in the portfolio.

Other links take the reader to related information.

Figure 15.10: An electronic portfolio is a great way to show off your materials and your experiences.

Purpose of the portfolio is clearly stated.

To see samples of online portfolios, go to
www.pearsonhighered.com/johnsonweb3/15.14
Where can materials for your portfolio be found? Go to
www.pearsonhighered.com/johnsonweb3/15.15

Creating a Professional Portfolio

401

You should keep your electronic portfolio separate from any personal websites you create. The materials you include in your electronic portfolio should be appropriate for discussion in a professional interview. (Your vacation photos from that Aspen skiing trip with your friends probably aren't appropriate.)

Interviewing Strategies

When you are called for an interview, an employer is already telling you that you are qualified for the position. Now you just need to compete with the other qualified candidates who are also interviewing for the position.

Preparing for the Interview

In many ways, interviewing is a game. It's not an interrogation. The interviewers are going to ask you some questions or put you in situations that will test your problem-solving abilities. To answer the interviewers' questions appropriately, you need to do some preparation.

RESEARCH THE COMPANY   Before the interview, find out as much as possible about the company and the people who will be interviewing you. The Internet, especially the company's website, is a great place to start. But you should also look for magazine and newspaper articles on the company in your university or local library. While researching the company, locate facts about the size of the company, its products, and its competitors. You should also be aware of major trends in the company's market.

The Interview

Interviewing is more like a game than an interrogation. Once you know how the game is played, you can win it.

DRESS APPROPRIATELY   The interview game begins with your appearance. There is an old saying: "Dress for the job you want, not for the job you have." When you are interviewing, you should be dressed in a suitably formal manner, avoiding flashy jewelry or too much cologne or perfume. Appropriate clothing usually depends on the kind of job you are seeking. During your interview, you want to be one of the best-dressed people in the room.

At the Interview

When you are at the interview, try to relax and present yourself as someone the interviewer would like to hire. Remember that each question from an interviewer is a move in the game, and there are always appropriate countermoves available.

Link

For more information on making a professional presentation, see Chapter 16, page 411.

GREET PEOPLE WITH CONFIDENCE   When you meet people at the interview, you should greet them with confidence. Most North Americans will expect you to shake their hand firmly and make eye contact. Let the interviewer indicate where you are going to sit. Then, set your briefcase and/or portfolio at your side on the floor. Don't put things on the interviewer's desk, unless he or she asks you to.

When time permits, write down the name and title of each interviewer you meet. You will need these names and titles when you are writing thank you e-mails and letters after the interview.

ANSWER QUESTIONS   Interviewers will usually work from a script of questions. However, most interviews go off the script as interesting topics come up. You should be prepared with answers to some of the following questions:

Link

For more strategies on answering questions, see Chapter 16, page 429.

> **"Tell me about yourself"**—Spend about two minutes talking about your work experience, education, and skills, relating them to the position. Don't start with, "I was born in New York in 1986...."
>
> **"What about this position attracted you?"**—Talk about the strengths of the company and what qualities of the position you find interesting.
>
> **"Why should we hire you?"**—Stress qualifications and skills appropriate to the job that set you apart from the other candidates.
>
> **"Where do you want to be in three to five years?"**—Without being too ambitious, talk about doing your job well and moving up in the company.
>
> **"What are your salary requirements?"**—This question is uncommon, but you should have a salary figure in your head in case it is asked. You don't want to fumble this question or ask for too little.
>
> **"What is your greatest strength?"**—Discuss a strong qualification, skill, or knowledge area relevant to the job.
>
> **"What is your greatest weakness?"**—Discuss something you would like to learn that would enhance your ability to do your job (e.g., a new language, more advanced computer skills, greater communication skills). The "weakness" question is not the time to admit your shortcomings. If you are a procrastinator or you have a bad temper, the interview is not the time to confess your sins. Also, answers like, "I just work too hard"

or "I am too committed to doing an excellent job" don't really fool anyone. Instead, mention something you want to learn to improve yourself.

USE YOUR PORTFOLIO   No doubt you have been told numerous times, "Show, don't just tell." In your interview, use your professional portfolio to back up your answers to interview questions. The materials in your portfolio should be used to reinforce your statements about your background, qualifications, and skills.

ASK QUESTIONS   As the interview comes to an end, the interviewer will usually ask if you have any questions. You should be ready to ask two or three insightful questions. Here are a few examples of questions that will demonstrate your interest in the company and the job:

Where do you see the company going in the next five years?

What can you tell me about your customers/clients?

What kinds of additional learning opportunities are available?

What happens in a normal day at this position?

AT A GLANCE

Interviewing Strategies

• Research the company.
• Dress appropriately.
• Greet people with confidence.
• Answer questions.
• Use your portfolio.
• Ask questions.
• Leave with confidence.

Avoid asking questions at this point about salary, vacation, and benefits. Usually these items are discussed after a job offer has been made.

After the interviewer answers each question, you might offer a follow-up statement that reinforces one of your strengths.

LEAVE WITH CONFIDENCE   When the interview is finished, thank the interviewers for their time and say that you are looking forward to hearing from them. Also, ask if they would like you to send them any other information. Then, shake each interviewer's hand firmly and go.

As soon as possible after the interview, find a place where you can write down everything you can remember about what you and the interviewers talked about. These notes may be helpful later, especially if a week or two lapses before you hear about the job. Your notes should mention any important discussion points that developed during the interview.

Writing Thank You Letters and/or E-Mails

After an interview, it is polite to write a thank you letter to the people who interviewed you. A basic thank you letter shows your appreciation for the interviewers' time while expressing continued interest in the job. A more sophisticated letter could reinforce one or more of your strengths, in addition to saying thank you and expressing continued interest in the job (Figure 15.11).

If you send a thank you through e-mail, follow it with a letter through the mail. An e-mail is nice for giving the interviewers immediate feedback, but the letter shows more professionalism.

If all goes well, you will be offered the position. At that point, you can decide if the responsibilities, salary, and benefits fit your needs.

GO TO THE NET

Want some strategies for answering tough questions? Go to
www.pearsonhighered.com/johnsonweb3/15.17

A Thank You Letter

Inside address

A thank you statement leads off the letter.

An important point is reinforced with details from the interview.

The signature includes the full name.

834 County Line Rd.
Hollings Point, Illinois 62905
May 15, 2009

Valerie Sims, Human Resources Manager
Sunny View Organic Products
1523 Cesar Chavez Lane
Sunny View, California 95982

Dear Ms. Sims:

Thank you for interviewing me for the Organic Agronomist position at Sunny View Organic Products. I enjoyed meeting you and the others at the company. Now, after speaking with you, I am more interested than ever in the position.

I noticed during the interview that my experience with benevolent insects was a recurring topic of discussion. This area of pest control is indeed very exciting, and my work with Professor George Roberts is certainly cutting edge. By paying attention to release times and hatching patterns, we have been able to maximize the effectiveness of predator insects. I would bring the latest research in this area to Sunny View's crops.

Again, thank you for the opportunity to interview with you for this position. I can be contacted at home (618-555-2993) or through e-mail (afranklin@unsb5.net). Please call or e-mail me if you need more information or would like to speak with me further about this position.

Sincerely,

Ann Franklin

Anne Franklin

Figure 15.11: A thank you letter can be used to reinforce an important point that came out during the interview.

The main point is clearly stated in the introduction.

The conclusion thanks the reader and offers contact information.

- From the beginning, you should adopt a professional approach to job seeking: Set goals and make a plan.

- Your job-searching plan should include job search engines, personal networking, professional networking, targeting, and classified advertisements.

- Two types of résumés are commonly used, the chronological résumé and the functional résumé. Either should summarize your background, experience, and qualifications.

- A well-designed résumé should use the basic principles of design; a scannable résumé incorporates keywords so that it can be sorted electronically.

- An effective application letter can be more individual and specifically targeted than the résumé. It shares the common features of a letter, including the appropriate "moves" for the introduction, body, and conclusion.

- An appropriate style for an application letter uses the "you" attitude, active voice, and nonbureaucratic language; you can also work in a theme that will make your résumé stand out.

- A professional portfolio is a helpful tool for an interview because it can demonstrate your qualifications and abilities; material for your "master" portfolio can be assembled from classroom work, volunteer projects, related work experience, awards, certificates, or letters of reference.

- When you go to your interview, be prepared with information about the company and appropriate questions to ask; dress appropriately for the position; greet people with confidence; answer questions thoughtfully; and obtain the name and title of each interviewer (and its correct spelling) for a follow-up note to be sent that day.

Individual or Team Projects

1. Find a sample résumé on the Internet. Write a one-page critique of the résumé for your instructor. Discuss its strengths and make suggestions for improvement. Pay special attention to the content, organization, style, and design of the résumé.

2. Imagine you are looking for an internship in your field. Write a one-page résumé that summarizes your education, work experience, skills, awards, and activities. Then, pay attention to issues of design, making sure the text uses principles of balance, alignment, grouping, consistency, and contrast.

3. Using an Internet job search engine, use keywords to find a job for which you might apply after college. Underline the qualifications required and the responsibilities of the position. Then, write a résumé and application letter suitable for the job. In a cover memo addressed to your instructor, discuss some of the reasons you would be a strong candidate for the position. Then, discuss some areas where you might need to take more courses or gain more experience before applying for the position.

4. Contact a human resources manager at a local company. Request an "informational interview" with this manager (preferably in person, but an e-mail interview is sufficient). If the interview is granted, ask him or her the best way to approach the company about a job. Ask what kinds of qualifications the company is usually looking for in a college graduate. Ask what you can do now to enhance your chances of obtaining a position at the company. After your interview, present your findings to your class.

5. Create a professional portfolio that you could use at an interview. Pull together examples of your work as well as any other relevant materials that you can find. What are the gaps in your portfolio? What kinds of materials should you try to obtain before you leave college? How can you go about filling in these gaps?

6. Make an electronic portfolio of your materials. Besides the materials available in your regular portfolio, what are some other links and documents you might include in your electronic portfolio?

Collaborative Project

With a group of people pursuing similar careers, develop some career goals and a job search plan for reaching those goals. Then, send each member of the group out to collect information. One group member should try out job search engines, while another should explore personal and professional networking opportunities. One member should create a list of potential employers in your area. Another should explore newspapers' online classified advertisements.

After your group has collected the information, write a small report for your class and instructor in which you discuss the results of your research. What did your group discover about the job market in your field? Where are the hottest places to find jobs? How can professional groups and personal networking help you make contacts with potential employers?

Individuals in your group should then choose one job that seems interesting. Each member of the group should write an application letter for that job and create a résumé that highlights qualifications and strengths.

Then the group should come up with five questions that might be asked at an interview, four questions that are usually asked at an interview, and one question that is meant to trip up an interviewee.

Finally, take turns interviewing each other. Ask your questions and jot down good answers and bad answers to the questions. Discuss how each member of the group might improve his or her interviewing skills based on your experiences in this project.

For additional technical writing resources, including interactive sample documents, document design tutorials and guidelines, and more, go to **www.mytechcommlab.com**.

CASE STUDY

The Lie

Henry Romero had wanted to be an architect his whole life. Even as a child, he was fascinated by the buildings in nearby Chicago. So, he spent years preparing himself to be an architect, winning top honors at his university in design.

He was thrilled when one of the top architectural firms, Goming and Cooper, announced it would be visiting his campus looking to interview promising new talent. This firm was certainly one of the top firms in the country. It had designed some of the most innovative buildings in recent years. Henry was especially interested in the firm's international projects, which would allow him to work and live abroad. Recently, in *Architecture Times,* he had read about the firm's successful relationships with companies in France.

So, Henry gave his résumé and a letter of application to the university's placement office personnel and told them to send it to Goming and Cooper. A month later, his placement counselor called him and said he had an interview with the firm. Henry was very pleased.

There was only one problem. On his résumé, Henry had put down that he "read and spoke French fluently." But, in all honesty, he had taken French for only a few years in high school and one year in college. He hadn't really mastered the language. If necessary, he could muddle his way through a conversation in French, but he was hardly fluent.

To make things worse, last week Henry's roommate, Paul, had a nightmare interview. He too had reported on his résumé that he spoke a language fluently—in this case, German. When he arrived at the interview, the interviewer decided to break the ice by talking in German. Paul was stunned. He stammered in German for a few minutes. Then the interviewer, clearly angry, showed him the door.

Henry was worried. What if the same thing happened to him? He knew the ability to speak French was probably one of the reasons he had received the interview in the first place. If the interviewer found out he was not fluent in French, there was a good chance the interview would go badly.

If you were Henry, how would you handle this touchy situation?

In this chapter, you will learn:

- The importance of being able to prepare and deliver public presentations.

- How to define the rhetorical situation (subject, purpose, audience, and context of use) for your presentation.

- Strategies for organizing the content of presentations.

- How to create an effective presenting style.

- How to create and use visuals in presentations.

- The importance of practice and rehearsal.

- How to work effectively with translators in cross-cultural situations.

If you don't like giving public presentations, you are not alone. Each year, surveys show that people fear speaking in public more than anything else—even more than death.

Yet giving public presentations is an essential part of most technical careers. More than likely, you will find yourself regularly giving presentations to clients, supervisors, and colleagues. Presenting information in public is a crucial skill in today's technical workplace.

Fortunately, computers have made presenting in public a bit easier. With the aid of computers, you can develop a polished presentation with professional visual aids that help reinforce your message. Presentation software like Microsoft PowerPoint, Open Office Impress, Corel Presentations, and Apple Keynote can help you organize and design your information for maximum effect.

Public Speaking Is More Important Than Ever

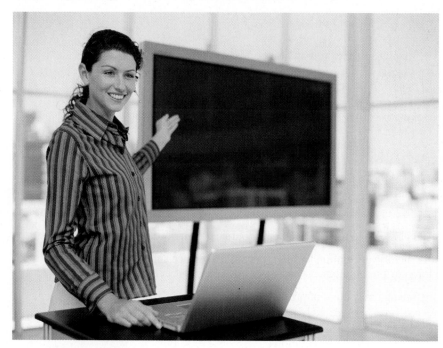

Public presentations are easier than ever with computers. You will find, though, that audiences now expect polished, professional presentations with plenty of visuals and visual appeal.

Keep in mind that computers have only increased the need for making public presentations. In today's computer-centered workplace, people are more visual, and they often prefer interactive presentations over written documents. As a result, public presentations are more common than ever because

- they are more visual than print documents.
- they require less time and effort from the audience.
- they allow people to interact directly with the speakers.

In fact, the need to present information publicly will likely only increase as computer-based technologies, like video streaming and videoconferencing, become regular features of the workplace. You will likely find yourself giving presentations to a digital camera for audiences that are hundreds and thousands of miles away. You might also find yourself giving "distance" presentations with Internet software that uses a wired whiteboard. The ability to give presentations is an important skill in today's technical workplace.

Planning and Researching Your Presentation

The preparation of a public presentation should follow a process, just like preparing written documents (Figure 16.1). Before getting up in front of an audience, you will need to

- plan and research your subject.
- organize your ideas.
- choose an appropriate presentation style.
- create graphics and slides.
- practice and rehearse your presentation.

Even if you are already an accomplished public speaker, it is important that you spend significant time on each stage of the process. After all, a presentation cannot succeed on content alone. You need to pay attention to issues of organization, style, design, and delivery.

The Preparation Process for a Presentation

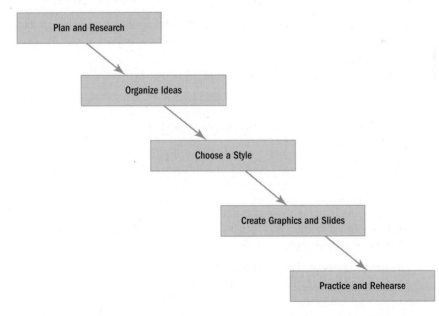

Figure 16.1: Just as with written documents, you should follow a process as you prepare a presentation.

As you begin planning your talk, remember that presentations can be either formal or informal:

Formal presentations—These presentations often include the use of a podium, speaking notes, and slides made with presentation software. Formal presentations include speeches, workshops, trainings, briefings, demonstrations, and panel discussions. They are made to clients, management, and colleagues.

Informal presentations—Most presentations at work are informal. At monthly meetings, you will be asked to report on your team's progress. If you have a new idea, you will need to pitch it to your boss. And, if your supervisor stops by and says, "Hey, in 10 minutes, could you come by my office to tell the vice president how the project is going?" you are about to make an informal public presentation.

In both formal and informal presentations, solid planning is the key to success. A good way to start the planning phase is to analyze the rhetorical situation of your presentation. Begin by asking some strategic questions about your subject, purpose, and audience and the context in which you will be giving your talk.

The Five-W and How Questions give you a good place to start.

Who will be in my audience?

What kind of information do the audience members need or want?

Why am I presenting this information to this audience?

Where will I be presenting?

When will I need to give my talk?

How should I present this information?

Your answers to these questions should help you start crafting your materials into something that will be interesting and informative to the audience.

Defining the Rhetorical Situation

Now that you have considered the Five-W and How Questions, you can think a little deeper about the situation in which you will be speaking (Figure 16.2). Consider closely your subject, purpose, audience, and the context of use.

SUBJECT   Identify what the audience needs to know, putting emphasis on information they require to take action or make a decision. You might also ask yourself what the audience does *not* need to know. That way, you can keep your presentation concise and to the point.

PURPOSE   You need to know exactly what you want to achieve in your presentation. Try to state your purpose in one sentence.

My goal is to persuade elected officials that global warming is a looming problem for our state.

For websites that discuss giving presentations, go to
www.pearsonhighered.com/johnsonweb3/16.1

Asking Questions About the Rhetorical Situation

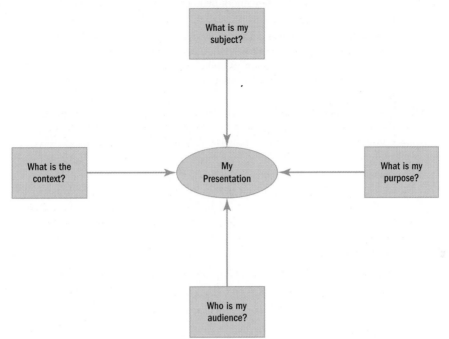

Figure 16.2:
Asking questions
about the subject,
purpose, audience,
and context will
help you anticipate
the information
you need and how
that information
should be
presented.

We need to demonstrate the G290 Robot Workstation to the CEO of Geocom Industries, showing her how it can be used to clean up toxic spills.

I need to motivate our technical staff to improve quality so we can meet the standards demanded by our new client.

If you need more than one sentence to express your purpose, you are probably trying to do too much in your presentation.

Link
For more help
defining your
purpose, go to
Chapter 2,
page 21.

AUDIENCE   Members of your audience will come to your presentation with various needs, values, and attitudes. You should anticipate these characteristics and shape your presentation to their specific requirements and interests.

>**Primary audience (action takers)**—For most presentations, the primary audience is the most important, because it is made up of people who will be making a decision or taking action.

>**Secondary audience (advisors)**—These members of the audience might advise the primary audience on what actions to take. They might be experts in your field or people who have information or opinions on your subject.

To learn more about analyzing audiences, go to
www.pearsonhighered.com/johnsonweb3/16.2

**Planning and
Researching Your
Presentation**

413

Tertiary audience (evaluators)—Others in the audience may have an interest in what you are saying. They might be journalists, lawyers, activists, or concerned citizens.

Gatekeepers (supervisors)—Your supervisors and others at your company will often need to see your presentation before you give it to an audience. They will be looking for accuracy and checking whether you are achieving your purpose and fulfilling the mission of the company.

Link

For more audience analysis techniques, go to Chapter 3, page 42.

CONTEXT OF USE   Context is always important in technical communication, but it is especially important in public presentations. You need to be fully aware of the physical, economic, ethical, and political factors that will shape your presentation and how your audience will react to it.

Physical context—Take time to familiarize yourself with the room in which you will be speaking and the equipment you will be using. You will need to adjust your presentation, including your visuals, to the size, shape, and arrangement of the room. Also, find out what kind of furniture and equipment you will have available.

Will you be using a podium or lectern?

Will you be sitting behind a table or standing out in the open?

Will there be a microphone?

Will a projector be available, and will you be able to use it?

Will you need to bring your own projector and computer?

Will other visual aids like whiteboards, flip charts, and large notepads be available?

Will the audience be eating, and will drinks be available?

When will the audience need breaks?

It is astonishing how many presentations fail because speakers are not prepared for the physical characteristics of the room. They show up with slides that can't be read because the room is too large, or they try to talk to a large audience with no public address system.

These sorts of problems might be someone else's fault, but it's your presentation. You will be the person who looks unprepared. Proper preparation for the physical context will help you avoid these problems.

Economic context—As always, money will be a central concern for your audience members. So, consider the microeconomic and macroeconomic factors that will influence how they will receive your presentation. Microeconomic issues might include budgetary concerns or constraints. Macroeconomic issues might include your audience's economic status, economic trends in the industry, or the state of the local or national economy.

Ethical context—Presentations almost always touch on ethical issues in one way or another. As you prepare, identify and consider any rights, laws, or issues of common concern that might shape your presentation.

Link

For more information on ethics, go to Chapter 5, page 92.

For worksheets to help you analyze your audience and the context of the presentation's use, go to **www.pearsonhighered.com/johnsonweb3/16.3**

Political context—Politics also play a role in presentations. In some cases, politics might simply involve the usual office politics that shape how people react to you and your subject. In other situations, larger political issues may come into play. National issues (energy policies, conservation, gun rights, privacy, security, etc.) will evoke different political responses among members of your audience.

Link

To learn more about analyzing context of use, go to Chapter 3, page 46.

Allotting Your Time

As the speaker, you have an unstated "contract" with the audience. According to this contract, your audience is allowing you a specific number of minutes. It's your responsibility to fill that time productively—*and not go over that amount of time.* Nothing annoys an audience more than a speaker who runs past the time allotted.

So, as you are planning your presentation, first determine how much total time you have to speak. Then, *scale* your presentation to fit the time you have allotted. Figure 16.3 shows how a few common time periods might be properly budgeted.

Allotting Your Presentation Time

	15-Minute Presentation	30-Minute Presentation	45-Minute Presentation	One-Hour Presentation
Introduction	1 minute	1–2 minutes	2 minutes	2–3 minutes
Topic 1	2 minutes	5 minutes	8 minutes	10 minutes
Topic 2	2 minutes	5 minutes	8 minutes	10 minutes
Topic 3	2 minutes	5 minutes	8 minutes	10 minutes
Topic 4	2 minutes	5 minutes	8 minutes	10 minutes
Conclusion	1 minute	1–2 minutes	2 minutes	2 minutes
Questions	2 minutes	5 minutes	5 minutes	10 minutes

Figure 16.3: When planning your presentation, carve up your time carefully to avoid going over that amount of time.

Of course, if you have fewer or more than four topics, you should make adjustments to the times allowed for each. Also, you might need to spend more time on one topic than another. If so, adjust your times accordingly.

There are two things you should notice about the times listed in Figure 16.3.

- Longer presentations do not necessarily allow you to include substantially longer introductions and conclusions. No matter how long your presentation is scheduled to run, keep your introductions and conclusions concise.
- You should not budget all the time available. Always leave yourself some extra time in case something happens during your talk or you are interrupted.

In the end, the unstated contract with your audience is that you will finish within the time scheduled. If you finish a few minutes early, you won't hear any complaints. However, if you run late, your audience will not be pleased.

Choosing the Right Presentation Technology

As you plan your presentation, it is a good idea to think about what presentation technology you will use for your talk. Will you use presentation software with a digital projector? Are you going to use a whiteboard? Are you going to make transparencies for an overhead projector? The kind of presentation technology you need depends on the type of presentation you will be making (Figure 16.4).

Presentation Technologies

Type of Presentation	Visuals
Presentation to a group of more than 10 people	Digital projector with computer
	Overhead projector with transparencies
	35-mm slide projector
	Whiteboard or chalkboard
Presentation to a group of fewer than 10 people	Digital projector with computer
	Overhead projector with transparencies
	35-mm slide projector
	Flip charts
	Large notepads
	Digital video on TV monitor (DVD or CD-ROM)
	Posters
	Handouts
	Computer screen
	Whiteboard or chalkboard

Figure 16.4: There are many different ways to present materials. You should choose the one that best fits your subject and audience.

Fortunately, presentation software like PowerPoint, Presentations, and Keynote makes it easy to create slides and use graphics in your presentations. These programs will help you create visually interesting presentations for a variety of situations.

Each kind of visual aid offers specific advantages and disadvantages. Here are some pros and cons of the more common types of visuals:

Digital projector with a computer—Most companies have a digital projector available for your use. The projector can display the slides from your computer screen onto a large screen. The advantages of digital projectors are their ease of use and their ability to create highly attractive, colorful presentations. The disadvantage is that the projected slides often dominate the room because the lights need to be turned down. As a result, the audience can become fixated on the slides and stop listening to what you are saying.

To find a selection of presentation software packages, go to
www.pearsonhighered.com/johnsonweb3/16.5

Using a Digital Projector

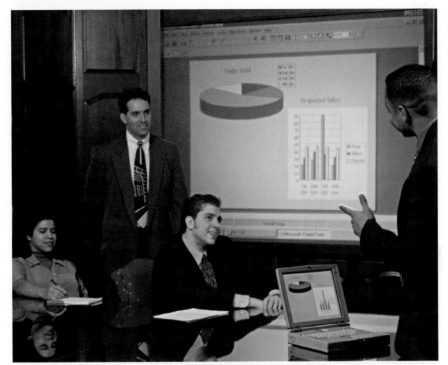

Digital projectors are increasingly common. They project your computer screen onto a large screen.

Overhead projector with transparencies—The overhead projector is the tried-and-true method for giving presentations. You can use an overhead projector to project slides onto a large screen. Transparencies for the projector can be made on a paper copier. The advantages of overhead projectors are that they are commonly available in workplaces and they are more reliable than digital projectors. The disadvantage is that presentations using overheads often seem more static and lifeless than ones made with digital projectors. The colors are not as sharp, and the pictures can be blurry.

Whiteboard, chalkboard, or large notepad—People often forget about the possibility of using a whiteboard, chalkboard, or large notepad in a room. But if you are giving a presentation that requires interaction with the audience, you can use these items to make visuals on the fly. The advantage is that you can create your visuals in front of the audience. Your listeners won't feel that they are receiving a canned presentation in which they have little input. The disadvantage is that you need to think on your feet. You need to find ways to translate your and the audience's comments into visuals on the board.

Want to know more about presentation technologies? Go to
www.pearsonhighered.com/johnsonweb3/16.6

Using a Flip Chart

Flip charts are low tech, but they are especially effective in presentations to a few people.

Flip charts—For small, more personal presentations, a flip chart is a helpful tool. As the speaker talks, he or she flips a new page forward or behind with each new topic. The advantage of flip charts is their closeness to the audience. You can give a flip chart presentation to a small group. The disadvantage of flip charts is that they are too small to be seen from a distance. If you have more than a handful of people in the audience, the flip chart won't work.

Posters—In some cases, you might be asked to present a poster. A poster usually includes about five to seven slides that describe a product or show a procedure. The slides are often used to summarize an experiment. The advantage of a poster is that everything covered in the talk is visually available to the audience. In some cases, the poster might be left alone on a display or a wall so readers can inspect it on their own time. Posters, however, have the same disadvantage as flip charts. They can be

GO TO THE NET

To see sample poster presentations, go to
www.pearsonhighered.com/johnsonweb3/16.7

used only in presentations to a handful of people. They are too hard to see from a distance.

Handouts—Handouts can be helpful in some cases, but they are not always appropriate. When used properly, they can reinforce points made in your presentation or provide data that won't be visible with a projector. Also, handouts made with presentation software can be formatted to leave room for note taking. Handouts, though, can also be very distracting. In a large room, handouts can take a few minutes to be passed around, causing the speaker to lose momentum. The speaker might also need to spend a few minutes making sure everyone has a copy. Meanwhile, people in the audience are often distracted by the handouts, looking at them rather than at the presentation.

These kinds of technology decisions are best made up front—while you are planning—because your choice of technology will often shape your decisions about the content, organization, style, and design/delivery of your information.

Remember Your Audience

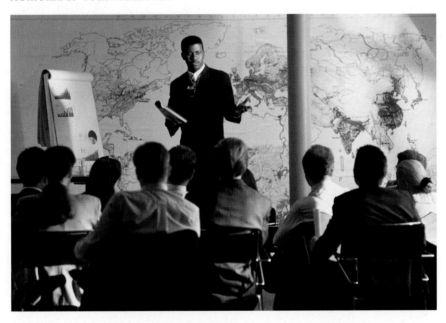

Always keep your audience in mind. Most audiences prefer a concise, to-the-point presentation.

Organizing the Content of Your Presentation

One problem with organizing public presentations is that you usually end up collecting more information than you can talk about in the time allowed. Of course, you cannot tell the audience everything, so you need to make some hard decisions about what they need to know and how you should organize that information.

Link

To help you collect content for your presentation, see Chapter 7, page 146.

Keep your purpose and audience foremost in your mind as you make decisions about what kind of content you will put in the presentation. You want to include only need-to-know information and cut out any want-to-tell information that is not relevant to your purpose or audience. As you make decisions about what to include or cut, you should keep the following in mind: *The more you say, the more they will forget.*

Audiences rarely complain about short presentations. As with wedding ceremonies, the longer your presentation, the less your listeners are going to pay attention. If you go on too long, they will simply stop listening altogether.

Building the Presentation

There is an old adage about public presentations that has been used successfully for years: *Tell them what you're going to tell them. Tell them. Tell them what you told them.*

In other words, like any document, your presentation should have a beginning (introduction), a middle (body), and an end (conclusion) (Figure 16.5).

The Introduction: Tell Them What You're Going to Tell Them

The beginning of your presentation is absolutely crucial. If you don't catch the audience's attention in the first few minutes, there is a good chance they will tune out and not listen to the rest of your presentation. Your introduction slide(s) for

Basic Pattern for a Presentation

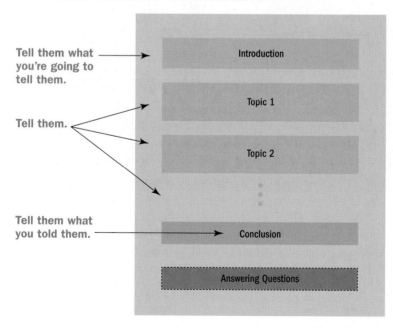

Figure 16.5: A presentation has a beginning, a middle, and an end. Usually time is left for questions at the end.

the presentation should present at least the subject, purpose, and main point of your presentation (one slide). As shown in Figure 16.6, you might also use a second slide to forecast the structure of the talk.

Like the introduction to a document, your presentation should begin by making up to six moves:

MOVE 1: DEFINE THE SUBJECT   Make sure the audience clearly understands the subject of your presentation. You might want to use a *grabber* to introduce your

Introduction Slides for a Presentation

Define the subject.

Offer background information.

Stress the importance of the subject to the audience.

State the purpose and main point of the presentation.

Introduction

- What Is Climate Change?
- When Did Our Climate Begin to Change?
- Why Is Climate Change Especially Important to People Living in the West?
- Answer: Our Ecosystem Is Too Fragile to Adjust to These Changes

Figure 16.6: An introduction builds a context for the body of the presentation.

Forecast the structure of the presentation.

Today's Presentation

- Causes of Climate Change
- Effects of Climate Change in the West
- Solutions to This Problem
- Recommendations

Want more ideas on how to start a presentation? Go to www.pearsonhighered.com/johnsonweb3/16.8

subject. A grabber states something interesting or challenging to capture the audience's attention. Some effective grabbers include the following:

A rhetorical question—"Have you ever thought about changing your career and becoming a professional chef?"

A startling statistic—"A recent survey shows that 73 percent of children aged 15 to 18 in the Braynard area have tried marijuana. Almost a third of Braynard teens are regular users of drugs."

A compelling statement—"Unless we begin to do something about global warming soon, we will see dramatic changes in our ecology within a couple decades."

An anecdote—"A few years ago, I walked into a computer store, only to find the place empty. I looked around and didn't find a salesperson anywhere. Now, I've always been an honest person, but it did occur to me that I could pocket thousands of dollars of merchandise without being caught."

A quotation—"William James, the famous American philosopher, once said, 'Many people believe they are thinking, but they are merely rearranging their prejudices.'"

A show of hands—"How many of you think children watch too much violent television? Raise your hands." Follow this question with an interesting, startling statistic or compelling statement.

But where do you find grabbers? The Internet is a source of endless material for creating grabbers. There are plenty of reference websites like Bartleby.com or Infoplease.com that will help you find quotations, statistics, and anecdotes.

Otherwise, you can use search engines like Google.com, Ask.com, or Yahoo.com to find interesting information for grabbers. Type in your subject and a keyword or phrase that sets a specific tone that you want your grabber to establish. The search engine will locate stories, quotes, statistics, and other information that you can use to create an interesting grabber.

MOVE 2: STATE THE PURPOSE OF YOUR PRESENTATION   In public presentations, you can be as blunt as you like about what you are trying to achieve. Simply tell the audience your purpose up front.

The purpose of this presentation is to prove to you that global warming is real, and it is having a serious impact on Nevada's ecology.

In this demonstration, our aim is to show you that the G290 Robot is ideal for cleaning up toxic spills.

MOVE 3: STATE YOUR MAIN POINT   Before moving into the body of your presentation, you should also state your main point. Your main point holds the

How do you set the right tone at the start of your presentation? For answers, go to
www.pearsonhighered.com/johnsonweb3/16.9
Need a good grabber? Go to
www.pearsonhighered.com/johnsonweb3/16.10

Giving Presentations with Your iPod, MP3, PDA, or Smartphone

One major hassle about giving presentations with digital projectors is lugging along the laptop that holds your presentation. Even if a laptop is provided with the projector, it's always risky to show up with only a flash drive, DVD, or CD that holds your presentation. After all, you never know whether the computer's hardware and/or software is compatible with your presentation until you arrive and plug it in. If something doesn't work, you're in trouble.

The solution? Why not use your iPod, MP3 player, PDA, or smartphone to store and give your presentation? Your player, PDA, or phone is portable (you were taking it anyway, right?) and you can hold it in your hand while you're talking. Also, depending on your player, you can add background music and video to your presentation and play it right through the projector or television (Figure A).

Using Your iPod or MP3 Player to Give a Presentation

Figure A:
Your iPod or MP3 player is a lightweight way to transport your presentation. It also eliminates some of the problems of connecting laptops to projectors or televisions.

An added advantage is that you can go over your slides and practice your presentation any time without firing up that laptop or even bringing it along to campus or on your trip.

How can you do this? You will need an iPod, MP3 player, PDA, or smartphone that has color photo or video capability and an AV port (not just a headphone port). You will also need an AV cable that can be plugged into a projector or television. This kind of cable usually plugs into the place where your headphones are plugged in, and it has three connectors at the other end.

Then, create your presentation in PowerPoint, Keynote, or any other presentation software. When you're done, select "Save As" and choose jpeg (photographs) or PDF

(continued)

(files) format. (For iPods, there is also a free software package called iPresent It by ZappTek that makes the process even easier and allows more options for music and video. This software works on Windows and Mac OS.)

Then, download your files into your iPod, MP3 player, PDA, or smartphone as photographs or PDF files. You can then use "Settings" to add transitions and background music if you want.

When you are ready to give your presentation, plug your AV cord into the projector or television, matching the colors of the three connectors to the colors of the ports. Your presentation should appear on the screen. (If not, turn off the projector and turn it back on.)

Then, while you do your presentation, use the forward and back buttons to move through your slides as you talk.

presentation together, because it is the one major idea that you want the audience to take away from your talk.

> Global warming is a serious problem for our state, and it is growing worse quickly. By switching to nonpolluting forms of energy, we can do our part to minimize the damage to our ecosystem.

> The G290 Robot gives you a way to clean up toxic spills without exposing your hazmat team to dangerous chemical or nuclear materials.

MOVE 4: STRESS THE IMPORTANCE OF THE SUBJECT TO THE AUDIENCE   At the beginning of any presentation, each of the audience members wants to know, "Why is this important to me?" So, tell them up front why your subject is important.

> Scientists predict that the earth's overall temperature will minimally rise a few degrees in the next 30 years. It might even rise 10 degrees. If they are correct, we are likely to see major ecological change on this planet. Oceans will likely rise a foot as the polar ice caps melt. We will also see an increase in the severity of storms like hurricanes and tornadoes. Here in Nevada, we will watch many of our deserts simply die and blow away.

> OSHA regulations require minimal human contact with hazardous materials. This is especially true with toxic spills. As you know, OSHA has aggressively gone after companies that expose their employees to toxic spills. To avoid these lawsuits and penalties, many companies are letting robot workstations do the dirty work.

MOVE 5: OFFER BACKGROUND INFORMATION ON THE SUBJECT   Providing background information on your subject is a good way to build a framework for the audience to understand what you are going to say. You can give the audience a little history on the subject or perhaps tell about your relationship with it.

AT A GLANCE

Opening Moves in a Presentation

- Define the subject.
- State the purpose of your presentation.
- State your main point.
- Stress the importance of the subject to the audience.
- Offer background information on the subject.
- Forecast the structure of the presentation.

Karen Paone

OWNER AND LEAD CONSULTANT, PAONE & ASSOCIATES, ALBUQUERQUE, NEW MEXICO

Paone & Associates is a communication consulting firm that trains employees to make effective presentations.

How can I overcome my fear of speaking in public?

The fear of public speaking routinely ranks ahead of death among the top three fears of most adults. Indeed, there's nothing quite as horrifying as the rush of adrenaline, the rising sense of panic in the pit of your stomach, increased heart rate, dry mouth, and complete loss of memory that assault you the very moment you realize your turn at the podium is imminent.

So what's the difference between these terrifying sensations and the energy, enthusiasm, and engaging style of a polished public speaker? In fact, there is very little difference between the physical reaction of an inexperienced individual presenting at a conference for the first time and an engaged, seasoned presenter. Both experience a heightened sense of awareness and both feel a strong sensation in the solar plexus. Why is one likely to succeed and the other to fail at presenting the information at hand?

- **Rehearsal**—The seasoned speaker creates a realistic rehearsal situation in which she either enlists a practice audience or visualizes her anticipated audience. She knows that simply thinking about the presentation is insufficient to prepare for a real audience. She stands in front of a practice audience or a mirror and runs the entire presentation at least three times. She reworks the rough spots until the entire presentation flows smoothly, anticipates questions, and practices responding with the answers.
- **Heightened awareness**—The seasoned presenter recognizes that the rush of adrenaline is very normal, and that a heightened sense of awareness is essential for an interesting presentation. She uses deep breathing, stretching, and smiling to focus that energy into a confident and enthusiastic stage presence.
- **Communication and connection**—Finally, the seasoned presenter knows that the secret to managing stage fright is focusing on communicating the content at hand in a sincere and personal way to each and every member of the audience. By focusing on connecting with each listener, the presenter's concern is with the audience rather than the way she or he looks, sounds, or feels.

The prescription for stage fright is to rehearse until you can deliver your presentation smoothly from start to finish. Recognize that the physical response to fear and excitement is essentially the same. What differs is how you use those sensations. And finally, when all of your energy is focused on connecting with your audience rather than observing yourself, you too will be perceived as a polished and successful public speaker.

MOVE 6: FORECAST THE STRUCTURE OF THE PRESENTATION   In your introduction, tell the audience how you have organized the rest of the presentation. If you are going to cover four topics, say something like,

> In this presentation, I will be going over four issues. First, I will discuss....Second, I will take a look at....Third, I will identify some key objectives that....And finally, I will offer some recommendations for....

Forecasting gives your audience a mental framework to follow your presentation. A major advantage to forecasting is that it helps the audience pay attention. If you say up front that you will be discussing four topics, the audience will always know where you are in the presentation. Audience members will stay more alert.

The Body: Tell Them

The body of your presentation is where you are going to do the heavy lifting. Start out by dividing your subject into two to five major topics that you want to discuss.

Why only two to five topics? Why not eight or twelve topics? Experience and research show that people can usually remember only five to seven items comfortably. So, if a presentation goes beyond five topics, the audience will start feeling overwhelmed or restless. If you have more than five topics, try to consolidate smaller topics into larger ones.

Link

For more help creating your introduction, go to Chapter 8, page 182.

Presenting the Content

Keep your presentation to two to five major points. You risk losing the audience if you try to cover more than five points.

If you have already written a document, you might follow its organizational structure. If you are starting from scratch, you can follow some of the basic organizational patterns listed below.

PROBLEM, NEED, SOLUTION   This pattern is most effective for proposing new ideas. After your introduction, offer a clear definition of the problem or opportunity you are discussing. Then, specify what is needed to solve the problem or take advantage of the opportunity. Finally, offer a solution/plan that achieves the objective.

CHRONOLOGICAL   When organizing material chronologically, divide the subject into two to five major time periods. Then, lead the audience through these time periods, discussing the relevant issues involved in each. In some cases, a three-part *past-present-future* pattern is a good way to organize a presentation.

SPATIAL   You might be asked to explain or demonstrate visual spaces, like building plans, organizational structures, or diagrams. In these cases, divide the subject into two to five zones. Then, walk the audience through these zones, showing each zone individually and discussing how it relates to the zones around it.

NARRATIVE   Audiences always like stories, so you might organize your presentation around the narrative pattern. Narratives typically (1) set a scene, (2) introduce a complication, (3) evaluate the complication, (4) resolve the complication, and (5) explain what was learned from the experience.

Link
For more information on presenting research, go to Chapter 23, page 662.

METHODS, RESULTS, DISCUSSION   This pattern is commonly used to present the results of research. This pattern (1) describes the research plan or methodology, (2) presents the results of the study, (3) discusses and interprets the results, and (4) makes recommendations.

CAUSES AND EFFECTS   This pattern is common for problem solving. Begin the body of the presentation by discussing the causes for the current situation. Then, later in the body, discuss the effects of these causes and their likely outcomes. You can also alternate between causes and effects. In other words, discuss a cause and its effect together. Then discuss another cause and its effect, and so on.

Link
To learn more about describing products or processes, turn to Chapter 19, page 517.

DESCRIPTION BY FEATURES OR FUNCTIONS   If you are demonstrating a product or process, divide your subject into its two to five major features or functions. Then, as you discuss each of these major features/functions, you can discuss the minor features/functions that are related to them.

COMPARISON AND CONTRAST   Usually this pattern is followed when the speaker is comparing something new or unfamiliar with something that the audience knows well. Choose two to five major points on which these two things can be compared and contrasted. Then, compare and contrast them point by point.

There are countless patterns available for organizing the body of your presentation. The ones shown in Figure 16.7 are some of the most common in technical

Common Patterns for Public Presentations

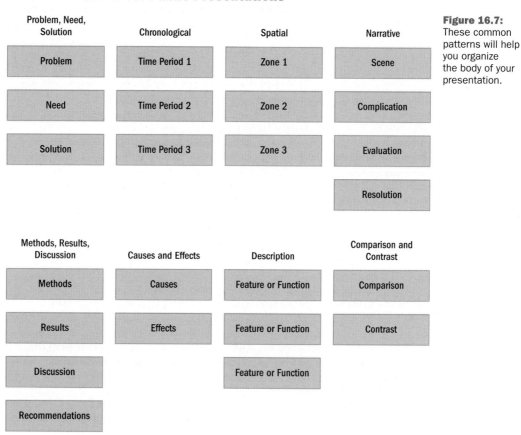

Figure 16.7: These common patterns will help you organize the body of your presentation.

Link

For more ideas about organizing body information, go to Chapter 8, page 186.

communication. These patterns are not formulas to be followed in lockstep. Rather, they can be manipulated to fit a variety of speaking situations.

The Conclusion: Tell Them What You Told Them

The conclusion is often the most important part of any presentation, and yet speakers consistently make mistakes at the ends of their talks. They often end the presentation by shrugging their shoulders and saying, "Well, that's all I have to say. Any questions?"

Your conclusion needs to do much more. Specifically, you want to summarize your key points, while leaving the people in your audience in a position to say *yes* to your ideas. But you don't have much time to do all these things. Once you signal that you are concluding, you probably have about 1 to 3 minutes to make your final points. If you go beyond a few minutes, your audience will become agitated and frustrated.

Like the introduction, a conclusion should make some standard moves.

GO TO THE NET

Not sure how to end your presentation? For ideas, go to
www.pearsonhighered.com/johnsonweb3/16.11

MOVE 1: SIGNAL CLEARLY THAT YOU ARE CONCLUDING   When you begin your conclusion, use an obvious transition such as, "In conclusion," "Finally," "To summarize my main points," or "Let me wrap up now." When you signal your conclusion, your audience will sit up and pay attention because they know you are going to tell them your main points.

MOVE 2: RESTATE YOUR KEY POINTS   Summarize your key points for the audience, including your overall main point (Figure 16.8). Minimally, you can simply list them and go over them one last time. That way, if your audience remembers anything about your presentation, it will be these most important items.

MOVE 3: RESTRESS THE IMPORTANCE OF YOUR SUBJECT TO THE AUDIENCE   Tell the people in your audience again why they should care about this subject. Don't tell them why it is important to you—that's assumed. Instead, answer the audience's "What's in it for me?" questions.

MOVE 4: CALL THE AUDIENCE TO ACTION   If you want people in the audience to do something, here is the time to tell them. Be specific about what action they should take.

MOVE 5: LOOK TO THE FUTURE   Briefly, offer a vision of the future, usually a positive one, that will result if they agree with your ideas.

MOVE 6: SAY THANK YOU   At the end of your presentation, don't forget to thank the audience. Saying, "Thank you" signals that you are really finished. Often, it will also signal the audience to applaud, which is always a nice way for a presentation to end.

MOVE 7: ASK FOR QUESTIONS   Once the audience has stopped applauding, you can ask for questions.

Preparing to Answer Questions

While preparing and researching your presentation, you should spend some time anticipating the kinds of questions you might be asked. Questions are an opportunity to interact with the audience and clarify your ideas. You will generally be asked three types of questions: elaboration, hostile, and heckling.

THE ELABORATION OR CLARIFICATION QUESTION   Members of the audience might ask you to expand on your ideas or explain some of your concepts. These questions offer you an opportunity to reinforce some of your key points. You should not feel defensive or threatened by these questions. The person asking the question is really giving you a chance to restate some of your main points or views.

When you receive one of these kinds of questions, start out by rephrasing the question for the audience. For example, "The question is whether global warming will have an impact that we can actually observe with our own eyes." Your rephrasing of

Conclusion Slide for a Presentation

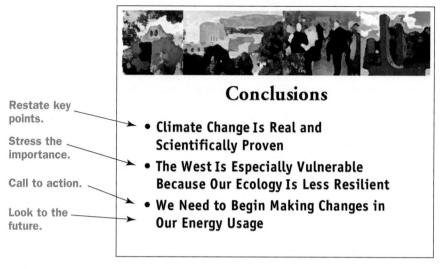

Figure 16.8:
The conclusion slide should drive home your main points and look to the future.

Restate key points.

Stress the importance.

Call to action.

Look to the future.

Conclusions

• **Climate Change Is Real and Scientifically Proven**

• **The West Is Especially Vulnerable Because Our Ecology Is Less Resilient**

• **We Need to Begin Making Changes in Our Energy Usage**

the question will allow you to shift the question into your own words, making it easier to answer.

Then, offer more information or reinforce a main point. For instance, "The answer is 'yes.' Long-time desert residents are already reporting that desert plants and animals are beginning to die off. One example is the desert willow...."

THE HOSTILE QUESTION   Occasionally, an audience member will ask you a question that calls your ideas into doubt. For example, "I really don't trust your results. Do you really expect us to believe that you achieved a precision of .0012 millimeters?"

Here is a good three-step method for deflecting these kinds of questions:

1. **Rephrase the question**—"The questioner is asking whether it is possible that we achieved a precision of .0012 millimeters."

2. **Validate the question**—"That's a good question, and I must admit that we were initially surprised that our experiment gave us this level of precision."

3. **Elaborate and move forward**—"We achieved this level of precision because...."

You should allow a hostile questioner only one follow-up remark or question. After giving the hostile questioner this second opportunity, do not look at that person. If you look elsewhere in the room, someone else in the audience will usually raise his or her hand and bail you out.

THE HECKLING QUESTION   In rare cases, a member of the audience will be there only to heckle you. He or she will ask rude questions or make blunt statements like, "I think this is the stupidest idea we've heard in 20 years."

For helpful question-and-answer strategies, go to
www.pearsonhighered.com/johnsonweb3/16.12

In these situations, you need to recognize that the heckler is *trying* to sabotage your presentation and cause you to lose your cool. Don't let the heckler do that to you. You simply need to say something like, "I'm sorry you feel that way. Perhaps we can meet after the presentation to talk about your concerns." Usually, at this point, others in the audience will step forward to ask more constructive questions.

It is rare that a heckler will actually come to talk to you later. That's not why he or she was there in the first place. If a heckler does manage to dominate the question-and-answer period, simply end your presentation. Say something like,

> Well, we are out of time. Thank you for your time and attention. I will stick around in the room for more questions.

Then, step away from the podium or microphone and walk off the stage. Find someone to shake hands with. Others who have questions can approach you one-on-one.

Answering Questions

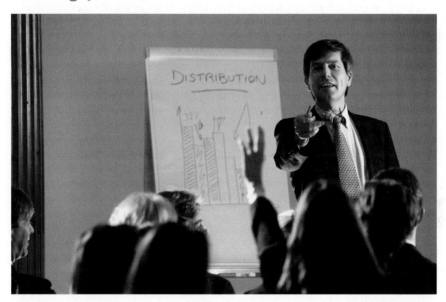

Prepare in advance for the kinds of questions you might be asked after your presentation.

Choosing Your Presentation Style

Your speaking style is very important. In a presentation, you can use style to add flavor to your information while gaining the trust of the audience. Poor style, on the other hand, can bore the audience, annoy them, and even turn them against you.

There are many ways to create an appropriate style for your presentation, but four techniques seem to work best for technical presentations. Each of these techniques will help you project a special tone in your speaking.

GO TO THE NET

Need to improve your style? Go to
www.pearsonhighered.com/johnsonweb3/16.13

Choosing Your
Presentation
Style

431

DEVELOP A PERSONA   In ancient Greek, the word *persona* meant "mask." So, as you consider your presentation style, think about the mask you want to wear in front of the audience. Then, step into this character. Put on the mask. You will find that choosing a persona will make you feel more comfortable, because you are playing a role, much like an actor plays a role. The people in the audience aren't seeing/judging *you*. They are seeing the mask you have chosen to wear for the presentation.

SET A THEME   A theme is a consistent tone you want to establish in your presentation. The best way to set a theme is to decide which one word best characterizes how you want the audience to *feel* about your subject. Then, use logical mapping to find words and phrases associated with that feeling (Figure 16.9).

As you prepare your talk, use these words regularly. When you give your presentation, the audience will naturally pick up on your theme. For example, let's say you want your audience to be "concerned." Map out words and phrases related to this feeling. Then, as you prepare your presentation, weave these words and phrases into your speech. If you do it properly, your audience will feel your concern.

SHOW ENTHUSIASM   If you're not enthusiastic about your subject, your audience won't be either. So, get excited about what you have to say. Be intense. Get pumped up. Show that you are enthusiastic about this subject (even if you aren't) and the audience members will be too.

Creating Your Presentation Style

AT A GLANCE

- Develop a persona.
- Set a theme.
- Show enthusiasm.
- KISS: Keep It Simple (Stupid).

A Theme in a Presentation

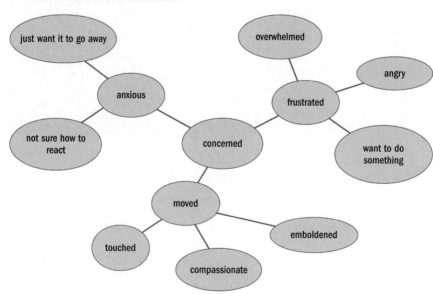

Figure 16.9:
By mapping around a key term and seeding your speech with related words, you can set a theme that creates a specific feeling in your audience.

To see websites that offer advice for presenting well, go to
www.pearsonhighered.com/johnsonweb3/16.14

KISS: KEEP IT SIMPLE (STUPID)   The KISS principle is always something to keep in mind when you are presenting, especially when it comes to style. Speak in plain, simple terms. Don't get bogged down in complex details and concepts.

In the end, good style is a choice. You may have come to believe that some people just have good style and others don't. That's not true. In reality, people with good style are just more conscious of the style they want to project.

Creating Visuals

In this visual age, you are really taking a risk if you try to present information without visuals. People not only want visuals, they *need* them to fully understand your ideas (Munter & Russell, 2002).

Designing Visual Aids

One of the better ways to design visual aids is to use the presentation software (PowerPoint, Keynote, or Presentations) that probably came bundled with your word-processing software. These programs are rather simple to use, and they can help you quickly create the visuals for a presentation. They also generally ensure that your slides will be well designed and readable from a distance.

The design principles discussed in Chapter 10 (balance, alignment, grouping, consistency, and contrast) work well when you are designing visual aids for public presentations. In addition to these design principles, here are some special considerations concerning format and font choices that you should keep in mind as you are creating your visuals:

FORMAT CHOICES

- Title each slide with an action-oriented heading.
- Put five or fewer items on each slide. If you have more than five points to make about a topic, divide the topic into two slides.
- Use left-justified text in most cases. Centered text should be used infrequently and right-justified text almost never.
- Use lists instead of paragraphs or sentences.
- Use icons and graphics to keep your slides fresh for the audience.

FONT CHOICES

- Use a simple typeface that is readable from a distance. Sans serif fonts are often more readable from a distance than serif fonts.
- Use a minimum of a 36-point font for headings and a minimum of a 24-point font for body text.
- Use color to keep slides interesting and to improve retention.
- Do not use ALL UPPERCASE letters because they are hard to read from a distance.

Overall, it is best to keep your slides as simple as possible (Figure 16.10). After all, if your audience needs to puzzle through your complex slides, they probably won't be listening to you.

Sample Slides from a Presentation

Cover Slide

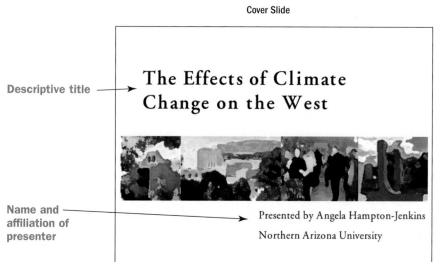

Descriptive title →

The Effects of Climate Change on the West

Presented by Angela Hampton-Jenkins

Northern Arizona University

Name and affiliation of presenter →

Figure 16.10: These two slides show good balance, simplicity, and consistency. Keep the slides simple so that the audience doesn't need to work too hard to read them.

Slide from Body of Presentation

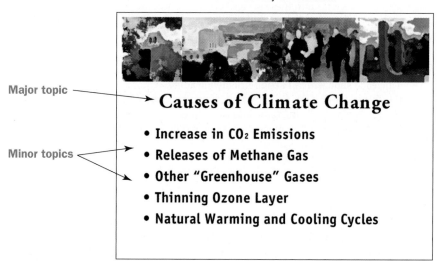

Major topic →

Causes of Climate Change

Minor topics →

- **Increase in CO_2 Emissions**
- **Releases of Methane Gas**
- **Other "Greenhouse" Gases**
- **Thinning Ozone Layer**
- **Natural Warming and Cooling Cycles**

Using Graphics

Graphics are also helpful, especially when you are trying to describe something to the audience. An appropriate graph, chart, diagram, picture, or even a movie will help support your argument (Figure 16.11). Chapter 11 in this book discusses the use of graphics in documents. Most of those same guidelines apply to presentations.

Using a Graphic on a Presentation Slide

The graphic tells a simple story.

The source of the graphic is clearly displayed.

Figure 16.11:
A good visual tells a simple story. The line graph in this visual is complex, but the story is obvious. Also, make sure you identify the source of the graphic and the information it displays.

Here are some guidelines that pertain specifically to using graphics in a presentation:

- Make sure words or figures in the graphic are large enough to be read from a distance.
- Label each graphic with a title.
- Keep graphics uncomplicated and limited to simple points.
- Keep tables small and simple. Large tables full of data do not work well as visuals because the audience will not be able to read them—nor will they want to.
- Use clip art or photos to add life to your slides.

Graphics, including clip art and photos, should never be used to merely decorate your slides. They should reinforce the content, organization, and style of your presentation.

Link

For more information on creating and using graphics, go to Chapter 11, page 275.

Slides to Avoid

All of us have probably been to a presentation in which the speaker created ineffective slides. He or she put up a transparency with a 12-point type font and minimal design (Figure 16.12). Or, the speaker put up a table or graph that was completely indecipherable because the font was too small or the graphic was too complex.

These kinds of slides are nothing short of painful. The only thing the audience wants from such a slide is for the speaker to remove it—as soon as possible. Always remember that we live in a visual culture. People are sensitive to bad design. So, take the time to properly create slides that enhance your presentation—not send the audience running for the door.

An Ineffective Slide

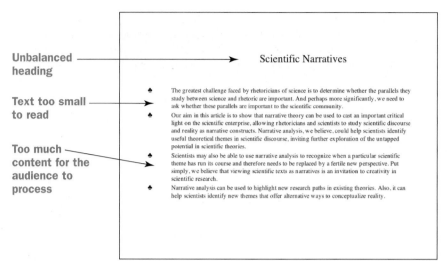

Unbalanced heading

Text too small to read

Too much content for the audience to process

Scientific Narratives

The greatest challenge faced by rhetoricians of science is to determine whether the parallels they study between science and rhetoric are important. And perhaps more significantly, we need to ask whether these parallels are important to the scientific community.

Our aim in this article is to show that narrative theory can be used to cast an important critical light on the scientific enterprise, allowing rhetoricians and scientists to study scientific discourse and reality as narrative constructs. Narrative analysis, we believe, could help scientists identify useful theoretical themes in scientific discourse, inviting further exploration of the untapped potential in scientific theories.

Scientists may also be able to use narrative analysis to recognize when a particular scientific theme has run its course and therefore needs to be replaced by a fertile new perspective. Put simply, we believe that viewing scientific texts as narratives is an invitation to creativity in scientific research.

Narrative analysis can be used to highlight new research paths in existing theories. Also, it can help scientists identify new themes that offer alternative ways to conceptualize reality.

Figure 16.12: An ineffective slide often says a great amount—if the audience can read it.

Delivering the Presentation

Why do people go to presentations, especially if a paper version of the talk is available?

People attend presentations because they want to see you perform the material. They want to see how you act and interact with them. They want you to put a human face on the material. For this reason, you should pay close attention to your delivery so the audience receives a satisfying performance.

The usual advice is to "be yourself" when you are presenting. Of course, that's good advice if you are comfortable talking to an audience. Better advice is to "be the person the audience expects." In other words, like an actor, play the role that seems to fit your material and your audience.

Body Language

The audience will pay close attention to your body language. So use your body to reflect and highlight the content of your talk.

DRESS APPROPRIATELY   How you dress should reflect the content and importance of your presentation. A good rule of thumb is to dress a level better than how you expect your audience to dress. For example, if the audience will be in casual attire, dress a little more formally. A female speaker might wear a blouse and dress pants or a nice skirt. Men might wear a shirt, tie, and dress pants. If the audience will be wearing suits and "power" dresses, you will need to wear an even *nicer* suit or an even *better* power dress.

STAND UP STRAIGHT   When people are nervous, they have a tendency to slouch, lean, or rock back and forth. To avoid these problems when you speak, keep your feet squarely under your shoulders, with your knees slightly bent. Keep your shoulders back and your head up.

DROP YOUR SHOULDERS   Under stress, people also have a tendency to raise their shoulders. Raised shoulders restrict your airflow and make the pitch of your voice go up. By dropping your shoulders, you will improve airflow and lower your voice. A lower voice sounds more authoritative.

USE OPEN HAND AND ARM GESTURES   For most audiences, open hand and arm gestures will convey trust and confidence. If you fold your arms, keep them at your sides, or put both hands in your pockets, you will convey a defensive posture that audiences will not trust.

Delivering the Presentation

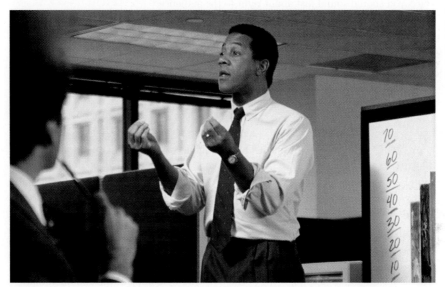

You can use your body and hands to highlight important parts of your presentation.

MAKE EYE CONTACT   Everyone in the audience should believe that you made eye contact with him or her at least once during your presentation. As you are presenting, make it a point to look at all parts of the room at least once. If you are nervous about making eye contact, look at audience members' foreheads instead. They will think you are looking them directly in the eye.

There are exceptions to these generally accepted guidelines about gestures and eye contact. In some cultures, like some Native American cultures, open gestures and eye contact might be considered rude and even threatening. If you are speaking to an unfamiliar audience, find out which gestures and forms of eye contact are appropriate for that audience.

MOVE AROUND THE STAGE   If possible, when you make important points, step toward the audience. When you make transitions in your presentation from one topic

Want more advice on delivering presentations? Go to
www.pearsonhighered.com/johnsonweb3/16.16

Delivering the Presentation   437

to the next, move to the left or right on the stage. Your movement across the stage will highlight the transition.

POSITION YOUR HANDS APPROPRIATELY Nervous speakers often strike a defensive pose by using their hands to cover specific parts of their bodies (perhaps you can guess which parts). Keep your hands away from these areas.

Voice, Rhythm, and Tone

A good rule of thumb about voice, rhythm, and tone is to *speak lower and slower than you think you should.*

Why lower and slower? When you are presenting, you need to speak louder than normal. As your volume goes up, the pitch of your voice will go up. So, your voice will seem unnaturally high (even shrill) to the audience. By consciously lowering your voice, you should sound just about right to the audience.

Meanwhile, nervousness usually encourages you to speak faster than you normally would. By consciously slowing down, you will sound more comfortable and more like yourself.

USE PAUSES TO HIGHLIGHT MAIN POINTS When you make an important point, pause for a moment. Your pause will signal to audience members that you just made an important point that you want them to consider and retain.

USE PAUSES TO ELIMINATE VERBAL TICS Verbal tics like "um," "ah," "like," "you know," "OK?" and "See what I mean?" are simply nervous habits that are intended to fill gaps between thoughts. If you have problems with a verbal tic (who doesn't?), train yourself to pause when you feel like using one of these sounds or phrases. Before long, you will find them disappearing from your speech altogether.

Using Your Notes

The best presentations are the ones that don't require notes. Notes on paper or index cards are fine, but you need to be careful not to keep looking at them. Nervousness will often lead you to keep glancing at them instead of looking at the audience. Some speakers even get stuck looking at their notes, glancing up only rarely at the audience. Looking down at your notes makes it difficult for the audience to hear you. You want to keep your head up at all times.

The following are some guidelines for making and using notes.

USE YOUR SLIDES AS MEMORY TOOLS You should know your subject inside out. So you likely don't need notes at all. Practice rehearsing your presentation with your slides alone. Eventually, you will be able to completely dispense with your written notes and work solely off your visual aids while you are speaking.

TALK TO THE AUDIENCE, NOT TO YOUR NOTES OR THE SCREEN Make sure you are always talking to the audience. It is sometimes tempting to begin looking at your notes while talking. Or, in some cases presenters end up talking to the screen. You should only steal quick glances at your notes or the screen. Look at the audience instead.

Need help improving your voice, rhythm, and tone? Go to
www.pearsonhighered.com/johnsonweb3/16.17

PUT WRITTEN NOTES IN A LARGE FONT   If you need to use notes, print them in a large font on the top half of a piece of paper. That way, you can quickly find needed information without effort. Putting the notes on the top half of the paper means you won't need to glance at the bottom of the piece of paper, restricting airflow and taking your eyes away from the audience.

USE THE NOTES VIEW FEATURE IN YOUR PRESENTATION SOFTWARE   Presentation software usually includes a Notes View feature that allows you to type notes under or to the side of slides (Figure 16.13). These notes can be helpful, but be wary of using them too much. First, they force you to look at the bottom of a sheet of paper, restricting your airflow. Second, if you are nervous, you may be distracted by these notes or start reading them to the audience.

IF SOMETHING BREAKS, KEEP GOING   If your overhead projector dies, your computer freezes, or you drop your notes all over the front row, don't stop your presentation to fumble around trying to fix the situation. Keep going. You need only acknowledge the problem: "Well, it looks like some gremlins have decided to sabotage my projector. So, I will move forward without it." Then do so.

Practicing and Rehearsing

The common advice for improving your ability to present in public is to "practice, practice, practice." This is good advice, but it doesn't tell the whole story. You also need to "rehearse, rehearse, rehearse."

Practice, Practice, Practice

Practicing your presentation is a good way to work out the kinks and revise your materials. While you are practicing, you should

- look for problems with the content, organization, and style of your presentation.
- edit and proofread your visuals and handouts.
- decide how you are going to deliver the information by paying attention to your body language and voice.

Practice is something you usually do at your desk or in your office. It involves talking through your presentation and fixing problems as you find them.

Rehearse, Rehearse, Rehearse

Once you have ironed out the kinks in your presentation, it is time to rehearse, rehearse, rehearse. Unlike practicing, rehearsing means giving the presentation from beginning to end *without stopping*. While practicing, you can stop to make corrections or changes to your presentation. When rehearsing, just keep talking, even if you make a mistake. You won't have time to make changes or correct mistakes when you are in front of an audience, so use rehearsal to practice getting past those weak parts of your presentation. The closer you can replicate the actual speaking situation, the better your rehearsal will be.

As you rehearse, you will notice smaller problems with your presentation. Don't stop to fix them. You can fix them when you have finished the rehearsal. Pretend you are in front of the audience and you can't stop to fix small problems.

The "Notes View" Feature

The copy of the slide is up here.

Causes of Climate Change
- Increase in CO_2 Emissions
- Releases of Methane Gas
- Other "Greenhouse" Gases
- Thinning Ozone Layer
- Natural Warming and Cooling Cycles

Causes of Climate Change

Increase in CO_2 Emissions—scientists at the Hanson Laboratories report that carbon dioxide emissions, especially from cars, is on the rise. We have seen an increase of 15 percent in emissions.

Releases of Methane Gas—methane is naturally occurring. Cows for example, belch methane gas. The problem is with larger and larger releases of methane gas due to industrialization.

Other Greenhouse Gases—a variety of other greenhouse gases, like nitrous oxide and halocarbons, need to be controlled.

The notes appear down here.

Thinning Ozone Layer—the thinning has slowed, but it still continues. As developing countries become more industrialized, they are using more chemicals that thin the ozone layer.

Natural Warming and Cooling Cycles—the earth does naturally warm and cool—creating "ice ages" and "hot periods." These changes are very gradual, though, and could only minimally account for the climate change we are experiencing now.

5

Figure 16.13: The Notes View feature in your presentation software allows you to put notes below or to the side of a slide. Your audience won't be able to see them, but you can print out each slide with your notes so that you don't need to keep looking up at your slides.

For some presentations, you might want to rehearse your presentation in front of a test audience. At your workplace, you can draft people into listening to your talk. These practice audiences might be able to give you some advice about improving your presentation. If nothing else, though, they will help you anticipate the reactions of the audience to your materials. Groups like Toastmasters also can provide a test audience (Figure 16.14). At Toastmasters meetings, people give presentations and work on improving their public speaking skills.

The more you rehearse, the more comfortable you are going to feel with your materials and the audience. Practicing will help you find the errors in your presentation, but rehearsal will help you put the whole package together into an effective presentation.

Evaluating Your Performance

Evaluation is an important way to receive feedback on your presentation. In some speaking situations, the audience will expect to evaluate your presentation. If an evaluation form is not provided, audience members will expect you to hand one out so they can give you and supervisors feedback on your performance. Figure 16.15 shows a typical evaluation form that could be used for a variety of presentations.

Places to Improve Your Presentation Skills

Figure 16.14: Toastmasters is a group that helps people practice and improve their public speaking skills. Local groups can be found in almost any large town.

To locate groups like Toastmasters, go to
www.pearsonhighered.com/johnsonweb3/16.18
For more practice and rehearsing tips, go to
www.pearsonhighered.com/johnsonweb3/16.19

Presentation Evaluation Form

Please answer these questions with one or more sentences.

Content

Did the speaker include more information than you needed to know? If so, where?

Where did the speaker not include enough need-to-know information?

What kinds of other facts, figures, or examples would you like to see included in the presentation?

Organization

In the introduction, was the speaker clear about the subject, purpose, and main point of the presentation?

In the introduction, did the speaker grab your attention effectively?

In the body, was the presentation divided into obvious topics? Were the transitions among these topics obvious to you?

In the conclusion, did the speaker restate the presentation's main point clearly?

Did the speaker leave enough time for questions?

Style and Delivery

Did the speaker speak clearly and loudly enough?

Figure 16.15:
An evaluation form is a good way to receive feedback on your presentation. Have your audience comment on your presentation's content, organization, style, and delivery.

442   Chapter 16
**Preparing
and Giving
Presentations**

To download this evaluation form, go to
www.pearsonhighered.com/johnsonweb3/16.20

Did the speaker move effectively, using hands and body language to highlight important points?

Did the speaker have any verbal tics (uh, um, ah, you know)?

Did the speaker look around the room and make eye contact?

Was the speaker relaxed and positive during the presentation?

Visuals

Did the visuals effectively highlight the speaker's points in the presentation?

Did the speaker use the visuals effectively, or did they seem to be a distraction?

Did the speaker use notes effectively? Was the speaker distracted by his or her notes?

Concluding Remarks

List five things you learned during this presentation.

What did you like about this presentation?

What did you not like about this presentation?

What suggestions can you offer to make this presentation better?

Working Cross-Culturally with Translators

When speaking to an international audience, you will likely need the services of a translator. A translator does more than simply convert one language into another. He or she will also modify your words to better capture your intent and adjust them to the cultural expectations of your audience. An effective translator will help you better express any subtle points while avoiding cultural taboos and gaffes.

Here are some strategies to help you work more effectively with a translator. These strategies can also be helpful when speaking in any cross-cultural situation.

KEEP YOUR SPEECH SIMPLE   The words and sentences in your speech should be as plain and simple as possible. Figures of speech, clichés, or complex sentences will be difficult to translate, especially when the audience is right in front of you.

AVOID JOKES   Translators cringe when speakers decide to tell jokes, because jokes rarely translate well into another culture. What is funny in one culture is often not funny in another culture. Meanwhile, jokes often rely on turns of phrase or puns that are impossible to translate. Translators have been known to tell the audience, "The speaker is now telling a joke that doesn't translate into our language. I will tell you

Working with a Translator

Translators are becoming increasingly important as technology and manufacturing become more international.

when to laugh." Then, as the speaker finishes the joke, the translator signals that the audience should laugh.

SPEAK SLOWLY   A translator will struggle to keep up with someone who is speaking at a faster-than-normal pace, leading to errors in translation. Meanwhile, the structure of some languages (e.g., German) can cause translation to take a little longer.

MINIMIZE SLANG, JARGON, AND SAYINGS   These words and phrases rarely translate easily into other languages, because they are culturally dependent. For example, if the speaker says, "Instead of doing another kickoff meeting, we just need to sit down and hammer out an agreement," the translator would struggle to translate three concepts in this sentence: "doing" a meeting, "kicking off" that meeting, and "hammering out" an agreement. The meanings of these words are dependent on the culture of the speaker and might have little meaning to the audience.

AVOID RELIGIOUS REFERENCES   In most cases, it is risky to include religious themes or terms in cross-cultural speeches. Even seemingly harmless phrases like, "God help us" or "Let's pray that doesn't happen" can translate in unexpected ways. Meanwhile, attempts to incorporate the sayings of a religious figure or scripture can be potentially insulting and even sacrilegious.

KNOW YOUR TRANSLATOR   Whenever possible, check your translator's level of fluency and understanding of your subject matter. One of your bilingual colleagues may be able to help you determine your translator's abilities. Also, you should hire a translator from your audience's specific culture. Just because a translator knows Spanish doesn't mean he or she can handle the dialects and colloquialisms in all Spanish-speaking cultures.

PROVIDE YOUR SPEECH, VISUALS, AND HANDOUTS IN ADVANCE   Giving your translator your speech ahead of time will greatly improve the accuracy of the translation, because your translator will have time to become familiar with the topic and anticipate ideas that are difficult to translate.

STAND WHERE YOUR TRANSLATOR CAN SEE YOUR FACE   A translator may have trouble hearing you correctly if you are turned away from him or her. Also, translators sometimes read lips or facial expressions to help them figure out difficult words or concepts.

For now, English speakers are fortunate that their language has become an international language of business and technology. Consequently, many people in your audience will be able to understand your speech without the help of a translator. Before too long, though, people from other cultures will expect business to be conducted in their language, too. At that point, translators will become even more critical in technical fields.

- Public presentations are an essential part of communicating effectively in technical workplaces.

- Planning is the key to successful public presentations. You need a firm understanding of the subject, purpose, audience, and context of your presentation.

- A well-organized presentation "tells them what you're going to tell them, tells them, and tells them what you told them."

- Visual aids are essential in presentations. Software programs are available to help you make these aids.

- People come to presentations because they want to see you *perform* the material.

- Good delivery means paying attention to your body language, appearance, voice, rhythm, tone, and effective use of notes.

- Practice is important, but so is rehearsal. Use practice to iron out the rough spots in the presentation, and then rehearse, rehearse, rehearse.

- Translators can help you appropriately present your message to a cross-cultural audience. Keeping your words simple and speaking at a normal pace will help translators more accurately reflect your intended meaning.

Individual or Team Projects

1. Using the Internet, locate a professional in your major or field of study and ask that person about the kinds of oral presentations he or she makes. Avoid simply asking, "Do you make oral presentations?" Instead, ask about informal presentations at meetings. Ask about product demonstrations. And, of course, ask about more formal presentations in which your interviewee needs to inform or persuade an audience. Write a one- to two-page memo to your instructor in which you report on your findings about oral presentations in your field.

2. Attend a public presentation at your campus or workplace. Instead of listening to the content of the presentation, pay close attention to the speaker's use of organization, style, and delivery. In a memo to your instructor, discuss some of the speaker's strengths and suggest places where he or she might have improved the presentation.

3. Using presentation software, turn a document you have written for this class into a short presentation with slides. Your document might be a set of instructions, a report, a proposal, or any other document. Your task is to make the presentation interesting and informative to the audience.

4. Pick a campus or workplace problem that you think should be solved. Then, write a 3-minute oral presentation in which you (a) identify the problem, (b) discuss

what is needed to solve the problem, and (c) offer a solution. As you develop your presentation, think about the strategies you might use to persuade the audience to accept your point of view and solution.

5. Interview one of your classmates for 10 minutes about his or her hometown. Then, from your notes alone make an impromptu presentation to your class in which you introduce your classmate's hometown to the rest of the class. In your presentation, you should try to persuade the audience members that they should visit this town.

Collaborative Projects

Group presentations are an important part of the technical workplace because projects are often team efforts. Therefore, a whole team of people often needs to make the presentation, with each person speaking about his or her part of the project.

Turn one of the collaborative projects you have completed in this class into a group presentation. Some of the issues you should keep in mind are the following:

- Who will introduce the project and who will conclude it?
- How will each speaker "hand off" to the next speaker?
- Where will each presenter stand while presenting?
- Where will the other team members stand or sit while someone else is presenting?
- How can you create a consistent set of visuals that will hold the presentation together?
- How will questions and answers be handled during and after the presentation?
- How can the group create a coherent, seamless presentation that doesn't seem to change style when a new speaker steps forward?

Something to keep in mind with team presentations is that members of the audience are often less interested in the content of the presentation than they are in meeting your team. So, think about some ways in which you can convey that personal touch in your team presentation while still talking effectively about your subject.

For additional technical writing resources, including interactive sample documents, document design tutorials and guidelines, and more, go to **www.mytechcommlab.com.**

The Coward

Jennifer Sandman is an architect who works for a small firm in Kansas City. The firm is completing its largest project ever, designing a building for a GrandMart superstore. Jennifer and her boss, a lead architect, Bill Voss, are in charge of the project.

At an upcoming city council meeting, they will need to give a public presentation on their plans for the new GrandMart. At the meeting, the mayor and city council will look over the plans and listen to comments from people in the audience. Failure to persuade the audience might mean the loss of the project as well as future business with GrandMart, an important client.

Everyone knows that the meeting will be contentious, with many angry residents arguing against the plans. And yet, GrandMart's surveys have shown that a silent majority would welcome the new store in the neighborhood. Meanwhile, GrandMart's marketing specialists are certain that the store would be profitable, because there are no competitors within 10 miles.

Jennifer has one problem—her boss is a coward. He scheduled a business trip to coincide with the meeting, and he promptly delegated the presentation to her. As her boss walked out the door, he said, "It will be a good growth opportunity for you. Just don't make the firm look bad."

For Jennifer, giving presentations isn't a problem. She is a whiz with PowerPoint, and she has given enough presentations to feel comfortable in front of people. Her concern, though, is the hostile people in the audience. There might even be some hecklers. These hostile people will never accept the idea that a GrandMart should be built in their neighborhood. Others in the audience, though, will be undecided. So, she wants to devise ways in her presentation to soften and deflect some of the hostile criticisms, while gaining the support of people in the community who want the store built.

If you were Jennifer, how would you craft your presentation to fit these difficult conditions? How would you prepare for the hostile and heckling questions? How would you design and deliver your presentation to avoid being seen as defensive?

In this chapter, you will learn:

- The role of correspondence in the technical workplace.
- The basic features of letters and memos.
- How to plan, organize, and draft letters and memos.
- Common patterns for letters and memos.
- How to choose an appropriate style for correspondence.
- How to design and format letters and memos.
- How to revise, review, and proofread letters and memos.

As e-mail has become more common in the workplace, the role of letters and memos has changed. Not long ago, letters and memos were the best way to communicate with others in writing. They were used to convey both informal and formal messages. Today, e-mail has replaced memos and letters as the preferred way to communicate informal messages.

Letters and memos now tend to communicate formal messages, such as

- results of important decisions.
- formal inquiries.
- refusals.
- new policies.
- transmittals of important documents or materials.

The increasingly formal nature of letters and memos ensures that your readers will take them seriously when these documents arrive on their desks. A paper letter or memo signals that the message is too important or too proprietary to be sent as an e-mail.

In the workplace, writing letters and memos will be a regular part of your job, but they can also be a drain on your time. The key is to learn how to write these documents quickly and efficiently, within the natural flow of your workday.

Letters and Memos Convey Important Information

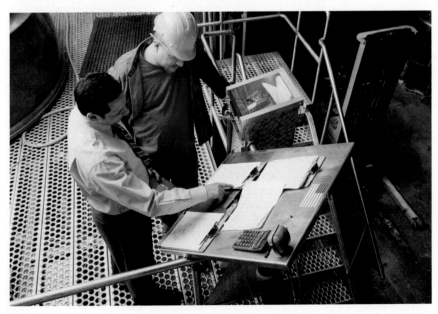

Letters and memos are usually used to convey formal messages, while e-mail is used for less formal messages.

GO TO THE NET

When is e-mail not appropriate? Go to
www.pearsonhighered.com/johnsonweb3/17.1

Letters and Memos   451

Basic Features of Letters and Memos

Letters and memos are similar in many ways. They look different because they use different *formats*. But, in the end, they tend to use the same kinds of content, organization, style, and design to put across their message. So, how are they different?

Letters are written to people *outside* the company or organization. Primarily, letters are used in formal situations in which an employee is acting as a representative of the company. Letters can be used to make requests or inquiries, accept or refuse claims, communicate important information, record agreements, and apply for jobs.

Memos are written to people *inside* the company or organization. They usually contain meeting agendas, policies, internal reports, and short proposals. When a message is too important or proprietary for e-mail, most people will send a memo instead. Memos are still more reliable than e-mails for information that should not be broadly released.

Letters and memos are also regularly used as *transmittal documents*. In the workplace, a letter or memo of "transmittal" is placed on top of another document to explain the document's purpose and clearly state who should receive it. A letter or memo of transmittal helps make sure your document reaches its intended readers.

Basic Pattern for Letters and Memos

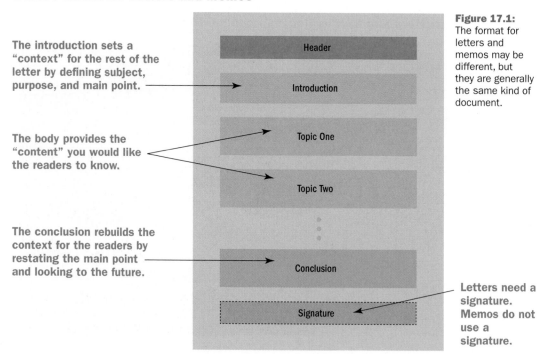

The introduction sets a "context" for the rest of the letter by defining subject, purpose, and main point.

The body provides the "content" you would like the readers to know.

The conclusion rebuilds the context for the readers by restating the main point and looking to the future.

Header

Introduction

Topic One

Topic Two

Conclusion

Signature

Figure 17.1: The format for letters and memos may be different, but they are generally the same kind of document.

Letters need a signature. Memos do not use a signature.

To see other sample letters and memos, go to
www.pearsonhighered.com/johnsonweb3/17.2

Despite their minor differences, letters and memos share many of the same basic features. As shown in Figure 17.1, both kinds of documents include the following elements, though they look different on the page:

- Header
- Introduction
- Body
- Conclusion

Letters and memos differ in formatting:

- The format for a letter usually includes: a letterhead, the date, an inside address, a greeting, and a closing with the writer's signature.
- The format for a memo usually includes: a header, the date, and lines for the addressee ("To:"), the sender ("From:"), and the subject ("Subject:").

Figure 17.2 shows a letter and a memo with basically the same content so you can see the differences in formatting.

Planning and Researching Letters and Memos

In the technical workplace, people don't do much planning before writing a letter or memo. They just write it. This approach works fine for everyday letters and memos, but, as the message becomes more important, you will want to prepare more thoroughly to write the correspondence.

When you begin writing a letter or memo on your computer, first consider how your readers will use the information you are providing. You might start by answering the Five-W and How Questions:

Who is the reader of my letter or memo?

Why am I writing to this person?

What is my point? What do I want my reader to do?

Where will the letter or memo be read?

When will the letter or memo be used?

How will the reader use this document now and in the future?

Determining the Rhetorical Situation

If the message is informal or routine, you might be ready to start typing right now. However, if your message is formal or especially important, you should explore the rhetorical situation in more depth.

SUBJECT   Pay attention to what your readers need to know to take action. Letters and memos should be as concise as possible, so include only need-to-know information. Strip away any want-to-tell information that will distract your readers from your main point.

Link

For more information about distinguishing between need-to-know and want-to-tell information, go to Chapter 7, page 159.

Letter and Memo

(a)

Letterhead

Kearney Engineering

3819 Washington St.
Austin, TX 78716
(512) 555-8200

Date

September 18, 2008

Inside Address

Tom Setter, CEO
Setter Industries, LLC
192 High Ave.
Aurora, IL 60321

Greeting

Dear Mr. Setter,

With this letter, I have enclosed our report reviewing your plans to convert the Aurora factory to biomass energy.

Overall, as our report shows, it looks like a biomass generator will work, but we have made some recommendations for improving the design. Specifically, we recommend you consider a design that more finely shreds wood by-products to improve energy yield. We believe this change would allow you to run your factory "off grid" during nonpeak hours of production.

Please look over the report as soon as possible. We would like to schedule a conference call with you in the next couple of weeks to discuss the report.

Thank you for giving us the opportunity to work with you. If you have any questions or comments, please call me at (512) 555-8200, ext. 541, or e-mail me at jamesw@kearneyeng.net.

Sincerely,

James Williams

James Williams, P.E.
Lead Engineer
Kearney Engineering

Closing with Signature

Figure 17.2:
Letters (a) and memos (b) are basically the same, except in their formatting. The main differences are that letters are written to readers outside the company, whereas memos are written to readers inside the company.

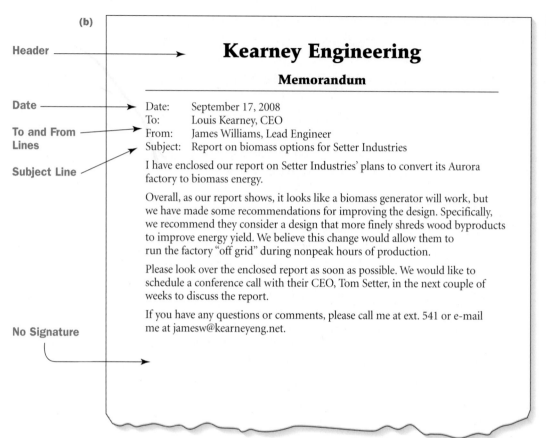

(b)

Header

Kearney Engineering

Memorandum

Date

To and From
Lines

Subject Line

Date: September 17, 2008
To: Louis Kearney, CEO
From: James Williams, Lead Engineer
Subject: Report on biomass options for Setter Industries

I have enclosed our report on Setter Industries' plans to convert its Aurora factory to biomass energy.

Overall, as our report shows, it looks like a biomass generator will work, but we have made some recommendations for improving the design. Specifically, we recommend they consider a design that more finely shreds wood byproducts to improve energy yield. We believe this change would allow them to run the factory "off grid" during nonpeak hours of production.

Please look over the enclosed report as soon as possible. We would like to schedule a conference call with their CEO, Tom Setter, in the next couple of weeks to discuss the report.

If you have any questions or comments, please call me at ext. 541 or e-mail me at jamesw@kearneyeng.net.

No Signature

PURPOSE   The purpose of your letter or memo should be immediately obvious to your readers. You should include a purpose statement in the first paragraph, perhaps even in the first sentence. Some key words for the purpose might include the following.

to inform   *to apologize*
to explain   *to discuss*
to complain   *to clarify*
to congratulate   *to notify*
to answer   *to advise*
to confirm   *to announce*
to respond   *to invite*

Your purpose statement might sound like one of the following:

> We are writing to inform you that we have accepted your proposal to build the Washington Street overpass.

> I would like to congratulate the Materials Team for successfully patenting the fusion polymer blending process.

> This memo explains and clarifies the revised manufacturing schedule for the remainder of this year.

Your purpose should be obvious to your readers as soon as they read the first paragraph. They should not need to guess why you are writing to them.

Link

For more guidance on defining a document's purpose, see Chapter 2, page 21.

READERS   Letters and memos can be written to individuals or to whole groups of people. Since these documents are often shared or filed, you need to anticipate all possible readers who might want a copy of your document.

Primary readers (action takers) are the people who will take action after they read your message. Your letter or memo needs to be absolutely clear about what you want these readers to do. It should also be tailored to their individual motives, values, and attitudes about the subject.

Secondary readers (advisors) are the people to whom your primary readers will turn if they need advice. They may be experts in the area, support staff, supervisors, or colleagues. You should anticipate these readers' concerns, but your focus should still be on the primary readers' needs.

Tertiary readers (evaluators) are any other people who may have an interest in your letter or memo. These readers may be more important than you expect. Letters and memos have a strange way of turning up in unexpected places. For example:

- That "confidential" memo you wrote to your research team might end up in your competitor's hands.
- The local newspaper might get a hold of a letter you sent to your company's clients explaining a problem with an important new product.

Link

For more information on analyzing readers, turn to Chapter 3, page 40.

Before sending any correspondence, you should think carefully about how the document would look if it were made public. Anticipate how it might be used against you or your company.

Gatekeeper readers (supervisors), such as your supervisor or legal counsel, may want to look over an especially important correspondence before it is sent out. You should always keep in mind that you are representing your company in your letters and memos. Your supervisor or the corporate lawyer may want to ensure that you are communicating appropriately with clients.

CONTEXT OF USE   Imagine all the different places your letter or memo may be used. Where will readers use this document now and in the future? Where will the document be kept (if at all) after it is read? Be sure to consider the physical, economic, political, and ethical factors that will influence how your readers will interpret and respond to your message. Put yourself in their place, imagining their concerns as they are reading your document.

Link

For strategies to help identify contextual issues, go to Chapter 3, page 46.

Organizing and Drafting Letters and Memos

Some messages require more time and care than others, but in most cases you should be able to generate letters and memos within the natural flow of your workday.

How can you write these documents efficiently? Keep in mind that the introductions and conclusions of these texts tend to make some predictable moves. If you memorize these moves, you can spend more time concentrating on what you need to say in the body of your document.

Introduction with a Purpose and a Main Point

In the introduction, a letter or memo should make at least three moves: (a) identify the *subject* of the letter, (b) state the *purpose,* and (c) state the *main point* (Figure 17.3). Depending on your message, you might also make two additional moves: (d) offer some *background information* and (e) stress the *importance of the subject.*

SUBJECT   Your *subject* should be stated or signaled in the first or second sentence of the introduction. Simply tell your readers what you are writing about, and *do not assume* they know what you are writing about.

> Recently, the Watson Project has been a source of much concern for our company.

> This memo discusses the equipment thefts that have occurred in our office over the last few months.

PURPOSE   Your *purpose* for writing should also be stated almost immediately in the first paragraph, preferably in the first or second sentence.

> Now that we have reached Stage Two of the Oakbrook Project, I would like to refine the responsibilities of each team member.

> The purpose of this letter is to inform you about our new transportation policies for low-level nuclear waste sent to the WIPP Storage Facility in New Mexico.

MAIN POINT   All letters and memos should have a *main point* that you want your readers to grasp or remember. In many cases, "the point" is something you want your readers to do when they are finished reading. In other words, state the *action* you want readers to take.

> We request the hiring of three new physician's assistants to help us with the recent increases in emergency room patients.

> Put bluntly, our subcontractors must meet ISO-9001 quality standards. It is our job to make sure that they comply.

It may seem odd to state your main point up front. Wouldn't it be better to lead up to the point, perhaps putting it in the conclusion? No. Most of your readers will only

Introduction, Body, Conclusion

Morris Blue Industries

Date: November 18, 2008
To: Hanna Marietta, Chief Executive Officer
From: Jason Santos, Corporate Health Officer
Subject: Bird Flu Contingency Plan

The subject is identified in the first sentence.

Background information is offered to remind the reader about the subject.

The body provides need-to-know information.

Last week, the Executive Board inquired about our company's contingency plans if a bird flu pandemic occurs. As the Board mentioned, the exposure of our overseas manufacturing operations, especially in the Asian Pacific region, puts our company at special risk. At this point, we have no approved contingency plan, but my team strongly believes we need to create one as soon as possible. In this memo, I will highlight important issues to consider, and my team requests a meeting with you to discuss developing a plan for Board approval.

The main point and purpose are clearly stated up front.

Despite the media hype, a bird flu pandemic is not imminent. A remote possibility exists that the H5N1 avian influenza virus could mutate into a form that can be transmitted among humans. To this point, though, only a small number of bird flu infections have occurred in humans. In these cases, birds have infected humans through close contact. The World Health Organization (WHO) reported in May 2008 that only 241 confirmed deaths had occurred worldwide, almost all in Asia. Human-to-human transmissions of bird flu are extremely rare.

This paragraph uses facts to inform the readers.

Nevertheless, the risk of a pandemic is real and the WHO recommends the immediate development of contingency plans. We recommend the following actions right now:

A. Develop a decision tree that outlines how our company will respond to a pandemic.
B. Design an alert system that notifies managers how to identify bird flu symptoms, when to be watchful, when to send employees home, and how to evacuate them.

This list makes important details easy to find.

Figure 17.3:
(continued)

C. Strengthen ties with local health authorities and law enforcement near our factories to speed the flow of information to local managers.

D. Create a training package for managers to educate them about bird flu and our company's response to a pandemic.

The Executive Board should also consider a) whether we want to procure stocks of antiviral drugs like Tamiflu and Relenza, b) whether our sick leave policies need to be adjusted to handle a pandemic, and c) how our medical insurance would cover prevention and recovery for employees. These issues will require legal counsel from each country in which we have employees.

Thank you for contacting me about this matter. We believe a contingency plan should be developed as soon as possible. To get things rolling, we would like to schedule an appointment with you to go over these issues in more depth. You or your assistant can reach me at ext. 2205 or e-mail me at tjackson@morrisblue.com.

A "thank you" signals the conclusion of the memo.

The conclusion restates the main point and action item, while looking to the future.

Contact information is provided.

scan your message. By putting your main point (the action item) up front, you will ensure that they do not miss it.

BACKGROUND INFORMATION   Writers often like to start their letters and memos with a statement that gives some background information or makes a personal connection to readers.

> Our staff meeting on June 28 was very productive, and I hope we all came away with a better understanding of the project. In this memo....

> When you and I met at the NEPSCORE Convention last October, our company was not ready to provide specifics about our new ceramic circuit boards. Now we are ready....

IMPORTANCE OF THE SUBJECT   In some cases, you might also want your introduction to stress the importance of the subject.

> This seems like a great opportunity to expand our network into the Indianapolis market. We may not see this opportunity again.

> If we don't start looking for a new facility now, we may find ourselves struggling to keep up with the demand for our products.

Introductions in letters and memos should be as concise as possible. At a minimum, the introduction should tell readers your subject, purpose, and main point.

Link

For more information on writing introductions, see Chapter 8, page 182.

Background information and statements about the importance of the subject should be used where needed.

Elements of a Letter or Memo

- Header
- Introduction—subject, purpose, main point, background information, importance of the subject
- Body—discussion topics, usually with one paragraph per topic
- Conclusion—thank you, main point (restated), and a look to the future

AT A GLANCE

Body That Provides Need-to-Know Information

The body is where you will provide your readers with the information they need to make a decision or take action. As shown in Figure 17.3, the body is the largest part of the memo or letter, and it will take up one or more paragraphs.

As you begin drafting the body of your document, divide your subject into the two to five major topics you need to discuss with your readers. These major topics will likely receive a paragraph or two of discussion.

If you are struggling to develop the content, you can use mapping to put your ideas on the screen or a piece of paper (Figure 17.4). Start out by putting the purpose statement in the center of the screen or top of a piece of paper. Then, branch out into two to five major topics. You can use mapping to identify any supporting information that will be needed for those topics.

As you are drafting the body of the document, keep looking back at your purpose statement in the introduction. Ask yourself, "What information do I need to provide to achieve that purpose?" Then, include any facts, examples, data, and reasoning that will help you support your argument.

Link

For more information on using logical mapping, go to Chapter 7, page 143.

Using Mapping to Generate Content

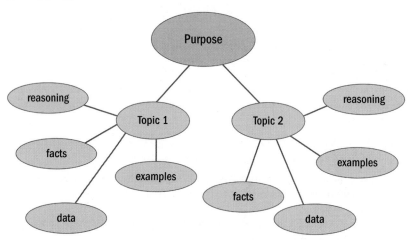

Figure 17.4: Using your purpose as a guide, identify the topics you will need to cover in your correspondence.

Conclusion That Restates the Main Point

The conclusion of your letter or memo should be short and to the point. Nothing essential should appear in the conclusion that has not already been stated in the introduction or body.

Conclusions in these documents tend to make three moves: *thank the readers, restate your main point,* and *look to the future.*

Need help writing an introduction? Go to
www.pearsonhighered.com/johnsonweb3/17.5

THANK THE READERS   Tell them that you appreciate their attention to your message. By thanking them at the end, you leave them with a positive impression as you conclude.

> Thank you for your time and attention to this important matter.

> We appreciate your company's efforts on this project, and we look forward to working with you over the next year.

RESTATE YOUR MAIN POINT   Remind your readers of the action you would like them to take.

> Time is short, so we will need your final report in our office by Friday, September 15, at 5:00.

> Please discuss this proposal right away with your team so that we can make any final adjustments before the submission deadline.

LOOK TO THE FUTURE   Try to end your letters and memos by looking forward in some way.

> When this project is completed, we will have taken the first revolutionary step toward changing our approach to manufacturing.

> If you have questions or comments, please call me at 555–1291 or e-mail me at sue.franklin@justintimecorp.com.

Your conclusion should run about one to three sentences. If you find yourself writing a conclusion that is more than a small paragraph, you probably need to trim the added information or move some of it into the body of the letter or memo.

Link
To learn more about writing conclusions, go to Chapter 8, page 196.

Types of Letters and Memos

In the technical workplace, letters and memos are used for a variety of purposes.

Inquiry Letters and Memos

The purpose of an *inquiry* is to gather information, especially answers to questions about important or sensitive subjects. In these situations, you could use e-mail, but a printed document is sometimes preferable because the recipients will view it as a formal request.

Here are some guidelines to follow when writing a letter or memo of inquiry:

- Clearly identify your subject and purpose.
- State your questions clearly and concisely.
- Limit your questions to five or fewer.
- If possible, offer something in return.
- Thank readers in advance for their response.
- Provide contact information (address, e-mail address, or phone number).

Figure 17.5 shows a typical letter of inquiry. Notice how the author of the letter is specific about the kinds of information she wants.

State the
subject and
purpose of the
letter.

State questions
clearly and
concisely.

Offer
something
in return.

Thank the
readers.

Provide
contact
information.

Arctic Information Associates
2315 BROADWAY, FARGO, ND 58102

February 23, 2009

Customer Service
Durable Computers
1923 Hanson Street
Orono, Maine 04467

Dear Customer Service:

My research team is planning a scientific expedition to the northern
Alaskan tundra to study the migration habits of caribou. We are looking
for a rugged laptop that will stand up to the unavoidable abuse that will
occur during our trip. Please send us detailed information on your Yeti
rugged laptop. We need answers to the following questions:

· How waterproof is the laptop?

· How far can the laptop fall before serious damage will occur?

· How well does the laptop hold up to vibration?

· Does the laptop interface easily with GPS systems?

· Can we receive a discount on a purchase of 20 computers?

Upon return from our expedition, we would be willing to share stories
about how your laptops held up in the Alaskan tundra.

Thank you for addressing our questions. Please respond to these inquiries
and send us any other information you might have on the Yeti rugged
laptop. Information can be sent to me at Arctic Information Associates,
2315 Broadway, Fargo, ND 58102. I can also be contacted at 701-555-
2312 or salvorman@arcticia.com.

Sincerely,

S Vorma

Sally Vorman, Ph.D.
Arctic Specialist

Figure 17.5:
A letter of inquiry
needs to be clear
about the
information it is
seeking. In this
letter, notice how
the writer has
listed her
questions in an
unmistakable way.

Response Letters and Memos

A response is written to answer an inquiry. A response letter should answer each of the inquirer's questions in specific detail. The amount of detail you provide will depend on the kinds of questions asked. In some situations, you may need to offer a lengthy explanation. In other situations, a simple answer or referral to the corporate website or enclosed product literature will be sufficient.

Here are some guidelines to follow when writing a response:

- Thank the writer for the inquiry.
- Clearly state the subject and purpose of the letter or memo.
- Answer any questions point by point.
- Offer more information, if available.
- Provide contact information (address, e-mail address, or phone number).

Figure 17.6 shows an example of a response letter. Pay attention to the author's point-by-point response to the questions in the original letter of inquiry (Figure 17.5).

Transmittal Letters and Memos

When sending documents or materials through the mail, you should include a letter or memo of transmittal. Also called "cover letters" or "cover memos," the purpose of these documents is to explain the reason the enclosed materials are being sent. For example, if you were sending a proposal to the vice president of your company, you would likely add a memo of transmittal like the one shown in Figure 17.7. Earlier in this chapter, the documents in Figure 17.2 also showed a letter and a memo of transmittal.

A transmittal letter or memo should do the following:

- Identify the materials enclosed.
- State the reason the materials are being sent.
- Summarize the information being sent.
- Clearly state any action requested or required of readers.
- Provide contact information.

You should keep your comments brief in transmittal letters or memos. Assume your readers will not read them closely. After all, readers are mostly interested in the enclosed materials, not your transmittal letter or memo.

Why should you include a letter or memo of transmittal in the first place? There are a few good reasons:

- If a document, such as a report, shows up in your readers' mail without a transmittal letter or memo, they may not understand why it is being sent to them and what they should do with it.
- Transmittal letters and memos give you an opportunity to make a personal connection with the readers.
- They also give you an opportunity to set a specific tone for readers, motivating them to respond positively to the document or materials you have enclosed.

An effective transmittal letter or memo welcomes your readers to the materials you have sent.

Response Letter

Figure 17.6:
A response letter should answer the inquirer's questions point by point and offer additional information, if available.

Durable Computers

1923 Hanson Street, Orono, Maine 04467

March 7, 2009

Sally Vorman, Arctic Specialist
Arctic Information Associates
2315 Broadway, Fargo, ND 58102

Dear Dr. Vorman:

Thank the readers for their inquiry. →
Thank you for your inquiries regarding our Yeti Rugged Laptop. This computer is one of our most durable products, and it is particularly suited to the kinds of arctic climates you will be experiencing.

State the subject and purpose of the letter. →
Here are the answers to your questions:

Answer the questions point by point. →
Waterproofing: The Yeti stands up well to rain and other kinds of moisture. It can be submersed briefly (a few seconds), but it cannot be left underwater for a sustained amount of time.

Damage Protection: The Yeti can be dropped from 20 feet onto concrete without significant damage. Its magnesium alloy casing provides maximum protection.

Vibration: The Yeti meets the tough US-MIL 810E standards, which pay close attention to vibration, especially across rough terrain in a vehicle.

GPS Compatibility: The Yeti is compatible with all GPS systems we are aware of.

Discounts: We offer a discount of 10% for orders of 10 or more Yetis.

Offer more information, if available. →
I am also enclosing some of our promotional literature on the Yeti, including the technical specifications. In these materials you will find the results of our endurance testing on the laptop.

Provide contact information. →
We would very much like to hear about your trip and your experiences with the Yeti. If you have any more questions or would like to place an order, please call me at 293-555-3422. Or, I can be e-mailed at garys@duracomps.net.

Sincerely,

Gary Smothers

Gary Smothers
Design Engineer
Durable Computers

Memo of Transmittal

Identify the enclosed materials.

State the reason materials are being sent.

Summarize the enclosed materials.

State the action item clearly.

Provide contact information.

Figure 17.7: A transmittal memo should be concise. Make sure any action items are clearly stated.

Rockford Services

MEMORANDUM

Date: May 8, 2009
To: Brenda Young, VP of Services
From: Valerie Ansel, Outreach Coordinator
cc: Hank Billups, Pat Roberts
Re: Outreach to Homeless Youth

Enclosed is the Proposal for the Rockford Homeless Youth Initiative, which you requested at the Board Meeting on February 16, 2009. We need you to look it over before we write the final version.

The proposal describes a broad-based program in which Rockford Services will proactively reach out to the homeless youth in our city. In the past, we have generally waited for these youths to find their way to our shelter on the west side of town. We always knew, though, that many youths are reluctant or unable to come to the shelter, especially the ones who are mentally ill or addicted to drugs. The program described in this proposal offers a way to reach out to these youths in a nonthreatening way, providing them a gateway to services or treatment.

Please look over this proposal. We welcome any suggestions for improvement you might have. We plan to submit the final version of this proposal to the Board on May 24th at the monthly meeting.

Thank you for your help. You can contact me by phone at 555-1242, or you can e-mail me at valansel@rockfordservices.org.

Enclosed: Proposal for the Rockford Homeless Youth Initiative

Claim Letters and Memos

In the technical workplace, products break and errors happen. For these situations, you may need to write a claim letter or memo, also called a complaint. The purpose of a claim letter or memo is to explain a problem and ask for amends. Here are some guidelines to follow when writing a claim:

- State the subject and purpose clearly and concisely.
- Explain the problem in detail.
- Describe how the problem inconvenienced you.
- State what you would like the receiver to do to address the problem.
- Thank your reader for his or her response to your request.
- Provide contact information.

Figure 17.8 shows a claim letter with these features.

A claim should always be professional in tone. You might be tempted to send an angry message when errors are made. Angry letters and memos might give you a temporary sense of satisfaction, but they are less likely to achieve your purpose—to have the problem fixed. If possible, you want to avoid putting readers on the defensive, because they may choose to ignore you or halfheartedly try to remedy the situation.

Adjustment Letters and Memos

If you receive a claim letter or memo, you may need to respond with an *adjustment* letter or memo. The purpose of an adjustment letter or memo is to respond to the issue described by the client, customer, or co-worker. These documents, though, need to do more than simply respond to the problem. They should also try to rebuild a potentially damaged relationship with the reader.

Here are some guidelines to follow when writing an adjustment letter:

- Express regret for the problem *without directly taking blame.*
- State clearly what you are going to do about the problem.
- Tell your reader when he or she should expect results.
- Show appreciation for his or her continued business with your company.
- Provide contact information.

Figure 17.9 shows an adjustment letter with these features.

Why shouldn't you take direct blame? Several factors might be involved when something goes wrong. So, it is fine to acknowledge that something unfortunate happened. For example, you can say, "We are sorry to hear about your injury when using the Zip-2000 soldering tool." But, it is something quite different to say, "We accept full responsibility for the injuries caused by our Zip-2000 soldering tool." This kind of statement could make your company unnecessarily liable for damages.

Ethically, your company may need to accept full responsibility for an accident. In these situations, legal counsel should be involved with the writing of the letter.

Refusal Letters and Memos

Refusals, also called "bad news" letters or memos, always need to be carefully written. In these documents, you are telling the readers something they don't want to

Claim Letter

Figure 17.8:
A claim letter should explain the problem in a professional tone and describe the remedy being sought.

State the subject and purpose of the letter.

Explain the problem in detail.

Describe how the problem inconvenienced you.

State what the reader should do.

Thank the reader for the anticipated response.

Provide contact information.

Outwest Engineering

2931 Mission Drive, Provo, UT 84601 (801) 555-6650

June 15, 2008

Customer Service
Optima Camera Manufacturers, Inc.
Chicago, IL 60018

Dear Customer Service:

We are requesting the repair or replacement of a damaged ClearCam Digital Camcorder (#289PTDi), which we bought directly from Optima Camera Manufacturers in May 2008.

Here is what happened. On June 12, we were making a promotional film about one of our new products for our website. As we were making adjustments to the lighting on the set, the camcorder was bumped and it fell ten feet to the floor. Afterward, it would not work, forcing us to cancel the filming, causing us a few days' delay.

We paid a significant amount of money for this camcorder because your advertising claims it is "highly durable." So, we were surprised and disappointed when the camcorder could not survive a routine fall.

Please repair or replace the enclosed camcorder as soon as possible. I have provided a copy of the receipt for your records.

Thank you for your prompt response to this situation. If you have any questions, please call me at 801-555-6650, ext. 139.

Sincerely,

Paul Williams

Paul Williams
Senior Product Engineer

Adjustment Letter

O   C   M

Optima Camera Manufacturers, Inc.
Chicago, IL 60018 312-555-9120

July 1, 2008

Paul Williams, Senior Product Engineer
Outwest Engineering Services
2931 Mission Drive
Provo, UT 84601

Dear Mr. Williams,

Express regret for the problem.

We are sorry that the ClearCam Digital Camcorder did not meet your expectations for durability. At Optima, we take great pride in offering high-quality, durable cameras that our customers can rely on. We will make the repairs you requested.

State what will be done.

After inspecting your camera, our service department estimates the repair will take two weeks. When the camera is repaired, we will return it to you by overnight freight. The repair will be made at no cost to you.

Tell when results should be expected.

We appreciate your purchase of a ClearCam Digital Camcorder, and we are eager to restore your trust in our products.

Show appreciation to the customer.

Thank you for your letter. If you have any questions, please contact me at 312-555-9128.

Provide contact information.

Sincerely,

Ginger Faust

Ginger Faust
Customer Service Technician

hear (i.e., "no"). Yet, if possible, you want to maintain a professional or business relationship with these customers or clients.

When writing a refusal, show your readers how you logically came to your decision. In most cases, you will not want to start out immediately with the bad news (e.g., "We have finished interviewing candidates and have decided not to hire you"). However, you also do not want to make readers wait too long for the bad news.

Here are some guidelines for writing a refusal:

- State your subject.
- Summarize your understanding of the facts.
- Deliver the bad news, explaining your reasoning.
- Offer any alternatives, if they are available.
- Express a desire to retain the relationship.
- Provide contact information.

Keep any apologizing to a minimum, if you feel you must apologize at all. Some readers will see your apology as an opening to negotiate or complain further. An effective refusal logically explains the reasons for the turndown, leaving your reader satisfied with your response—if a bit disappointed. Figure 17.10 shows a sample refusal letter with these features.

Using Style in Letters and Memos

The style of a letter or memo can make a big difference. One thing to keep in mind as you consider your style in a letter or memo is this: *All letters and memos are personal.* They make a one-to-one connection with readers. So, even if you are writing a memo to the whole company or sending out a form letter to your company's customers, you are still making a personal, one-to-one connection with each of those readers. People will read your correspondence as a message written to them individually.

The personal nature of these documents has its advantages and disadvantages:

- The advantage is that your readers will view your letter or memo as something sent directly to them. They will assume that they are interacting with a person (you), not just a faceless company.
- A potential disadvantage is that readers tend to react to these documents in an emotional way. If your correspondence sounds indifferent, rude, or uncaring, your readers are likely to become irritated or upset. Readers respond very negatively to memos or letters that they perceive to be angry, insulting, or condescending.

Strategies for Developing an Appropriate Style

Since letters and memos are personal documents, their style needs to be suited to their readers and contexts of use. Here are some strategies for projecting the appropriate style in your letter or memo:

- Use the "you" style.
- Create an appropriate tone.
- Avoid bureaucratic phrasing.

Refusal Letter

O C M

Optima Camera Manufacturers, Inc.
Chicago, IL 60018 312-555-9120

July 1, 2008

Paul Williams, Senior Product Engineer
Outwest Engineering Services
2931 Mission Drive
Provo, UT 84601

Dear Mr. Williams,

State the subject.

We are sorry that the ClearCam Digital Camcorder did not meet your expectations for durability. At Optima, we take great pride in offering high-quality, durable cameras that our customers can rely on.

Summarize what happened.

According to the letter you sent us, the camcorder experienced a fall and stopped working. After inspecting your camcorder, we have determined that we will need to charge for the repair. According to the warranty, repairs can only be made at no cost when problems are due to manufacturer error. A camcorder that experienced a fall like the one you described is not covered under the warranty.

Deliver the bad news, explaining your reasoning.

We sent your camcorder to the service department for a repair estimate. After inspecting your camera, they estimate the repair will take two weeks at a cost of $156.00. When it is repaired, we will return it to you by overnight freight.

Offer alternatives.

If you would like us to repair the camcorder, please send a check or money order for $156.00. If you do not want us to repair the camcorder, please call me at 312-555-9128. Upon hearing from you, we will send the camcorder back to you immediately.

Provide contact information.

Again, we are sorry for the damage to your camcorder. We appreciate your purchase of a ClearCam Digital Camcorder, and we are eager to retain your business.

Express a desire to retain the relationship.

Sincerely,

Ginger Faust

Ginger Faust
Customer Service Technician

Enclosed: Warranty Information

Figure 17.10:
A refusal letter should deliver the bad news politely and offer alternatives if available. You should strive to maintain the relationship with the person whose request is being refused.

USE THE "YOU" STYLE   When you are conveying neutral or positive information, you should use the word *you* to address your readers. The "you" style puts the emphasis on readers rather than on you, the author.

> Well done. Your part of the project went very smoothly, saving us time and money.

> We would like to update your team on the status of the Howards Pharmaceutical case.

> You are to be congratulated for winning the Baldrige Award for high-quality manufacturing.

In most cases, negative information should not use the "you" style, because the readers will tend to react with more hostility than you expect.

> **Offensive:** Your lack of oversight and supervision on the assembly line led to the recent work stoppage.

> **Improved:** Increased oversight and supervision will help us avoid work stoppages in the future.

> **Offensive:** At our last meeting, your ideas for new products were not fully thought through. In the future, you should come more prepared.

> **Improved:** Any ideas for new products should be thoroughly considered before they are presented. In the future, we would like to see presenters more prepared.

Don't worry about whether your readers will notice that you are criticizing them. Even without the "you" style, they will figure out that you are conveying negative information or criticisms. By avoiding "you" in these negative situations, though, you will create a constructive tone and avoid an overly defensive reaction from your readers.

Link

For more advice about choosing an appropriate style, see Chapter 9, page 226.

CREATE A TONE   Think about the image you want to project. Put yourself into character as you compose your message. Are you satisfied, hopeful, professional, pleased, enthusiastic, or annoyed? Write your message with that tone in mind.

Mapping is an especially good way to project a specific tone in your letter or memo (Figure 17.11). For example, perhaps you want to argue that you are an "expert." Put the word "expert" in the middle of the screen or a piece of paper. Then, write down words associated with this word.

Once you have mapped out the tone you want to create, you can then weave these words into your letter or memo. If the words are used strategically, your readers will subconsciously sense the tone you are trying to create.

AVOID BUREAUCRATIC PHRASING   When writing correspondence, especially a formal letter, some people feel a strange urge to use phrasing that sounds bureaucratic:

> **Bureaucratic:** Pursuant to your request, please find the enclosed materials.

> **Nonbureaucratic:** We have included the materials you requested.

Do you notice how the bureaucratic phrasing in the first sentence only makes the message harder to understand? It doesn't add any information. Moreover, this phrasing depersonalizes the letter, undermining the one-to-one relationship between writer and reader.

Mapping a Tone

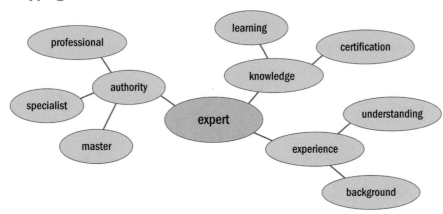

Here are a few other bureaucratic phrases and ways they can be avoided:

Bureaucratic Phrase	Nonbureaucratic Phrase
Per your request	As you requested
In lieu of	Instead of
Attached, please find	I have attached
Enclosed, please find	I have enclosed
Contingent upon receipt	When we receive
In accordance with your wishes	As you requested
In observance with	According to
Please be aware that	We believe
It has come to our attention	We know
Pursuant to	In response to
Prior to receipt of	Before receiving

A simple guideline is not to use words and phrases that you would not use in everyday speech. If you would not use words like *lieu, contingent,* or *pursuant* in a conversation, you should not use them in a letter or memo.

Designing and Formatting Letters and Memos

Letters and memos are usually rather plain in design. In the workplace, they typically follow standardized formats and templates that prescribe how they will look.

GO TO THE NET

Want more advice about avoiding bureaucratic language? Go to **www.pearsonhighered.com/johnsonweb3/17.9**

Huiling Ding
Assistant Professor, Clemson University
Dr. Huiling Ding researches proposal writing and Chinese rhetorical practices.

How should letters be written to Chinese readers?

To effectively communicate with your Chinese business partners through letters, it is extremely important to understand the cultural background of business communication practices in China and to have a strong sense of audience.

Chinese businesses work in a conservative manner, which requires acknowledgement and respect of status differences while stressing mutual face-saving in business transactions. You should address your partners with their titles and their surnames (e.g., Dr., Mr., or Ms.) to show your respect for their status. Don't address them with their first names unless you have their permission to do so. Meanwhile, since Chinese women never adopt their husbands' name after getting married, always be careful *not* to address somebody with the title "Mrs."

Face-saving is highly stressed in Chinese culture, so you should always avoid saying "no" directly in your letter. If you decide to turn your business partners down, tell them you will consider their request. Then, let them wait. Afterward, follow up with a nice message apologizing for not considering their product or service because of fierce competition. Indicate the intention of collaborating with them in the future. Don't mention problems in quality or service.

Living in a high-context culture, Chinese people prefer to conduct business through established and trustworthy personal contacts rather than strangers. Therefore, your letter should convey a personal interest in your business partners instead of using standard businesslike tones. Here are a few guidelines:

- Use the letter to establish yourself as a friendly and credible partner.
- Adopt a friendly tone, i.e., using *I/we* instead of *you* to show your respect for your audience.
- Use active voice rather than passive voice to avoid the impression of being arrogant. For instance, say, "I will appreciate the information" instead of "the information will be appreciated."

Finally, to communicate effectively, you need to consider the English proficiency of your audience. Please keep in mind that in most cases, your partner may have adequate reading skills but little practice in writing. Therefore, keep your language simple and your message crystal clear. Repeat important information so that your partner will not miss it because of the lack of professional vocabulary or poor reading skills.

Most companies have premade word-processing templates for letters and memos that you can download on your computer. These templates allow you to type your letter or memo directly into a word-processing file. When you print out the document, the letterhead or memo header appears at the top.

Formatting Letters

As stated earlier in this chapter, letter formats typically include some predictable features: a header (letterhead), an inside address, a greeting, the message, and a closing with a signature (Figure 17.12).

LETTERHEAD   Companies typically have letterhead available as a premade word-processor template or as stationery. Letterhead includes the company name and address. If letterhead is not available, you should enter your return address followed by the date. Do not include your name in the return address.

> 1054 Kellogg Avenue, Apt. 12
> Hinsdale, Illinois 60521
> December 19, 2008

The return address is best set along the left margin of the letter.

INSIDE ADDRESS   The address of the person to whom you are sending the letter (called the *inside address*) should appear two lines below the date or return address.

> George Falls, District Manager
> Optechnical Instruments
> 875 Industrial Avenue, Suite 5
> Starkville, New York 10034

The inside address should be the same as the address that will appear on the letter's envelope.

GREETING   Include a greeting two lines below the inside address. It is common to use the word "Dear," followed by the name of the person to whom you are sending the letter. A comma or colon can follow the name, although in business correspondence a colon is preferred.

If you do not know the name of the person to whom you are sending the letter, choose a gender-neutral title like "Human Resources Director," "Production Manager," or "Head Engineer." A generic greeting like "To Whom It May Concern" is inappropriate because it is too impersonal. With a little thought, you can usually come up with a neutral title that better targets the reader of your letter.

Also, remember that it is no longer appropriate to use gender-biased terms like "Dear Sirs" or "Dear Gentlemen." You will offend at least half the receivers of your letters with these kinds of gendered titles.

Formatting a Letter

Letterhead with return address →

GeneriTech
1032 Hunter Ave., Philadelphia, PA, 19130

⊤ 1 blank line

March 29, 2009

⊤ 1 blank line

Inside address →

George Falls, District Manager
Optechnical Instruments
875 Industrial Avenue, Suite 5
Starkville, New York 10034

⊤ 1 blank line

Greeting →

Dear Mr. Falls:

⊤ 1 blank line

Message →

I enjoyed meeting with you when we visited Starkville last week. I am writing to follow up on our conversation about building a wireless network for your business. A wireless network would certainly avoid many of the Internet access problems your company is experiencing in its current facility. It would also save you the wiring costs associated with a more traditional system.

With this letter, I have enclosed a variety of materials on GeneriTech's wireless systems. You will find that there is a range of options to suit your company's needs. I have tried to include only brochures and white papers that you would find useful. I have also included testimonials written by a few of our customers about their wireless systems and their experiences with our company.

I will call you in a couple of weeks to see if you have any questions about our wireless products. If you want answers sooner, please call me at 547-555-5282 or e-mail me at lhampton3712@generitechsystems.com. I can answer your questions.

Again, thank you for meeting with us last week. I look forward to speaking with you again.

⊤ 1 blank line

Sincerely,

Lisa Hampton

⊤ 3 blank lines

Closing with signature →

Lisa Hampton
Senior Engineer, Wireless Division

MESSAGE   The message should begin two lines below the greeting. Today, most letters are set in *block format,* meaning the message is set against the left margin with no indentation. In block format, a space appears between each paragraph.

CLOSING WITH SIGNATURE   Two lines below the message, you should include a closing with a signature underneath. In most cases, the word "Sincerely," followed by a comma, is preferred. Sometimes writers will be more creative and say "Best Wishes," "Respectfully," or "Cordially." Avoid being cute with the closing. Phrases like "Your Next Employee" or "Respectfully Your Servant" are unnecessarily risky. Just use the word "Sincerely" in almost all cases.

Your signature should appear next, with your name and title typed beneath it. To save room for your signature, you should leave three blank lines between the closing and your typed name.

Sincerely,

Lisa Hampton

Lisa Hampton
Senior Engineer, Wireless Division

If you are sending the letter electronically, you can create an image of your signature with a scanner. Then, insert the image in your letter.

Formatting Envelopes

Once you have finished writing your letter, you will need to put it in an envelope. Fortunately, with computers, putting addresses on envelopes is not difficult. Your word-processing program can capture the addresses from your letter (Figure 17.13). Then,

Formatting for an Envelope

Return address

GeneriTech
1032 Hunter Ave., Philadelphia, PA 19130

Recipient address

George Falls, District Manager
Optechnical Instruments
875 Industrial Avenue, Suite 5
Starkville, NY 10034

Figure 17.13: An envelope includes a return address and a recipient address.

with the Envelopes and Labels function (or equivalent), you can have the word processor put the address on an envelope or label. Most printers can print envelopes.

An envelope should have two addresses, the *return address* and the *recipient address*. The return address is printed in the upper left-hand corner of the envelope, a couple lines from the top edge of the envelope. The recipient address is printed in the center of the envelope, about halfway down from the top edge of the envelope.

If your company has premade envelopes with the return address already printed on them, printing an envelope will be easier. You will need to add only the recipient address.

Formatting Memos

Memos are easier to format than letters because they include only a header and message.

HEADER   Most companies have stationery available that follows a standard memo format (Figure 17.14). If memo stationery is not available, you can make your own by typing the following list:

> Date:
> To:
> cc:
> From:
> Subject:

The "Subject" line should offer a descriptive and specific phrase that describes the content of the memo. Most readers will look at the subject line first to determine if they want to read the memo. If it is too generic (e.g., "Project" or "FYI"), they may not read the memo. Instead, give them a more specific phrase like "Update on the TruFit Project" or "Accidental Spill on 2/2/08."

The "cc" line (optional) includes the names of any people who will receive copies of the memo. Often, copies of memos are automatically sent to supervisors to keep them informed.

If possible, sign your initials next to your name on the "From" line. Since memos are not signed, these initials serve as your signature on the document.

MESSAGE   Memos do not include a "Dear" line or any other kind of greeting. They just start out with the message. The block style (all lines set against the left margin and spaces between paragraphs) is preferred, though some writers indent the first line of each paragraph.

Longer memos should include headings to help readers identify the structure of the text. In some cases, you might choose to include graphics to support the written text.

It is important to remember that memos do *not* include a closing or signature. When your conclusion is complete, the memo is complete. No closing or signature is needed.

Link

For more ideas about designing documents, go to Chapter 10, page 232.

Sample Memo

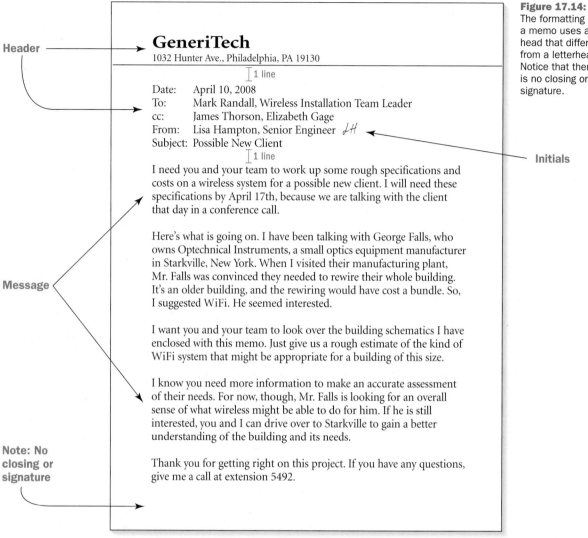

Header

GeneriTech
1032 Hunter Ave., Philadelphia, PA 19130

⊺ 1 line

Date: April 10, 2008
To: Mark Randall, Wireless Installation Team Leader
cc: James Thorson, Elizabeth Gage
From: Lisa Hampton, Senior Engineer *LH*
Subject: Possible New Client

⊺ 1 line

Message

I need you and your team to work up some rough specifications and costs on a wireless system for a possible new client. I will need these specifications by April 17th, because we are talking with the client that day in a conference call.

Here's what is going on. I have been talking with George Falls, who owns Optechnical Instruments, a small optics equipment manufacturer in Starkville, New York. When I visited their manufacturing plant, Mr. Falls was convinced they needed to rewire their whole building. It's an older building, and the rewiring would have cost a bundle. So, I suggested WiFi. He seemed interested.

I want you and your team to look over the building schematics I have enclosed with this memo. Just give us a rough estimate of the kind of WiFi system that might be appropriate for a building of this size.

I know you need more information to make an accurate assessment of their needs. For now, though, Mr. Falls is looking for an overall sense of what wireless might be able to do for him. If he is still interested, you and I can drive over to Starkville to gain a better understanding of the building and its needs.

Thank you for getting right on this project. If you have any questions, give me a call at extension 5492.

Note: No closing or signature

Figure 17.14: The formatting of a memo uses a head that differs from a letterhead. Notice that there is no closing or signature.

Initials

Want more help with formatting your letter or memo? Go to
www.pearsonhighered.com/johnsonweb3/17.11

Making Letter and Memo Templates

Personal letters and memos looked rather bland before computers were available, because they used only minimal design features. Only companies could afford letterhead or memo headers.

Today, you can use the Templates feature in your word-processing program to make your own headers for letters and memos. These templates are especially handy when you are applying for jobs.

Using Existing Templates

Almost all word-processing software, like Microsoft Word and Corel WordPerfect, includes a variety of letter and memo templates that you can immediately adapt to your own needs. When you select "New" or "New Document" in your word-processing program, you will be given a choice of different templates you can use (Figure A).

Using a Premade Template

Selection of template options →

Figure A: The template feature of your word-processing program allows you to use professional-looking headers in your letters and memos.

Preview of the template

At this point, you can select the template that most closely reflects the kind of document you need to write. The template will allow you to enter information such as an inside address or greeting. The computer will take care of the formatting of the document, though you should always check the format to make sure it looks appropriate after you have added your text.

Making Your Own Template

You can also make your own templates for letters and memos. In a blank document on your word processor, design a header that reflects your personality. You can add an icon or logo, while including your address (Figure B).

Link
To learn how to download graphics off the Internet, go to Chapter 11, page 295.

(continued)

Personalized Letter Template

James Chosuke
1923 W. Lincolnway, Ames, IA 50010
515-555-2315

Figure B: You can quickly make up personalized templates for your letters and memos for work. For your personal use, you can design a header that fits your personality.

When your template is complete, select "Save As." In the box that appears on your screen, you will find an option to save the file as a "template." Once you save the file as a template, the word-processing program will add it to the other templates so you can use it in the future.

From then on, you can use your template whenever you want. These templates are especially handy when you are doing a job search and need to make numerous form letters.

Revising, Editing, and Proofreading

You should always leave some time for revising and proofreading your letter or memo before you send it out. Since these documents are one-on-one forms of communication, readers often become especially annoyed at smaller mistakes.

After you have finished drafting and formatting your letter or memo, reconsider the rhetorical situation in which the document will be used. Pay attention to any information or wording that might annoy or offend the readers.

Have I included only need-to-know information?

Is the purpose of the document stated or obvious in the introduction?

Does the document have a clear point?

Is it clear in the introduction what I want my readers to do?

How might secondary or tertiary readers use this information?

In what other contexts might the document be used?

Is the style appropriately formal or informal? Plain or persuasive?

Have I properly formatted the document?

Link

For more advice about proofreading, go to Chapter 12, page 313.

If the message is especially important, let others look over your letter or memo before you send it out. Your immediate supervisor would be a good person to ask for a critical review. In some cases, you may want to leave the document alone for a couple of hours. That way, you can revise it with a more critical eye.

Revising and proofreading your work is very important. Since letters and memos tend to be formal, you want them to represent your best work. Also, keep in mind that letters and memos are often filed, and they have a nasty habit of reappearing at unexpected moments (like performance reviews, court cases, or important meetings). You want them to reflect your true thoughts and your best work.

- Writing letters and memos can be a drain on your time. You should learn how to write them quickly and efficiently.

- Letters and memos are essentially the same kind of document, called a *correspondence*. They use different formats, but they tend to achieve the same purposes. Letters are used for messages that go outside the company. Memos are for internal messages.

- Letters and memos are always personal. They are intended to be a one-to-one communication to a reader, even if they are sent to more than one person.

- Letters and memos share the same basic features: header, introduction, informative body, and conclusion. They differ mostly in *format,* not content or purpose.

- The introduction should identify the subject, purpose, and point of the correspondence. If you want the readers to take action, put that action up front.

- The body of the correspondence should give the readers the information they need to take action or make a decision.

- The conclusion should thank the readers, restate your main point, and look to the future.

- To develop an appropriate style, use the "you" style, create a tone, and avoid bureaucratic phrasing.

- Use standard formats for letters and memos. These formats will make the nature of your message easier to recognize.

- Be sure to revise your work as necessary and proofread it when you are finished.

Individual or Team Projects

1. Write a memo to an instructor in which you request a letter of reference. Your memo should be polite, and it should offer suggestions about what the instructor should include in the letter of reference.

2. Find a sample letter or memo on the Internet. In a memo to your instructor, discuss why you believe the letter is effective or ineffective. Discuss how the content, organization, style, and design are effective/ineffective. Then, make suggestions for improvement. Sample letters and memos are available at www.pearsonhighered.com/johnsonweb3/17.2.

3. Think of something that bothers you about your college campus. To a named authority, write a letter in which you complain about this problem and discuss how it has inconvenienced you in some way. Offer some suggestions about how the problem might be remedied. Be sure to be tactful.

Want to make your own templates? Go to
www.pearsonhighered.com/johnsonweb3/17.12

Exercises and Projects   481

4. Imagine you are the college administrator who received the complaint letter in Exercise 3. Write a response letter to the complainant in which you offer a reasonable response to the writer's concerns. If the writer is asking for a change that requires a great amount of money, you may need to write a letter that refuses the request.

5. Create a letter template that you can use for your personal correspondence. It should include your name and address, as well as a symbol or logo that adds a sense of style and character to your correspondence.

Collaborative Project

With a group, choose three significantly different cultures that interest you. Then, research these cultures' different conventions, traditions, and expectations concerning letters and memos. You will find that correspondence conventions in countries such as Japan or Saudi Arabia are very different from those in the United States. The Japanese often find American correspondence to be blunt and rude. Arabs often find American correspondence to be bland (and rude, too).

Write a brief report, in memo form, to your class in which you compare and contrast these three different cultures' expectations for correspondence. In your memo, discuss some of the problems that occur when a person is not aware of correspondence conventions in other countries. Then, offer some solutions that might help the others in your class become better intercultural communicators.

Present your findings to the class.

Revision Challenge

The memo shown in Figure A needs to be revised before it is sent to its primary readers. Using the concepts and strategies discussed in this chapter, analyze the weaknesses of this document. Then, identify some ways it could be improved through revision.

- What information in the memo goes beyond what readers need to know?
- How can the memo be reorganized to highlight its purpose and main point?
- What is the "action item" in the memo, and where should it appear?
- How can the style of the memo be improved to make the text easier to understand?
- How might design be used to improve the readers' understanding?

An electronic version of this memo is available at www.pearsonhighered.com/johnsonweb3/17.13. There, you can also download the file for revision.

> For additional technical writing resources, including interactive sample documents, document design tutorials and guidelines, and more, go to **www.mytechcommlab.com**.

ChemConcepts, LLC

Memorandum

Date: November 14, 2008
To: Laboratory Supervisors
cc: George Castillo, VP of Research and Development
From: Vicki Hampton, Safety Task Force
Re: FYI

It is the policy of the ChemConcepts to ensure the safety of its employees at all times. We are obligated to adhere to the policies of the State of Illinois Fire and Life Safety Codes as adopted by the Illinois State Fire Marshal's Office (ISFMO). The intent of these policies is to foster safe practices and work habits throughout companies in Illinois, thus reducing the risk of fire and the severity of fire if one should occur. The importance of chemical safety at our company does not need to be stated. Last year, we had four incidents of accidental chemical combustion in our laboratories. We needed to send three employees to the hospital due to the accidental combustion of chemicals stored or used in our laboratories. The injuries were minor and these employees have recovered; but without clear policies it is only a matter of time before a major accident occurs. If such an accident happens, we want to feel assured that all precautions were taken to avoid it, and that its effects were minimized through proper procedures to handle the situation.

In the laboratories of ChemConcepts, our employees work with various chemical compounds that cause fire or explosions if mishandled. For example, when stored near reducing materials, oxidizing agents such as peroxides, hydroperoxides and peroxyesters can react at ambient temperatures. These unstable oxidizing agents may initiate or promote combustion in materials around them. Of special concern are organic peroxides, the most hazardous chemicals handled in our laboratories. These

(continued)

compounds form extremely dangerous peroxides that can be highly combustible. We need to have clear policies that describe how these kinds of chemicals should be stored and handled. We need policies regarding other chemicals, too. The problem in the past is that we have not had a consistent, comprehensive safety policy for storing and handling chemicals in our laboratories. The reasons for the lack of such a comprehensive policy are not clear. In the past, laboratories have been asked to develop their own policies, but our review of laboratory safety procedures shows that only four of our nine laboratories have written safety policies that specifically address chemicals. It is clear that we need a consistent safety policy that governs storage and handling of chemicals at all of our laboratories.

So, at a meeting on November 3, it was decided that ChemConcepts needs a consistent policy regarding the handling of chemical compounds, especially ones that are flammable or prone to combustion. Such a policy would describe in depth how chemicals should be stored and handled in the company's laboratories. It should also describe procedures for handling any hazardous spills, fires, or other emergencies due to chemicals. We are calling a mandatory meeting for November 28 from 1:00–5:00 in which issues of chemical safety will be discussed. The meeting will be attended by the various safety officers in the company, as well as George Castillo, VP of Research and Development. Before the meeting, please develop a draft policy for chemical safety for your laboratory. Make fifteen copies of your draft policy for distribution to others in the meeting. We will go over the policies from each laboratory, looking for consistencies. Then, merging these policies, we will draft a comprehensive policy that will be applicable throughout the corporation.

The Nastygram

Shannon Phillips is the testing laboratory supervisor at the Rosewood Medical Center. It is her laboratory's job to test samples, including blood samples, drawn from patients. After the samples are tested, one of her technicians returns a report on the sample to the doctor.

Last week, Shannon's laboratory was in the process of transferring over to a new facility. It was a complex move because the samples needed to be delivered on rolling carts to the new laboratory. There was no other way to make the transfer.

Then, the accident happened. One of the technicians who was pushing a cart of blood samples left it alone for a moment. The cart rolled down an incline, into the lobby, and turned over. Samples were strewn all over the floor.

The hazardous materials team cleaned up the mess, but the samples were all destroyed. In an e-mail, Shannon notified the doctors who had ordered the tests that they would need to draw new samples from their patients. With her apologies, she tried her best to explain the situation and its remedy.

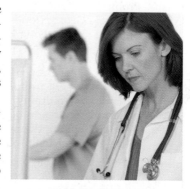

Most doctors were not pleased, but they understood the situation. However, Dr. Alice Keenan, the director of the AIDS Center at the hospital, was quite angry about the accident. She wrote the memo shown in Figure B (next page) to Shannon.

Needless to say, Shannon was quite upset about the memo. She had done everything she could to avoid the accident, and now a doctor was going to ask the board of directors to fire her. At a minimum, Shannon might need to fire her technician, who had never made a mistake up to this point. She was also quite angry about the tone of the letter. It seemed so unprofessional.

If you were Shannon, how might you respond to this situation?

Rosewood Medical Center

712 Hospital Drive, Omaha, Nebraska 68183

Memorandum

Date: February 13, 2008
To: Shannon Phillips, Laboratory Supervisor
From: Dr. Alice Keenan, MD, Director of AIDS Center
cc: George Jones, CEO Rosewood Medical Center
Subject: Your Idiotic Blunder

I cannot believe you morons in the lab ruined the blood samples we sent from our patients. Do you realize how difficult it is to draw these samples in the first place? Our patients are already in great discomfort and in a weakened state. So, asking them for a second blood draw within only a few days is a big problem.

Not to mention the hazardous situation created by this idiotic blunder. Do you realize that someone in the lobby could have come into contact with these samples and perhaps contracted HIV? Are you too obtuse to realize what kind of horrible consequences your mistake might have caused? You would have made us liable for huge damages if we were sued (which we would be).

Someone down there should be fired over this mistake! In my opinion, Shannon, you should be the first to go. I'm very concerned about your leadership abilities if you could allow something like this to happen. If I don't hear about a firing or your resignation in the next week, I'm going to bring a formal complaint to the Board of Directors at their next meeting.

Are you people all idiots down there? I've never heard of something this stupid happening at this hospital.

Figure B: Sometimes it is difficult to respond professionally to a nasty memo or letter like this one.

In this chapter, you will learn:

- How definitions are used in technical communication.

- How to write three kinds of definitions: parenthetical, sentence, and extended.

- How to organize and draft each kind of definition.

- How to compose an extended definition.

- How to use plain style to make descriptions clear and readable.

- How to revise and edit definitions to ensure consistency.

Accurate technical definitions are essential in technical communication. Any new term or concept needs to be clearly defined before it can be used properly in a technical workplace. For example, think about all the new words that have come into the English language since the invention of the computer, such as *hypertext, software, Internet, information technology, bits, bytes, monitor,* and *server,* among numerous other new terms. Some of these words are completely new, like *hypertext* and *bytes,* taking on their own unique meanings. Other words, like *server* and *hardware,* have meanings adapted to the world of computers.

Each technical discipline has a vocabulary of specialized words that it uses.

- A mechanical engineer must be able to define words like *velocity, torque,* and *viscosity.*
- A medical professional needs to define words like *cerebellum, hepatitis, West Nile virus,* and *ankylosing spondyliti.*
- An anthropologist needs to understand the definitions of words like *flaking station, reflexivity,* and *karyotype.*

Knowing how to define the terms in your discipline is an important part of succeeding in that discipline.

It is also important to recognize that different disciplines have their own distinct meanings for words. For example, consider the meanings of the word *field* in four different disciplines.

Discipline	Definition	Sample Usage
Physics	A region where a given effect (e.g., electricity, magnetism, or gravitation) exists	"The magnetic field influences the direction of the compass."
Medicine	An area of professional expertise	"He is an expert in the skin cancer field."
Geology	An area of study outside the office or laboratory	"She confirmed her theory with observations in the field."
Agronomy	An open area used for farming or pasture	"They let the cattle graze in the field."

GO TO THE NET

For websites that offer definitions of computer terms, go to
www.pearsonhighered.com/johnsonweb3/18.1

Fortunately, computers and the Internet have made defining terms easier. At your fingertips are countless resources, like online dictionaries and glossaries, that can help you find precise definitions for the terms you use in your documents and presentations.

You will also occasionally need to write your own definitions for new terms or technical concepts. Technical words are a natural part of a discipline's vocabulary, so they are unavoidable. Defining these terms is crucial to effective technical communication.

Types of Technical Definitions

There are three types of definitions: *parenthetical*, *sentence*, and *extended*.

Parenthetical definitions—A parenthetical definition explains or clarifies the meaning of a word or phrase within the sentence itself.

Parenthetical definitions

A butterfly's thorax (<u>body</u>) has three segments that bear four wings and six legs.

In 1929, Earnest Lawrence developed the first workable design for a cyclotron, <u>a device that accelerates protons to high energies,</u> helping scientists better explore the strange universe of subatomic particles.

A parenthetical definition usually appears in parentheses or is set off by commas or dashes.

Sentence definitions—These definitions are whole sentences in which a term is defined by naming its "category" and the "distinguishing features" that differentiate it from its category.

Category

Distinguishing characteristics

An ion is an atom that has a positive charge because one or more electrons have been stripped from it.

Sentence definitions are often embedded in documents when a word or concept needs to be defined in exact terms. They are also commonly found in dictionaries and glossaries.

Extended definitions—Extended definitions can fill a small paragraph or even run as long as several pages. In extended definitions, complex terms are defined very precisely by using techniques such as analogies, comparisons, contrasts, examples, negation, and graphics. Figure 18.1, for example, shows an extended definition of a laser from Bell Labs.

As you will notice in this extended definition of lasers, many words could be even further defined. In this way, a definition can be extended indefinitely.

Three Types of Definitions
• Parenthetical definition—a word or phrase
• Sentence definition—a full sentence
• Extended definition—more than a sentence

AT A GLANCE

For websites with quick sentence definitions, go to **www.pearsonhighered.com/johnsonweb3/18.2**

An Extended Definition

Word history →

What Is a Laser?

The word *laser* is an acronym for light amplification by stimulated emission of radiation, although common usage today is to use the word as a noun—laser—rather than as an acronym—LASER.

Sentence definition →

A laser is a device that creates and amplifies a narrow, intense beam of coherent light.

Atoms emit radiation. We see it every day when the "excited" neon atoms in a neon sign emit light. Normally, they radiate their light in random directions at random times. The result is incoherent light—a technical term for what you would consider a jumble of photons going in all directions.

Similarities and differences

The trick in generating coherent light—of a single or just a few frequencies going in one precise direction—is to find the right atoms with the right internal storage mechanisms and create an environment in which they can all cooperate—to give up their light at the right time and all in the same direction.

Exciting atoms or molecules

In a laser, the atoms or molecules of a crystal, such as ruby or garnet—or of a gas, liquid, or other substance—are excited in what is called the *laser cavity* so that more of them are at higher energy levels than are at lower energy levels. Reflective surfaces at both ends of the cavity permit energy to reflect back and forth, building up in each passage. (See figure below)

Division into parts

Graphic →

Mirror

Flash Lamp

Ruby Rod

Semi-transparent Mirror

In a ruby laser, light from the flash lamp, in what is called "optical pumping," excites the molecules in the ruby rod, and they bounce back and forth between two mirrors until coherent light escapes from the cavity.

Figure 18.1: An extended definition starts with a sentence definition. Then it uses a variety of rhetorical tools to expand and sharpen the meaning of the term.

To see samples of extended definitions, go to
www.pearsonhighered.com/johnsonweb3/18.3

If a photon whose frequency corresponds to the energy difference between the excited and ground states strikes an excited atom, the atom is stimulated as it falls back to a lower energy state to emit a second photon of the same (or a proportional) frequency, in phase with and in the same direction as the bombarding photon.

This process is called *stimulated emission*. The bombarding photon and the emitted photon may then each strike other excited atoms, stimulating further emission of photons, all of the same frequency and phase. This process produces a sudden burst of coherent radiation as all the atoms discharge in a rapid chain reaction.

Wide range of sizes and uses

History

First built in 1960, lasers now range in size from semiconductor lasers as small as a grain of salt to solid-state and gas lasers as large as a storage building. The light beam produced by most lasers is pencil-thin and maintains its size and direction over very large distances.

Lasers are widely used in industry for cutting and boring metals and other materials, in medicine for surgery, and in communications, scientific research, and holography. They are an integral part of such familiar devices as bar-code scanners used in supermarkets, scanners, laser printers, and compact disk players.

Examples

Planning and Researching Technical Definitions

In most cases, writing parenthetical and sentence definitions does not require a great amount of preparation. However, if you are writing an extended definition, you will need to spend some time doing research and collecting information.

Planning

Whatever kind of definition you are writing, you should start by gaining a full understanding of the term you are defining and the context in which that term is being used. A good way to start defining a term is to ask the Five-W and How Questions.

> *Who needs this definition, and what is their familiarity with terms in this field?*
>
> *What amount of detail will be needed to accurately define this term for these readers?*
>
> *Where will this definition be used in the document and elsewhere?*
>
> *When will this definition be used?*
>
> *Why is this definition needed?*
>
> *How might this definition be used in the document and elsewhere?*

Your answers to these questions will help you decide what kinds of information you should include in your definition. Expert readers in your own discipline probably don't need an overly detailed definition of a term. They simply want to know how you are using the term in your document. Nonexpert readers, on the other hand, will probably need a more detailed definition.

Once you have briefly answered the Five-W and How Questions, you are ready to start thinking about the subject, purpose, readers, and context of use for your definition.

SUBJECT   Think about what facts and details are needed to write an accurate definition of this term. What information does your reader need to know to understand the term?

Link

For more information on defining your purpose, go to Chapter 2, page 21.

PURPOSE   Write down *why* you are defining this term. Are you offering a new definition of this term? Are you trying to distinguish your definition of the term from other definitions of it? Are you striving for a certain level of accuracy? Your purpose statement might say something like

> My purpose is to define the word *butterfly* in this report so that my readers will understand the differences between true butterflies (*Papilionoidea*) and other winged insects that are mistakenly called butterflies.

Link

For more reader analysis strategies, go to Chapter 3, page 39.

READERS   When considering your readers, your first task should be to assess their level of expertise and what kinds of information they will need. After all, experts and nonexperts require different kinds of information to take action or make a decision.

Using Online Sources for Definitions

The Internet is an amazing source for definitions. Your word processor also includes a thesaurus that you can use to help you write definitions.

Online Dictionaries

Numerous dictionaries are available on the Internet. A favorite online dictionary is the *Oxford English Dictionary (OED)*, which has been an ongoing project in print form since 1857. This dictionary identifies the original usage of English words and keeps track of new words. The *OED* also keeps tabs on English words used throughout the world, including North America, Australia, South Africa, India, and New Zealand, among other countries. Currently, it defines over half a million words.

The print version of the *OED* is a massive 20 volumes, but you can now look up definitions at the *OED* online through the Internet if your university or company is a subscriber. To check if you have access, go to the website (www.oed.com) and see if it allows you to enter the site. If you can't, you might want to ask your librarian how you can access the online version.

If you can access the *OED*, go to the website and click on the link "Enter OED Online." You will then be able to enter any English word in the website's search engine. Hit Return, and the definition will appear on your screen, along with a wealth of historical information on how the word has been used (Figure A).

HELP

OED Online

Figure A: It is simple and fun to use the online *OED*. Type the word in the box in the upper right-hand corner, and the dictionary will find the definition for you.

Source: From Oxford English Dictionary *(1998)*, edited by J. Simpson & E. Weiner. By permission of Oxford University Press. www.oup.co.uk.

Want to see some online dictionaries, including the *OED*? Go to **www.pearsonhighered.com/johnsonweb3/18.4**

Planning and Researching Technical Definitions

In Figure A, for example, the word "solstice" was entered. Initially, as shown, the dictionary will offer you the definition and the history of the word. The buttons at the top of the page (Pronunciation, Spellings, Etymology, Quotations, and Date Chart) allow you to gather further information.

One limitation to the *OED* is its conservative nature. You probably won't find the latest technical terms or slang in this dictionary. Before adding a new word, the *OED* typically waits for a word to become a regular part of the English language. Other online dictionaries are available if you want definitions of current or trendy words.

Primary readers (action takers)—These readers need to know enough information to make informed choices about your subject. You need to offer enough information for them to be able to understand the term without overwhelming them with unnecessary details.

Secondary readers (advisors)—These readers may be experts in your field who are advising the primary readers. They will mainly be looking for accuracy in your definition.

Tertiary readers (evaluators)—Your tertiary readers might be just about anyone else who is interested in your document. They might include lawyers, auditors, accountants, or reporters who will be paying close attention to how you define your terms.

Gatekeepers (supervisors)—Your supervisors will be concerned about how you define any terms. They will be looking for accuracy and clarity.

Link

For more ideas about defining the context of use, go to Chapter 3, page 46.

CONTEXT OF USE   Definitions are used in a variety of places and times. Think about the places where your readers might use or need your definition. Do you need to define your terms up front? Or, should you put most definitions in a glossary in the appendix? Also, don't forget that definitions can involve political, ethical, and legal considerations. How you define a word can make a big difference in a lawsuit.

Researching

When researching your definition, you should draw information from online, print, and empirical sources. Here are a few strategies that are especially applicable to writing definitions.

Do background research—Start out by using print and online dictionaries to gather existing definitions. Locate information about the origins and history of the word. The Help box in this chapter describes how to use online sources to find information about words.

Find examples of usage—Using a variety of sources, gather sentences in which the word is used. Put the word into just about any Internet search

Need help defining the rhetorical situation? Go to
www.pearsonhighered.com/johnsonweb3/18.5

Researching Usage of a Word

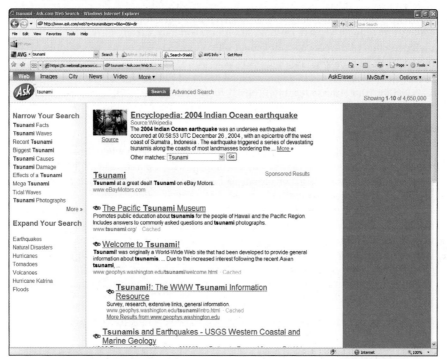

Figure 18.2: Running a simple Internet search is a good way to research how a word is used. Here, the word "tsunami" was put into Ask.com.

engine, like Ask.com, shown in Figure 18.2. You will find many sentences in which the word is used. Also, you should take notes on how people use the word in everyday usage.

Compare and contrast—Identify similarities and differences between your subject and other things. You can make direct comparisons and note contrasts between similar things (e.g., "Keller Hall is larger and more modern than Ortega Hall"). Or, you can compare and contrast dissimilar things (e.g., "The upright granite rocks in this valley stand tall, unlike the smaller sandstone rocks found farther down the river").

Collect visuals—Collect pictures and illustrations of your subject. You can even make your own visuals, using a camera, scanner, or drawings you made yourself.

Link

For more information about doing research, turn to Chapter 7, page 142.

The key to researching a definition is not to rely solely on the definitions you find in dictionaries. Dictionary definitions are rather generic and static. They rarely capture the full evolution and usage of a word, especially in technical disciplines. Instead, do research of your own to gain a fuller understanding of the word you are trying to define.

GO TO THE NET

To find a dictionary specific to your field, go to
www.pearsonhighered.com/johnsonweb3/18.6

Organizing and Drafting Technical Definitions

When defining a term, you first need to gain a thorough understanding of it and the contexts in which it is used. In some situations, you may already have a rather firm understanding of the concept but cannot offer a precise definition. For example, almost anyone knows what a biological *virus* is, but few people would be able to offer a precise definition. So, as you begin organizing and drafting, think about the kinds of information your readers need to properly understand your term.

Parenthetical Definitions

Parenthetical definitions use a word or phrase to define a term when it is first used in a technical document. The easiest way to come up with a parenthetical definition is to look in a dictionary, glossary, or thesaurus for a synonym (an alternative word that has almost the same meaning).

Traditional paper-based dictionaries are helpful at this point, but there are also many dictionaries and glossaries on the Internet, like Merriam-Webster online (www.merriam-webster.com), Dictionary.com, the *Oxford English Dictionary* (www.oed.com), and Bartleby.com. Moreover, each field or discipline usually has an online dictionary or glossary that you can use. You can find them with an Internet search engine.

In these dictionaries, locate the word you are seeking to define. Then, use a word or phrase from that definition to define your term:

> If the circuit board is heated beyond 500°C, delamination (layer separation) will likely occur.

> In disbelief, we realized Kudzu (*Pueraria lobata*) was growing in this arid environment.

> We determined that the patient had gastritis, an inflammation of the mucous membrane of the stomach.

> The ketch (i.e., a multimasted sailboat) moved slowly out to sea.

Avoid using words or phrases that are equally unfamiliar to your readers to define a term.

> We will need to develop gateways—application-specific interfaces that link all seven layers of the OSI Model—when they are dissimilar at any or all levels.

Nonexperts might not know the meaning of "application-specific interface." As a result, this parenthetical definition would not be helpful.

Sentence Definitions

Sentence definitions can appear in a variety of places in a document. They are especially helpful when a term needs to be exactly defined.

> **Within the text**—Occasionally, you will need to use a word that might be unfamiliar to your readers. In these situations, the sentence that follows the use of the word should be a definition.

Unlike bacteria, a virus is not really alive. A virus is genetic material (DNA or RNA) protected by a protein coating, but it is unable to replicate without a host cell.

J. J. Thompson, using cathode rays, developed the concept of an electron. Today, an electron is defined as an elementary particle with a negative charge of about 1.6×10^{-19} coulomb.

Glossary—In highly technical documents, especially ones written for non-experts, a glossary of terms can be placed toward the end of the document, usually in an appendix. A glossary contains terms that readers may want to refer to as they are reading the document.

Margin notes—In some documents, you might choose to put sentence definitions in the margins to help readers understand the meanings of technical terms.

Sentence definitions follow a rather standard three-part pattern, which includes (1) the item being defined, (2) the category of similar things to which the item belongs, and (3) any distinguishing features that separate this item from other items in its category.

Term	Category	Distinguishing Features
The chromosphere	is a region of the sun's atmosphere	that is close to the sun, thousands of miles thick, and composed chiefly of hydrogen gas.
Oxacillin	is a semisynthetic penicillin	that is used to control infections caused by penicillin-resistant staphylococci.
A solstice	is a time of the year	when the sun, having reached the tropical points, is farthest from the equator and appears to stand still; i.e., about June 21 (the summer solstice) and December 22 (the winter solstice).

You might want to use a diagram like the one in Figure 18.3 to help you sort out these three parts of the sentence definition. Put the name of the item in the smallest quarter circle. Then, in the middle area, note the possible categories into which the item might be classified. In the largest area, write down the features that distinguish the item from other similar items in these categories.

For samples of glossaries, go to
www.pearsonhighered.com/johnsonweb3/18.7
Need more help writing a sentence definition? Go to
www.pearsonhighered.com/johnsonweb3/18.8

Diagram for Sentence Definitions

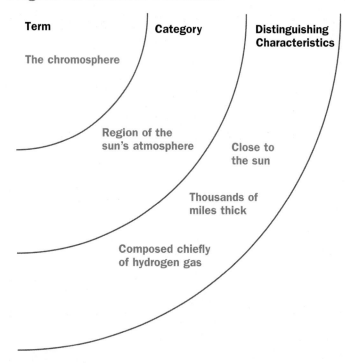

Term

Category

Distinguishing
Characteristics

The chromosphere

Region of the
sun's atmosphere

Close to
the sun

Thousands of
miles thick

Composed chiefly
of hydrogen gas

Figure 18.3:
A diagram like
this one can
help you write a
good sentence
definition.

Again, dictionaries, available both online and in print, are a good place to start developing a sentence definition. Using dictionaries, you can compare a few different definitions and then devise your own. Remember, though, that if you copy a definition directly from a dictionary, you should cite the dictionary as a source. For example, the following is a cited sentence definition used in a text:

> According to *Merriam-Webster's Collegiate Dictionary,* Eleventh Edition, a gene is "a specific sequence of nucleotides in DNA or RNA that is located usually on a chromosome and that is the functional unit of inheritance controlling the transmission and expression of one or more traits by specifying the structure of a particular polypeptide and especially a protein or controlling the function of other genetic material." (520)
>
> Source: By permission. From Merriam-Webster's Collegiate® Dictionary, 11th Edition © 2008 by Merriam-Webster Incorporated (www.Merriam-Webster.com).

Here, the page number is included so readers can check the dictionary as a source.

Extended Definitions

Extended definitions are usually used for terms that need to be explained with utmost precision. These kinds of definitions are often found in larger documents such as reports and websites, where they are used to explain a term thoroughly.

This diagram can be downloaded at
www.pearsonhighered.com/johnsonweb3/18.9

Extended Definition

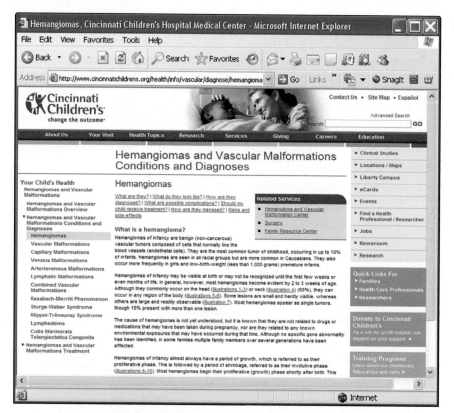

Source: Cincinnati Children's Hospital Medical Center, http://www.cincinnatichildrens.org/
health/info/vascular/diagnose/hemangioma.htm.

Figure 18.4:
Extended
definitions are
commonly found
on professional
websites.

They are also commonly found in guidebooks, handbooks, and websites that are devoted to specific disciplines. For example, Figure 18.4 shows an extended definition of a medical condition.

Even a common word sometimes needs to be exactly defined. Lawyers and insurance agents, for example, often need exact definitions of common terms.

> What is "negligence"? In its broadest sense, negligence means carelessness. But not all carelessness will result in a legal claim for negligence. To recover money damages for negligence, the plaintiff must prove that the defendant had a duty under the law to be careful, that the defendant wasn't careful, that the defendant's carelessness caused the plaintiff's injury, and that the plaintiff suffered damages, like medical expenses and pain and suffering, because of the defendant's carelessness.
>
> Source: U.S. Public Health Service, http://oep.osophs.dhhs.gov/dmat/training/phsfr/cha02/12cha02.htm.

As these examples show, most extended definitions start out with a sentence definition. Then you can use a variety of techniques to extend the definition.

GO TO THE NET

More examples of extended definitions can be found at
www.pearsonhighered.com/johnsonweb3/18.10

**Organizing and
Drafting Technical
Definitions**

499

Extending a Definition

Link

For more help using logical mapping to develop new ideas, go to Chapter 7, page 146.

Drafting an extended definition is like drafting any other document. You are directly or indirectly claiming that a term can be defined a particular way. Then you will need to offer support to back up that claim.

To help you start extending a definition, you might try using logical mapping to explore the many ways in which the word could be used or defined. As shown in Figure 18.5, start out with your sentence definition in the middle of the screen or a sheet of paper. Then, use logical mapping to explore all the different ways you might define the term.

You can extend a definition by discussing or using some of the following techniques:

WORD ORIGIN (ETYMOLOGY)   A term's origin can often provide some interesting insights into its meaning.

> According to the *Oxford English Dictionary,* the word *cricket* comes from variations of the French word *criquer,* "to creak, rattle, crackle."

> The word *hysterical* includes the root *hyster-,* which comes from the Greek word for "womb." Hysteria was believed to be a women's malady that was related to a disorder with the uterus.

Extending a Sentence Definition

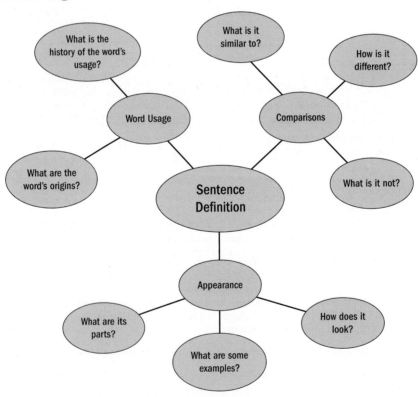

Figure 18.5: You can extend a definition by asking questions about the usage of the word or the appearance of the subject and by making comparisons to other things.

Would you like to have a word of the day e-mailed to you automatically? Go to **www.pearsonhighered.com/johnsonweb3/18.11**

The *OED* is probably the most authoritative source of word origins in English, but most online dictionaries will provide etymological information (see the Help box in this chapter). You might also look for technical dictionaries that serve your field.

WORD HISTORY   The history of a word's usage can also offer interesting insights into its current meaning.

> The origin of the word *geek* is somewhat disputed. *Geek* actually comes from the old Dutch/German word *geck,* which dates to the sixteenth century. It originally meant "dupe, fool, or simpleton." However, some word historians argue that the American use of the word originally referred to carnival performers who bit the heads off animals, like snakes or chickens. This meaning gives a whole new sense to what it means to be a geek or geeked up.

> The word *sabotage* first meant "making noise or destroying things with shoes" (*sabot* means "shoe" in French). For example, an audience would disrupt musical performances by making noise with *sabots,* or workers would destroy machinery by striking it with their shoes. Since then, the word has evolved to mean disruption of military operations or manufacturing by deliberately damaging equipment.

EXAMPLES   You might also include an example to clarify a term's usage. The example usually offers a specific usage of the word in a context familiar to readers. The following uses an example to define *computer virus:*

> Perhaps playing on instinctive fears of illness, the concept of a computer "virus" sends shudders through our society. For instance, in 1992 the media whipped up a frenzy about the Michelangelo virus, which was supposed to wipe out up to a quarter of American computer hard drives on March 6, the famed artist's birthday. Companies spent millions of dollars and valuable time preparing for this attack. In the end only about 10,000 computers were affected, a relatively small number.

NEGATION   Often, you can define something by showing what it is not. Health officials, especially, seem to focus on defining terms by telling people what a disease is *not.*

> According to the Saskatchewan Government Health website, HIV (human immunodeficiency virus) cannot be transmitted through casual contact with other people, like hugging, shaking hands, or eating meals prepared by people infected with HIV. It is not transmitted through sweat, saliva, or tears, so you cannot get the virus from computer keyboards, public phones, swimming pools, water fountains, or other similar places (for more details, see http://www.health.gov.sk.ca/hiv_aids).

> St. John's wort is not a stimulant, and it won't cure all kinds of depression. Rather, it is a mild sedative. This is why St. John's wort has been found useful for treating patients suffering from mild depression with high levels of anxiety.

DIVISION INTO PARTS   There is a fine line between defining something and describing it. In your extended definition, you might describe your subject by breaking it down into parts.

AT A GLANCE

Techniques for Extending a Definition

- Word origin—What is the etymology of the word?
- Word history—How has this word been used in the past?
- Example—What is an example of the word's usage?
- Negation—What does *not* define this word?
- Division into parts—How can my subject be divided?
- Similarities and differences—What things can be compared and contrasted with my subject?
- Analogy—What is something completely different from my subject and yet similar in some important ways?
- Graphics—How can my subject be illustrated?

For websites that offer word origins, go to
www.pearsonhighered.com/johnsonweb3/18.12
For technical dictionaries that offer word histories, go to
www.pearsonhighered.com/johnsonweb3/18.13

Organizing and Drafting Technical Definitions

501

According to the American Orchid Society, the orchid family is a group of flowering plants that has many different species and hybrids. Nevertheless, all orchids have some characteristics in common. Let's consider the common characteristics in the orchid family's flowers, leaves, and roots.

Flowers: The flowers of all orchids follow a similar pattern. There is a drum-shaped container, called the whorl, with three leaf-like sepals. Within the sepals is another whorl that contains three petals. One of these petals, called the labellum or lip, is different from the other two in color or size. The center of the flower includes the column, where the stamens and pistil are fused together.

Leaves: The leaves of an orchid can be folded or pleated, and they can be pliant or tough. The major veins of an orchid are usually parallel to each other.

Roots: The roots are typically adapted to the environment in which the plant grows. Most orchids have roots that protect interior tissues from water loss.

Source for information: The American Orchid Society, http://www.aos.org.

Link

For more help dividing things into parts, go to Chapter 19, page 525.

SIMILARITIES AND DIFFERENCES   You can compare your subject to things that are similar, showing their common characteristics and their differences.

Orangutans are similar to the other great apes (gorillas, chimpanzees, and bonobos) in several ways. Like other apes, they primarily eat fruit and plants, though they will eat meat occasionally. They have prehensile hands and feet. They can stand but are most comfortable walking on all fours. Orangutans are different from other apes because they tend to be solitary animals. Also, their distinctive reddish color makes them stand out. Other apes tend to be black in color.

Link

To learn more about using comparison, go to Chapter 8, page 189.

ANALOGY   It is helpful to compare your subject to something completely different but with some similar qualities.

Imagine your body is a city. It has a circular system of arteries and veins that are like roads in a city. These arteries and veins contain white blood cells, which act like police patrolling for viruses and bacteria. The brain is the government, directing the city where it needs to go. And yes, your body has a mouth for importing the things it needs and its own sewage system for releasing things when it has used them.

Link

To find out how to use analogies effectively, go to Chapter 9, page 223.

Just about anything can be compared to anything else, some things with more success than others.

GRAPHICS   A drawing, picture, diagram, or other kind of graphic is especially useful for describing something. The graphic should not stand alone. Rather, it should be used to support the written text. For example, the drawing of an x-ray tube in Figure 18.6 should not stand alone. Instead, it should support written text that offers further information on all these parts.

Any graphic used to define a term should label the parts of the subject. Make sure the labels on the graphic correspond directly to the terms you use in the written text.

Link

For more ideas on how to use graphics, go to Chapter 11, page 275.

For more examples of definition strategies, go to
www.pearsonhighered.com/johnsonweb3/18.14
Interested in analogies? Go to
www.pearsonhighered.com/johnsonweb3/18.15

Using a Graphic in a Definition

Cathode   Filament   Anode   Target

X-Ray Tube

Figure 18.6: A graphic can be used to support the written text, but a graphic alone cannot replace the written text.

In an extended definition, you can use a variety of techniques to define a term. For example, the definition of a tsunami in Figure 18.7 uses several of the strategies discussed in this chapter.

Using Style and Design in Technical Definitions

Technical definitions are almost always written in plain style, and their design is typically not flashy.

Keeping the Style Plain and Simple

When writing and revising a definition, you want to provide your readers with the straightforward meaning of the word. So, there is little need to be overly persuasive. Here are some suggestions for style in definitions:

Use only words that will be familiar to the readers—If your readers are experts, a specialized vocabulary is fine, perhaps even preferred. If your readers are not experts, use common words and define any terms being used in specialized ways.

Keep sentences short—Parenthetical and sentence definitions should use the least amount of words possible to provide an accurate definition. Extended definitions should avoid long sentences that go beyond breathing length.

Use definitions within definitions—When you use a specialized word within a definition, also include a parenthetical definition to define it.

Keep it visual—Where possible, use words and phrases that allow readers to visualize the subject. Use color, texture, and shapes to define it. Add a graphic. Use analogies and similes. Show examples.

An Extended Definition

Sentence definition →

Similarities and differences →

Graphics →

Negation

Figure 18.7:
An extended definition uses a variety of techniques to explain the meaning of a term. In this definition of the word *tsunami*, the author uses examples, shows similarities and differences, and includes an analogy.

Source: International Institute for Geo-Information Science and Earth Observation.

504   Chapter 18
Technical Definitions

For more advice on writing plainly, go to
www.pearsonhighered.com/johnsonweb3/18.16

In almost all cases, the style of a definition should not stand out. Using plain style is probably the best approach whether you are writing a parenthetical, sentence, or extended definition.

Link

To learn more about using plain style, go to Chapter 9, page 215.

Designing for Clarity

Since most definitions are embedded within a larger document, they tend to adopt the design of the larger document. However, there are situations, such as white papers or specifications, where extended definitions need to take on their own design (Figures 18.7 and 18.8).

As you consider the definition's design, think about its readers, purpose, and context of use. Will the extended definition be used to define a new product or concept to experts? Will it be sent out to people around the country to clarify a subtle change in

Good Style and Design in an Extended Definition

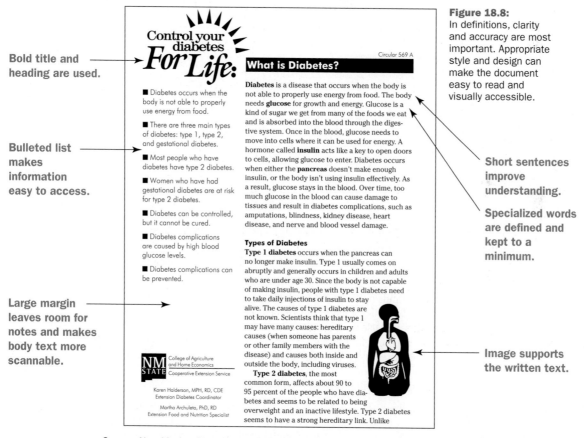

Bold title and heading are used.

Bulleted list makes information easy to access.

Large margin leaves room for notes and makes body text more scannable.

Short sentences improve understanding.

Specialized words are defined and kept to a minimum.

Image supports the written text.

Figure 18.8: In definitions, clarity and accuracy are most important. Appropriate style and design can make the document easy to read and visually accessible.

Source: New Mexico State University Extension Office.

(continued)

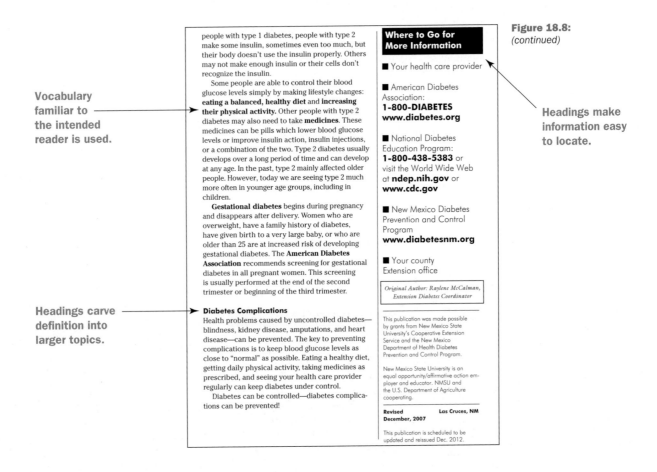

Figure 18.8:
(continued)

Vocabulary familiar to the intended reader is used.

people with type 1 diabetes, people with type 2 make some insulin, sometimes even too much, but their body doesn't use the insulin properly. Others may not make enough insulin or their cells don't recognize the insulin.

Some people are able to control their blood glucose levels simply by making lifestyle changes: **eating a balanced, healthy diet** and **increasing their physical activity**. Other people with type 2 diabetes may also need to take **medicines**. These medicines can be pills which lower blood glucose levels or improve insulin action, insulin injections, or a combination of the two. Type 2 diabetes usually develops over a long period of time and can develop at any age. In the past, type 2 mainly affected older people. However, today we are seeing type 2 much more often in younger age groups, including in children.

Gestational diabetes begins during pregnancy and disappears after delivery. Women who are overweight, have a family history of diabetes, have given birth to a very large baby, or who are older than 25 are at increased risk of developing gestational diabetes. The **American Diabetes Association** recommends screening for gestational diabetes in all pregnant women. This screening is usually performed at the end of the second trimester or beginning of the third trimester.

Headings carve definition into larger topics.

Diabetes Complications
Health problems caused by uncontrolled diabetes—blindness, kidney disease, amputations, and heart disease—can be prevented. The key to preventing complications is to keep blood glucose levels as close to "normal" as possible. Eating a healthy diet, getting daily physical activity, taking medicines as prescribed, and seeing your health care provider regularly can keep diabetes under control.

Diabetes can be controlled—diabetes complications can be prevented!

Where to Go for More Information

■ Your health care provider

■ American Diabetes Association:
1-800-DIABETES
www.diabetes.org

■ National Diabetes Education Program:
1-800-438-5383 or visit the World Wide Web at **ndep.nih.gov** or **www.cdc.gov**

■ New Mexico Diabetes Prevention and Control Program
www.diabetesnm.org

■ Your county Extension office

Original Author: Raylene McCalman, Extension Diabetes Coordinator

This publication was made possible by grants from New Mexico State University's Cooperative Extension Service and the New Mexico Department of Health Diabetes Prevention and Control Program.

New Mexico State University is an equal opportunity/affirmative action employer and educator. NMSU and the U.S. Department of Agriculture cooperating.

Revised   **Las Cruces, NM**
December, 2007

This publication is scheduled to be updated and reissued Dec. 2012.

Headings make information easy to locate.

the meaning of a word? Will it be put on a website to help the public understand your company's products or services?

If you are given a free hand to design the text, you might consider some of the following design features:

- a larger title that stands out from the body text
- a typeface and font size that reflect your readers' characteristics and the contexts in which they will use the document
- headings that break the definition into larger blocks of information
- a format with two or more columns that allows you to put graphics or explanatory information in a separate column
- a graphic or picture that offers a helpful image of the subject

Link

For more about designing documents, go to Chapter 10, page 232.

Technical definitions are rarely known for their innovative design. So, you should look for examples of previous definitions in your company literature for guidance on design issues. Also, if you are writing a definition that will be embedded in a larger document, pay close attention to how that document is designed. Your definition should blend in seamlessly with the document's overall design.

Revising, Editing, and Proofreading

Accuracy is very important in definitions, so you should leave plenty of time to revise and edit your work. Make sure others, especially experts, have a chance to offer commentary on the definition.

One thing to keep in mind as you revise is your readers' level of expertise in your field. After all, if you use words to define your subject that are unfamiliar to readers, you are likely making the subject harder to understand.

Amy Jamison
SPECIALIST IN AFRICAN LANGUAGE AND CULTURE, MICHIGAN STATE UNIVERSITY
Amy Jamison is earning her Ph.D. in African studies.

How are new technical concepts brought into African languages?

When a new technical concept enters an African language and culture, more than simple translation is involved. Rather, the definition of a new word is shaped by many forces, including governments, private corporations, popular culture, and academic institutions. The goal is to make the technology accessible to people who are often far removed from technical advances.

One such society is the African nation of Tanzania, where Swahili is the official language. Though some people speak English, people in Tanzania are more comfortable using Swahili to do business. As a result, proper transmission of technological concepts is necessary for maintaining Tanzanian identity and facilitating communication.

For example, the Swahili word for "icons" on a computer screen is *nembo*. Originally, *nembo* meant "tribal scars." After independence, it also was used to refer to political party emblems. The similarity of emblems and icons made the word a good choice as computers became part of African society. These kinds of historical transformations help to explain the emergence of new words and definitions in Swahili.

In the 1960s and 1970s, the newly independent government of Tanzania began to coin new Swahili words for technological innovations in order to develop the language and show its capacity for expressing modern concepts. This practice slowed down in the 1990s, just as advances in telecommunications and computers were increasing.

Popular coinage of technical terms in Swahili proliferated, filling the gap. Corporations also saw the need for new terminology, further contributing to the development of Swahili technical language.

Recently, government and academic institutions have returned to creating new words and definitions while helping software and telecommunications companies translate technical manuals.

Overall, the development of a Swahili technical language has been a complex process, beyond simple translation of words and definitions from one language to another. Politics, commerce, and the need for accessibility have each played a role.

AT WORK

GO TO THE NET
To see well-designed and badly designed definitions, go to
www.pearsonhighered.com/johnsonweb3/18.17

Revising for Conciseness and Visual Detail

While revising, look for places where you can cut out any information that goes beyond your readers' need to know. Definitions can be extended endlessly, so you need to scale the amount of information to your readers' needs.

Then, look for ways to use the senses, especially the visual, to bring your definition to life. Look for places where you can add color or texture to your writing. Add one or more graphics to help readers understand what you are defining. With these visual techniques, you will help readers gain a full three-dimensional understanding of the subject.

Always look for sentences that are too long or complex. It is tempting to use long sentences to "get it just right." That's fine in a draft, but the final version should use simple, plain sentences that readers can grasp quickly.

Editing for Accuracy and Consistency

Often, sentence definitions end up in a glossary, while extended definitions can be placed in an appendix. When readers take the time to look up these definitions, they will expect them to be absolutely accurate and consistent. So, as you are revising and proofreading, pay close attention to the precision and predictability of the definitions.

Accuracy is essential. If you have some doubts about whether a word is being defined properly, it is worth your time to double-check its meaning. Inaccuracies in definitions can lead to large problems down the line.

CHAPTER REVIEW

- Reliable, accurate technical definitions are essential in technical communication. They ensure clarity in technical documents.

- There are three types of definitions: parenthetical, sentence, and extended.

- To determine the rhetorical situation for a definition, you can use the Five-W and How Questions. Then, you should review your subject, purpose, readers, and context of use.

- Organizing and drafting a parenthetical definition involves finding a word or phrase to clarify the meaning of a term.

- Organizing and drafting a sentence definition involves determining the term, category, and distinguishing features of the object being defined.

- Drafting an extended definition may involve research into the background and meaning of a term, and it could involve using different techniques such as analogies or graphics to provide an extra dimension to your definition.

- A plain and simple style will make your definition clear.

- Revise and edit your definitions to make sure they are accurate, concise, and consistent. Use visual language and graphics where appropriate to illustrate the definition.

Individual or Team Projects

1. By yourself or with a team, list ten technical words that are commonly used in your major or field. Then, using dictionaries, glossaries, and Internet search engines, find sentence definitions for each of these words. Finally, turn this list of definitions into a "glossary" that you can give to your class or hand in to your instructor.

2. Pick a word that interests you. Then, go to the *Oxford English Dictionary* or another dictionary and find the history of this word. Pay special attention to its etymology and when it was first used in the English language. Then, write a memo to your instructor in which you discuss the word, its usage, and its history.

3. Here is a list of words and phrases with some interesting word histories. Pick a few and hunt down their history on the Internet. To your class, present your favorite word, showing how its history reflects (or doesn't reflect) its usage today.

   America
   blackbox
   blurb
   boondoggle
   bug (as in computer bug)
   catch-22
   cybernetics
   doughnut
   the full monty
   glitch
   ground zero
   hermetic seal
   jeep
   jinx
   Murphy's Law
   on the fritz
   quark
   scapegoat
   spam (as in e-mail spam)
   the whole nine yards

4. Choose a word that is common in your major or field. Then, write an extended technical definition of that word for a nonexpert reader. Start out by defining your subject, purpose, readers, and context of use. Then, using various sources, research the meaning of the word. Write an extended definition in which you use examples, analogies, and similarities and differences, among other techniques, to expand the definition of the word.

5. Of some alarm to language purists are the abbreviated words used in instant messaging and e-mails. Terms like *IMO* ("in my opinion"), *brb* ("be right back"), *lol* ("laughing out loud"), *rtfm* ("read the * manual"), *ttfn* ("ta ta for now"), and

ICQ ("I seek you") are commonly used codes in these electronic formats. With your team, produce a small glossary of these words, explaining what they mean and how they are used. Your glossary should be written for people who are new to instant messaging or e-mail.

Collaborative Projects

Your group has been assigned to develop a glossary of terms that are commonly used on your college campus. The glossary is being written for new students and their parents, as well as for hapless college administrators and faculty who often don't know what their students are talking about. Your glossary will be put on the college's website so people will have quick access to the words.

When your group is finished with its glossary, combine it with the glossaries from other groups in the class. Certainly, some of the words will overlap with the words defined by other groups. Probably other groups' definitions for these words will be somewhat different from yours. Negotiating these differences, develop a master list of definitions that the whole class can agree to. Pay special attention to where people disagree about the meanings of words. Discuss why these disagreements come about.

Revision Challenge

The following technical definition in Figure A was written for a communication company's website. Its purpose was to help explain why solar flares occasionally affect the company's operations. The definition was aimed primarily at the company's clients. However, the company also wanted to provide information about communication technology to schoolchildren and other people interested in satellites.

This draft is probably a couple revisions from complete. Using the concepts discussed in this chapter, identify places where this document could be strengthened. Then, develop a revision strategy for improving this website text.

- What content would you like to see added to this definition?
- Is there any content that goes beyond the readers' need to know?
- Does the definition have a clear introduction, body, and conclusion?
- Where and how could the style be improved to make it more readable?
- Do you think other graphics might be helpful?

An electronic version of this definition is available at www.pearsonhighered.com/johnsonweb3/18.20. There, you can also download the file for revision.

Websites that discuss word histories are available at
www.pearsonhighered.com/johnsonweb3/18.19

Figure A:
This document is a draft of some information that will appear on a website. Can you make this text more readable to a broader group of readers, including children?

Solar Flares

For those of us who work with satellites, solar flares are a particularly interesting problem. They are mostly unpredictable phenomena that we cannot foresee, and we are just now beginning to understand how they affect the delicate hardware that we send up into space.

2003/10/28 06:24 UT

Source: SOHO, ESA/NASA.

(continued)

Solar flares happen when a large amount of magnetic energy is discharged from the sun's atmosphere. When a solar flare occurs, observers on Earth and those using observation tools like the SOHO satellite witness a huge burst coming from the sun. In some cases, the sun also emits an incredible burst of x-rays, which are a form of light. Other particles like heavy nuclei, electrons, and protons are released too. This is called a coronal mass ejection. The picture included here shows one of the larger bursts that we have seen recently, which occurred on October 21, 2003. This flare disturbed many of the earth's communication systems and power systems when its emitted electromagnetic waves slammed into the earth's atmosphere and magnetosphere. Fortunately, it takes awhile for the solar storm to reach the earth, so we had time to put our satellites on standby.

The effects of a solar flare are significant. Here on earth, the effects are not immediately noticeable. Your radio or mobile phone may pick up some static. False alarms might be triggered on alarms, like your car. But up in space, the effects can be pretty serious. The emitted x-rays can charge up components on satellites, causing damage. Probably the best known effect of solar flares are the colorful auroras which are visible in the sky, especially in northern parts of the country.

Even though the effects are usually not immediately noticeable here on earth, they are still important. The high-energy particles that are released during a solar radiation storm can expose airplane passengers to radiation that is equivalent to ten chest x-rays. Also, navigational systems, especially ones close to earth or on earth can be bothered, creating orientation problems. Weather is affected by solar flares, too.

Another interesting thing about solar flares is that they tend to follow eleven-year cycles. The last "solar maximum" was in 1989. During those periods there are many sunspots and coronal mass ejections. Satellites often need to be shut down to avoid damage. Even the SOHO satellite, which is used to observe the sun for the NOAA, cannot continue to operate during the largest of solar storms. It needs to be put on standby during the height of a solar storm.

Fast Food High School

The people of Edwardsville, like those in many communities, were growing concerned about their children's obesity problems. A recent study commissioned by the Edwardsville school board reported that 34 percent of the children in the school district were overweight or obese. High school students, especially, were carrying extra pounds. Nearly 41 percent of the students at Edwardsville High School were overweight.

At a meeting in March, the school board members decided to do something about this problem. They voted to remove the candy and soda machines from the schools, and they decided that only "nutritious food" would be served in school cafeterias.

The school board then hired Sally Mayer, a nutritionist, to assess the high school cafeteria's current food offerings and make recommendations for changes to the menu. Her report to the school board was due in two months.

Her first visit to the school confirmed that the food served in the cafeteria was typical fast food. She noted that students had a selection of pizza, burgers, hot dogs, and fried chicken strips. Overwhelmingly, students ate french fries and potato chips as sides, even though fruits and vegetables like apples, oranges, bananas, carrots, and celery were available. The cafeteria also had a salad bar, but only a few students were using it.

After observing lunch at the cafeteria, Sally met with the principal, Tom Young. He seemed like a person who genuinely cared for the students, but he was skeptical about changing their eating habits. He said, "We provide a choice. I don't think it's the school's job to regulate what kids eat. That's their parents' job. If we offer healthy food, students just won't eat it."

But Tom had another reason for resisting the move to nutritious foods—the expense. The cafeteria was a revenue source for the school. He told Sally that the profits from the cafeteria funded the after-school athletic intramural programs. The additional costs of nutritious foods would wipe out those revenues and possibly the intramural programs.

"Ironically," Tom pointed out, "the change to so-called 'nutritious food' would eliminate the kinds of physical activities and exercise these kids need to stay fit."

Nevertheless, Tom knew he had to make changes to satisfy the school board. He said, "Much of the solution depends on how you define 'nutritious food.' We

(continued)

can make some small changes, like buying leaner meats and adding to the salad bar. Maybe we won't offer french fries a couple days a week. But, Sally, you're going to need to meet us halfway. We need you to define nutritious food in a way that's realistic."

Sally drove back to her office. Tom was right—her whole report hinged on her definition of nutritious food. If she followed a strict definition, there was a possibility that (1) the students would not eat the food, (2) the food would be expensive, and (3) the after-school intramural athletic programs would be cut. However, a loose definition, leading only to the minor changes Tom described, would not address the current nutrition problems in the school.

If you were Sally, how might you define "nutritious food" in your report? How would this definition influence your assessment of the school cafeteria's current offerings and your recommendations?

Want more information on fast food and school cafeterias? Go to
www.pearsonhighered.com/johnsonweb3/18.21

In this chapter, you will learn:

- How descriptions are used in technical workplaces.

- Common features of descriptions.

- How to determine the rhetorical situation for a description.

- Strategies for partitioning objects, places, or processes into major and minor parts.

- Techniques for organizing and drafting descriptions.

- The use of plain style to make descriptions understandable.

- How to use page layout and graphics to highlight and illustrate important concepts.

- The importance of revising and editing descriptions for accuracy.

In any technical career, you will find that the ability to accurately describe things, places, or processes is essential. In the workplace, technical descriptions are called a variety of different names, but they are used in similar ways:

- A medical researcher needs to describe how a new chemotherapy treatment attacks cancerous cells in laboratory mice.
- An astrophysicist needs to describe a newly discovered comet that will orbit the sun in the coming year.
- A mechanical engineer needs to describe a new microprocessor for a patent application.
- An architect needs to illustrate his ideas for a butterfly pavilion that will be the centerpiece of a city's new biological park.

There are a few types of technical descriptions, written for various purposes in the technical workplace:

Technical description—Manufacturers use technical descriptions to describe their products for patents, quality control, and sales. These descriptions are often used to establish an archetype, or ideal, against which future products can be measured and tested.

Patents—An application for a patent requires a detailed technical description of how to make and use the new invention.

Specifications (often referred to as the "specs")—Engineers write specifications to describe a product in great detail, providing exact information about features, dimensions, power requirements, capacities, and other qualities. As discussed in Chapter 20, specifications are also used to describe, step-by-step, how a product is assembled or a task is completed.

Field notes—Naturalists, anthropologists, sociologists, and others use field notes to help them accurately describe people, animals, and places.

Observations—Scientists and medical personnel need to make detailed observations. For example, medical doctors need to describe their patients' conditions in great detail. These observations help them keep track of changes in their patients' symptoms.

Look around and you will find that descriptions appear in almost every technical document, including experimental reports, user's manuals, reference materials, proposals, marketing literature, magazine articles, and conference presentations. Knowing how to write technical descriptions will likely be an important part of your career.

To see examples of technical descriptions, go to
www.pearsonhighered.com/johnsonweb3/19.1

Basic Features of Technical Descriptions

A technical description can stand alone or be part of a larger document. A stand-alone technical description is a separate document and will generally have the following features:

- specific and precise title
- introduction with an overall description
- description of features, functions, or stages of a process
- use of senses, similes, analogies, and metaphors
- graphics
- conclusion that shows the thing, place, or process in action

When a technical description needs to stand alone as a separate document, it tends to follow a pattern like the one shown in Figure 19.1.

Technical Description

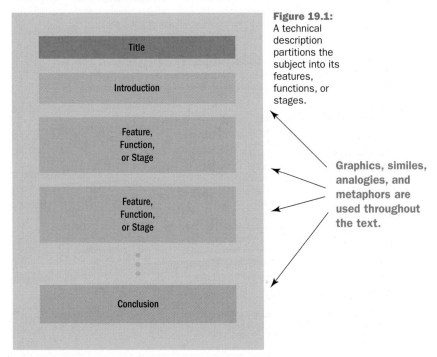

Figure 19.1: A technical description partitions the subject into its features, functions, or stages.

Graphics, similes, analogies, and metaphors are used throughout the text.

If your description is embedded in a larger document, such as a report or proposal, you will likely need to adjust the content, organization, style, and design to suit the larger document. The technical description in Figure 19.2, for instance, was placed on the Segway website.

Figure 19.2: This extended description of the Segway effectively partitions the subject by features and describes each feature in depth. In this well-written example, a conclusion was not included. This absence saves space but leaves the description lacking closure.

Definition of the product →

Overall description →

Importance stressed →

DISCOVER THE SEGWAY HT EVOLUTION.

The Segway™ Human Transporter (HT) is the first of its kind—a self-balancing, personal transportation device designed to go anywhere people do. It gives people everywhere the ability to move faster and carry more, allowing them to commute, shop, and run errands more efficiently while also having fun. It makes businesses more productive by allowing workers greater versatility, mobility and carrying capacity. It does it all by harnessing some of the most advanced, thoroughly tested technology ever created.

MAJOR COMPONENTS OF THE SEGWAY

Forecasting of the structure of the description →

The Segway™ Human Transporter (HT) includes dozens of individual parts that have been carefully designed, tested and manufactured to work in synchronous harmony with one another. Segway HT components highlighted below include the controller boards, balance sensory assembly, motors, gearbox, tires and wheels, and batteries.

Major feature →

CONTROLLER BOARDS

Two sophisticated controller boards from Delphi Electronics provide both brains and brawn for the system. Delphi Electronics was chosen as a partner based on their track record in the production of high-volume, high-quality automotive electronics for such demanding applications as airbag modules.

← Metaphor is used to personify the product.

Minor features →

Each board contains a Texas Instruments digital signal processor, monitoring the entire Segway HT system and checking 100 times per second for any faults or conditions that might require immediate response. It reads the information from the BSA to determine if the rider is leaning forward or backward, and instantly uses this information to deliver power from the batteries to the motors through a set of 12 high-power, high-voltage field-effect transistors (FETs). These calculations take place 100 times a second, and the motors are adjusted at up to 1,000 times per second, responding far more quickly than the human body is capable of perceiving. Although each board is capable of operating the Segway HT after a failure (each board, in fact, powers electrical circuits on both motors), under normal conditions they share the load.

Controller Boards

Balance Sensor Assembly

← Illustration shows minor parts.

Major feature →

BALANCE SENSOR ASSEMBLY

The balance sensor assembly (BSA), supplied by Silicon Sensing Systems, is an elegantly designed, extremely robust, and yet incredibly sensitive piece of equipment.

Source: © 2004 Segway Inc. All Rights Reserved.

Clearly labeled diagram

This small cube, 3 inches on a side, is packed with five solid-state, vibrating-ring, angular-rate sensors ("gyroscopes") that use the Coriolis effect to measure rotation speed. These tiny rings are electromechanically vibrated in such a way that when they are rotated, a small force is

Balance Sensor Assembly
(Top view)

Tilt sensor

Vibrating ring
(gyroscope)

Details add realism to the text.

generated that can be detected in the internal electronics of the sensor. Each "gyro" is placed at a unique angle that allows it to measure multiple directions. Segway's onboard computers constantly compare the data from all five gyros to determine if any of the five is supplying faulty data—in this condition, it can compensate and use data from the remaining sensors to continue balancing through a controlled safety shutdown. Two tilt sensors filled with an electrolyte fluid provide a gravity reference in the same way your inner ear does for your own sense of balance. The BSA is monitored by two independent microprocessors and is split into two independent halves for redundancy. Even the communication between sides is performed optically to avoid electrical faults on one side propagating to the other.

Analogy is used to clarify a complex concept.

Major feature

MOTORS

The Segway HT's motors are unique in a number of respects. Produced by Pacific Scientific, a division of Danaher, they are the highest-power motors for their size and weight ever put into mass production. Each motor is capable of maintaining a power output of 1.5 kilowatts—that's 2 horsepower!

Definitions are used throughout the text to explain concepts.

The motors use brushless servo technology, meaning there are no contacts to wear, arc, and reduce performance. The magnets are constructed of an incredibly powerful rare-earth material: neodymium-iron-boron. Each motor is constructed with two independent sets of windings, each driven by a separate board and motor. Under normal conditions, both sets of windings work in parallel, sharing the load. In the event of a failure, the motor is designed to instantly disable the faulty side and use the remaining winding to maintain control of

Motors
(Cross sectional)

B side windings

Magnets

A side windings

the Segway HT until it can be brought to a stop. The motor is carefully balanced to operate up to 8,000 rpm, allowing it to produce very high power levels in a small package. Feedback from the motor back to the Segway HT is provided by redundant, noncontact analog hall sensors that sense the positions of magnets with no moving parts other than the motor shaft itself.

(continued)

Major feature

GEARBOX

The Segway HT's gearbox, a joint effort between Axicon Technologies and Segway, is constructed more like a precision Swiss watch than a traditional gear drive from an automobile. A two-stage reduction system provides a 24:1 reduction, allowing the motor to operate at powerful, efficient speeds throughout the full range of speeds of the Segway HT. Each gear is cut to a helical profile, which creates a spiral engagement to minimize noise and increase the load capability of the gears. The number of teeth on each gear is chosen to produce noninteger gear ratios. This means that the teeth will mesh in a different location each revolution, maximizing the life of the gearbox.

A simile makes a comparison to something familiar.

Definition explains technical term.

Our engineers were so obsessed with the details on the Segway HT that they designed the meshes in the gearbox to produce sound exactly two musical octaves apart—when the Segway HT moves, it makes music, not noise.

Metaphor adds depth.

The gearbox is pre-assembled and lubricated, and is designed to require no maintenance over the life of the Segway HT. It has been tested for thousands of miles under severe conditions.

Major feature

TIRES AND WHEELS

Michelin was one of the earliest partners in the Segway HT design project. The tires on the Segway HT have been designed specifically for this product, using a unique tread compound (a silica-based compound instead of traditional carbon-based materials), giving enhanced traction and minimized marking on indoor floors, and a specially engineered tubeless construction that allows low pressure for comfort and traction while minimizing rolling resistance for long range.

Minor features

The tire is mounted on an equally unique wheel design: the wheel is constructed of a sophisticated engineering-grade thermoplastic chosen in partnership with GE Plastics. This composite material allows light weight and excellent durability, and actually reduces noise transmitted from the drive system. The wheel is molded around a forged steel hub, eliminating fasteners that can loosen over time. The attachment between this hub and the transmissions incorporates a unique (patent pending) taper and hex design, which allows a single nut to attach or remove the wheel while retaining the security of multiple-bolt systems more typically seen in conventional applications.

Major feature

BATTERIES

The Segway HT uses twin NiMH battery packs, designed in partnership with SAFT (a division of Alcatel), running at a nominal 72 volts. These nickel-metal hydride cells deliver the highest power of any currently available chemistry, optimized to maintain the Segway HT's balance under severe conditions. These are not your cell phone batteries!

Minor features

Each pack consists of an array of high-capacity cells and a custom-designed circuit board that constantly monitors the temperature and voltage of the pack in multiple locations. This assembly is enclosed in another unique application of GE thermoplastics—the battery box is sealed using a vibration welding technique that makes the outside of the pack a single, continuous structure—sealed from moisture and strong enough to survive the most extreme tests our durability engineers could throw at it.

Batteries
High-capacity
cells

The internal electronics in the battery incorporate "smart" charging—the customer need only plug the Segway HT into the wall and the battery will choose the appropriate charge rate based on temperature, voltage, and level of charge. The batteries will quick charge, then automatically transition into a balance and maintenance charge mode. The Segway HT customer does not need to worry about memory or timing their charges—just plug it in.

Under normal operation, the Segway HT carefully monitors both batteries and automatically adjusts to drain the batteries evenly. In the unlikely event of a battery failure, the system is designed to use the second battery to operate the machine and allow it to continue balancing until it is brought to a safe stop.

Planning and Researching Technical Descriptions

During the planning and researching phase, you should identify what kinds of information your readers need to know, how they will use that information, and the contexts in which they will use it. Then, you need to research the subject to collect the content for your description.

Planning

Technical descriptions are written for a variety of people and uses. So, as you begin planning your description, it is important that you first have a good understanding of the situation in which your description will be used. A good way to start is to first consider the Five-W and How Questions that will be important in your description.

> *Who might need this description?*
>
> *Why is this description needed?*
>
> *What details and facts should the description include?*
>
> *Where will the description be used?*
>
> *When will the description be used?*
>
> *How will this description be used?*

Link

For more help defining your subject, see Chapter 7, page 143.

Once you have briefly answered these questions, you are ready to start defining the subject, purpose, readers, and context of your description.

SUBJECT Technical descriptions tend to be written about three types of subjects: objects, places, or processes. As you research your subject, define its boundaries and major characteristics.

PURPOSE Ask yourself what your description should achieve. Do you want it to provide exact detail? Or, are you just trying to familiarize your readers with the subject?

In one sentence, write down the purpose of your description. Here are some verbs that might help you write that sentence:

to describe	*to represent*
to illustrate	*to clarify*
to show	*to reveal*
to depict	*to explain*
to characterize	*to portray*

Link

For more information on defining a document's purpose, see Chapter 2, page 21.

Your purpose statement might say something like the following:

> The purpose of this description is to show how a fuel cell generates power.
>
> In this description, I will explain the basic features of the International Space Station.

You should be able to write your purpose statement in one sentence. If it goes beyond one sentence, you likely need to be more specific about what you are trying to achieve.

Need help defining the rhetorical situation? Go to
www.pearsonhighered.com/johnsonweb3/19.2

READERS   Technical descriptions tend to be written for readers who are unfamiliar with the subject. So, your job is to help them understand it by using terms and images that they will find familiar. You will also need to adjust the detail and complexity of your description to suit their specific interests and needs.

Primary readers (action takers) are readers who most need to understand your description. What information do they need to know to make a decision about your product, place, or process?

Secondary readers (advisors) will likely be experts in your area. They may be engineers, technicians, or scientists who are advising the primary readers about the strengths and weaknesses of the product, place, or process. How much technical detail and accuracy will these readers require to feel satisfied with your description?

Tertiary readers (evaluators) could include just about anyone who has an interest in the product, place, or process you are describing. Your technical description may be used by auditors, lawyers, reporters, or concerned citizens.

Gatekeeper readers (supervisors) may want to check your description for accuracy. Descriptions, especially specifications and observations, need to be exact. Your supervisors may want to review your materials for exactness and correctness.

Link

For more information on analyzing readers, see Chapter 3, page 40.

CONTEXT OF USE   Imagine the places where your description might be used. Will the description be embedded in a report or proposal? Will it be placed in a larger document's appendix to provide additional details? Will salespeople use the description to promote the product or service? Will the description be published in a magazine? Will it be used as a specification to establish the ideal measurements of the product or service?

Also, imagine your primary readers in a likely context, using your description. What physical, economic, ethical, and political factors will influence how they interpret the text?

Link

For more ideas about analyzing the context of use, see Chapter 3, page 46.

Addressing ISO 9000/ISO 14000 Issues

One important issue involving context of use is whether your technical description needs to conform to ISO 9000 or ISO 14000 standards. These voluntary standards are accepted internationally and developed by the International Organization for Standardization (ISO). ISO 9000 standards involve quality management systems, while ISO 14000 involve environmental management systems. Many high-tech companies, especially ones working for the U.S. government, follow these quality management and environmental management standards. Figure 19.3 shows an introduction to these standards drawn from the ISO website (www.iso.org).

The ISO standards cannot be discussed in sufficient depth here, but you should be aware that they exist. If your company follows ISO standards, any descriptions you write will need to reflect and conform to these standards.

Worksheets are available to help you define readers and contexts of use. Go to
www.pearsonhighered.com/johnsonweb3/19.3

Planning and Researching Technical Descriptions   523

ISO 9000 and ISO 14000

Certification is hard to achieve, but necessary in many fields.

Environmental management standards are becoming a great concern.

Source: International Organization for Standardization, http://www.iso.org/iso/iso_catalogue/management_standards/iso_9000_iso_14000.htm.

Figure 19.3: The ISO standards are crucial to maintaining quality and consistency in national and international manufacturing.

Researching

In most cases, doing research for a technical description is primarily experiential. In other words, you will likely need to personally observe the object, thing, or process you are describing. Here are some strategies that are especially applicable to writing descriptions.

DO BACKGROUND RESEARCH Before making direct observations, you should know as much as possible about your subject. On the Internet, use search engines to find as much information as you can. Then, collect print sources, like books, documents, and other literature. These materials should help you understand your subject better so you can make more informed observations.

USE YOUR SENSES While making direct observations, use all of your available senses. As much as possible, take notes about how something looks, sounds, smells, feels, and tastes. Pay special attention to colors and textures, because they will add depth and vividness to your description.

TAKE MEASUREMENTS When possible, measure qualities like height, width, depth, and weight. If you cannot gain exact measurements, make estimates. In some cases,

To learn more about ISO guidelines, go to
www.pearsonhighered.com/johnsonweb3/19.4

you can compare the subject to something of similar size (e.g., "The rough-legged hawk was about the size of a large crow").

DESCRIBE MOTION AND CHANGE   Pay attention to your subject's movements. Look for patterns and note places where your subject deviates from those patterns. Also, take special note of any changes in your subject. Over time, you may notice that it evolves or transforms in some way.

DESCRIBE THE CONTEXT   Take copious notes about the surroundings of the subject. Pay attention to how your subject acts or interacts with the things and people around it.

COLLECT VISUALS   If available, collect graphics that illustrate your subject, or create them yourself. You can make drawings or take pictures of your subject.

Link

For more information on doing research, go to Chapter 7, page 142.

ASK SUBJECT MATTER EXPERTS (SMEs)   If possible, find experts who can answer your questions. SMEs can help fill in any gaps in your understanding of your subject, while pointing out subtle features that you might miss.

While researching, collect as much information as you can. When you are finished researching, you should determine how much your readers already know about the subject and how much they need to know. You can then prioritize your notes to suit their needs.

Partitioning the Subject

To describe something, you need to fully understand it. You need to go beneath the visible surface to discover how it works and how it affects the people who use it.

- If you are describing a product or artifact, take it apart if you can. Then, look at the object from a variety of different angles.
- If you are describing a place, go there and look around. Pay attention to what makes this place different or unique. Spend some time observing how people interact with the place.
- If you are describing a process, note its larger stages. Pay close attention to how each stage leads to and affects the following stage.

Once you are familiar with the subject, you can start describing it in words and images. The first question you should ask is, "How can I partition it?" Or, to put the question more simply, "How can I break it down into its major features, functions, or stages?" Your answer to this question will determine your *partitioning strategy* for describing your subject.

By features—You might partition the subject by separately describing its parts or features. For example, a description of a computer might describe it part by part by first partitioning it into a monitor, keyboard, external disk drives, and a central processing unit (CPU).

By functions—You might partition the subject by noting how its different parts function. A description of the International Space Station, for example,

Want to learn how to closely observe things? Go to
www.pearsonhighered.com/johnsonweb3/19.5

Partitioning the Subject 525

Chris Peterson
CAD TECHNICIAN, LARON, INC., PHOENIX, AZ

Laron, Inc., is an engineering company that specializes in heavy equipment.

How does computer-aided drafting (CAD) help write descriptions?

"A picture is worth a thousand words." It just doesn't get any better than that. A drawing can show a potential client your expertise for his or her particular needs. With the newer 3D CAD programs, such as Autodesk's® Inventor® or SolidWorks®, you can produce awesome pictures. For video presentations, you can use CAD to walk your clients through anything, from their new home or new plant, to a tour through a new machine. When new video games come out, like PS2® or Xbox®, what I hear kids (including big kids) say is "the graphics are awesome." CAD is a way to make your descriptions "awesome."

In order to impress your clients, you must master the CAD programs as well as your product. You can have the greatest talent to produce a "mousetrap," but if you lack the skills to present it, you will look like all the others, or worse! You may also be the greatest illustrator, but lack of understanding of the product you are promoting will stick out like a sore thumb.

Another place CAD is very helpful is for "internal clients" such as machinists, welders, carpenters, and pipe fitters. Here again you must have a good knowledge of your discipline as well as the CAD program. There is no substitute for knowledge of what you are trying to convey to the client.

might partition it function by function into research, power generation, infrastructure, habitation, and docking sections.

By stages of a process—You could partition the subject chronologically by showing how it is assembled or how it works. A description of Hodgkin's disease, for example, might walk readers step-by-step through detection, diagnosis, staging, and remission stages.

At this point, logical mapping can help you break down (partition) your subject into major and minor parts (Figure 19.4).

To use logical mapping to help you describe something, follow these steps:

1. Put the name of your subject in the middle of your screen or a sheet of paper.

2. Write down the two to five major parts in the space around it.

3. Circle each major part.

4. Partition each major part into two to five minor parts.

Link

For more help using logical mapping, see Chapter 7, page 143.

For model descriptions that use each partitioning strategy, go to
www.pearsonhighered.com/johnsonweb3/19.6

Partitioning with Logical Mapping

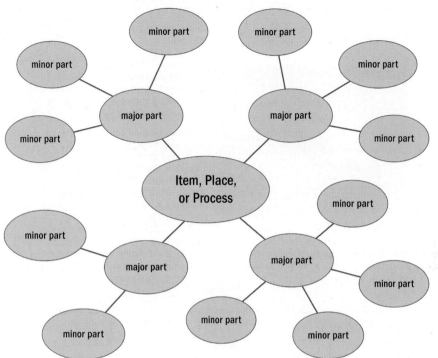

For example, in Figure 19.5, NASA's description of the Mars Exploration Rover partitions the subject into

Core structure

Suspension system

Navigation system

Instrument deployment device

Batteries

Radioscope heater

Computer

In this example, notice how the description carves the subject into its major parts. Then, each of these major parts is described in detail by paying attention to its minor parts.

Mars Exploration Rover

At the heart of each Mars Exploration Rover spacecraft is its rover. This is the mobile geological laboratory that will study the landing site and travel to examine selected rocks up close.

The Mars Exploration Rovers differ in many ways from their only predecessor, Mars Pathfinder's Sojourner Rover. Sojourner was about 65 centimeters (2 feet) long and weighed 10 kilograms (22 pounds). Each Mars Exploration Rover is 1.6 meter (5.2 feet) long and weighs 174 kilograms (384 pounds).

Sojourner traveled a total distance equal to the length of about one football field during its 12 weeks of activity on Mars. Each Mars Exploration Rover is expected to travel six to 10 times that distance during its three-month prime mission. Pathfinder's lander, not Sojourner, housed that mission's main telecommunications, camera and computer functions. The Mars Exploration Rovers carry equipment for those functions onboard and do not interact with their landers any further once they roll off.

On each Mars Exploration Rover, the core structure is made of composite honeycomb material insulated with a high-tech material called aerogel. This core body, called the warm electronics box, is topped with a triangular surface called the rover equipment deck. The deck is populated with three antennas, a camera mast and a panel of solar cells. Additional solar panels are connected by hinges to the edges of the triangle. The solar panels fold up to fit inside the lander for the trip to Mars, and deploy to form a total area of 1.3 square meters (14 square feet) of three-layer photovoltaic cells. Each layer is of different materials: gallium indium phosphorus, gallium arsenide and germanium. The array can produce nearly 900 watt-hours of energy per martian day, or sol. However, by the end of the 90-sol mission, the energy generating capability is reduced to about 600 watt-hours per sol because of accumulating dust and the change in season. The solar array repeatedly recharges two lithium-ion batteries inside the warm electronics box.

Doing sport utility vehicles one better, each rover is equipped with six-wheel drive. A rocker-bogie suspension system, which bends at its joints rather than using any springs, allows rolling over rocks bigger than the wheel diameter of 26 centimeters (10 inches). The distribution of mass on the vehicle is arranged so that the center of mass is near the pivot point of the rocker-bogie system. That enables the rover to tolerate a tilt of up to 45 degrees in any direction without overturning, although onboard comput-

Source: NASA, http://www.jpl.nasa.gov/news/press_kits/merlandings.pdf.

Figure 19.5: This description of the Mars Exploration Rover shows how a subject can be partitioned into major and minor parts.

Definition of subject

Overall description of subject

Major part: Core Structure

Minor parts

Major part: Suspension System

Minor parts →

ers are programmed to prevent tilts of more than 30 degrees. Independent steering of the front and rear wheels allows the rover to turn in place or drive in gradual arcs.

Major part:
Navigation
System →

The rover has navigation software and hazard-avoiding capabilities it can use to make its own way toward a destination identified to it in a daily set of commands. It can move at up to 5 centimeters (2 inches) per second on flat hard ground, but under automated control with hazard avoidance, it travels at an average speed about one-fifth of that.

Minor parts →

Two stereo pairs of hazard-identification cameras are mounted below the deck, one pair at the front of the rover and the other at the rear. Besides supporting automated navigation, the one on the front also provides imaging of what the rover's arm is doing. Two other stereo camera pairs sit high on a mast rising from the deck: the panoramic camera included as one of the science instruments, and a wider-angle, lower-resolution navigation camera pair. The mast also doubles as a periscope for another one of the science instruments, the miniature thermal emission spectrometer.

Major part:
Instrument
Deployment
Device →

The rest of the science instruments are at the end of an arm, called the "instrument deployment device," which tucks under the front of the rover while the vehicle is traveling. The arm extends forward when the rover is in position to examine a particular rock or patch of soil.

Major part:
Batteries →

Batteries and other components that are not designed to survive cold martian nights reside in the warm electronics box. Nighttime temperatures may fall as low as minus 105°C (minus 157°F). The batteries need to be kept above minus 20°C (minus 4°F) for when they are supplying power, and above 0°C (32°F) when being recharged. Heat in-

Minor parts →

side the warm electronics box comes from a combination of electrical heaters, eight radioisotope heater units and heat given off by electronics components.

Major part:
Radioisotope
Heater →

Each radioisotope heater unit produces about one watt of heat and contains about 2.7 grams (0.1 ounce) of plutonium dioxide as a pellet about the size and shape of the eraser on the end of a standard pencil. Each pellet is encapsulated in a metal cladding of platinum-rhodium alloy and surrounded by multiple layers of carbon-graphite composite material, making the complete unit about the size and shape of a C-cell battery.

Minor parts →

This design of multiple protective layers has been tested extensively, and the heater units are expected to contain their plutonium dioxide under a wide range of launch and orbital-reentry accident conditions. Other spacecraft, including Mars Pathfinder's Sojourner rover, have used radioisotope heater units to keep electronic systems warm and working.

Major part:
Computer →

The computer in each Mars Exploration Rover runs with a 32-bit Rad 6000 microprocessor, a radiation-hardened version of the PowerPC chip used in some models of Macintosh computers, operating at a speed of 20 million instructions per second. Onboard

Minor parts →

memory includes 128 megabytes of random access memory, augmented by 256 megabytes of flash memory and smaller amounts of other non-volatile memory, which allows the system to retain data even without power.

Organizing and Drafting Technical Descriptions

With your subject partitioned into major and minor features, you are ready to start organizing and drafting your description. There are many different ways to describe your subject, but it is best to follow an organizational pattern that demonstrates an obvious logic that readers will immediately recognize. Figure 19.6 shows a basic pattern that might be followed.

Specific and Precise Title

The title for your description can be long or short. In most cases, a concise title like "Mars Exploration Rover" would be fine. However, this title could suit a variety of documents about the rover, including a report, a proposal, or a magazine article. If you want your title to clearly identify the purpose of the document, you might write something more exact, such as

> Description of the Mars Exploration Rover

Here are some other sample titles:

> Specifications for the XC-9000 Microprocessor
>
> How Does a Fuel Cell Work?
>
> Lung Cancer: Profile of a Killer

A good title for a description is a title that cannot be used for anything else.

Introduction with an Overall Description

Like any document, your description should begin with an introduction that sets a framework, or context, for the rest of the document (Figure 19.6). Typically, the introduction will include some or all of the following features:

DEFINITION OF SUBJECT   Often, descriptions begin with a sentence definition of the subject. A sentence definition includes three parts: the term, the class in which the subject belongs, and the characteristics that distinguish the subject from the other members in its class.

> The International Space Station is a multinational research facility that will house six state-of-the-art laboratories in orbit.

Link

For more information on writing definitions, see Chapter 18, page 496.

> Hodgkin's disease is a type of cancer that starts in the lymph nodes and other organs that are the body's system for making blood and protecting against germs.

The definition of the subject should appear early in the introduction, preferably in the first sentence.

Possible Outline for a Description or Specification

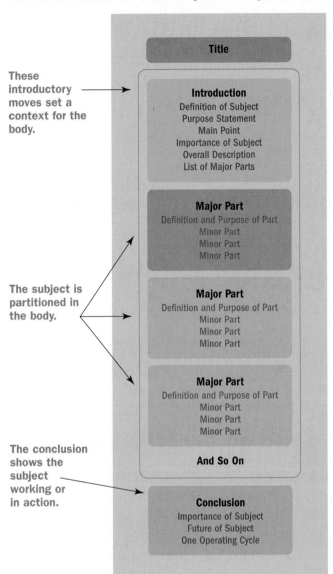

Figure 19.6:
A generic outline of a technical description. The document includes an introduction, body, and conclusion. The body partitions the subject.

These introductory moves set a context for the body.

Title

Introduction
Definition of Subject
Purpose Statement
Main Point
Importance of Subject
Overall Description
List of Major Parts

Major Part
Definition and Purpose of Part
Minor Part
Minor Part
Minor Part

The subject is partitioned in the body.

Major Part
Definition and Purpose of Part
Minor Part
Minor Part
Minor Part

Major Part
Definition and Purpose of Part
Minor Part
Minor Part
Minor Part

And So On

The conclusion shows the subject working or in action.

Conclusion
Importance of Subject
Future of Subject
One Operating Cycle

PURPOSE STATEMENT Directly or indirectly state that you are describing something.

> This description of the International Space Station will explain its major features, highlighting its research capabilities.

> In this article, we will try to demystify Hodgkin's disease, so you can better understand its diagnosis and treatment.

MAIN POINT Give your readers an overall claim that your description will support or prove.

> Building the International Space Station is an incredible engineering feat that will challenge the best scientists and engineers in many different nations.

> To fight Hodgkin's disease, you first need to understand it.

IMPORTANCE OF THE SUBJECT Your readers may or may not be familiar with the object, place, or process you are describing. For readers who are unfamiliar with it, you might want to include a sentence or paragraph that stresses the importance of your subject.

> The ISS, when completed, will provide scientists and other researchers an excellent platform from which to study space.

> Hodgkin's disease is one of the most acute forms of cancer, and it needs to be aggressively treated.

OVERALL DESCRIPTION OF THE SUBJECT Usually, descriptions start by offering an overall look at the item being described.

> From a distance, the International Space Station looks like a large collection of white tubes with two rectangular solar panels jutting out like ears from its side.

> Hodgkin's disease spreads through the lymphatic vessels to other lymph nodes. It enlarges the lymphatic tissue, often putting pressure on vital organs and other important parts of the body.

An overall description should give your readers an image of the subject as a whole. When you begin describing the smaller features of the subject, this whole image will help your readers visualize how the parts fit together.

LIST OF THE MAJOR FEATURES, FUNCTIONS, OR STAGES In many descriptions, especially longer descriptions, the introduction will list the major features, functions, or stages of the subject.

Link

For more information on writing introductions, see Chapter 8, page 182.

> The International Space Station includes five main features: modules, nodes, trusses, solar power arrays, and thermal radiators.

> Once Hodgkin's has been detected, doctors will usually (1) determine the stage of the cancer, (2) offer treatment options, and (3) make a plan for remission.

You can then use this partitioning scheme to organize the body of your description.

To see some interesting descriptions from NASA, go to
www.pearsonhighered.com/johnsonweb3/19.8

Description by Features, Functions, or Stages in a Process

The body of your description will be devoted to describing the subject's features, functions, or stages.

Address each major part separately, defining it and describing it in detail. Within your description of each major part, identify and describe the minor parts.

Definition of major part ——→ **Modules** are pressurized cylinders of the habitable space on board the Station. They may

Minor parts ——→ contain research facilities, living quarters, and any vehicle operational systems and equipment the astronauts may need to access.

If necessary, each of these minor parts could then be described separately. In fact, you could extend your description endlessly, teasing out the smaller and smaller features of the subject.

Figure 19.7 shows a description of a subject by "stages in a process." In this description of a fuel cell, the author walks readers through the energy generation process, showing them step-by-step how the fuel cell works.

Description by Senses, Similes, Analogies, and Metaphors

The key to a successful technical description is the use of vivid detail to bring your subject to life—to make it seem real. A well-written description helps the readers visualize the thing, place, or process you are illustrating for them. To add this level of detail, you might consider using some of the following techniques:

DESCRIPTION THROUGH SENSES   Humans experience the world through their five senses: sight, hearing, smell, touch, and taste. Consider each of these senses separately, asking yourself, "How does it look?" "How does it sound?" "How does it smell?" "How does it feel?" and "How does it taste?"

> A visit to a Japanese car manufacturing plant can be an overwhelming experience. Workers in blue jumpsuits seem to be in constant motion. Cars of every color—green, yellow, red—are moving down the assembly line with workers hopping in and out. The smell of welding is in the air, and you can hear the whining hum of robots at work somewhere else in the plant.

SIMILES   A simile describes something by comparing it to something familiar to the readers ("A is like B").

> The mixed-waste landfill at Sandia Labs is like a football field with tons of toxic chemical and nuclear waste buried underneath it.

Similes are especially helpful for nonexpert readers, because they make the unfamiliar seem familiar.

ANALOGIES   Analogies are like similes, but they work on two parallel levels ("A is to B as C is to D").

> Circuits on a semiconductor wafer are like tiny interconnected roads crisscrossing a city's downtown.

For more advice on using the senses, go to
www.pearsonhighered.com/johnsonweb3/19.9

A Technical Description: Stages in a Process

Definition of subject

Overall description of subject

Steps in the process

Numbers refer to diagram.

Subject is shown at work in the conclusion.

Look to the future

Fuel Cell Technology
How It Works

A fuel cell produces electricity by means of an electro-chemical reaction much like a battery. But there is an important difference. Rather than extracting the chemical reactants from the plates inside the cells, a fuel cell uses hydrogen fuel and oxygen extracted from the air to produce electricity. As long as these substances are fed into the fuel cell, it will continue to generate electric power.

Different types of fuel cells work with different electrochemical reactions. The following is a basic description of how a phosphoric acid fuel cell generates electric power.

1. Hydrogen gas is extracted from natural gas or other hydrocarbon fuels and permeates the anode. Oxygen from the air permeates the cathode.

2. Aided by a catalyst in the anode, electrons are stripped from the hydrogen. Hydrogen ions pass into the electrolyte.

3. Electrons cannot enter the electrolyte. They travel through an external circuit, producing electricity.

4. Electrons travel back to the cathode where they combine with hydrogen ions and oxygen to form water.

A fuel cell provides DC (direct current) voltage that can be used to power motors, lights or other electrical appliances. To supply electricity for homes, businesses, and buildings, however, the direct current must be changed into AC (alternating current). A device called an "inverter" makes this conversion.

Hydrogen needed by a fuel cell can be extracted from a variety of fuels. Natural gas—a chemical combination of carbon and hydrogen atoms—is perhaps the most common fuel, but other hydrocarbon fuels can also be used. For example, some fuel cells operate on gases released from wastewater digesters or from landfills. In the future, gas made from coal or biomass might be candidate fuels. Some types of fuel cells extract the hydrogen in a separate fuel processor called a "reformer"; other fuel cells incorporate reforming inside the cell stack itself.

Diagram illustrates the process.

Figure 19.7: A description of a process. In this description the subject has been partitioned into major and minor stages.

Source: U.S. Department of Energy, http://www.fe.doe.gov/coal_power/fuelcells/fuelcells_howitworks.shtml.

METAPHORS   Metaphors are used to present an image of the subject by equating two different things ("A is B"). For example, consider these two common metaphors:

Link

For more information on using similes, analogies, and metaphors, see Chapter 9, page 223.

> The heart is a pump: it has valves and chambers, and it pushes fluids through a circulation system of pipes called arteries and veins.
>
> Ants live in colonies: a colony will have a queen, soldiers, workers, and slaves.

The use of senses, similes, analogies, and metaphors will make your description richer and more vivid. Readers who are unfamiliar with your subject will especially benefit from these techniques, because concepts they understand are being used to describe things they don't understand.

Conclusion

The conclusion of a technical description should be short and concise. Conclusions often portray one working cycle of the object, place, or process.

> When the International Space Station is completed, it will be a center of activity. Researchers will be conducting experiments. Astronomers will study the stars. Astronauts will be working, sleeping, exercising, and relaxing. The solar power arrays will pump energy into the station, keeping it powered up and running.
>
> Fighting Hodgkin's disease is difficult but not impossible. After detection and diagnosis, you and your doctors will work out treatment options and staging objectives. If treatment is successful, remission can continue indefinitely.

In the conclusion, you are putting your subject in motion, showing your readers how it works when it is operating.

Using Style in Technical Descriptions

Most technical descriptions are written in the plain style. However, there are times when you want your description to be more persuasive, such as in sales literature or a proposal. Here are some suggestions for using style in descriptions:

Keep your words simple—The words you use should be familiar to your readers. If they are not, you should define any words they may not know. Jargon should be avoided with nonexpert readers.

Keep sentences short, within breathing length—Sentences that are too long only cloud your readers' abilities to visualize the subject. Your readers should be able to comfortably read each sentence in one breath.

Use subject alignment and given/new techniques to weave sentences together—Aligning your subjects and the given/new will make your text smoother and easier to follow.

To see examples of similes, analogies, and metaphors in descriptions, go to
www.pearsonhighered.com/johnsonweb3/19.10

Using Style in Technical Descriptions   535

Link

To learn more about subject alignment and given/new techniques, turn to Chapter 9, page 218.

Subject Alignment

Here, subject alignment is being used to keep a consistent perspective.

<u>RFID</u> stands for **R**adio **F**requency **ID**entification, a technology that uses tiny computer chips smaller than a grain of sand to track items at a distance. <u>RFID "spy chips"</u> have been hidden in the packaging of Gillette razor products and in other products you might buy at a local Wal-Mart, Target, or Tesco. <u>Each tiny chip</u> is hooked up to an antenna that picks up electromagnetic energy beamed at it from a reader device. When it picks up the energy, <u>the chip</u> sends back its unique identification number to the reader device, allowing the item to be remotely identified. <u>Spy chips</u> can beam back information anywhere from a couple of inches to up to 20 or 30 feet away.

Source: CASPIAN, www.spychips.com/what-is-rfid.html

Given/New

The given/new technique links sentences together by using something from the previous sentence early in the next sentence.

The golden eagle is one of the <u>larger raptors</u> found in the Midwestern United States. <u>Its large size</u> makes this eagle almost unmistakable, and you will often see it <u>perched</u> on fence posts and telephone poles. <u>From its perch,</u> the eagle will swoop down to seize an unsuspecting rodent or small mammal.

Use the senses to add color, texture, taste, sound, and smell—Use all the senses, not just sight, to add detail to your description. The style of your document will improve if you give readers the full sensory experience.

Use similes, analogies, and metaphors to add a visual dimension—These techniques will help you give your readers an overall image of the subject.

In most cases, the best style for a technical description is an unnoticed style. Your goal is to use plain style techniques to improve your readers' understanding.

Designing Technical Descriptions

The design of your technical description will depend on the context in which it will be used. A description used in your company's sales literature, for example, will usually be more colorful than a technical specification kept in your company's files. So, as you consider the design of your description, think carefully about where and how the document will be used.

Designing a Page Layout

The page layout of your description likely depends on your company's previous documentation or established style. If you are given a free hand to design the text, you might consider

Link

For more information on designing documents, see Chapter 10, page 232.

- using a two-, three-, or four-column format.
- using lists to highlight minor parts of an object, place, or process.
- using a sidebar to focus on a particular part or function of the subject.
- using headings to show the organization of the content.
- placing data or measurements in a table.

You don't need to restrict yourself to a simple one-column format. Use your imagination. Figure 19.8, for example, shows how columns and tables can be used to pack

To learn more about improving your style, go to
www.pearsonhighered.com/johnsonweb3/19.11

A Specification

The use of reverse type highlights the main point.

Photo shows product.

This table highlights the features of the product.

OUR MOST POPULAR HANDHELD IS BETTER THAN EVER.

TUNGSTEN | *E2*

Affordable and packed with productivity for wherever business takes you.

Wherever your life is going, this is the perfect handheld to take along for the ride. The sleek, stylishly designed Tungsten™ E2 handheld from palmOne brings your entire world to life. Your calendar, contacts, documents, presentations, photos and videos look sharper and more colorful. But just as important, its new flash memory keeps all that information safe—even if you don't have time to recharge.

Of course, it wouldn't be a palmOne handheld if we didn't push the boundaries of what a handheld can do. That's why we included Bluetooth® wireless technology for wireless connectivity. Now working and communicating will be that much easier. Taking work to go? View and even edit spreadsheets and word processing documents right on your handheld. Plus you can sync your calendar and contacts from Outlook! And don't forget to grab your MP3 tunes? You're going places.

palmOne

KEY FEATURES

Brighter, richer color display	See your information clearly indoors and out. Brighter display, better color saturation brings photos and videos to life.
Non-volatile, flash memory	There's more than enough room to hold your calendar, contacts, applications, photos, and even your spreadsheets or presentations. And because it's flash memory, the information on your handheld is protected—even if you're on the go and don't have time to recharge.
Built-In Bluetooth®	Stay connected. With built-in Bluetooth® wireless technology, you can synchronize with your desktop without wires getting in the way.² Use your Tungsten™ E2 handheld with a compatible phone to send email and text messages, or to check news headlines on the Web?
Documents To Go®	Productivity in your pocket. The Tungsten™ E2 comes with Documents To Go, which lets you carry Word, Excel, and PowerPoint files-so you can be more productive wherever you are. And with a simple conversion step, you can even view Acrobat PDFs.⁵

1 Windows compatible only.
2 Expansion card required for MP3s. Sold separately.
3 Requires a compatible Bluetooth enabled Windows computer.
4 Works within wireless service coverage area only and requires compatible GSM mobile phone with Bluetooth wireless technology, data service from a mobile service provider and an ISP, sold separately.
5 PDFs require conversion step.

Figure 19.8: The design of this specification allows readers to quickly gain access to the information they need. Notice how information is presented in easy-to-access blocks.

The two-column design makes the text easy to scan.

Company name and logo are framed with white space.

Source: Palm, Inc.

(continued)

These annotations help readers understand the parts of the product.

This table lists the specifications of the product.

The layout balances the design features of the page.

a solid amount of information into one place, while still presenting the information in an attractive way.

Using Graphics

Graphics are helpful in technical descriptions. Pictures, illustrations, and diagrams help the readers visualize your subject and its parts.

Using your computer, you can collect or create a wide range of graphics. Many free-use graphics are widely available on the Internet. (Reminder: Unless the site specifies that the graphics are free to use, you must ask permission to use them.) If you cannot find graphics on the Internet, you can use a digital camera to take photographs that can be downloaded into your text. You can also use a scanner to digitize pictures, illustrations, and diagrams.

Here are some guidelines for using graphics in a technical description:

Link

For more information on using graphics, see Chapter 11, page 275.

- If possible, use a title and number with each graphic so it can be referred to in the written text.
- Include captions that explain what the graphic shows.
- Label specific features in the graphic.
- Reference the graphic in the written text; e.g., "As shown in Figure 3 . . ." or "(See Figure 3)."
- Place the graphic on the page where it is referenced or soon afterward.

It is not always possible to include titles, numbers, and captions with your graphics. In these situations, all graphics should appear immediately next to or immediately after the places where they are discussed.

Even simple graphics are helpful in technical descriptions. In almost all cases, you should plan to include some kind of picture, illustration, or diagram to support your written text.

As we move into a visual age, readers will expect technical descriptions to be more and more visual. For example, Scott McCloud, a well-known graphic artist, was brought in to illustrate the technical description of Google's Internet browser, Chrome (Figure 19.9).

Revising, Editing, and Proofreading

You should save plenty of time to revise and edit your description. Often, what seems obvious or straightforward to you will not be obvious to your readers. Revising and editing will help you locate and clarify unclear parts of your description.

Revising for Conciseness

At this point, you should look for ways to make your description more concise. You should

- comb through the text, looking for places where you go beyond need-to-know information.
- find places where you have included too many details for your readers. Cut out this excessive information.
- shorten any long sentences to make them easier to read.

Need graphics? Go to
www.pearsonhighered.com/johnsonweb3/19.13

Revising, Editing, and Proofreading

539

Technical Descriptions Becoming More Visual

Figure 19.9: As our culture becomes more visual, even technical descriptions need to be as visual as possible. Scott McCloud drew these images for Google's Chrome browser.

Using Digital Photography in Descriptions

Digital cameras and scanners offer an easy way to insert visuals into your descriptions (Figure A). These digitized pictures are inexpensive, alterable, and easily added to a text. Moreover, they work well in print and on-screen texts. Here are some photography basics to help you use a camera more effectively.

Resolution—Digital cameras will usually allow you to set the resolution at which you shoot pictures. If you are using your camera to put pictures on the web, you should use the 640 × 480 pixel setting. This setting will allow the picture to be downloaded quickly, because the file is smaller. If you want your picture to be a printed photograph, a minimum 1280 × 1024 pixel setting is probably needed. Online pictures are usually best saved in jpeg format, while print photos should be saved as tiff files.

ISO sensitivity—ISO is the amount of light the camera picks up. A high ISO number like 400 is good for shooting in dark locations or snapping something that is moving quickly. A low ISO number like 100 produces a higher-quality shot. You might leave your camera's ISO setting on "auto" unless you need to make adjustments.

Shutter speed—Some digital cameras and conventional cameras will allow you to adjust your shutter speed. The shutter speed determines how much light is allowed into the camera when you push the button. Shutter speeds are usually listed from 1/1000 of a second to 1 second. Slower speeds (like 1 second) capture more detail but risk blurring the image, especially if your subject is moving. Faster speeds (like 1/1000 second) will not capture as much detail, but they are good for moving subjects.

Cropping—Once you have downloaded your picture to your computer, you can "crop" the picture to remove things you don't want. Do you want to remove an old roommate from your college pictures? You can use the cropping tool to cut him or her out of the picture. Most word processors have a cropping tool that you can use to carve away unwanted parts of your pictures (Figure A).

Retouching—One of the main advantages of digital photographs is the ease with which they can be touched up. Professional photographers make ample use of programs like Adobe Photoshop to manipulate their photographs. If the picture is too dark, you can lighten it up. If the people in the photo have "red-eye," you can remove that unwanted demonic stare.

(continued)

Cropping a Digital Photograph

Figure A: Using the cropping tool, you can focus the photograph on the subject. Or, you can eliminate things or people you don't want in the picture.

The cropping tool lets you frame the part of the picture you want.

The toolbar offers a variety of options for altering the picture.

One of the nice things about digital photography is that you can make any photograph look professional. With a digital camera, since the "bullets are free," as photographers say, you can experiment freely.

When revising, try to boil your subject down to the fewest words possible, while still meeting the purpose of the description.

Editing and Proofreading for Accuracy

Above all, descriptions need to be accurate. As you are editing your description,

- check the accuracy of any figures.
- double-check measurements for exactness.
- confirm units of measurement.
- look up any words that you are not completely sure about.

Proofreading is always important, but it is even more important if your description will be used as a sales tool or in a proposal. A typo or misspelling might be forgivable in your company's in-house materials, but your company's reputation is at stake when written materials are being shown to customers or clients. You want these materials to be error free.

Worksheets to help you revise your descriptions are available at **www.pearsonhighered.com/johnsonweb3/19.15**

You should show your description to your supervisor or colleagues before you finish it. Your familiarity with the subject might cause you to overlook problems in your text. So, letting someone with a fresh perspective look over your materials might help you locate trouble spots or errors.

- Technical descriptions are written to describe objects, places, phenomena, and processes. They are important documents in all technical workplaces.

- Basic features of a technical description include a title, introduction, body, graphics, and conclusion.

- An object, place, or process can be partitioned according to its features, functions, or stages.

- Technical descriptions tend to be written in the "plain style," meaning words and sentences are kept simple, direct, and concise.

- To add a visual element to the description, use your senses, similes, analogies, and metaphors to describe the subject.

- Graphics are crucial in technical descriptions, because the purpose of the description is to allow the readers to "see" the object, place, or process. You can use pictures, illustrations, and diagrams to add graphics to your text.

- The design of the description will depend on how it is being used. In sales literature, the design will probably be colorful or ornate. A specification for the company's files, on the other hand, might be rather plainly designed.

Individual or Team Projects

1. Find a technical description on the Internet or in your workplace or home. First, determine the rhetorical situation (subject, purpose, readers, context) for which the description was written. Then, study its content, organization, style, and design. Write a two-page memo to your instructor in which you offer a critique of the description. What do you find effective about the description? What could be improved?

2. Your company sells a variety of different products, listed below. Choose one of these items and write a one-page technical description. Your description should be aimed at a potential customer who might purchase one of these products:

   plasma-screen television
   DVD player
   MP3 player
   bicycle

clock radio
telescope
washing machine
baby stroller
toaster
coffeemaker
video camera

3. Professionals and researchers in your field of study (major) likely use tools, machines, or processes that are unfamiliar to nonexperts. Choose one of these items and write a technical description for an audience of nonexperts. In your introduction, make sure you define the subject you are describing and stress its importance to your field.

4. Pick a building on your campus and write a technical description of it. Before drafting your description, though, identify some readers (e.g., new students, visitors, alumni) who might actually have a use for your description. Then, write and design your description in a way that suits their needs.

5. Find a common process that you can describe. Then, describe that process, walking your readers through its stages. In your description, you should define any jargon or technical terms that may be unfamiliar to your readers.

6. Study the description for the Segway or Mars Exploration Rover in this chapter. Write a two-page memo critiquing the description you choose. Do you think the description includes enough information? How might its content, organization, style, and design be improved?

Collaborative Projects

Your group has been assigned to describe a variety of renewable energy sources that might be used in your state. These energy sources might include solar, wind, geothermal, biomass generators, and fuel cells. While keeping the energy needs and limitations of your region in mind, offer a brief description of each of these renewable energy sources, showing how it works, its advantages, and its disadvantages.

In a report to your state's energy commissioner, describe these energy sources and discuss whether you think they offer possible alternatives to nonrenewable energy sources.

For additional technical writing resources, including interactive sample documents, document design tutorials and guidelines, and more, go to **www.mytechcommlab.com.**

Sample descriptions are available at
www.pearsonhighered.com/johnsonweb3/19.16

Revision Challenge

Figure A shows a fact sheet from the Occupational Safety and Health Administration (OSHA). The description of flooding and flooding cleanup in this document is fine for an office environment. However, the size of the document makes it not particularly portable into areas that have been flooded.

Revise and redesign this document so that it fits on a 3×5 inch card that will be given out to first responders. This "Quickcard" will be easier to carry, and it can be stored in pockets and small storage areas. It could also be laminated, so it would hold up to severe conditions, like those found in flood zones.

Using the principles discussed in this chapter, analyze the content, organization, style, and design of this document. Then, revise this fact sheet so that it would be suitable for the kinds of emergency situations in which it would be used.

To help you revise, you can download a version of this document at www.pearsonhighered.com/johnsonweb3/19.17.

A Description That Could Be Revised

OSHA FactSheet

Flood Cleanup

Flooding can cause the disruption of water purification and sewage disposal systems, overflowing of toxic waste sites, and dislodgement of chemicals previously stored above ground. Although most floods do not cause serious outbreaks of infectious disease or chemical poisonings, they can cause sickness in workers and others who come in contact with contaminated floodwater. In addition, flooded areas may contain electrical or fire hazards connected with downed power lines.

Floodwater

Floodwater often contains infectious organisms, including intestinal bacteria such as E. coli, Salmonella, and Shigella; Hepatitis A Virus; and agents of typhoid, paratyphoid and tetanus. The signs and symptoms experienced by the victims of waterborne microorganisms are similar, even though they are caused by different pathogens. These symptoms include nausea, vomiting, diarrhea, abdominal cramps, muscle aches, and fever. Most cases of sickness associated with flood conditions are brought about by ingesting contaminated food or water. Tetanus, however, can be acquired from contaminated soil or water entering broken areas of the skin, such as cuts, abrasions, or puncture wounds. Tetanus is an infectious disease that affects the nervous system and causes severe muscle spasms, known as lockjaw. The symptoms may appear weeks after exposure and may begin as a headache, but later develop into difficulty swallowing or opening the jaw.

Floodwaters also may be contaminated by agricultural or industrial chemicals or by hazardous agents present at flooded hazardous waste sites. Flood cleanup crew members who must work near flooded industrial sites also may be exposed to chemically contaminated floodwater. Although different chemicals cause different health effects, the signs and symptoms most frequently associated with chemical poisoning are headaches, skin rashes, dizziness, nausea, excitability, weakness, and fatigue.

Pools of standing or stagnant water become breeding grounds for mosquitoes, increasing the risk of encephalitis, West Nile virus or other mosquito-borne diseases. The presence of wild animals in populated areas increases the risk of diseases caused by animal bites (e.g., rabies) as well as diseases carried by fleas and ticks.

Protect Yourself

After a major flood, it is often difficult to maintain good hygiene during cleanup operations. To avoid waterborne disease, it is important to wash your hands with soap and clean, running water, especially before work breaks, meal breaks, and at the end of the work shift. Workers should assume that any water in flooded or surrounding areas is not safe unless the local or state authorities have specifically declared it to be safe. If no safe water supply is available for washing, use bottled water, water that has been boiled for at least 10 minutes or chemically disinfected water. (To disinfect water, use 5 drops of liquid household bleach to each gallon of water and let it sit for at least 30 minutes for disinfection to be completed.) Water storage containers should be rinsed periodically with a household bleach solution.

If water is suspected of being contaminated with hazardous chemicals, cleanup workers may need to wear special chemical resistant outer clothing and protective goggles. Before entering a contaminated area that has been flooded, you should don plastic or rubber gloves, boots, and other protective clothing needed to avoid contact with floodwater.

Source: Occupational Safety and Health Administration
www.osha.gov/OshDoc/data_Hurricane_Facts/floodcleanup.pdf.

Figure A:
This fact sheet is somewhat long-winded and not easy to access in an emergency situation. Try turning it into a "Quickcard" that fits on a 3 × 5 inch card. You can use both sides of the card.

To download the text for revision, go to
www.pearsonhighered.com/johnsonweb3/19.17

Decrease the risk of mosquito and other insect bites by wearing long-sleeved shirts, long pants, and by using insect repellants. Wash your hands with soap and water that has been boiled or disinfected before preparing or eating foods, after using the bathroom, after participating in flood cleanup activities, and after handling articles contaminated by floodwater. In addition, children should not be allowed to play in floodwater or with toys that have been in contact with floodwater. Toys should be disinfected.

What to Do If Symptoms Develop

If a cleanup worker experiences any of the signs or symptoms listed above, appropriate first aid treatment and medical advice should be sought. If the skin is broken, particularly with a puncture wound or a wound that comes into contact with potentially contaminated material, a tetanus vaccination may be needed if it has been five years or more since the individual's last tetanus shot.

Tips to Remember

- Before working in flooded areas, be sure that your tetanus shot is current (given within the last 10 years). Wounds that are associated with a flood should be evaluated for risk; a physician may recommend a tetanus immunization.

- Consider all water unsafe until local authorities announce that the public water supply is safe.

- Do not use contaminated water to wash and prepare food, brush your teeth, wash dishes, or make ice.

- Keep an adequate supply of safe water available for washing and potable water for drinking.

- Be alert for chemically contaminated floodwater at industrial sites.

- Use extreme caution with potential chemical and electric hazards, which have great potential for fires and explosions. Floods have the strength to move and/or bury hazardous waste and chemical containers far from their normal storage places, creating a risk for those who come into contact with them. Any chemical hazards, such as a propane tank, should be handled by the fire department or police.

- If the safety of a food or beverage is questionable, throw it out.

- Seek immediate medical care for all animal bites.

This is one in a series of informational fact sheets highlighting OSHA programs, policies or standards. It does not impose any new compliance requirements. For a comprehensive list of compliance requirements of OSHA standards or regulations, refer to Title 29 of the Code of Federal Regulations. This information will be made available to sensory impaired individuals upon request. The voice phone is (202) 693-1999; teletypewriter (TTY) number: (877) 889-5627.

For more complete information:

OSHA Occupational
Safety and Health
Administration

U.S. Department of Labor
www.osha.gov
(800) 321-OSHA

DSTM 9/2005

In the Vapor

Linda Galhardy, a computer engineer at Gink, rushed to finish her technical description of FlashTime, a new handheld video game console her team had developed. She and her team had suffered through two long months of 18-hour days with no weekends off in order to get to this point. They were exhausted.

But, they succeeded in developing a prototype of the console, and they had created a couple innovative games. The purpose of Linda's technical description was to update Gink's management on the project and persuade them to begin planning for production. If everything worked out, the game console could be released in October.

This mad rush all started when Gink's management read on gaming blogs that CrisMark, Gink's main competitor, was only six months away from developing a new handheld gaming device that would revolutionize the market. According to the bloggers, CrisMark's console had a small screen, much like a Nintendo DS, but it also allowed the gamer to project video onto any blank wall. According to reports, CrisMark had figured out a way to keep that video from shaking or turning with the controller. If true, this handheld game would revolutionize the market and take away a significant chunk of Gink's marketshare.

Fortunately, Gink had some smart engineers like Linda, and it had a reputation for more reliable and innovative consoles and games. Consumers tempted to buy CrisMark's console would probably be willing to wait a couple extra months for a similar console from Gink. Gink's engineers and designers would have time to catch up.

Gink's management was impressed with Linda's presentation, including her technical description and the prototype. The marketing division was brought in immediately to begin working up the advertising and public relations campaign. Linda was congratulated and told to wait for management's answer.

A week later, Linda's supervisor, Thomas Hale, sent her an instant message with a link to a popular consumer electronics blog. On the front page of the blog was a picture of Linda's prototype and a story taken from "leaked" sources. The story was surprisingly accurate.

She was furious, but she didn't know whom to blame. Did the marketing department leak the information? A week later, several consumer electronics blogs reported that CrisMark was abandoning its attempt to develop their revolutionary new handheld console. Anonymous sources said CrisMark abandoned the project when they heard Gink was ahead of them in developing the new product.

Soon afterward, Linda's boss, Thomas, called her in his office and told her the FlashTime project was being "slow-tracked."

"You mean we're being killed slowly," Linda said angrily.

"Well," replied Thomas, "we need to make it look like we weren't just floating vaporware. So, you and a couple designers are going to stay on the project part-time. A year from now, we'll quietly pull the plug."

"That's garbage. We can do this, Thomas!" Linda said loudly. "We can make this console!"

Thomas said in a reassuring voice, "This thing was never going to get made anyway, Linda. Management just wanted to get something into the blogs to head off CrisMark. It worked. As soon as our 'leaked' story hit the blogs, CrisMark's stock dropped like a rock. They had to cancel the project to concentrate on their current products. Management just wanted to scare them off."

Linda couldn't believe what she was hearing. "So all that work was just to create vaporware to scare off CrisMark?"

Thomas nodded. "Yeah, and it worked. Congratulations."

If you were Linda, what would you do at this point? Would you play along? Would you try to convince management to continue the project? Would you leak the truth to the blogs? Would you strike out on your own?

CHAPTER
20

Instructions and Documentation

In this chapter, you will learn:

- The importance of instructions, specifications, and procedures in the technical workplace.

- The basic features of instructions, specifications, and procedures.

- How to plan and research instructions, specifications, and procedures.

- About the needs of cross-cultural readers.

- How to organize and draft instructions, specifications, and procedures.

- How style and design can be used to highlight and reinforce written text.

- Strategies for revising, editing, and proofreading instructions, specifications, and procedures.

More than likely, you have read and used countless sets of instructions in your lifetime. Instructions are packaged with the products we buy, such as phones, cameras, and televisions. In the technical workplace, documentation helps people complete simple and complex tasks, such as downloading software, building an airplane engine, drawing blood from a patient, and assembling a computer motherboard.

Instructions and other kinds of documentation are among the least noticed but most important documents in the technical workplace. Three types of documentation are commonly written and used in this arena:

Instructions—Instructions describe how to perform a specific task. They typically describe how to assemble a product or do something step-by-step.

Specifications—Engineers and technicians write specifications (often called the "specs") to describe in exact detail how a product is assembled or how a routine process is completed.

Procedures/Protocols—Procedures and protocols are written to ensure consistency and quality in a workplace. In hospitals, for example, doctors and nurses are often asked to write procedures that describe how to handle emergency situations or care for a specific injury or illness. Similarly, scientists will use protocols to ensure consistent methods in the laboratory.

Good Documentation Is Essential in the Technical Workplace

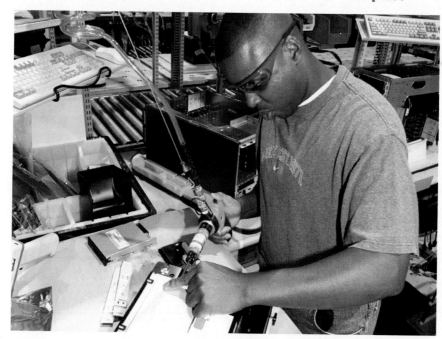

Instructions, procedures, and specifications are important, though often unnoticed, documents.

Link

For more information on writing descriptions, go to Chapter 19, page 530.

To avoid confusion in this chapter, the word *documentation* will be used as general term to mean instructions, procedures, and specifications. When the chapter discusses issues that are specific to instructions, procedures, or specifications, those terms will be used.

Basic Features of Documentation

Documentation tends to follow a consistent step-by-step pattern, whether you are describing how to make coffee or how to assemble an automobile engine (Figure 20.1).

Here are the basic features of most forms of documentation:

- specific and precise title
- introduction with background information
- list of materials, parts, tools, and conditions required

Basic Organization for Documentation

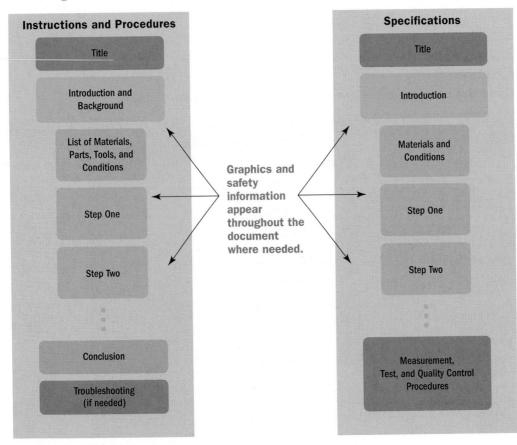

Instructions and Procedures

- Title
- Introduction and Background
- List of Materials, Parts, Tools, and Conditions
- Step One
- Step Two
- Conclusion
- Troubleshooting (if needed)

Graphics and safety information appear throughout the document where needed.

Specifications

- Title
- Introduction
- Materials and Conditions
- Step One
- Step Two
- Measurement, Test, and Quality Control Procedures

Figure 20.1: Instructions, procedures, and specifications use some or all of these parts.

- sequentially ordered steps
- graphics
- safety information
- measurement, test, and quality control procedures (for specifications)
- conclusion that signals the completion of the task

The content, style, and design of your documentation will change to suit the readers and the contexts in which the text will be used. For example, Figure 20.2 shows survival instructions from an entertaining and useful book called *Worst-Case Scenarios*. These instructions use simple text and visuals to explain how to jump out of a moving car—in case you ever need to.

A Set of Instructions

Specific and precise title

Brief introduction

Additional notes to clarify steps

Figure 20.2: Here is a rather simple set of instructions. Notice how the steps make up the bulk of the document.

Graphic visually reinforces steps.

Source: From the Worst-Case Scenario Survival Handbook™ *by Joshua Piven and David Borgenicht. Copyright © 1999 by Quirk Productions, Inc. Used with permission of Chronicle Books LLC, San Francisco. Visit ChronicleBooks.com.*

Figure 20.3 shows an example of a protocol used at a hospital. This protocol works like most sets of instructions. A numbered list of steps explains what to do if someone is experiencing chest pain. The protocol includes helpful visuals, like the gray boxes and pictures of a doctor, to signal situations where a doctor should be called.

Figure 20.4, on page 557, is a testing specification used by civil engineers. The numbers and indentation help engineers follow the procedure consistently.

A Procedure

Header shows identification number of procedure.

Title of emergency procedure

Brief introduction defines medical condition.

Steps of the procedure

Gray areas signal situations in which a doctor needs to be involved.

WEST VIRGINIA EMS SYSTEM

**EMT-Paramedic
Treatment Protocol
4202**

**Chest Pain/Discomfort Acute
Coronary Syndrome (ACS)** | **Page 1 of 3**

A. Indications for this protocol include one or more of the following:

    1. Male over 25 years of age or female over 35 years of age, complaining of substernal chest pain, pressure or discomfort unrelated to an injury.

    2. History of previous ACS/AMI with recurrence of "similar" symptoms.

    3. Any patient with a history of cardiac problems who experiences lightheadedness or syncope.

    4. Patients of any age with suspected cocaine abuse and chest pain.

B. Perform **MAMP (4201).**

C. Obtain 12 lead ECG, if available and causes no delay in treatment or transport.

D. If patient has no history of allergy to aspirin **and** has no signs of active bleeding (i.e., bleeding gums, bloody or tarry stools, etc.), then administer 4 (four) 81 mg chewable aspirin orally (324 mg total). Note: May be administered prior to establishment of IV access.

E. If blood pressure > 90 systolic and patient has **not** taken *Viagra* or *Levitra* within last 24 hours (or *Cialis* within the last 48 hours):

    1. Administer nitroglycerine 0.4 mg (1/150 gr) SL. Note: May be administered prior to establishment of IV access.

    2. Repeat every 3-5 minutes until pain is relieved.

    3. If blood pressure falls below 90 systolic or decreases more than 30 mm Hg below patient's normal baseline blood pressure, then discontinue dosing and **contact MCP** to discuss further treatment.

West Virginia Office of Emergency Medical Services - State ALS Protocols
4602 Stroke.doc   Finalized 12/1/01, Revised 9/11/07

Source: West Virginia EMS System, 2007.

Figure 20.3: Procedures like this one are used for training. They also standardize care.

554   Chapter 20
**Instructions and
Documentation**

Header is retained from first page.

**EMT-Paramedic
Treatment Protocol
4202**

**Chest Pain/DiscomfortAcute
Coronary Syndrome (ACS)**

Page 2 of 3

Pages are clearly numbered.

Steps of the instructions

F. If blood pressure <90 systolic and/or patient is experiencing severe bradycardia or tachycardia treat according to appropriate protocol. Further treatment **per MCP orders**. If patient has taken *Viagra* or *Levitra* within last 24 hours (*Cialis* within 48 hours), nitroglycerine should only be given **by MCP order**.

G. Transport.

Explains when to call for more help.

H. Contact **Medical Command.**

I. If chest pain persists:

1. Morphine sulfate 2 mg slow IV **per order of Medical Command.**

2. Repeat morphine sulfate 2 mg **per order of Medical Command** after five minutes if pain persists and BP is over 100 systolic.

3. Administer additional nitroglycerine 0.4 mg (1/150 gr.) sublingual **per order of Medical Command.**

Icons signal when a doctor should be consulted.

J. Treat dysrhythmias according to specific protocols.

K. If transport time permits, complete **AHA Fibrinolytic Checklist (STEMI)** (see section L on next page).

Closes with instructions and states where more information can be found if necessary.

West Virginia Office of Emergency Medical Services - State ALS Protocols
4602 Stroke.doc   Finalized 12/1/01, Revised 9/11/07

(continued)

Figure 20.3
(continued)

Design helps reader make decisions about what to do.

This flowchart and checklist help the user make decisions.

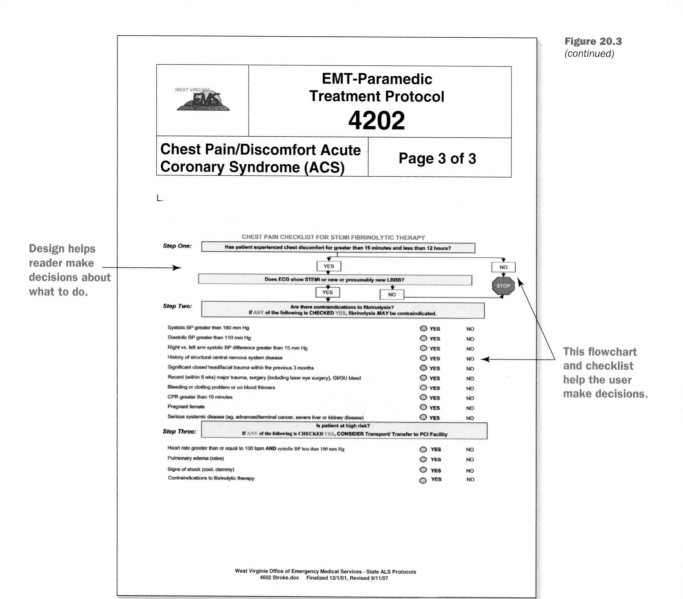

EMT-Paramedic
Treatment Protocol
4202

Chest Pain/Discomfort Acute Coronary Syndrome (ACS) | **Page 3 of 3**

L.

CHEST PAIN CHECKLIST FOR STEMI FIBRINOLYTIC THERAPY

Step One: Has patient experienced chest discomfort for greater than 15 minutes and less than 12 hours?

YES   NO

Does ECG show STEMI or new or presumably new LBBB?

YES   NO   STOP

Step Two: Are there contraindications to fibrinolysis?
If ANY of the following is CHECKED YES, fibrinolysis *MAY* be contraindicated.

Systolic BP greater than 180 mm Hg	YES	NO
Diastolic BP greater than 110 mm Hg	YES	NO
Right vs. left arm systolic BP difference greater than 15 mm Hg	YES	NO
History of structural central nervous system disease	YES	NO
Significant closed head/facial trauma within the previous 3 months	YES	NO
Recent (within 6 wks) major trauma, surgery (including laser eye surgery), GI/GU bleed	YES	NO
Bleeding or clotting problem or on blood thinners	YES	NO
CPR greater than 10 minutes	YES	NO
Pregnant female	YES	NO
Serious systemic disease (eg, advanced/terminal cancer, severe liver or kidney disease)	YES	NO

Step Three: Is patient at high risk?
If ANY of the following is CHECKED YES, CONSIDER Transport/ Transfer to PCI Facility

Heart rate greater than or equal to 100 bpm AND systolic BP less than 100 mm Hg	YES	NO
Pulmonary edema (rales)	YES	NO
Signs of shock (cool, clammy)	YES	NO
Contraindications to fibrinolytic therapy	YES	NO

West Virginia Office of Emergency Medical Services - State ALS Protocols
4602 Stroke.doc   Finalized 12/1/01, Revised 9/11/07

A Specification

**INDIANA DEPARTMENT OF TRANSPORTATION
OFFICE OF MATERIALS MANAGEMENT**

**DRY FLOW TESTING
OF
FLOWABLE BACKFILL MATERIALS
ITM No. 217-07T**

1.0 SCOPE.

Introduction explains the purpose of the specifications.

1.1 This test method covers the procedure for the determination of the flow time of dry flowable backfill materials for the purpose of verifying changes in sand sources for an approved Flowable Backfill Mix Design (FBMD).

1.2 The values stated in either acceptable English or SI metric units are to be regarded separately as standard, as appropriate for a specification with which this ITM is used. Within the text, SI metric units are shown in parentheses. The values stated in each system may not be exact equivalents; therefore, each system shall be used independently of the other, without combining values in any way.

1.3 This ITM may involve hazardous materials, operations, and equipment. This ITM may not address all of the safety problems associated with the use of the test method. The user of the ITM is responsible for establishing appropriate safety and health practices and determining the applicability of regulatory limitations prior to use.

2.0 REFERENCES.

2.1 AASHTO Standards.

M 231   Weighing Devices Used in the Testing of Materials

T 304   Uncompacted Void Content of Fine Aggregate

T 248   Reducing Samples of Aggregate to Testing Size

3.0 TERMINOLOGY. Definitions for terms and abbreviations shall be in accordance with the Department's Standard Specifications, Section 101, except as follows.

3.1 Dry flow time. The time to for a specified sample size of dry flowable materials to flow through a specified funnel

Source: Indiana Department of Transportation, http://www.in.gov/indot/div/M&T/itm/pubs/217_testing.pdf.

(continued)

4.0   SIGNIFICANCE AND USE.

4.1   This ITM is used to determine the time of dry flow of loose uncompacted flowable backfill material through a flow cone. The flowable backfill material includes sand or sand and fly ash mixture. The test result is done to ensure that an alternate sand shall have the same flow characteristic as the sand in the approved FBMD.

4.2   The dry flow cone test characterizes the state of flow of dry materials on any sand of known grading that may provide information about the sand or sand and fly ash mixture angularity, spherical shape, and surface texture.

4.3   Other test procedures or test methods exist for various flow cones with different dimensions and cone tip forms and sizes that may or may not have a correlation with the AASHTO T 304 flow cone.

5.0   APPARATUS.

5.1   Cylindrical measure, in accordance with AASHTO T 304, except the nominal 100-mL cylindrical measure is replaced by a one quart glass jar

The testing apparatus is clearly described.

5.2   Metal spatula, with a blade approximately 4 in. (100 mm) long, and at least ¾ in. (20 mm) wide, with straight edges. The end shall be cut at a right angle to the edges. (The straight edge of the spatula blade is used to strike off the fine aggregate.)

This "nested" numbering system is commonly used with specifications for easy reference.

5.3   Timing device, such as a stop watch, with an accuracy to within ± 0.1 seconds

5.4   Balance, Class G2, conforming to the requirements of AASHTO M 231

5.5   Sample splitter, in accordance with AASHTO T 248 for fine aggregate

6.0   SAMPLE PREPARATION.

6.1   The sample may consist of sand or a sand and fly ash mixture proportioned according to the FBMD. The sample shall be oven dried at 230 ± 9° F (110 ± 5° C) for 24 h. Upon completion of the drying, the sample shall be split to a sample size of approximately 1,500 g using a small sample splitter for fine aggregate in accordance with AASHTO T 248.

6.2   The dry sample of sand or sand and fly ash mixture shall be thoroughly mixed with the spatula until the sample appears to be homogenous.

7.0   PROCEDURE.

    7.1   Place the dry sample into the one quart glass jar and put the lid on. Agitate the glass jar to mix the dry sample for 30 seconds.

    7.2   Place a finger at the end of the funnel to block the opening of the funnel.

    7.3   Pour and empty the dry sample of sand or sand and fly ash mixture from the glass jar into the Mason jar.

    7.4   Level the dry sample in the Mason jar with a spatula.

    7.5   Place the empty glass jar directly centered under the funnel.

    7.6   Remove the finger and allow the dry sample to fall freely into the glass jar, and start timing the dry flow.

    7.7   Record the time T_1 of the dry flow to an accuracy of ±0.1 second.

    7.8   Repeat 7.1 through 7.6 for times T_2 and T_3.

8.0   CALCULATIONS.

    8.1   Calculate the average dry flow time of the dry flowable backfill materials as follows:

$$T_{average} = \frac{T_1 + T_2 + T_3}{3}$$

where:

    $T_{average}$ = Average dry flow time, s
    T_1 = Dry flow time on first trial, s
    T_2 = Dry flow time on second trial, s
    T_3 = Dry flow time on third trial, s

9.0   REPORT.

    9.1   Report the average dry flow time to within ± 0.1 seconds.

The testing procedure is explained step-by-step.

(continued)

The illustration helps the engineer build the testing apparatus.

Planning and Researching Documentation

When you are asked to write documentation for a product or procedure, you should first consider the situations in which it might be used. You also need to research the process you are describing so that you fully understand it and can describe it in detail.

Planning

When planning, first gain a thorough grasp of your subject, your readers, and their needs. A good way to start is to answer the Five-W and How Questions that define the rhetorical situation:

> *Who might use these instructions?*
>
> *Why are these instructions needed?*
>
> *What should the instructions include?*

Where will the instructions be used?

When will the instructions be used?

How will these instructions be used?

Once you have answered these questions, you are ready to define the rhetorical situation that will shape how you write the text.

SUBJECT   Give yourself time to use the product or follow the process. Try to identify any unexpected difficulties or dangers. Tinker with the product or process, looking for places where your readers might have trouble or make mistakes.

PURPOSE   Take a moment to consider and compose your purpose statement, limiting yourself to one sentence. That way, you will be very clear about what you are trying to accomplish. Some key verbs for the purpose statement might include the following:

to instruct	*to guide*
to show	*to lead*
to illustrate	*to direct*
to explain	*to train*
to teach	*to tutor*

For example, here are a few purpose statements that might be used in a set of instructions:

> The purpose of these instructions is to show you how to use your new QuickTake i700 digital video camera.

> These procedures will demonstrate the suturing required to complete and close up a knee operation.

> These specifications illustrate the proper use of the Series 3000 Router to trim printed circuit boards.

A version of your purpose statement will likely appear in the introduction of your instructions.

Link

For more information on defining a document's purpose, see Chapter 2, page 21.

READERS   Of course, it is difficult to anticipate all the types of people who might use your documentation. But people who decide to use a specific set of instructions, a set of specifications, or a procedure usually have common characteristics, backgrounds, and motivations that you can use to make your documentation more effective.

Primary readers (action takers) are people who will use your documentation to complete a task. What is their skill level? How well do they understand the product or process? What is their age and ability? Typically, instructions are written for the primary readers who have minimal experience with and knowledge of the task.

Secondary readers (advisors) are people who might supervise or help the primary readers complete the task. What is the skill level of these secondary readers? Are they training/teaching the primary readers? Are they helping them assemble the product or complete the process?

Tertiary readers (evaluators) often use documentation to ensure quality. Auditors and quality experts will review procedures and specifications closely when evaluating products or processes in a technical workplace. Also, sets of instructions can be used as evidence in lawsuits. So, these kinds of readers need to be considered and their needs anticipated.

Gatekeeper readers (supervisors) will need to look over your documentation before it is sent out with a product or approved for use in the workplace. These gatekeeper readers may or may not be experts in your subject. They will be checking the accuracy, safety, and quality of your instructions.

Link

For more information on analyzing readers, see Chapter 3, page 40.

Try not to overestimate your readers' skills and understanding. If your documentation is specific and uses simple language, novice users will appreciate the added help. Experienced users may find your documentation a bit too detailed, but they can just skim information they don't need. In most cases, you are better off giving your readers more information than they need.

CONTEXT OF USE   Put yourself in your readers' place for a moment. When and where will your readers use the documentation? In their living rooms? At a workbench? At a construction site? In an office cubicle? At night? Each of these different places and times will require you to adjust your documentation to your readers' needs.

Depending on the context of use, instructions can follow a variety of formats. They may be included in a user's manual, or they could be part of a poster explaining how to accomplish a task. Increasingly, documentation is being placed on websites for viewing or downloading. Figure 20.5, for example, shows how the documentation from *Worst-Case Scenarios* are portrayed on a website.

Context of use also involves safety and liability issues. If users of the documentation are at risk for injury or an accident, you are ethically obligated to warn them about the danger and tell them how to avoid it. You should try to anticipate all the ways users might injure themselves or experience a mishap while following the documentation you are providing. Then, use warning statements and cautions (discussed later in this chapter) to help them avoid these problems.

Researching

Once you have defined the rhetorical situation, you should spend some time doing research on the task you are describing. Research consists of gaining a thorough understanding of your subject by considering it from several angles (Figure 20.6). Here are a few research strategies that are especially useful when writing documentation.

DO BACKGROUND RESEARCH   You should research the history and purpose of the product or process you are describing. If the product or process is new, find out why it was developed and study the documents that shaped its development. If the product or process is not new, determine whether it has evolved or changed. Also, collect any prior instructions, procedures, or specifications that might help you write your own documentation.

For document planning worksheets, go to
www.pearsonhighered.com/johnsonweb3/20.2
For reader analysis worksheets, go to
www.pearsonhighered.com/johnsonweb3/20.3

Instructions on a Website

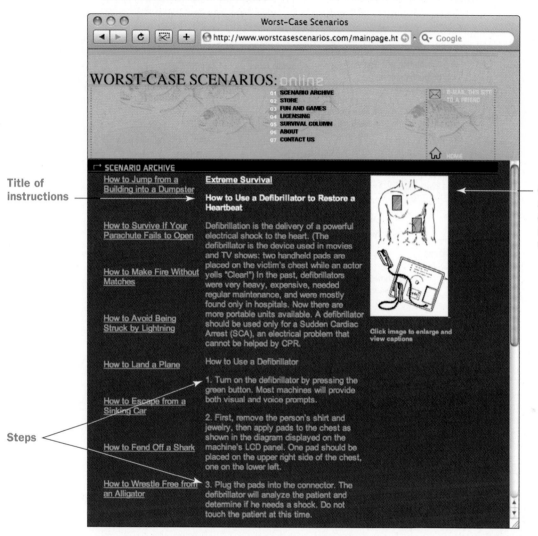

Title of instructions

Steps

Figure 20.5: Instructions are increasingly found on websites. Here are some survival tips on the *Worst-Case Scenarios* website.

Graphic offers a visual to enhance reader's understanding.

MAKE OBSERVATIONS Observe people using the product. If, for example, you are writing instructions for using a coffeemaker, observe someone making coffee with it. Pay attention to his or her experiences, especially any mistakes. Your notes from these experiments will help you anticipate some of the situations and problems your readers will experience.

To read about blunders in which writers did not anticipate their readers' needs, go to
www.pearsonhighered.com/johnsonweb3/20.4

Researching for Documentation

Figure 20.6: When doing research to write documentation, you should study your subject from a few different perspectives.

ASK SUBJECT MATTER EXPERTS (SMEs)   Interview the experts who are very familiar with the product or have used the procedure. They may be able to give you some insight or pointers into how the product is actually used or how the procedure is completed. They might also be able to point out trouble spots where nonexperts might have problems.

Link

For more information on doing research, go to Chapter 7, page 143.

USE YOUR SENSES   As you work with the product or follow the process, pay attention to all your senses. Where appropriate, take notes about appearance, sounds, smells, textures, and tastes. These details will add depth to your documentation. They will also help your readers determine if they are following the directions properly.

DESCRIBE MOTION AND CHANGE   Pay special attention to the movements of your subject and the way it changes as you complete the steps. Each step will lead to some kind of motion or change. By noting these motions and changes, you will be better able to describe them.

COLLECT VISUALS   If available, collect graphics that illustrate the steps you are describing in your documentation. If necessary, you can create your own photographs with a digital camera, or you can use drawings to illustrate your subject.

Planning for Cross-Cultural Readers and Contexts

As world trade continues to expand, documentation needs to be written for cross-cultural readers. As you profile your readers, you should anticipate how your documentation should be adjusted for cross-cultural readers.

Verbal Considerations

First, figure out how the written parts of your documentation will fit the needs of cross-cultural readers.

TRANSLATE THE TEXT   If the documentation is being sent to a place where another language is spoken, you should have it translated. Then, include both the translated and English versions with the product.

USE BASIC ENGLISH   If the documentation might be sent to people who are not fluent in English, you should use basic words and phrases. Avoid any jargon, idioms, and metaphors that will be understood only by North Americans (e.g., "senior citizens," "bottom line," or "back to square one"). Companies that do business in cross-cultural markets often maintain lists of basic words to use in documentation.

CHECK MEANINGS OF NAMES AND SLOGANS   Famously, names and slogans don't always translate well into other languages. For example, Pepsi's "Come alive with Pepsi" was translated into, "Pepsi brings your ancestors back from the grave" in Taiwan. In Spanish, Chevrolet's popular car the "Nova" means "It doesn't go." Coors's slogan "Turn it loose" translated into "Suffer from diarrhea" in some Spanish-speaking countries. So, check with people who are familiar with the target culture and its language to see if your names and slogans have meanings other than those intended.

Design Considerations

Then, do some research to figure out the best way to present your documentation visually to your target readers.

USE ICONS CAREFULLY   Some symbols commonly used in North America can be offensive to other cultures. A pointing finger, for example, is offensive in some Central and South American countries. An "OK" sign is offensive in Arab nations. Dogs are "unclean" animals in many parts of the world, so cartoon dogs often do not work well in documentation for multicultural readers. To avoid these problems, minimize your use of animals, human characters, or body parts whenever possible. Make sure that all the icons you use do not confuse or offend your readers in some unintended way.

USE IMAGES CAREFULLY   Images can convey unintended messages to the readers, especially in high-context cultures like those of Asia, parts of Africa, and the Middle East. For instance, how people are dressed in photographs can signal respect or disrespect. Obvious displays of emotion in professional settings can be seen as rude. In

Link

For more information on working with cross-cultural readers, go to Chapter 3, page 53.

some conservative Middle Eastern countries, meanwhile, photographs of people are used for identification only. So, before moving forward with your documentation, you might let someone from the target culture look over the images for any unintended meanings.

Organizing and Drafting Documentation

Like other technical documents, documentation should include an introduction, body, and conclusion. The introduction typically offers background information on the task being described. The body describes the steps required to complete the task. The conclusion usually offers readers a process for checking their work.

Specific and Precise Title

The title of your documentation should clearly describe the specific task the reader will complete.

Not descriptive	RGS-90x Telescope
Descriptive	Setting Up Your RGS-90x Telescope
Not descriptive	Head Wound
Descriptive	Procedure for Treating a Head Wound

Introduction

The length of the introduction depends on the complexity of the task and your readers' familiarity with it. If the task is simple, your introduction might be only a sentence long. If the task is complex or your readers are unfamiliar with the product or process, your introduction may need to be a few paragraphs long.

Introductions should include some or all of the following moves.

STATE THE PURPOSE   Simple or complex, all documentation should include some kind of statement of purpose.

> These instructions will help you set up your RGS-90x telescope.

> The Remington Medical Center uses these procedures to treat head wounds in the emergency room.

STATE THE IMPORTANCE OF THE TASK   You may want to stress the importance of the product or perhaps the importance of doing the task correctly.

> Your RGS-90x telescope is one of the most revolutionary telescope systems ever developed. You should read these instructions thoroughly so that you can take full advantage of the telescope's numerous advanced features.

> Head wounds of any kind should be taken seriously. The following procedures should be followed in all head wound cases, even the cases that do not seem serious.

To read about some other funny translation problems, go to
www.pearsonhighered.com/johnsonweb3/20.6

DESCRIBE THE NECESSARY TECHNICAL ABILITY   You may want to describe the necessary technical background that readers will need to use the product or to complete the task. Issues like age, qualifications, education level, and prior training are often important considerations that should be mentioned in the introduction.

> With advanced features similar to those found in larger and more specialized telescopes, the RGS-90x can be used by casual observers and serious astronomers alike. Some familiarity with telescopes is helpful but not needed.

> Because head wounds are usually serious, a trained nurse should be asked to bandage them. Head wounds should never be bandaged by trainees without close supervision.

IDENTIFY THE TIME REQUIRED FOR COMPLETION   If the task is complex, you may want to estimate the time the readers will need to complete all the steps.

> Initially, setting up your telescope should take about 15 to 20 minutes. As you grow more familiar with it, though, setup should take only 5 to 10 minutes.

> Speed is important when treating head wounds. You may have only a few minutes before the patient goes into shock.

MOTIVATE THE READER   An introduction is a good place to set a positive tone. Add a sentence or two to motivate readers and make them feel positive about the task they are undertaking.

> With push-button control, automatic tracking of celestial objects, and diffraction-limited imaging, an RGS-90x telescope may be the only telescope you will ever need. With this powerful telescope, you can study the rings of the planet Saturn or observe the feather structure of a bird from 50 yards away. This telescope will meet your growing interests in astronomical or terrestrial viewing.

> Head wounds of any kind are serious injuries. Learn and follow these procedures so you can effectively treat these injuries without hesitation.

Procedures and specifications often also include motivational statements about the importance of doing the job right, as companies urge their employees to strive for the highest quality.

List of Parts, Tools, and Conditions Required

After the introduction, you should list the parts, tools, and conditions required for completing the task.

LIST THE PARTS REQUIRED   This list should identify all the necessary items required to complete the task. Your parts list will allow the readers to check whether all the parts were included in the package (Figure 20.7). Other items not included with the package, like adhesive, batteries, and paint, should be mentioned at this point so readers can collect these items before following the steps.

A Parts List

Figure 20.7:
A parts list for a set of instructions. This feature usually begins by asking readers to check whether all the parts were included.

The opening encourages readers to check the kit's parts.

The parts are listed and numbered so that they can be checked against the graphic.

The graphic illustrates the parts in the kit.

Source: Stewart-MacDonald, http://www.stewmac.com.

IDENTIFY TOOLS REQUIRED   Nothing is more frustrating to readers than discovering midway through a set of instructions or a specification that they need a tool that was not previously mentioned. The required tools should be listed up front so readers can gather them before starting.

SPECIFY SPECIAL CONDITIONS   If any special conditions involving temperature, humidity, or light are required, mention them up front.

> Paint is best applied when temperatures are between 50°F and 90°F.

> If the humidity is above 75 percent, do not solder the microchips onto the printed circuit board. High humidity may lead to a defective joint.

Sequentially Ordered Steps

The steps are the centerpiece of any form of documentation, and they will usually make up the bulk of the text. These steps need to be presented logically and concisely, allowing readers to easily understand them and complete the task.

Want to see some sample introductions for instructions? Go to
www.pearsonhighered.com/johnsonweb3/20.7

As you carve the task you are describing into steps, you might use logical mapping to sort out the major and minor steps (Figure 20.8). First, put the overall task you are describing on the left-hand side of the screen or a sheet of paper. Then, break the task down into its major and minor steps. You might also, as shown in Figure 20.8, make note of any necessary hazard statements or additional comments that might be included.

Once you have organized the task into major and minor steps, you are ready to draft your instructions.

USE COMMAND VOICE   Steps should be written in *command voice,* or imperative mood. To use command voice, start each step with an action verb.

1. Place the telescope in an upright position on a flat surface.

2. Plug the coil cord for the Electronic Controller into the HBX port (see Figure 5).

Using Logical Mapping to Identify Steps in a Task

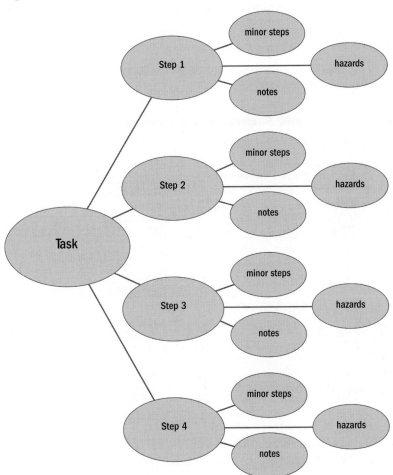

Figure 20.8: With logical mapping, the task is broken down into major and minor steps. Places where notes and hazard statements might appear are also noted.

In most steps, the verb should come first in the sentence. This puts the action up front, while keeping the pattern of the steps consistent. The "you" in these sentences is not stated, but rather implied ("*You* place the telescope in an upright position").

STATE ONE ACTION PER STEP   Each step should express only one action (Figure 20.9). You might be tempted to state two smaller actions in one step, but your readers will appreciate following each step separately.

Ineffective

2. Place the telescope securely on its side as shown in Figure 4 and open the battery compartment by simultaneously depressing the two release latches.

Revised

2. Place the telescope securely on its side as shown in Figure 4.

3. Open the battery compartment by simultaneously depressing the two release latches.

However, when two actions must be completed at the same time, you should put them in the same sentence.

6. Insert a low-power eyepiece (e.g., 26mm) into the eyepiece holder and tighten the eyepiece thumbscrew.

Instructions with Sequentially Ordered Steps

Steps are clearly numbered.

Graphics are used to clarify written text.

Notes are used to explain steps.

Graphics support the text.

Figure 20.9: Each step should express one action. Putting the steps in a list makes them easier to follow.

Source: Nikon, Cool Pix 885 *guide.*

You should state two actions in one step only when the two actions are dependent on each other. In other words, completion of one action should require the other action to be handled at the same time.

KEEP THE STEPS CONCISE   Use concise phrasing to describe each step. Short sentences are preferred so readers can remember each step while they work.

    7. Adjust the focus of the telescope with the focusing knob.

    8. Center the observed object in the lens.

If your sentences seem too long, consider moving some information into a follow-up "Note" or "Comment" that elaborates on the step.

NUMBER THE STEPS   In most kinds of documentation, steps are presented in a numbered list. Start with the number 1 and mark each step sequentially with its own number. Notes or warnings should not be numbered, because they do not state steps to be followed.

Incorrect

    9. Aim the telescope with the electronic controller.

    10. Your controller is capable of moving the telescope in several different directions. It will take practice to properly aim the telescope.

There is no action in step 10 above, so a number should not be used.

Correct

    9. Aim the telescope with the electronic controller.

    Your controller is capable of moving the telescope in several different directions. It will take practice to properly aim the telescope.

An important exception to this "number only steps" guideline involves the numbering of procedures and specifications. Procedures and specifications often use an itemized numbering system in which lists of cautions or notes are "nested" within lists of steps.

10.2.1 Putting on Clean Room Gloves, Hood, and Coveralls

    10.2.1.1 Put on a pair of clean room gloves so they are fully extended over the arm and the coverall sleeves. Glove liners are optional.

    10.2.1.2 Put a face/beard mask on, completely covering the mouth and nose.

    10.2.1.2.1 *Caution: No exposed hair is allowed in the fab.*

    10.2.1.2.2 *Caution: Keep your nose covered at all times while in the fab.*

    10.2.1.2.3 *Note: Do not wear the beard cover as a face mask. A beard cover should be used with a face mask to cover facial hair.*

    10.2.1.3 Put on coveralls.

Cautions and notes receive numbers in some procedures and specifications.

In specifications, comments and hazard statements receive a number. The purpose for this advanced numbering scheme is to make items in the documentation easier to reference.

In some cases, you may also want to use *paragraph style* to describe the steps (Figure 20.10). In these situations, you can use headings or sequential transitions to

Paragraph Style Instructions

Introduction to instructions

Paragraphs slow the readers down.

Headings highlight steps.

List style is used when appropriate.

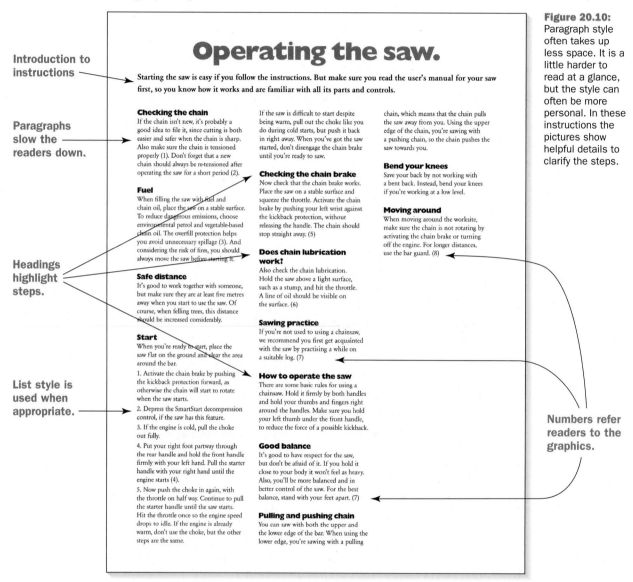

Operating the saw.

Starting the saw is easy if you follow the instructions. But make sure you read the user's manual for your saw first, so you know how it works and are familiar with all its parts and controls.

Checking the chain
If the chain isn't new, it's probably a good idea to file it, since cutting is both easier and safer when the chain is sharp. Also make sure the chain is tensioned properly (1). Don't forget that a new chain should always be re-tensioned after operating the saw for a short period (2).

Fuel
When filling the saw with fuel and chain oil, place the saw on a stable surface. To reduce dangerous emissions, choose environmental petrol and vegetable-based chain oil. The overfill protection helps you avoid unnecessary spillage (3). And considering the risk of fires, you should always move the saw before starting it.

Safe distance
It's good to work together with someone, but make sure they are at least five metres away when you start to use the saw. Of course, when felling trees, this distance should be increased considerably.

Start
When you're ready to start, place the saw flat on the ground and clear the area around the bar.
1. Activate the chain brake by pushing the kickback protection forward, as otherwise the chain will start to rotate when the saw starts.
2. Depress the SmartStart decompression control, if the saw has this feature.
3. If the engine is cold, pull the choke out fully.
4. Put your right foot partway through the rear handle and hold the front handle firmly with your left hand. Pull the starter handle with your right hand until the engine starts (4).
5. Now push the choke in again, with the throttle on half way. Continue to pull the starter handle until the saw starts. Hit the throttle once so the engine speed drops to idle. If the engine is already warm, don't use the choke, but the other steps are the same.

If the saw is difficult to start despite being warm, pull out the choke like you do during cold starts, but push it back in right away. When you've got the saw started, don't disengage the chain brake until you're ready to saw.

Checking the chain brake
Now check that the chain brake works. Place the saw on a stable surface and squeeze the throttle. Activate the chain brake by pushing your left wrist against the kickback protection, without releasing the handle. The chain should stop straight away. (5)

Does chain lubrication work?
Also check the chain lubrication. Hold the saw above a light surface, such as a stump, and hit the throttle. A line of oil should be visible on the surface. (6)

Sawing practice
If you're not used to using a chainsaw, we recommend you first get acquainted with the saw by practising a while on a suitable log. (7)

How to operate the saw
There are some basic rules for using a chainsaw. Hold it firmly by both handles and hold your thumbs and fingers right around the handles. Make sure you hold your left thumb under the front handle, to reduce the force of a possible kickback.

Good balance
It's good to have respect for the saw, but don't be afraid of it. If you hold it close to your body it won't feel as heavy. Also, you'll be more balanced and in better control of the saw. For the best balance, stand with your feet apart. (7)

Pulling and pushing chain
You can saw with both the upper and the lower edge of the bar. When using the lower edge, you're sawing with a pulling

chain, which means that the chain pulls the saw away from you. Using the upper edge of the chain, you're sawing with a pushing chain, so the chain pushes the saw towards you.

Bend your knees
Save your back by not working with a bent back. Instead, bend your knees if you're working at a low level.

Moving around
When moving around the worksite, make sure the chain is not rotating by activating the chain brake or turning off the engine. For longer distances, use the bar guard. (8)

Numbers refer readers to the graphics.

Source: Husqvarna.

Figure 20.10: Paragraph style often takes up less space. It is a little harder to read at a glance, but the style can often be more personal. In these instructions the pictures show helpful details to clarify the steps.

highlight the steps. Numerical transitions ("first," "second," "third," "finally") are best in most cases. In shorter sets of steps, you might use transitions like "then," "next," "5 minutes later," and "finally" to mark the actions. In Figure 20.10, headings are used to mark transitions among the major steps.

Figure 20.10:
(continued)

Pictures depict actions.

Safety equipment is used in pictures.

Close-ups help readers follow instructions.

Numbers reflect numbers in written text.

Readers often find paragraph style harder to follow because they cannot easily find their place again in the list of instructions. In some cases, though, paragraph style takes up less space and sounds more friendly and conversational.

AT A GLANCE

Writing Effective Steps

- Use command voice.
- State one action per step.
- Keep the steps concise.
- Number the steps.
- Add comments, notes, or examples.
- Provide feedback.
- Refer to the graphics.

ADD COMMENTS, NOTES, OR EXAMPLES   After each step, you can include additional comments or examples that will help readers complete the action. Comments after steps might include additional advice or definitions for less experienced readers. Or, comments might provide troubleshooting advice in case the step did not work out.

3. Locate a place to set your telescope.

Finding a suitable place to set up your telescope can be tricky. A paved area is optimal to keep the telescope steady. If a paved area is not available, find a level place where you can firmly set your telescope's tripod in the soil.

Comments and examples are often written in the "you" style to maintain a positive tone.

PROVIDE FEEDBACK   After a difficult step or group of steps, you might offer a paragraph of feedback to help readers assess their progress.

When you finish these steps, the barrel of your telescope should be pointed straight up. The tripod should be stable so that it does not teeter when touched. The legs of the tripod should be planted firmly on the ground.

REFER TO THE GRAPHICS   In the steps, refer readers to any accompanying graphics. A simple statement like, "See Figure 4" or "(Figure 4)" will notify readers that a graphic is available that illustrates the step. After reading the step, they can look at the graphic for help in completing the step properly.

In some cases, graphics are not labeled. In these situations, the graphic should appear immediately next to the step or below it so readers know which visual goes with each step.

Safety Information

Safety information should be placed early in the documentation and in places where the reader will be completing difficult or dangerous steps. A common convention in technical writing is to use a three-level rating for safety information and warnings: *Danger, Warning,* and *Caution.*

DANGER   Signals that readers may be at risk for serious injury or even death. This level of warning is the highest, and it should be used only when the situation involves real danger to the readers.

Danger: Do not remove grass from beneath your riding lawn mower while the engine is running (even if the blade is stopped). The blade can cause severe injury. To clear out grass, turn off the lawn mower and disconnect the spark plug before working near the blade.

WARNING Signals that the reader may be injured if the step is done improperly. To help readers avoid injury, warnings are used frequently.

> *Warning:* When heated, your soldering iron will cause burns if it touches your skin. To avoid injury, always return the soldering iron to its holder between uses.

CAUTION Alerts readers that mistakes may cause damage to the product or equipment. Cautions should be used to raise readers' awareness of difficult steps.

> *Caution:* The new oil filter should be tightened by hand only. Do not use an oil filter wrench for tightening, because it will cause the filter to seal improperly. If the filter is too tight, oil will leak through the filter's rubber gasket, potentially leading to major damage to your car's engine.

Safety information should tell your readers the following three things: (1) the hazard, (2) the seriousness of the hazard, and (3) how to avoid injury or damage. As shown in Figure 20.11, safety information should appear in two places:

- If a hazard is present throughout the procedure, readers should be warned before they begin following the steps. In these cases, danger and warning statements should appear between the introduction and the steps.
- If a hazard relates to a specific step, a statement should appear prominently before that step. It is important for readers to see the hazard statement *before* the step so that they can avoid damage or injury.

AT A GLANCE

Labeling Hazards

- Danger—Risk of serious injury or death is possible.
- Warning—Injury likely if step is handled improperly.
- Caution—Damage to the product or equipment is possible.

You can use symbols to highlight safety information. Icons are available to reinforce and highlight special hazards such as radioactive materials, electricity, or chemicals. Figure 20.12 shows a few examples of icons commonly used in safety information.

In our litigious culture, the importance of safety information should not be underestimated. Danger, warning, and caution notices will not completely protect your company from lawsuits, but they will give your company some defense against legal action.

Conclusion That Signals Completion of Task

When you have listed all the steps, you should offer a closing that tells readers that they are finished with the task. Closings can be handled a few different ways.

SIGNAL COMPLETION OF THE TASK Tell readers that they are finished with the steps. Perhaps you might offer a few comments about the future.

> Congratulations! You are finished setting up your RGS-90x telescope. You will now be able to spend many nights exploring the night skies.

> When completed, the bandaging of the head wound should be firm but not too tight. Bleeding should stop within a minute. If bleeding does not stop, call an emergency room doctor immediately.

To see good and bad examples of instructions, go to
www.pearsonhighered.com/johnsonweb3/20.8

**Organizing and
Drafting
Documentation**

575

Placement of Hazard Statements

Warning statements are prominently displayed.

Symbols draw attention to warnings.

Boxes are used to capture the readers' attention.

Figure demonstrates proper use of the machine, including use of appropriate safety devices like glasses, earmuffs, and gloves.

OPERATING INSTRUCTIONS

OPERATING TIPS

⚠ **WARNING:** Dress properly to reduce the risk of injury when operating this unit. Do not wear loose clothing or jewelry. Wear eye and ear/hearing protection. Wear heavy, long pants, boots and gloves. Do not wear short pants, sandals or go barefoot.

1. Move the cultivator to the work area prior to starting the engine. The cultivator may be transported by pushing it on wheels or carrying it by the shaft tube grip.

⚠ **WARNING:** To prevent serious personal injury, never pick-up or carry the unit while the engine is running.

2. Start the unit per Starting Instructions.
3. With the engine running and the tines off the ground, depress the throttle control to increase the engine speed.
4. Holding both of the handlebar grips firmly, slowly lower the cultivator until the tines make contact with the ground (Fig. 13).
5. As cultivating action begins, pull back on the cultivator so that the tines can penetrate the ground.
6. Once the ground has been broken, continue at a moderate pace until you are familiar with the controls and the handling of the cultivator.
7. Pull the cultivator backwards to improve the depth of cultivation and reduce your effort.
8. If the tines are digging too deep or not deep enough, adjust the tines per Adjusting Tine Depth.

— Fig. 13 —

ADJUSTING TINE DEPTH

Tine adjustment will vary depending on the type of soil being cultivated and how it will be used. Generally, adjusting the tines to break the soil 4 to 6 inches is recommended for most gardens. Adjust the tines as follows:

1. Stop the engine and disconnect the plug wire.
2. Loosen (do not remove) the two wing nuts on the tine guard (Fig. 14).
3. Slide the wheel bracket assembly down for shallower and up for deeper tine penetration.
4. Once the tines are in the desired position, tighten the wing nuts, making sure that the carriage bolts are seated properly through the bracket.
5. If the tine depth is not correct, repeat steps 2 to 4.

Up

Down

— Fig. 14 —

Transporting the Unit

⚠ **WARNING:** To prevent serious personal injury, always stop the engine when operation is delayed or when transporting the unit from one location to another.

1. Stop the engine.
2. Slide the wheel bracket assembly all the way down.
3. Tilt the unit back until the tines clear the ground.
4. Push or pull the unit to the next location to be cultivated.

12

A close-up graphic shows how to accomplish important tasks.

Source: Ryobi, 2000.

Figure 20.11: Hazard statements need to be prominent in the page design. In this user's manual, the warnings stand out because boxes and symbols draw attention to them.

GO TO THE NET

Want to learn more about using safety symbols? Go to
www.pearsonhighered.com/johnsonweb3/20.9

Safety Symbols

Source: From Geoffrey Peckham, "Safety Symbols," Compliance Engineering Magazine, *http://www.ce-mag.com/archive/02/03/peckham.html (Figure 1). Used with permission, Hazard Communication Systems, LLC. safetylabel.com.*

Figure 20.12: Here are a few examples of ISO and IEC symbols used on safety signs (hot surface, laser, radiation).

DESCRIBE THE FINISHED PRODUCT   You might describe the finished product or provide a graphic that shows how it should look.

> When you have completed setting up your telescope, it should be firmly set on the ground and the eyepiece should be just below the level of your eyes. With the Electronic Controller, you should be able to move the telescope horizontally and vertically with the push of a button. Figure 5 shows how a properly set-up telescope should look.

> Your bandaging of the patient's head should look like Figure B. The bandaging should be neatly wound around the patient's head with a slight overlap in the bandage strips.

OFFER TROUBLESHOOTING ADVICE   Depending on the complexity of the task, you might end your documentation by anticipating some of the common problems that could occur. Simple tasks may require only a sentence or two of troubleshooting advice. More complex tasks may require a table that lists potential problems and their remedies (Figure 20.13). Depending on your company's ability to provide customer service, you may also include a web address or a phone number where readers can obtain additional help.

Using Style in Documentation

People often assume that technical documentation should be dry and boring. But documentation can be—and sometimes should be—written in a more interesting style. There are ways, especially with consumer products, that you can use style to reflect readers' attitudes as they follow the steps.

For example, instructions for using a telescope should reflect readers' enthusiasm for their new ability to see into space. A procedure for bandaging a head wound should set a reassuring tone for a nurse who is learning the procedure.

To download safety symbols, go to
www.pearsonhighered.com/johnsonweb3/20.10

Figure 20.13: Troubleshooting guides are often provided in a table format with problems on the left and solutions on the right. Note the positive, constructive tone in this table.

TROUBLESHOOTING

Your XM Reference Tuner is designed and built to provide trouble-free performance without the need for service. If it does not appear to be functioning correctly, please follow these troubleshooting tips.

1. Make sure all connectors are properly attached to the rear panel of the XM Reference Tuner and to your home audio system.
2. Make sure that the unit is plugged into a standard 120V home current. You may also plug the Power Cord of the XM Reference Tuner into the switched AC Outlets of your pre-amp/receiver, if available.
3. Make sure you have activated your XM Satellite Radio subscription. You must activate your XM Satellite Radio service in order to use this product. To activate your XM Satellite Radio service, make sure you have a major credit card and your XM Satellite Radio ID Number handy and contact XM at http://activate.xmradio.com. You can also activate your service by calling 1-800-852-9696.

If you see this on the display,	You should:
NO SIGNAL Cause: The XM signal is out of range.	Reposition your High-Gain Antenna.
ANTENNA	Make sure your antenna is securely attached to the rear of the unit, that the antenna cable is unkinked and undamaged. Turn the XM Reference Tuner off and then back on to reset this message.
OFF AIR Cause: The channel selected is not currently broadcasting.	Tune in to another channel.

If this happens,	You should:
You can only receive XM channels 0 and 1. Cause: Your XM Reference Tuner is not activated.	Contact XM as described in the ACTIVATING YOUR XM SYSTEM section.
The audio sounds distorted.	Reduce the audio output level as described in the **Menu Commands and Settings Line Level Out** section.
The audio level is too low.	Increase the audio output level as described in the **Menu Commands and Settings Line Level Out** section.
You cannot tune in to a channel. (See "Channel Skip/Add.")	1. Make sure the channel has not been "skipped." 2. Verify that you are authorized to receive that channel as part of your subscription with XM and that you have not asked XM to block that channel on your XM Reference Tuner.
No song title or artist name displays.	Nothing is wrong with your receiver. This is normal on many talk and news channels where there is no song playing. Also, this information may not yet be in the database for some music channels.

Source: Polk Audio, Satellite Radio XRt 12 Tuner owner's guide.

How can you improve the style of your documentation? First, look at your original analysis of your readers and the contexts in which your document will be used. Pay attention to your readers' needs, values, and attitudes. Try to identify the emotions and attitudes that shape how they will be reading and using the instructions. Will they be enthusiastic, frustrated, happy, apprehensive, or excited?

To see examples of troubleshooting guides, go to
www.pearsonhighered.com/johnsonweb3/20.11

On-Screen Documentation

One of the major changes brought about by computers is the availability of on-screen documentation. Today, many user's manuals and instructions for products are available through a website, provided on a CD-ROM, or included with the Help feature in a software package.

The advantages of online documentation are numerous, and the disadvantages are few. The greatest advantage is reduced cost. After all, with some products like software, the accompanying user's manual costs more to print than the software itself. By putting the manual on a website, CD-ROM, or online Help, a company can save thousands of dollars almost immediately. Also, online documentation can be updated regularly to reflect changes in the product or revisions to the documentation.

Companies are also putting specifications and procedures online, usually on the company's intranet. That way, all employees can use their computers to easily call up the document they need.

Several options are available for on-screen documentation:

Online Help—Increasingly, the online Help features that come with software packages are being used to present instructions (Figure A). Online Help features allow readers to access the instructions more quickly, because they do not need to hunt around for the user's manual. As more documents move online, it is likely that instructions will increasingly be offered as online Help. To write instructions as online Help, you will need Help-authoring software like RoboHelp and DoctoHelp. These programs simplify the writing of Help features.

Online Help

Icons are used for easy navigation.

Major topics are clearly identified.

Current topic is highlighted for users.

Figure A: Online Help features are another place where instructions are commonly found, especially for software programs. These instructions are from the online Help feature for Mozilla Firefox.

Users can type in what they are looking for here.

Source: Mozilla Firefox, 2006.

(continued)

CD-ROM—Increasingly, computers have a CD burner as a standard feature. To make a CD, you can use web development software such as MacroMedia Dreamweaver, Expression Web Designer, or Adobe GoLive to create the files. Then, burn the files to a CD the same way you would save to a disk. Browser programs like Firefox, Safari, or Explorer will be able to read these multimedia documents.

Website—You can put your documentation on a website for use through the Internet or a company intranet. These files can be created with web development software such as Dreamweaver or Adobe GoLive.

Portable document format (PDF)—Software programs such as Adobe Acrobat can turn your word-processing files into PDFs, which retain the formatting and color of the original document. PDFs can be read by almost any computer, and they store information efficiently. They can also be password protected so that readers cannot tamper with the text. PDFs can be placed on a website for easy downloading.

Identify a word that best reflects readers' feelings as they are using your documentation. Then, use logical mapping to come up with some words that are associated with that word (Figure 20.14).

Mapping a Tone for Instructions

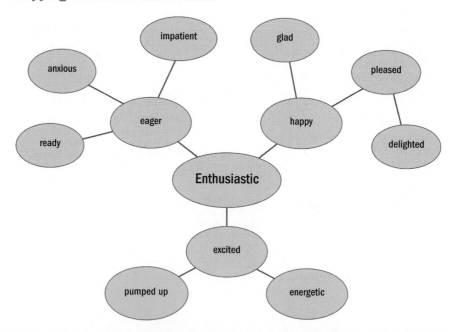

Figure 20.14: An enthusiastic tone can be created by finding synonyms associated with that word.

GO TO THE NET

Want to know more about Help-authoring software? Go to
www.pearsonhighered.com/johnsonweb3/20.13

After you have found words that are associated with the appropriate tone, use them in the introduction, notes, and conclusion. These words, when used strategically, will reflect your readers' attitudes or emotions.

If your readers have a negative attitude (perhaps they are annoyed that they need to read instructions), you can use antonyms to counteract their feelings. For example, to sooth annoyed readers, words like *satisfy, pleasure, please, delight,* and *fulfill* can be used to counteract their negativity. Don't overuse them, though, because angry readers may detect your attempt to soothe them.

Increasingly, companies are seeing documentation as their main contact with customers or clients. So, they are using style to make their documentation more responsive to readers' needs, values, and attitudes. Ideally, your documentation will leave customers with a positive image of the company.

Link

For more ideas about improving style, go to Chapter 9, page 207.

Designing Documentation

The design of your documentation will often determine whether it is usable (and whether it is used at all). If the page layout and images are helpful and attractive, readers will probably be more willing to use the documentation.

Page Layout

Documentation can be designed a variety of different ways. You might choose to incorporate the graphics within the page layout, as shown in Figure 20.15. Or, perhaps you might use a two-column format that allows you to put text on the left and graphics on the right. You can also use boxes and lines to highlight important information. At a minimum, headings should clearly show the levels of information in the text.

As you design your documentation, keep in mind your readers' specific needs and the situations in which they will be using the text. Design should not be used simply to decorate the text. Instead, it should enhance readers' ability to access the information. Design should make the steps more understandable. If you are not sure how to design your instructions, study sample texts from your home or workplace. You can use these examples as models for designing your own documentation.

Link

For more help on document design, see Chapter 10, page 232.

Graphics

With computers, a variety of ways are available for you to add graphics to your documentation. You can add illustrations and diagrams (Figure 20.15). Or, you can use a digital camera or scanner to add graphics to your text.

When including graphics, it is best to number and title them. Then, refer to the graphics by number in the written steps. That way, readers will be able to quickly locate them. Each graphic should appear next to or below the step that refers to it. Graphics should also appear on the same page as the step that refers to them. When using instructions, readers should never need to turn the page to see a graphic. Their hands might be busy, not allowing them to turn the page.

Design Is Important in Documentation

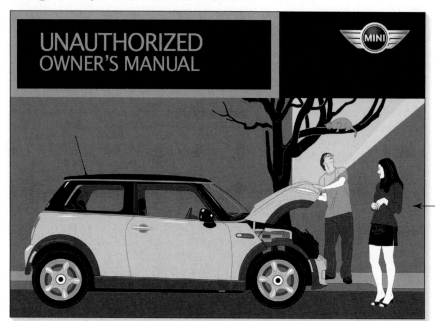

Figure 20.15: In this "unauthorized" user's guide for the Cooper Mini, the instructions use graphics to illustrate the text while making the information attractive.

Images of people help users identify with the product.

Headings make the text easy to scan.

BEST PLACES TO STASH STUFF IN YOUR MINI

First things first. This book fits snugly inside the case provided for your factory-authorized owner's manual, and is designed to be stowed away in the glove box. Or conveniently left about on the coffee table.

1. Coin Caddy
Technically engineered as the beverage holder. Reinvented by resourceful you as the handy loose change receptacle.

⚠CAUTION: When parking in low-lit urban areas, cover all exposed coinage with cup.

2. Toll Ticket Cranny
Your MINI's dashboard console features four vertical slits (two within driver's reach, two for co-pilot assistance) for securing toll receipts and parking stubs.

3. The Glove Box
Re-engineered by MINI designers. Because no one really keeps gloves in there anymore. Use it as a CD box. A toolbox. Or turn up the air conditioning, and it's a refrigerated icebox.

Hazard statements are highlighted with an icon.

Fig. 5

4. Cubby Space
Two center console bins for maps, cell phone, spare change, loose French fries. One per door for sunglasses (optimists). Collapsible umbrellas (pessimists).

⚠ **Cooling/Heating Feature**
Turns your MINI's glove box into a mini refrigerator for chilling contents to 50° Fahrenheit. For emergency roadside candy bars and spare hero sandwiches. Your MINI goes great with mayo. Or activate the heater to warm contents from soup to nuts. (See Fig. 6.)

Numbers and pointers highlight places discussed in text.

BEST PLACES TO STASH STUFF IN YOUR MINI

Fig. 6 Cooling/Heating Glove Box

Strategic Golf Club Placement

With both rear seats folded down, your MINI can accommodate four golf bags comfortably. Unfortunately, this leaves no room to accommodate a foursome. Recommended: with one rear seat folded down, fit one to two bags lying angled diagonally, and one to two golf buddies seated vertically.

How to Fit a Bike:

1. Adjust passenger's side seat to furthest forward non-reclining position.
2. Remove front wheel. OF BICYCLE!
3. Lay the bike on the side opposite the rear derailleur rotating the handlebars counterclockwise until they are parallel with the cargo area floor.
4. Place front wheel in cargo area and close hatch.

Instructions are easy to find in this numbered list.

Fig. 7Use floor space as parking citation crumple zone.

(side tabs: BASICS · STASH · CUSTOMIZE · ROMANCE · MUSIC · GOOD MOJO · TROUBLE)

BEST PLACES TO STASH STUFF IN YOUR MINI

Secret Jogging Key Nook

The undercarriage of your MINI features a handy hideaway for stashing cumbersome key sets when you take a break from motoring for a jog, bike ride, or skinny-dipping romp in the country. Place your hand just forward of either rear wheel near the rocker panel. Reach up and under the wheel well. You'll find a flat 4"X4" secret spot no one knows about but you. And tens of thousands of fellow U.S.MINI owners.

⚠ NOTE: Always perform a quick 360° visual scan to make sure no one is watching you. (See Fig. 8 & 9.)

A little humor can be refreshing.

Additional Storage Information

Pizza Capacity:The passenger's side floor accommodates four large pies stacked.

Fig.8 & 9 The jogging key nook.

Close-ups show users more detail.

Link

For more
information on
using graphics,
see Chapter 11,
page 275.

In some cases, graphics do not need to be numbered and titled. In these situations, the graphics should be tightly grouped with the text so readers can clearly see the relationship between the graphics and the text. For example, Figure 20.9 on page 570 shows an example of graphics tightly grouped with instructions.

Revising, Editing, and Proofreading

In many ways, the documentation reflects the quality of the product, service, and manufacturer. If the documentation looks unprofessional or has errors, readers will doubt the quality of the company behind it. So, you should leave yourself plenty of time for revising, editing, and proofreading. Also, leave yourself some time to user-test your documentation on some real readers.

Revising for Content, Style, and Design

When you are finished drafting, spend time critically studying the document's content, style, and design.

- The content needs to be complete and include all the information readers need to complete the task. Look for places where steps are missing or unclear. Identify steps that would be clearer with a follow-up note or comment. Identify places where hazard statements are needed.
- The style should be concise and clear. Keep sentences, especially commands, short and to the point. Where possible, replace complex words with simpler, plainer terms. Meanwhile, use words that reflect readers' attitudes or emotions as they will be using your documentation.
- The design should enhance readers' ability to follow the steps. Make sure you have used headings, notes, and graphics consistently. The page layout should also be consistent from the first page to the last.

Link

For more advice
on revision and
editing, see
Chapter 12,
page 306.

As you revise, ask yourself whether the content, style, and design are appropriate for the readers you defined in your reader profile. Your documentation should be usable by the least experienced, least knowledgeable person among your readers.

User-Testing with Sample Readers

User-testing your documentation on actual people can be done informally or formally, as discussed in Chapter 12 of this book.

- *Informal user-testing* can be used for in-house or less important forms of documentation. You should ask at least your supervisor or a few co-workers to look over your documentation before making it available to others. Often, your colleagues can help you identify missing content, ambiguous sentences, and ineffective graphics.
- *Formal user-testing* involves finding real users of the documentation and observing them as they try to follow the steps. Often, formal user-testing includes videotaping the subjects as they use the documentation. You can

Amanda Jervis
REQUIREMENTS ENGINEER, VIDIOM SYSTEMS CORPORATION,
BOULDER, COLORADO
Vidiom Systems makes software and other products for interactive television.

What is the best way to prepare to write instructions for a product?

Writing documentation often means writing about something that has just been conceived, has not yet been created, and certainly has never been written about. No library or website can help you with this assignment. You must rely on your own research skills to create your own story about this product or service.

How do you invent content for a set of instructions when you don't have sources to rely on?

- Talk to developers, marketing specialists, and everyone in the company to learn their vision of the product.
- Interview the programmers in-depth, asking them to explain the product at a simple level; copy down any diagrams they draw.
- Request all marketing and sales materials—PowerPoint presentations, flyers— and ask to attend a sales pitch.
- Attend all company meetings about the product, even developer reviews, and listen, listen, listen.

Finally, write up your information and give it to a project manager or developer for review. You'll always get great information when you give people an opportunity to point out that you're wrong.

---

then identify places where they stumble. Then, you can revise to fix those problems.

Chapter 12 discusses four different kinds of tests that would be helpful while user-testing your documentation: (1) read-and-locate tests, (2) understandability tests, (3) performance tests, and (4) safety tests. These tests would strengthen the usability of your documentation.

Link

For more information on user-testing documents, see Chapter 12, page 322.

Editing and Proofreading

Finally, you should carefully edit and proofread your documentation. The text should be clearly written and error free, whether the document is being sent with a product, included in a book of procedures, or placed in the specifications file.

You should challenge the content, organization, style, and design. Be your own harshest critic. Then, while proofreading, pay close attention to grammar and spelling. Flawless documentation will strengthen your readers' faith in you and your company.

- Documentation describes step-by-step how to complete a task.

- Basic features of documentation include a specific and precise title; an introduction; a list of parts, tools, and conditions required; sequentially ordered steps; graphics; safety information; a conclusion; and troubleshooting information.

- Determine the rhetorical situation by asking the Five-W and How Questions and analyzing the document's subject, purpose, readers, and context of use.

- Organize and draft your documentation step-by-step, breaking tasks down into their major and minor actions.

- Safety information should (1) identify the hazard, (2) state the level of risk (Danger, Warning, or Caution), and (3) offer suggestions for avoiding injury or damage.

- The style of your documentation does not need to be dry and boring. Try to use a style that reflects or counters readers' attitudes as they follow the steps.

- Graphics offer important support for the written text. They should be properly labeled by number and inserted on the page where they are referenced.

- While revising, user-testing your documentation with sample readers is an effective way to work out any bugs and locate places for improvement. Your observations of these sample readers should help you revise the document.

Individual or Team Projects

1. Find an example of documentation in your home or workplace. Using concepts discussed in this chapter, develop a set of criteria to evaluate its content, organization, style, and design. Then, write a two-page memo to your instructor in which you analyze the documentation. Highlight any strengths and make suggestions for improvements.

2. In your home or workplace, find an ineffective set of instructions. First, identify its weaknesses in content, organization, style, and design. Then, revise the instructions to make them easier to use. Write a cover memo to your instructor in which you discuss the ways you revised and improved the set of instructions.

3. Turn the instructions for playing Klondike at the end of this chapter into a numbered list and use graphics to support the text. You can make minor changes to the wording of the text; however, try to keep the written text intact as much as possible.

4. On the Internet or in your home, find information on first aid (handling choking, treating injuries, using CPR, handling drowning, treating shock, dealing with alcohol or drug overdoses). Then, turn this information into a text that is specifically aimed at college students living on campus. You should keep in mind that these

readers will be reluctant to read this text—until it is actually needed. So, write and design it in a way that will be both appealing before injuries occur and highly usable when an injury has occurred.

5. The Case Study at the end of this chapter presents a difficult ethical decision. Pretend you are Mark Harris in this situation. As Mark, write a memo to your supervisor in which you express your concerns about the product. Tell your supervisor what you think the company should do about the problem.

6. Choose a culture that is quite different from your own. Through the Internet and your library, research how documentation is written and designed in that culture. Look for examples of documentation designed for people of that culture. Then, write a memo to your instructor in which you explain how documentation is different in that target culture.

Collaborative Projects

Have someone in your group bring to class an everyday household appliance (toaster, blender, hot air popcorn popper, clock radio, portable CD player, etc.). With your group, write and design documentation for this appliance that would be appropriate for 8-year-old children. Your documentation should keep the special needs of these readers in mind. The documentation should also be readable and interesting to these readers, so they will actually use it.

Revision Challenge

These instructions for playing Klondike are technically correct; however, they are hard to follow. Can you use visual design to revise these instructions to make them more readable?

Playing Klondike (Solitaire)

Many people know Klondike simply as solitaire, because it is such a widely played solitaire game. Klondike is not the most challenging form of solitaire, but it is very enjoyable and known worldwide.

To play Klondike, use one regular pack of cards. Dealing left to right, make seven piles from 28 cards. Place one card on each pile, dealing one fewer pile each round. When you are finished dealing, the pile on the left will have one card, the next pile on the left will have two cards, and so on. The pile farthest to the right will have seven cards. When you are finished dealing the cards, flip the top card in each pile faceup.

You are now ready to play. You may move cards among the piles by stacking cards in decreasing numerical order (king to ace). Black cards are placed on red cards and red cards are placed on black cards. For example, a red four can be placed on a black five. If you would like to move an entire stack of faceup cards, the bottom card being moved must be placed on a successive card of the opposite color. For example, a faceup stack with the jack of hearts as the bottom card can be moved only to a pile with a black queen

For sample sets of effective and ineffective instructions, go to
www.pearsonhighered.com/johnsonweb3/20.16
For first aid sites on the web, go to
www.pearsonhighered.com/johnsonweb3/20.17

Exercises and Projects   587

showing on top. You can also move partial stacks from one pile to another as long as the bottom card you are moving can be placed on the top faceup card on the pile to which you are moving it. If a facedown card is ever revealed on top of a pile, it should be turned face up. You can now use this card. If the cards in a pile are ever completely removed, you can replace the pile by putting a king (or a stack with a king as the bottom faceup card) in its place.

The rest of the deck is called the "stock." Turn up cards in the stock one by one. If you can play a turned-up card on your piles, place it. If you cannot play the card, put it in the discard pile. As you turn up cards from the stock, you can also play the top card off the discard pile. For example, let us say you have an eight of hearts on top of the discard pile. You turn up a nine of spades from the stock, which you find can be played on a ten of diamonds on top of one of your piles. You can then play the eight of hearts on your discard pile on the newly placed nine of spades.

When an ace is uncovered, you may move it to a scoring pile separate from the seven piles. From then on, cards of the same suit may be placed on the ace in successive order. For example, if the two of hearts is the top faceup card in one of your piles, you can place it on the ace of hearts. As successive cards in the suit are the top cards in piles or revealed in the stock, you can place them on your scoring piles.

When playing Klondike properly, you may go through the stock only once (variations of Klondike allow you to go through the stock as many times as you like, three cards at a time). When you are finished going through the stock, count up the cards placed in your scoring piles. The total cards in these piles make up your score for the game.

For additional technical writing resources, including interactive sample documents, document design tutorials and guidelines, and more, go to **www.mytechcommlab.com.**

GO TO THE NET

Check your solitaire instructions against ones on the net. Go to
www.pearsonhighered.com/johnsonweb3/20.18
To find information on product liability, go to
www.pearsonhighered.com/johnsonweb3/20.19

The Flame

Mark Harris was recently hired as a design engineer at Fun Times Toy Manufacturers, Inc. In a way, the job was a dream come true. Mark liked designing toys, and he had two small children at home.

When he began working at the company, he inherited a project from the previous engineer, who had taken a job with another company. The project was an indoor campfire made out of a gas stove. The stove produced a flame about a foot tall in a pile of fake logs. The campfire was designed to be used outside, but it could be used inside with special precautions.

The prior engineer on the project had already built a prototype. All Mark needed to do was write the instructions for the user's manual. So, he took the campfire home to try it out with his kids.

At home, the campfire started up easily in his living room, but Mark immediately noticed that the flame was somewhat dangerous. As his kids tried to roast marshmallows over the fire, he started to become a little nervous. He wondered if the couch and rug were at risk of catching on fire. Certainly, he could imagine situations where the campfire could be used incorrectly, causing burns or fire.

The next day, he stopped by his supervisor's office to express his concerns. His supervisor, though, waved off the problems. "Yeah, the previous guy mentioned some problems, too. All you need to do is fill the user's manual with warnings and warning symbols. That should protect us from lawsuits."

As Mark walked out the door, his supervisor said, "Listen, we need that thing out the door in a couple weeks. Our manufacturers in Mexico have already retooled their plants, and we're ready to go. You need to get that manual done right now."

Mark went back to his office a bit concerned. Wouldn't his company still be negligent if people or their property were harmed? Would warnings and warning symbols really protect them from negligence? And even if the company couldn't be sued, would it be ethical to produce a potentially harmful product? What would you do if you were in Mark's place?

CHAPTER
21
Proposals

In this chapter, you will learn:

- The purpose of proposals and their uses in the workplace.
- The basic features and types of proposals.
- How to plan and do research for a proposal.
- How to organize and draft the major sections in a proposal.
- Strategies for using plain and persuasive style to make a proposal influential.
- How document design and graphics can enhance a proposal.
- Techniques for revising, editing, and proofreading proposals.

P roposals are the lifeblood of the technical workplace. No matter your field, you will be asked to write proposals that describe new projects, present innovative ideas, offer new strategies, and promote services.

The purpose of a proposal is to present your ideas and plans for your readers' consideration. Here are just a few examples of how proposals are used in the technical workplace:

- An electronic engineer would use a proposal to describe a new kind of plasma-screen television that his team wants to develop.
- A manager would use a proposal to argue for the use of robots to automate the assembly line at her factory.
- A civil engineer would use a proposal to present a plan for a monorail system in the downtown of a city.
- A marine biologist would use a proposal to request funding for her study of the effects of sonar on blue whale migrations.

Computers have increased the speed and competitiveness of proposal development. They have heightened the sophistication of proposals, allowing writers to use graphics, color, and even video to enhance the persuasiveness of their ideas.

Effective proposal writing is a crucial skill in today's technical workplaces. Almost all projects begin with proposals, so you need to master this important genre to be successful.

Basic Features of Proposals

Proposals are used "internally" or "externally" in a technical workplace. *Internal* proposals are used within companies to plan or propose new projects or products. *External* proposals are used to offer services or products to clients outside the company. Both external and internal proposals tend to have the following features:

- Introduction
- Description of the current situation
- Description of the project plan
- Review of qualifications
- Discussion of costs and benefits
- Graphics
- Budget

Figure 21.1 shows how these features are usually arranged in a typical proposal. As with other technical documents, though, you should not mechanically follow the pattern described here. The proposal genre is not a formula. It is only a model that offers a guideline for writing. You should alter this pattern to suit the needs of your proposal's subject, purpose, readers, and context of use.

Proposals are also classified as *solicited* or *unsolicited,* depending on whether they were requested or not.

Solicited proposals are proposals requested by the readers. For example, your company's management might ask your team to submit a proposal for

GO TO THE NET
To see sample proposals, go to
www.pearsonhighered.com/johnsonweb3/21.1

Basic Features
of Proposals   591

Basic Pattern for a Proposal

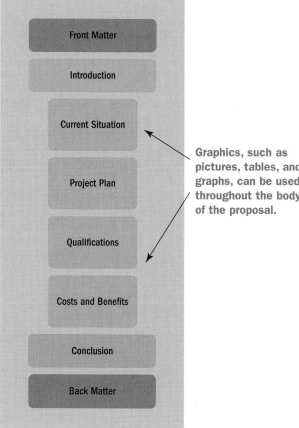

Figure 21.1: Proposals tend to address these areas, though not always in this order.

Graphics, such as pictures, tables, and graphs, can be used throughout the body of the proposal.

a new project. Or, your team might be asked to write a proposal that answers a request for proposals (RFP) sent out by a client.

Unsolicited proposals are proposals not requested by the readers. For example, your team might prepare an unsolicited internal proposal to pitch an innovative new idea to the company's management. Or, your team might use an unsolicited external proposal as a sales tool to offer your company's clients a product or service.

Figure 21.2 shows a solicited proposal written for internal purposes. In this example, a team within the company is pitching a plan to overhaul the company's website. Notice how the proposal is used to persuade management to agree to the team's ideas.

To learn more about different kinds of proposals, go to
www.pearsonhighered.com/johnsonweb3/21.2

Figure 21.2:
This small proposal is an internal proposal that is pitching a new idea to a manager. After a brief introduction, it describes the current situation and offers a plan for solving a problem. It concludes by highlighting the benefits of the plan.

Internal proposals are often written in memo format.

Background information signals that the proposal was solicited.

This section describes the current situation.

The main point of the proposal is stated up front.

JumperCom

Date: April 10, 2008
To: Jim Trujillo, VP of Operations
From: Sarah Voss, Lambda Engineering Team Leader
Re: Cutting Costs

At our meeting on April 4th, you asked each project team to come up with one good idea for cutting costs. Our team met on April 7th to kick around some ideas. At this meeting we decided that the best way to cut costs is to expand and enhance the company's website.

Our Current Website

When we developed the current website in Spring 2002, it served our company's purposes quite well. For its time, the website was attractive and interactive.

Six years later, our website is no longer cutting edge—it's obsolete. The website

- looks antiquated, making our company seem out of touch
- does not address our customers' questions about current products
- does not address our customers' needs for product documentation
- is not a tool that our salespeople can use to provide answers and documentation to the customers
- does not answer frequently asked questions, forcing clients to call our toll-free customer service lines for answers to simple questions.

As a result, our outdated website is causing a few important problems. First, we are likely losing sales because our customers don't see us as cutting edge. Second, we are wasting hundreds of thousands of dollars on printed documents that the customers throw away after a glance. And, third, we are unnecessarily spending many more thousands of dollars on customer service representatives and tollfree phone lines. *A conservative estimate suggests that our outdated website could be costing us around $400,000 each year.*

(continued)

This section offers a plan for the readers' consideration.

The plan is described step by step.

Renovating the Website

We believe a good way to cut costs and improve customer relations is to renovate the website. We envision a fully interactive site that customers can use to find answers to their questions, check on prices, and communicate with our service personnel. Meanwhile, our sales staff can use the website to discuss our products with clients. Instead of lugging around printed documents, our salespeople would use their laptop computers to show products or make presentations.

Renovating the site will require four major steps:

Step One: Study the Potential Uses of Our Website

With a consultant, we should study how our website might be better used by customers and salespeople. The consultant would survey our clients and salespeople to determine what kind of website would be most useful to them. The consultant would then develop a design for the website.

Step Two: Hire a Professional Web Designer to Renovate the Site

We should hire a professional web designer to implement our design, because modern websites are rather complex. A professional would provide us with an efficient, well-organized website that would include all the functions we are seeking.

Step Three: Train One of Our Employees to Be a Webmaster

We should hire or retrain one of our employees to be the webmaster of the site. We need someone who is working on the site daily and making regular updates. Being the webmaster for the site should be this employee's job description.

Step Four: User-Test the New Website with Our Customers and Salespeople

Once we have created a new version of the website, we should user-test it with our customers and salespeople. Perhaps we could pay some of our customers to try out the site and show us where it could be improved. Our salespeople will certainly give us plenty of feedback.

2

At the end of this process, we would have a fully functioning website that would save us money almost immediately.

Costs and Benefits of Our Idea

Renovating the website would have many advantages:

- The new website will save us printing costs. We estimate that the printing costs at our company could be sliced in half—perhaps more—because our customers would be able to download our documents directly from the website, rather than ask us to send these documents to them. That's a potential savings of $300,000.
- The new website will provide better service to our customers. Currently, our customers go to the website first when they have questions. By providing more information in an interactive format, we can cut down dramatically on calls to our customer service center. We could save up to $120,000 in personnel costs and long-distance charges.
- Currently our sales staff will find the website a useful tool when they have questions. When products change, salespeople will immediately see those changes reflected on the website. As a result, more sales might be generated because product information will be immediately available on-line.

A quick estimate shows that a website renovation would cost us about $40,000. We would also need to shift the current webmaster's responsibilities from part time to full time, costing us about $20,000 per year more. The savings, though, are obvious. For an initial investment of $60,000 and a yearly investment of $20,000 thereafter, we will minimally save about $400,000 a year.

Thank you for giving us this opportunity to present our ideas. If you would like to talk with us about this proposal, please call me at 555-1204, or you can e-mail me at sarahv@jumpercom.net.

3

Proposal concludes by discussing costs and benefits of the plan.

Another kind of proposal is the grant proposal. Researchers and nonprofit organizations prepare grant proposals to obtain funding for their projects. For example, one of the major funding sources for grants in science and technology is the National Science Foundation (NSF). Through its website, the NSF offers funding opportunities for scientific research (Figure 21.3).

The National Science Foundation Home Page

Recent research funded by the NSF

Examples of funded projects

Figure 21.3: The National Science Foundation (NSF) website offers information on grant opportunities. The home page, shown here, discusses some of the recent research projects that have received grants.

Source: National Science Foundation, http://www.nsf.gov/.

Planning and Researching Proposals

Because proposals are difficult to write, it is important that you follow a reliable writing process that will help you develop your proposal's content, organization, style, and design. An important first step in this process is to start with a planning and researching phase. During this phase, you will define the rhetorical situation and start collecting the content for the proposal.

Want to learn more about the NSF? Go to
www.pearsonhighered.com/johnsonweb3/21.3

Planning

A good way to start planning your proposal is to analyze the situations in which it will be used. Begin by answering the Five-W and How Questions:

Who will be able to say yes to my ideas, and what are their characteristics?

Why is this proposal being written?

What information do the readers need to make a decision?

Where will the proposal be used?

When will the proposal be used?

How will the proposal be used?

Once you have answered these questions, you are ready to start thinking in-depth about your proposal's subject, purpose, readers, and context of use.

Link

For more help on defining need-to-know information, go to Chapter 7, page 159.

SUBJECT   Define exactly what your proposal is about. Where are the boundaries of the subject? What information do your readers expect you to include in the proposal? What need-to-know information must readers have if they are going to say yes to your ideas?

PURPOSE   Clearly state the purpose of your proposal in one sentence. What should the proposal achieve? What do you want the proposal to do? By stating your purpose in one sentence, you will focus your writing efforts while making it easier for readers to understand what you are trying to accomplish.

Some key words for your purpose statement might include the following action verbs:

to persuade	*to present*
to convince	*to propose*
to provide	*to offer*
to describe	*to suggest*
to argue for	*to recommend*
to advocate	*to support*

A purpose statement might look something like this:

Link

To read more about defining your purpose, go to Chapter 2, page 21.

The purpose of this proposal is to recommend that our company change its manufacturing process to include more automation.

In this proposal, our aim is to persuade the state of North Carolina to develop a multi-modal approach to protect itself from stronger hurricanes, which may be caused by climate change.

READERS   More than any other kind of document, proposals require you to fully understand your readers and anticipate their needs, values, and attitudes.

Primary readers (action takers) are the people who can say yes to your ideas. They need good reasons and solid evidence. They also hold values

 Need help defining your purpose? Go to
www.pearsonhighered.com/johnsonweb3/21.4

Planning and Researching Proposals

597

and attitudes that will shape how they interpret your ideas. Meanwhile, keep in mind that economic issues are always important to primary readers, so make sure you carefully consider any money-related issues that might influence them.

Secondary readers (advisors) are usually experts in your field. They won't be the people who say yes to your proposal, but their opinions will be highly valued by your proposal's primary readers. You need to satisfy these advisors by offering enough technical information to demonstrate your understanding of the current situation and the soundness of your project.

Tertiary readers (evaluators) can be just about anyone else who might have an interest in the project. These readers might include lawyers, journalists, and community activists, among others. You need to anticipate these readers' concerns, especially because tertiary readers can often undermine the project if you are not careful.

Gatekeepers (supervisors) are the people at your own company who will need to look over your proposal before it is sent out. Your immediate supervisor is a gatekeeper, but you will likely also need to let other gatekeepers, such as the company's accountants, lawyers, and technical advisors, look over the proposal before it is sent.

Link

For more strategies for analyzing your readers, see Chapter 3, page 40.

CONTEXT OF USE   The document's context of use will also greatly influence how your readers will interpret the ideas in your proposal.

Physical context concerns the places your readers may read or use your proposal. Will readers look over your proposal at their desks, on their laptops, or in a meeting? Where will they discuss it?

Economic context involves the financial issues that will shape readers' responses to your ideas. How much money is available for the project? What economic trends will shape how your readers perceive the project? What are the financial limitations of the project?

Ethical context involves the ethical decisions that you and your readers will need to make. Where does the proposal touch on ethical issues? How might these ethical issues be resolved so they don't undermine the project? What are the legal issues involved with the proposal?

Link

For more help defining the context of use, turn to Chapter 3, page 46.

Political context concerns the people your proposal will affect. Who stands to gain or lose if your proposal is accepted? How will the proposal change relationships that are already in place? Would any larger political trends shape how the proposal is written or interpreted?

Something to keep in mind is that proposals, especially external proposals, are de facto legal contracts. They are legal documents that can be brought into court if a dispute occurs. So, you need to make sure that everything you

Worksheets are available to help you analyze your readers and contexts of use. Go to
www.pearsonhighered.com/johnsonweb3/21.5

say in the proposal is accurate and truthful, because the proposal may be used in a court case to prove (or disprove) that your company completed the promised work to the level proposed.

Researching

After defining your proposal's rhetorical situation, you should start collecting and creating the content of your document (Figure 21.4). Chapter 7 of this book describes how to do research, so research strategies won't be fully described here. However, here are some research strategies that are especially applicable for writing proposals:

Doing Research on Your Subject

Figure 21.4:
A logical map, like this one, might help you research your subject from a variety of directions. When researching the background of a proposal, you should collect as much information as possible.

DO BACKGROUND RESEARCH   The key to writing a persuasive proposal is to fully understand the problem you are trying to solve. First, you might go to the Internet to find as much information about your subject as you can. Second, locate print sources, such as books, reports, news articles, and brochures, on your subject. Third, interview, survey, and observe people who have a stake in the plan or project you are developing. Find out their views.

ASK SUBJECT MATTER EXPERTS (SMEs)   Spend time interviewing experts who know a great amount about your subject. They can probably give you insight into the problem you are trying to solve and suggest some potential solutions. They might also tell you what has worked (and not worked) in the past.

PAY ATTENTION TO CAUSES AND EFFECTS   All problems have causes, and all causes create effects. In your observations of the problem, try to identify the causes that are behind that problem. Then, try to identify some of the effects of the problem.

FIND SIMILAR PROPOSALS   On the Internet or at your workplace, you can probably locate proposals that have dealt with similar problems in the past. These samples can help you frame the problem and better understand the causes and effects. They might also give you some insight into how similar problems have been solved in the past.

Link

To learn more about doing research, turn to Chapter 7, page 142.

COLLECT VISUALS   Proposals are persuasive documents, so they often include plenty of graphics, such as photographs, charts, illustrations, and graphs. Collect any materials, data, and information that will help you add a visual dimension to your proposal. If appropriate, you might use a digital camera to take pictures to be added to the document.

Organizing and Drafting Proposals

Writing the first draft of a proposal is always difficult, because proposals describe the future—a future that you are trying to envision for your readers and yourself. Consequently, you will need to use your imagination to help create and describe the future you have in mind.

A good way to draft your proposal is to write it one section at a time. Think of the proposal as four or five separate mini-documents that could stand alone. When you finish drafting one section, move on to the next.

Writing the Introduction

As with all documents, the proposal's introduction sets a context, or framework, for the body of the document. A proposal's introduction will usually include up to six moves:

> **Move 1:** Define the *subject.*
>
> **Move 2:** State the *purpose.*
>
> **Move 3:** State the *main point.*
>
> **Move 4:** Stress the *importance of the subject.*
>
> **Move 5:** Offer *background information* on the subject.
>
> **Move 6:** Forecast the *organization of the document.*

Link

For additional help on writing introductions, see Chapter 8, page 182.

These moves can be made in just about any order, depending on your proposal, and they are not all required. Minimally, your proposal's introduction should clearly identify your *subject, purpose,* and *main point.* The other three moves are helpful, but they are optional. Figure 21.5 shows a sample introduction that uses all six moves. Your proposal's introduction should be concise. Usually introductions for proposals run about one to three paragraphs.

Figure 21.5:
This introduction makes all six "moves." As a result, it is somewhat lengthy. Nevertheless, this introduction prepares readers to understand the information in the body of the proposal.

Growth and Flexibility with Telecommuting

A Proposal to Northside Design

Offers background information.

Founded in 1979, Northside Design is one of the classic entrepreneurial success stories in architecture. Today, this company is one of the leading architectural firms in the Chicago market with over 50 million dollars in annual revenue. With growth, however, comes growing pains, and Northside now faces an important decision about how it will manage its growth in the near future. The right decision could lead to more market share, increased sales, and even more prominence in the architectural field. However, Northside also needs to safeguard itself against over-extension in case the Chicago construction market unexpectedly begins to recede.

Defines the subject.

Stresses the importance of the subject.

> "Northside needs to safeguard itself against over-extension in case the Chicago construction market unexpectedly begins to recede."

To help you make the right decision, this proposal offers an innovative strategy that will support your firm's growth while maintaining its flexibility. Specifically, we propose Northside implement a telecommuting network that allows selected employees to work a few days each week at home. Telecommuting will provide your company with the office space it needs to continue growing. Meanwhile, this approach will avoid a large investment in new facilities and disruption to the company's current operations.

States the purpose of the proposal.

States the main point.

Forecasts the body of the proposal.

In this proposal, we will first discuss the results of our research into Northside's office space needs. Second, we will offer a plan for using a telecommuting network to free up more space at Northside's current office. Third, we will review Insight Systems' qualifications to assist Northside with its move into the world of telecommuting. And finally, we will go over some of the costs and advantages of our plan.

1

Describing the Current Situation

The aim of the *current situation* section—sometimes called the *background* section—is to define the problem your plan will solve. In this section, you should help readers understand the current situation by clearly defining the problem, its causes, and its effects.

You should accomplish three things in this section of the proposal:

- Define and describe the problem.
- Discuss the causes of the problem.
- Discuss the effects of the problem if nothing is done.

For example, let us say you are writing a proposal to improve safety at your college or workplace. Your current situation section would first define the problem by proving there is a lack of safety and showing its seriousness. Then, it would discuss the causes and effects of that problem.

MAPPING OUT THE SITUATION   Logical mapping is a helpful technique for developing your argument in this section. Here are some steps you can follow to map out the content:

1. Write the problem in the middle of your screen or piece of paper. Put a circle around the problem.

2. Write down the two to five major causes of that problem. Circle them, and connect them to the problem.

3. Write down some minor causes around each major cause, treating each major cause as a separate problem of its own. Circle the minor causes and connect them to the major causes.

Figure 21.6 illustrates how your logical map for the current situation section might look.

DRAFTING THE CURRENT SITUATION SECTION   Your logical map of the current situation should help you define the basic content of this section. Now you are ready to turn your map into paragraphs and sentences. To begin, the current situation section should include an opening, a body, and a closing:

Opening—Identify and define the problem you will describe.

Body—Discuss the *causes* of the problem, showing how these causes brought about the problem.

Closing—Discuss the *effects* of not doing anything about the problem.

The length of the current situation section depends on your readers' familiarity with the problem. If readers are new to the subject, then several paragraphs or even pages might be required. However, if they fully understand the problem already, maybe only a paragraph or two are needed.

Figure 21.7 follows one possible pattern for organizing this section of the proposal. The current situation section can be organized

Mapping Out the Current Situation

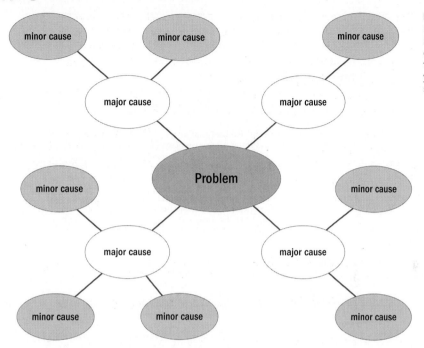

Figure 21.6:
Logical mapping helps you figure out what caused the problem that you are trying to solve.

many different ways. What is important is that the section define the problem and discuss its causes and effects.

Describing the Project Plan

A proposal's *project plan* section offers a step-by-step method for solving the problem. Your goal is to tell your readers *how* you would like to handle the problem and *why* you would handle it that way. In this section, you should do the following:

- Identify the solution.
- State the objectives of the plan.
- Describe the plan's major and minor steps.
- Identify the deliverables or outcomes.

As you begin drafting this section, look back at your original purpose statement for the proposal, which you wrote during the planning phase. Now, imagine a solution that might achieve that purpose. Perhaps you already have a solution in mind. If not, brainstorm with your team members to determine what kinds of solutions might solve the problem you described in the current situation section.

Example Current Situation Section

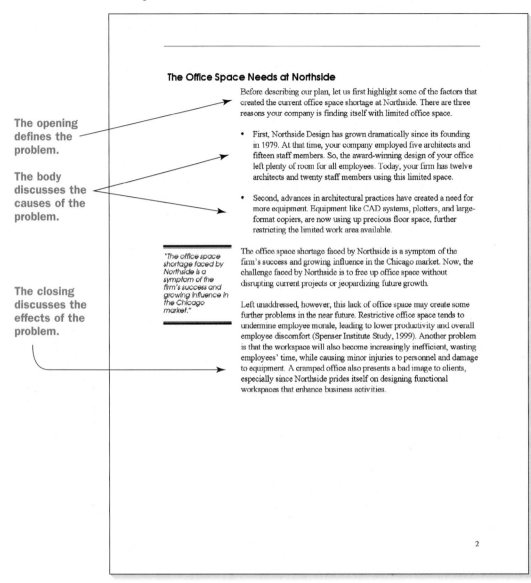

The opening defines the problem.

The body discusses the causes of the problem.

The closing discusses the effects of the problem.

The Office Space Needs at Northside

Before describing our plan, let us first highlight some of the factors that created the current office space shortage at Northside. There are three reasons your company is finding itself with limited office space.

- First, Northside Design has grown dramatically since its founding in 1979. At that time, your company employed five architects and fifteen staff members. So, the award-winning design of your office left plenty of room for all employees. Today, your firm has twelve architects and twenty staff members using this limited space.

- Second, advances in architectural practices have created a need for more equipment. Equipment like CAD systems, plotters, and large-format copiers, are now using up precious floor space, further restricting the limited work area available.

"The office space shortage faced by Northside is a symptom of the firm's success and growing influence in the Chicago market."

The office space shortage faced by Northside is a symptom of the firm's success and growing influence in the Chicago market. Now, the challenge faced by Northside is to free up office space without disrupting current projects or jeopardizing future growth.

Left unaddressed, however, this lack of office space may create some further problems in the near future. Restrictive office space tends to undermine employee morale, leading to lower productivity and overall employee discomfort (Spenser Institute Study, 1999). Another problem is that the workspace will also become increasingly inefficient, wasting employees' time, while causing minor injuries to personnel and damage to equipment. A cramped office also presents a bad image to clients, especially since Northside prides itself on designing functional workspaces that enhance business activities.

2

Figure 21.7:
The current situation section includes an opening, a body, and a closing. The causes of the problem are discussed mainly in the body paragraphs, while the effects are usually discussed at the end of the section.

GO TO THE NET

To see documents that analyze problems and causes, go to
www.pearsonhighered.com/johnsonweb3/21.9

MAPPING OUT THE PROJECT PLAN   When you have identified a possible solution, you can again use logical mapping to turn your idea into a plan:

1. Write your solution in the middle of your screen or a sheet of paper. Circle this solution.

2. Write down the two to five major steps needed to achieve that solution. Circle them and connect them to the solution.

3. Write down the minor steps required to achieve each major step. Circle them and connect them to the major steps.

As shown in Figure 21.8, your map should illustrate the basic steps of your plan.

DRAFTING THE PROJECT PLAN SECTION   Your project plan section should have an opening, a body, and a closing. This section will describe step-by-step how you will achieve your project's purpose.

Mapping Out a Project Plan

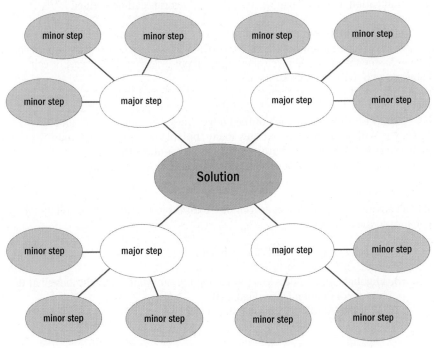

Figure 21.8: Logical mapping will help you figure out how to solve the problem. In a map, like the one shown here, you can visualize your entire plan by writing out the major and minor steps.

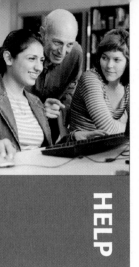

Roughing Out a Budget with a Spreadsheet

The budget is one of the most important but least understood parts of a proposal. At almost all companies, accountants and bookkeepers will prepare the final budget for projects, but they need the project managers to tell them what funds are needed and provide estimates of how much those items will cost. In other words, you will need to rough out budgets for your projects.

Here is a quick way to use a spreadsheet to create a rough budget for your project:

1. Open your spreadsheet program.
2. Create a new file.
3. List categories for the project's expenses in Column A, as shown in Figure A.
4. Type "Cost/hour," meaning cost per hour, in Column B
5. Type "Hours" in Column C
6. Type "Total Cost" in Column D

Most projects can be broken down into the seven budget categories shown in Figure A (Management, Labor, Facilities and Equipment, Materials, Travel, Communication, Profit). Then, each of these categories can be divided further.

Under each of these categories, list out all of the expenses you anticipate for the project. If you don't know how much something will cost, such as renting equipment or purchasing materials, then you should search out prices on the Internet or call someone who rents or sells that item.

At this point in the budgeting process, you don't need to get everything correct. Just try to think of as many of the project's expenses as you can. Then, collect estimates for each of the budget lines. For management and labor, you will need to figure out how many "billable hours" you and your co-workers will spend on the project. Usually, accountants at your company can calculate costs per hour for each employee, including managers.

When you are finished, you will need to sum the "Total Costs" column and have the full amount of the project appear in the "Total Project Costs" cell. To add up the column, do the following:

1. Calculate the "total cost" for each item.
2. Enter each "total cost" into Column D.
3. Highlight all numbers to be added in Column D and highlight the cell at the bottom of Column D (see Figure B).
4. Click the "AutoSum" button in the toolbar.

Creating a Rough Budget

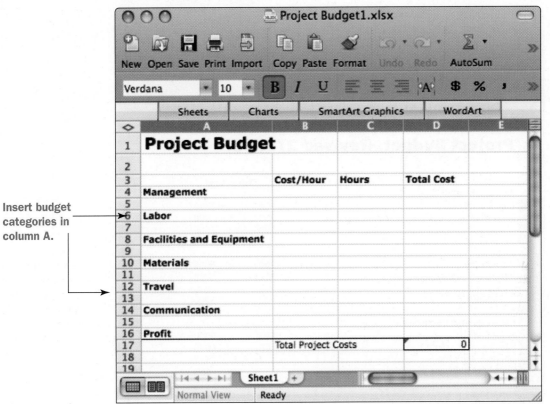

Insert budget categories in column A.

At this point, the spreadsheet program should add up all the project's costs and put the total at the bottom of the "Total Costs" column (see Column D in Figure B).

Of course, your budget will only be a rough estimate at this point, but it should give you an overall sense of the costs of the project. There are several budget items like benefits and depreciation that you probably won't be able to calculate yourself. Nevertheless, your company's accountant or cost estimator will appreciate your start on the budgeting process.

(continued)

Adding in Budget Items

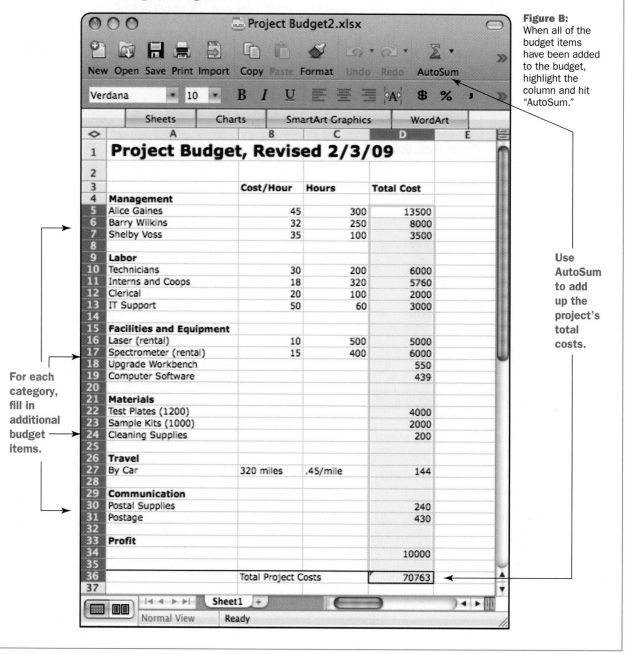

Figure B: When all of the budget items have been added to the budget, highlight the column and hit "AutoSum."

For each category, fill in additional budget items.

Use AutoSum to add up the project's total costs.

Project Budget2.xlsx

New Open Save Print Import Copy Paste Format Undo Redo AutoSum

Verdana   10   B I U $ %

Sheets | Charts | SmartArt Graphics | WordArt

	A	B	C	D	E
1	**Project Budget, Revised 2/3/09**				
2					
3		Cost/Hour	Hours	Total Cost	
4	**Management**				
5	Alice Gaines	45	300	13500	
6	Barry Wilkins	32	250	8000	
7	Shelby Voss	35	100	3500	
8					
9	**Labor**				
10	Technicians	30	200	6000	
11	Interns and Coops	18	320	5760	
12	Clerical	20	100	2000	
13	IT Support	50	60	3000	
14					
15	**Facilities and Equipment**				
16	Laser (rental)	10	500	5000	
17	Spectrometer (rental)	15	400	6000	
18	Upgrade Workbench			550	
19	Computer Software			439	
20					
21	**Materials**				
22	Test Plates (1200)			4000	
23	Sample Kits (1000)			2000	
24	Cleaning Supplies			200	
25					
26	**Travel**				
27	By Car	320 miles	.45/mile	144	
28					
29	**Communication**				
30	Postal Supplies			240	
31	Postage			430	
32					
33	**Profit**				
34				10000	
35					
36			Total Project Costs	70763	
37					

Sheet1 +

Normal View   Ready

For more advice about writing project plans in proposals, go to **www.pearsonhighered.com/johnsonweb3/21.10**

GO TO THE NET

The Project Plan

AT A GLANCE

- Identify the solution.
- State the objectives of the plan.
- Describe the plan's major and minor steps.
- Identify the deliverables or outcomes.

Opening—Identify your overall solution to the problem. You can even give your plan a name to make it sound more real (e.g., the "Restore Central Campus Project"). Your opening might also include a list of project objectives so the readers can see what goals your plan is striving to achieve.

Body—Walk the readers through your plan step-by-step. Address each major step separately, discussing the minor steps needed to achieve that major step. It is also helpful to tell readers *why* each major and minor step is needed.

Closing—Summarize the final *deliverables,* or outcomes, of your plan. The deliverables are the goods and services that you will provide when the project is finished. Tell the readers what the end results of your plan will be.

As shown in Figure 21.9, the project plan section balances the plan's steps with reasons why these steps are needed.

In most proposals, the project plan is the longest section of the document. This section needs to clearly describe your plan. Moreover, it needs to give your readers good reasons to believe your plan will work, while offering specific outcomes or results (deliverables).

Describing Qualifications

The qualifications section presents the credentials of your team or company, striving to prove that you are qualified to carry out the project plan. Minimally, the aim of the qualifications section is to show that your team or company is able to do the work. Ideally, however, you also want to prove that your team or company is *best qualified* to handle the project.

As you begin drafting this section, keep the following saying in mind: *What makes us different makes us attractive.* In other words, pay attention to the qualities that make your team or company different from your competitors. What are your company's strengths? What makes you better than the others?

In the qualifications section, you do not need to discuss every aspect of your team or company. Rather, you should offer just enough information to demonstrate that your team or your company is best qualified and able to handle the proposed project.

A typical qualifications section offers information on three aspects of your team or company:

Description of personnel—Short biographies of managers who will be involved in the project; demographic information on the company's workforce; description of support staff.

Description of organization—Corporate mission, philosophy, and history of the company; corporate facilities and equipment; organizational structure of the company.

Previous experience—Past and current clients; a list of similar projects that have been completed; case studies that describe past projects.

Figure 21.9:
An effective project plan section includes an opening, a body, and a closing. The opening states the solution and offers some objectives. The body walks the readers through the plan's steps. The closing identifies the major deliverables of the plan.

Objectives →

Our Plan: Flexibility and Telecommuting

Managing Northside's limited office space requires a solution that allows the company to grow but does not sacrifice financial flexibility. Therefore, we believe a successful solution must meet the following objectives:

- minimize disruption to Northside's current operations
- minimize costs, preserving Northside's financial flexibility
- retain Northside's current office on Michigan Avenue
- foster a dynamic workplace that will be appealing to Northside's architects and staff.

Our Objectives:
- *minimize disruption*
- *minimize costs*
- *retain Northside's current office*
- *foster a dynamic workplace*

To meet these objectives, Insight Systems proposes to collaborate with Northside to develop a telecommunication network that allows selected employees to work at home.

← Solution to problem

The primary advantage of telecommuting is that it frees up office space for the remaining employees who need to work in the main office. Telecommuting will also avoid overextending Northside's financial resources, so the firm can quickly react to the crests and valleys of the market.

Our plan will be implemented in four major phases. First, we will study Northside's telecommuting options. Second, we will design a local area network (LAN) that will allow selected employees to telecommute from a home office. Third, we will train Northside's employees in telecommuting basics. And finally, we will assess the success of the telecommuting program after it has been implemented.

Phase One: Analyzing Northside's Telecommuting Needs

We will start out by analyzing the specific workplace requirements of Northside's employees and management. The results of this analysis will allow us to work closely with Northside's management to develop a telecommuting program that fits the unique demands of a dynamic architecture firm.

3

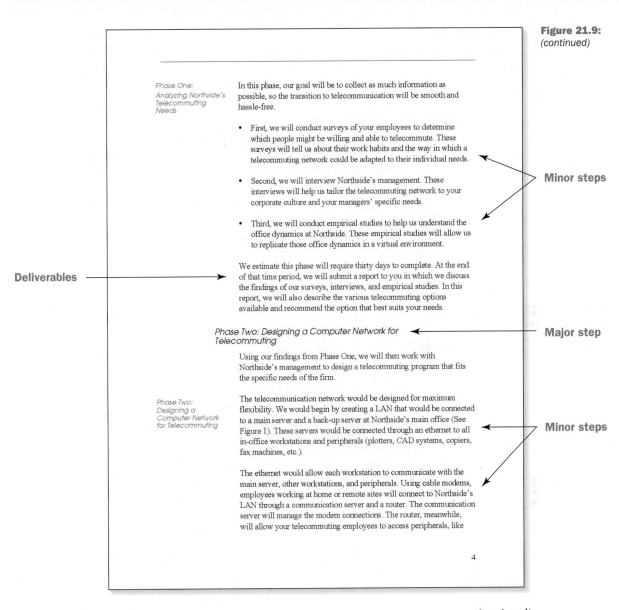

Phase One:
Analyzing Northside's
Telecommuting
Needs

In this phase, our goal will be to collect as much information as possible, so the transition to telecommunication will be smooth and hassle-free.

- First, we will conduct surveys of your employees to determine which people might be willing and able to telecommute. These surveys will tell us about their work habits and the way in which a telecommuting network could be adapted to their individual needs.

- Second, we will interview Northside's management. These interviews will help us tailor the telecommuting network to your corporate culture and your managers' specific needs.

- Third, we will conduct empirical studies to help us understand the office dynamics at Northside. These empirical studies will allow us to replicate those office dynamics in a virtual environment.

Minor steps

We estimate this phase will require thirty days to complete. At the end of that time period, we will submit a report to you in which we discuss the findings of our surveys, interviews, and empirical studies. In this report, we will also describe the various telecommuting options available and recommend the option that best suits your needs.

Deliverables

Phase Two: Designing a Computer Network for Telecommuting

Major step

Using our findings from Phase One, we will then work with Northside's management to design a telecommuting program that fits the specific needs of the firm.

Phase Two:
Designing a
Computer Network
for Telecommuting

The telecommunication network would be designed for maximum flexibility. We would begin by creating a LAN that would be connected to a main server and a back-up server at Northside's main office (See Figure 1). These servers would be connected through an ethernet to all in-office workstations and peripherals (plotters, CAD systems, copiers, fax machines, etc.).

Minor steps

The ethernet would allow each workstation to communicate with the main server, other workstations, and peripherals. Using cable modems, employees working at home or remote sites will connect to Northside's LAN through a communication server and a router. The communication server will manage the modem connections. The router, meanwhile, will allow your telecommuting employees to access peripherals, like

4

(continued)

the plotters and copiers, through the ethernet. The router will also allow Northside's main office to connect easily with future branch offices and remote clients.

To ensure the security of the LAN, we will equip the network with the most advanced security hardware and software available. The router (hardware) will be programmed to serve as a "firewall" against intruders. We will also install the most advanced encryption and virus software available to protect your employees' transmissions.

Figure 1: The Local Area Network

The graphic illustrates a complex concept.

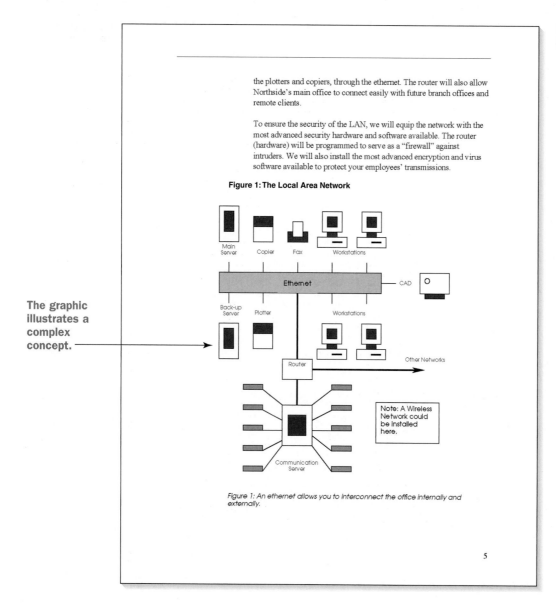

Figure 1: An ethernet allows you to interconnect the office internally and externally.

5

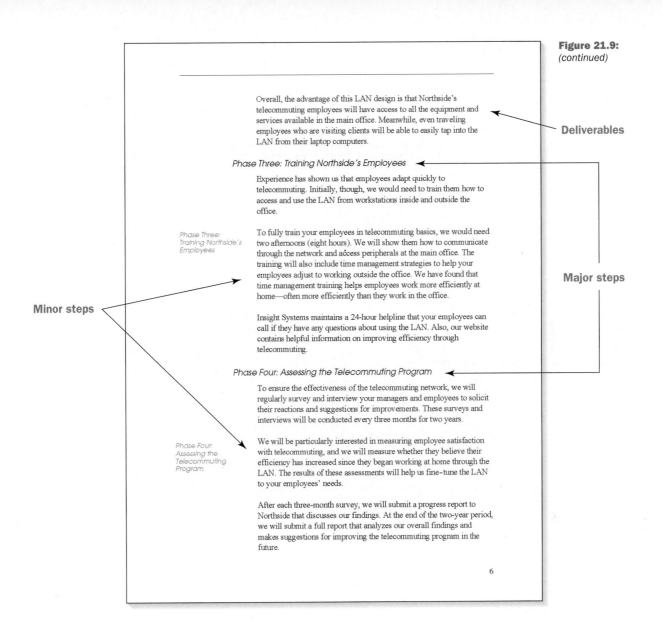

Overall, the advantage of this LAN design is that Northside's telecommuting employees will have access to all the equipment and services available in the main office. Meanwhile, even traveling employees who are visiting clients will be able to easily tap into the LAN from their laptop computers.

Deliverables

Phase Three: Training Northside's Employees

Experience has shown us that employees adapt quickly to telecommuting. Initially, though, we would need to train them how to access and use the LAN from workstations inside and outside the office.

Phase Three: Training Northside's Employees

To fully train your employees in telecommuting basics, we would need two afternoons (eight hours). We will show them how to communicate through the network and access peripherals at the main office. The training will also include time management strategies to help your employees adjust to working outside the office. We have found that time management training helps employees work more efficiently at home—often more efficiently than they work in the office.

Major steps

Minor steps

Insight Systems maintains a 24-hour helpline that your employees can call if they have any questions about using the LAN. Also, our website contains helpful information on improving efficiency through telecommuting.

Phase Four: Assessing the Telecommuting Program

To ensure the effectiveness of the telecommuting network, we will regularly survey and interview your managers and employees to solicit their reactions and suggestions for improvements. These surveys and interviews will be conducted every three months for two years.

Phase Four: Assessing the Telecommuting Program

We will be particularly interested in measuring employee satisfaction with telecommuting, and we will measure whether they believe their efficiency has increased since they began working at home through the LAN. The results of these assessments will help us fine-tune the LAN to your employees' needs.

After each three-month survey, we will submit a progress report to Northside that discusses our findings. At the end of the two-year period, we will submit a full report that analyzes our overall findings and makes suggestions for improving the telecommuting program in the future.

6

Qualifications Section

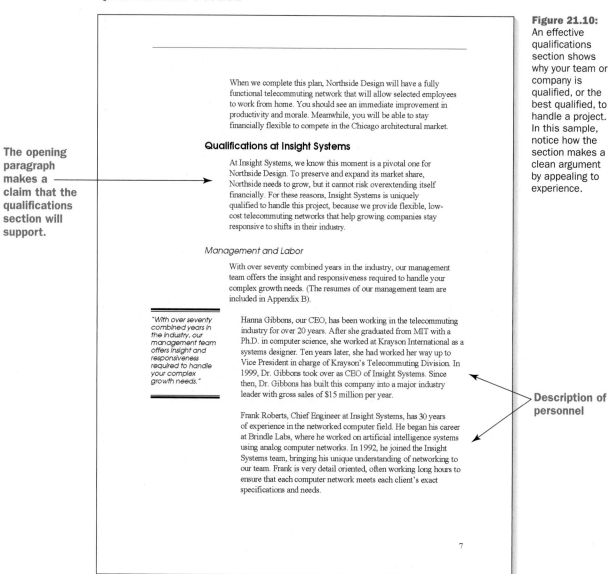

When we complete this plan, Northside Design will have a fully functional telecommuting network that will allow selected employees to work from home. You should see an immediate improvement in productivity and morale. Meanwhile, you will be able to stay financially flexible to compete in the Chicago architectural market.

Qualifications at Insight Systems

The opening paragraph makes a claim that the qualifications section will support.

At Insight Systems, we know this moment is a pivotal one for Northside Design. To preserve and expand its market share, Northside needs to grow, but it cannot risk overextending itself financially. For these reasons, Insight Systems is uniquely qualified to handle this project, because we provide flexible, low-cost telecommuting networks that help growing companies stay responsive to shifts in their industry.

Management and Labor

With over seventy combined years in the industry, our management team offers the insight and responsiveness required to handle your complex growth needs. (The resumes of our management team are included in Appendix B).

"With over seventy combined years in the industry, our management team offers insight and responsiveness required to handle your complex growth needs."

Hanna Gibbons, our CEO, has been working in the telecommuting industry for over 20 years. After she graduated from MIT with a Ph.D. in computer science, she worked at Krayson International as a systems designer. Ten years later, she had worked her way up to Vice President in charge of Krayson's Telecommuting Division. In 1999, Dr. Gibbons took over as CEO of Insight Systems. Since then, Dr. Gibbons has built this company into a major industry leader with gross sales of $15 million per year.

Description of personnel

Frank Roberts, Chief Engineer at Insight Systems, has 30 years of experience in the networked computer field. He began his career at Brindle Labs, where he worked on artificial intelligence systems using analog computer networks. In 1992, he joined the Insight Systems team, bringing his unique understanding of networking to our team. Frank is very detail oriented, often working long hours to ensure that each computer network meets each client's exact specifications and needs.

7

Figure 21.10: An effective qualifications section shows why your team or company is qualified, or the best qualified, to handle a project. In this sample, notice how the section makes a clean argument by appealing to experience.

Figure 21.10 shows a sample qualifications section that includes these three kinds of information about qualifications. Pay attention to how this section does more than describe the company—it makes an argument that the bidders are uniquely qualified to handle the project.

For more help writing a qualifications section, go to
www.pearsonhighered.com/johnsonweb3/21.11

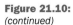

Lisa Miller, Insight System's Senior Computer Engineer, has successfully led the implementation of thirty-three telecommuting systems in companies throughout the United States. Earning her computer science degree at Iowa State, Lisa has won numerous awards for her innovative approach to computer networking. She believes that clear communication is the best way to meet her clients' needs.

Our management is supported by one of most advanced teams of high technology employees. Insight Systems employs twenty of the brightest engineers and technicians in the telecommunications industry. We have aggressively recruited our employees from the most advanced universities in the United States, including Stanford, MIT, Illinois, Iowa State, New Mexico, and Syracuse. Several of our engineers have been with Insight Systems since it was founded.

Corporate History and Facilities

Insight Systems has been a leader in the telecommuting industry from the beginning. In 1975, the company was founded by John Temple, a pioneer in the networking field. Since then, Insight Systems has followed Dr. Temple's simple belief that computer-age workplaces should give people the freedom to be creative.

"Insight Systems earned the coveted '100 Companies to Watch' designation from Business Outlook Magazine."

Recently, Insight Systems earned the coveted "100 Companies to Watch" designation from *Business Outlook Magazine* (May 2007). The company has worked with large and small companies, from Vedder Aerospace to the Cedar Rapids Museum of Fine Arts, to create telecommuting options for companies that want to keep costs down and productivity high.

Insight Systems' Naperville office has been called "a prototype workspace for the information age." (*Gibson's Computer Weekly*, May 2006). With advanced LAN systems in place, only ten of Insight System's fifty employees actually work in the office. Most of Insight Systems' employees telecommute from home or on the road.

Experience You Can Trust

Our background and experience give us the ability to help Northside manage its needs for a more efficient, dynamic office space. Our key to success is innovation, flexibility, and efficiency.

Description of organization

8

You should never underestimate the importance of the qualifications section in a proposal. In the end, your readers will not accept the proposal if they do not believe that your team or company has the personnel, facilities, or experience to do the work. In this section, your job is to persuade them that you are uniquely or best qualified to handle the project.

Costs and Benefits

The costs and benefits section summarizes the advantages of saying yes to the proposal while also telling readers how much the project will cost.

Start out this section by making an obvious transition. Say something like, "Let us conclude by summarizing the costs and benefits of our plan." This kind of transition will wake up your readers, because they will realize that you are going to be highlighting the most important points in the proposal.

Early in this section, tell them the costs. A good strategy for handling expenses is to simply state the costs up front without apology or a sales pitch.

> As shown in our budget, this renovation will cost $287,000.

> We anticipate that the price for retooling your manufacturing plant will be $5,683,000.

Immediately after this statement of the costs, you should describe the significant benefits of saying yes to the plan. Fortunately, you have already identified many of these benefits earlier in the proposal. They appeared in two other sections: the project plan section and the qualifications section.

Project plan section—You identified some deliverables when you described the project plan. Summarize those deliverables for the readers and discuss the benefits of obtaining them.

Qualifications section—When describing your team's or company's qualifications, you also showed that your organization is uniquely or best qualified to handle the project. You can now briefly remind your readers of the benefits of working with your company (e.g., qualified people, high-quality products and services, superior customer service, excellent facilities).

In most cases, your discussion of the benefits should not add new content to the proposal. Instead, this section should summarize the important benefits you discussed in earlier sections of the proposal. In Figure 21.11, for example, you will notice that this proposal's costs and benefits section really doesn't add anything new to the proposal, except the cost of the project.

Why should you discuss the benefits at all, especially if you already mentioned them in the proposal? There are two main reasons:

- If you summarize the benefits at this point, readers will know exactly what they are going to receive if they say yes to your ideas.
- The costs are usually a bitter pill for most readers. Just telling them the price without reminding them of the benefits would make the price of the project hard to swallow.

At this point in the proposal, your readers will be measuring the costs of your plan against its benefits. By discussing costs and benefits at the same time, you can show them how the benefits outweigh these costs.

To see discussions of costs and benefits, go to
www.pearsonhighered.com/johnsonweb3/21.12

Costs and Benefits Section

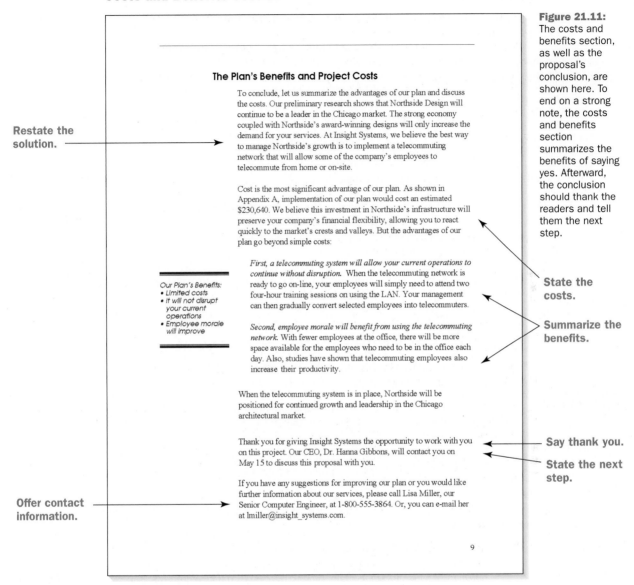

Figure 21.11: The costs and benefits section, as well as the proposal's conclusion, are shown here. To end on a strong note, the costs and benefits section summarizes the benefits of saying yes. Afterward, the conclusion should thank the readers and tell them the next step.

Restate the solution.

Offer contact information.

State the costs.

Summarize the benefits.

Say thank you.

State the next step.

The Plan's Benefits and Project Costs

To conclude, let us summarize the advantages of our plan and discuss the costs. Our preliminary research shows that Northside Design will continue to be a leader in the Chicago market. The strong economy coupled with Northside's award-winning designs will only increase the demand for your services. At Insight Systems, we believe the best way to manage Northside's growth is to implement a telecommuting network that will allow some of the company's employees to telecommute from home or on-site.

Cost is the most significant advantage of our plan. As shown in Appendix A, implementation of our plan would cost an estimated $230,640. We believe this investment in Northside's infrastructure will preserve your company's financial flexibility, allowing you to react quickly to the market's crests and valleys. But the advantages of our plan go beyond simple costs:

Our Plan's Benefits:
- *Limited costs*
- *It will not disrupt your current operations*
- *Employee morale will improve*

First, a telecommuting system will allow your current operations to continue without disruption. When the telecommuting network is ready to go on-line, your employees will simply need to attend two four-hour training sessions on using the LAN. Your management can then gradually convert selected employees into telecommuters.

Second, employee morale will benefit from using the telecommuting network. With fewer employees at the office, there will be more space available for the employees who need to be in the office each day. Also, studies have shown that telecommuting employees also increase their productivity.

When the telecommuting system is in place, Northside will be positioned for continued growth and leadership in the Chicago architectural market.

Thank you for giving Insight Systems the opportunity to work with you on this project. Our CEO, Dr. Hanna Gibbons, will contact you on May 15 to discuss this proposal with you.

If you have any suggestions for improving our plan or you would like further information about our services, please call Lisa Miller, our Senior Computer Engineer, at 1-800-555-3864. Or, you can e-mail her at lmiller@insight_systems.com.

9

Conclusion

The conclusion of a proposal should be concise, perhaps only one or two paragraphs.

Here at the end of the proposal is a good place to restate the main point of your proposal one last time. Of course, you have already told readers your solution at least a couple of times. Tell them again. This last repetition will leave them with a clear statement of what you want to achieve.

Also, you might thank your readers for their consideration of your ideas. Much like a public speaker signaling the end of a speech by thanking listeners for their attention, you can end your proposal by thanking readers for considering your plan. Thanking your readers will end your proposal on a positive note.

Finally, when concluding, you should leave your readers with a clear sense of what they should do when they finish reading your proposal. Should they call you? Should they wait for you to call them? Should they set up a meeting with you? A well-written conclusion makes their next move obvious for them.

In Figure 21.11, for example, the conclusion has been added to the end of the costs and benefits section. The authors thank the readers for their consideration and ask them to call when they are finished looking over the proposal.

AT A GLANCE

Concluding a Proposal

- Restate the proposal's main point (the solution).
- Say thank you.
- Describe the next step.
- Provide contact information.

Link

For more information on writing conclusions, go to Chapter 8, page 196.

Using Style in Proposals

Proposals are designed to both educate and persuade readers, so they tend to use a mixture of plain and persuasive style.

Link

For more strategies for using plain and persuasive style, see Chapter 9, page 206.

Plain style—Use plain style in places where description is most important, such as the current situation section, the project plan section, and the qualifications section.

Persuasive style—Use persuasive style in places where readers are expected to make decisions, such as the proposal's introduction and the costs and benefits section.

One persuasive style technique, called "setting a tone," is particularly effective when writing a proposal.

1. Determine how you want your proposal to sound. Do you want it to sound exciting, innovative, or progressive? Choose a word that best reflects the tone you want your readers to hear as they are looking over your proposal.

2. Put that word in the middle of your screen or sheet of paper. Circle it.

3. Write words associated with that word around it. Circle them also.

4. Keep mapping farther out until you have filled the page or screen.

Mapping to Set a Tone

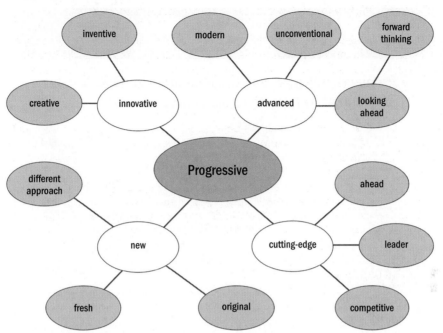

Figure 21.12: Logical mapping can help you develop a tone for your proposal. Weaving these words into a text would make the document sound "progressive."

Figure 21.12 shows a logical map of the word *progressive*. As shown in this map, you can quickly develop a set of words that are related to this word.

Once you have finished mapping out the tone, you can weave these words into your proposal to create a theme. As discussed in Chapter 9 of this book, if you use these words carefully and strategically in your proposal, your readers will hear this tone as they are reading your document. But, be sure not to overuse these words. Setting a tone is like adding spices in food. Used well, the spices will give the food an interesting flavor. When overused, their flavor will be too strong.

Designing Proposals

Computers have changed the way proposals are designed and delivered. Not long ago, it was common for proposals to include only minimal design with sparse graphics. Today, desktop publishing gives you a full range of page design options, including graphics, headers and footers, columns, and photographs, among other design elements. Increasingly, proposals are also being delivered in multimedia formats with video and sound.

Figure 21.13, for example, shows the first page of a well-designed proposal. Notice how the design of the proposal sets a professional, progressive tone for the whole document—even before you start reading.

Want more help developing your proposal's style? Go to www.pearsonhighered.com/johnsonweb3/21.13

Sample Proposal Design

Header adds color and anchors the top of the page. →

Title of proposal is bold and easy to read. →

Headings provide access points into text. →

Two-column format makes text more scannable. →

A Proposal for Upgrading the National-Scale Soil Geochemical Database for the United States

Figure 21.13: The page design of this proposal sets a professional tone while helping readers scan for important information. The tone is set with positive words and phrasings, while the graphics and headings make the information easy to access.

Graphic supports text while adding color. →

Source: U.S. Geological Survey, 2003.

To see samples of well-designed and badly designed proposals, go to
www.pearsonhighered.com/johnsonweb3/21.14

When designing your proposal, you should consider three components: graphics, page design, and medium:

Graphics—In proposals, it is common to include charts, graphs, maps, illustrations, photographs, and other kinds of graphics. You should look for places in your proposal where graphics can be used to reinforce important points.

Page design—Page layouts for proposals vary from simple to elaborate. At a minimum, you should use headings, lists, and graphics. More elaborate page layouts might include multiple columns, margin comments, pull quotes, and sidebars. Choose a page design that suits your readers and the context in which they will use the proposal.

Medium—The appropriate medium is also an important choice with the advent of computers. Paper is still the norm for most proposals, but increasingly companies are using CD-ROMs, websites, and presentation software to deliver their ideas.

In the rush to get the proposal out the door, you might be tempted to skip the design phase of the writing process. You might even convince yourself that visual design doesn't matter to readers. Don't fool yourself. Good design is very important in proposals. Design makes the proposal more attractive while helping readers locate important information in the document.

Link
For more information on using graphics, see Chapter 11, page 275.

Link
For more information on page and screen layout, see Chapter 10, page 233.

Link
For more information on developing oral presentations with presentation software, see Chapter 16, page 416.

Revising, Editing, and Proofreading

You and your team should always leave plenty of time to revise, edit, and proofread your proposal. At this point in the writing process, you can dramatically strengthen the content, organization, style, and design of the document.

Your time revising and editing will be well spent. After all, in today's competitive technical workplace, a carefully revised and edited proposal often determines the difference between success and failure.

As you revise, edit, and proofread your proposal, consider the following three areas:

Review the rhetorical situation—When you have completed a draft, look back at your original notes on the proposal's subject, purpose, readers, and context of use. In some cases your subject and purpose might have evolved as you drafted the proposal. Or, perhaps you now have a better understanding of your readers and the context in which the proposal will be used.

Edit for ethics—Proposals almost always involve ethical issues of some kind. Ask yourself whether the proposal stretches the truth in places. Does it hide important information? Are the graphics drawn ethically? Unethical proposals can lead to lawsuits or lost clients, so make sure your proposal anticipates any ethical issues that might be involved.

Link
For more information on ethics, see Chapter 5, page 92.

Revision worksheets for proposals are available at
www.pearsonhighered.com/johnsonweb3/21.15
Revising, Editing, and Proofreading
621

Jane Perkins
COMMUNICATIONS CONSULTANT, A.T. KEARNEY, CHICAGO, ILLINOIS
A.T. Kearney is a consulting firm that works in high-tech industries.

What makes a proposal successful?

Writing proposals is critical work at A.T. Kearney, a global management consulting firm made up primarily of MBAs with engineering backgrounds. It is how we obtain the majority of our business. Our success in writing proposals depends on key aspects of our proposal process:

- *Web-based processes for data and expertise.* We have formal processes in place to proactively gather important data that is necessary for providing potential clients with decision-making information. We also rely on an informal and collaborative network for tapping into expertise throughout the firm. These processes help us write proposals that include the necessary level of detail that persuades stakeholders.
- *A cohesive proposal and selling process.* For many opportunities, we develop a series of proposal documents, which are components of a negotiation/selling process, often extending over months. For example, an opportunity might begin with a formal RFP to down-select competitors. The RFP often specifies providing both on-line and print versions with strict requirements on content, length, and software application. For the next round, we would probably prepare a PowerPoint presentation for the client buyers. This round would open the door for refinements on the scope and initiate negotiations. The proposal process might then conclude with a proposal letter to summarize the aspects we had discussed.
- *Innovation and internal education.* Our proposal writing spurs much innovation and forward thinking, as we respond to evolving client needs and opportunities, and we build on recent project work. Proposal teams are not only educating clients through proposals but also important, internal stakeholders, such as those responsible for ensuring financials and quality delivery of the project. The use of eRooms and collaborative applications facilitates this learning process. Additionally, expert reviews at specified stages of the proposal process are part of our culture—leadership insights and approvals help the proposal team, while the latest thinking is spread across the firm.

As you can see, a well-written proposal is only part of a much larger process.

Link

For more information on editing and proofreading, see Chapter 12, page 306.

Proofread—You might be surprised how many proposals are rejected simply because of grammatical mistakes, typos, and misspellings. The importance of good proofreading makes sense. After all, why would readers put their money and faith in you if your proposal demonstrates carelessness and lack of quality? You need to proofread your text and graphics carefully for any errors.

Solid revision and editing are often the difference between proposals that succeed and those that fail. Make sure you reserve plenty of time for revising and editing.

CHAPTER REVIEW

- Proposals are documents that present ideas or plans for consideration.

- Proposals can be internal or external and solicited or unsolicited.

- Proposals usually include five sections: (1) introduction, (2) current situation, (3) project plan, (4) qualifications, and (5) costs and benefits.

- When planning the proposal, start out by defining the Five-W and How Questions. Then define the rhetorical situation (subject, purpose, readers, context of use).

- When organizing and drafting the proposal, each major section should be written separately with an opening, a body, and a closing.

- Proposals should use a combination of plain and persuasive style to motivate readers.

- Design is absolutely essential in professional proposals. You should look for places where page layout and graphics will improve the readability of your proposal.

- Don't skimp on the revision, editing, and proofreading of the proposal. A proposal that demonstrates poor quality in writing usually signals inattention to quality by the bidder (you).

EXERCISES AND PROJECTS

Individual or Team Projects

1. Analyze the proposal "'Meaningful' Outdoor Experiences for Students," at the end of this chapter. Study the content, organization, style, and design of the proposal. Does it cover the four areas (current situation, project plan, qualifications, costs and benefits) that were discussed in this chapter? Where does it include more or less information than you would expect in a proposal?

   Write a two-page memo to your teacher in which you discuss whether you believe the proposal was effectively written and designed. Discuss the content, organization, style, and design in specific details. Highlight the proposal's strengths and suggest ways in which it could be improved.

2. Find a Request for Proposals (RFP) at www.fedbizopps.gov or in the classifieds of your local newspaper. Analyze the RFP according to the Who, What, Where, When, Why, and How Questions. Then, prepare a presentation for your class in which you (a) summarize the contents of the RFP, (b) discuss why you believe the RFP was sent out, and (c) explain what kinds of projects would be suitable for this RFP.

3. Find a proposal that demonstrates weak style and/or design. Using the style and design techniques discussed in this chapter and in Chapters 9, 10, and 12, revise the document so that it is more persuasive and more visual. Locate places where

Exercises and Projects   623

blocks of text could be turned into lists, and identify places where graphics would reinforce the text.

Collaborative Project: Improving Campus

As students, you are aware—perhaps more so than the administrators—of some of the problems on campus. Write a proposal that analyzes the causes of a particular problem on campus and then offers a solution that the administration might consider implementing. The proposal should be written to a *named* authority on campus who has the power to say yes to your proposal and put it into action.

Some campus-based proposals you might consider include the following:

- improving lighting on campus to improve safety at night.
- improving living conditions in your dorm or fraternity/sorority.
- creating a day-care center on campus.
- creating an adult commuter room in the student union.
- improving campus facilities for the handicapped.
- improving security in buildings on campus.
- creating a university public relations office run by students.
- increasing access to computers on campus.
- improving the parking situation.
- reducing graffiti and/or litter on campus.
- improving food service in the student union.
- helping new students make the transition to college.
- changing the grading system at the university.
- encouraging more recycling on campus.
- reducing the dependence on cars among faculty and students.

Use your imagination to come up with a problem to which you can offer a solution. You don't need to be the person who implements the program. Just offer some guidance for the administrators. In the qualifications section of your proposal, you will likely need to recommend that someone else should do the work.

For additional technical writing resources, including interactive sample documents, document design tutorials and guidelines, and more, go to **www.mytechcommlab.com**.

Revision Challenge

This proposal in Figure A is written well, but it could be strengthened in organization, style, and design. Read through the proposal and identify places where you could make the proposal stronger.

To find sample proposals for analysis, go to
www.pearsonhighered.com/johnsonweb3/21.16

Sample Proposal

Figure A: This proposal is well written, but not perfect. How might you revise this document to make its content, organization, style, and design even stronger?

Project Summary

Organization:	**Monterey Bay High School**
Applicant Name:	Ted Watson
P.I.:	Ariel Jane Lichwart
Address:	590 Foam Street, Monterey, CA 93940
Telephone Number:	Ted: (831) 244-7843, Ariel: (831) 244-7842
Email Address:	Ted: Twatson@mbhs.edu, Ariel: Ajl@mbhs.edu
Partner:	Carmel River Watershed Council

Area of Interest:	**"Meaningful" Outdoor Experiences for Students**

Project Title:	A Meaningful Outdoor Experience for Students: To Monitor the Health of the Carmel River and Carmel Watershed

Project Period:	09/01/2003 – 10/31/2004

Project Objectives:

— To provide a meaningful outdoor experience for all of the 650 students at Monterey Bay High School;

— To assess the environmental quality and health of the Carmel River and the impact these problems have on the entire watershed and the larger ecological system;

— To teach the students how to monitor fish traps, conduct water quality tests, measure stream flow, operate plant surveys, perform bird density measurements (point counts), sample invertebrate and plankton tows, and perform herpetology surveys along the River;

— To assist local agencies that are concerned with the health of the Carmel River, to maintain existing populations and to assist in restoring those that are in decline through habitat restoration work, community-based education programs and increased local involvement in environmental projects;

— To raise the students' social awareness, stimulate observation, motivate critical thinking, and develop problem-solving skills.

Summary of Work:	To provide high school students with Meaningful Outdoor Experiences by assessing the environmental quality and health of the Carmel River and Watershed.

Total Federal Funds:	$50,000
Cost Sharing:	$25,643
Total Project Costs:	$75,643

Source: National Marine Sanctuaries, http://sanctuaries.noaa.gov/news/bwet/.

(continued)

Grant Proposal

Purpose of Project

Monterey Bay High School is requesting $50,000 from NOAA for Environmental Education Projects in the Monterey Bay Watershed to provide high school students with **Meaningful Outdoor Experiences** by assessing the environmental quality and health of the Carmel River and Watershed. The Carmel River Watershed Council will be our collaborative working partner during the course of the grant.

How

The project will take place over a one-year period of time, beginning September 1, 2003. It will consist of three phases: the preparation phase, the action phase, and the reflection phase. The preparation and reflection stages will each be 3 months long; the action phase will take place over a 6 month period.

Project Objectives

— To provide a meaningful outdoor experience for all of the 650 students at Monterey Bay High School;

— To assess the environmental quality and health of the Carmel River and the impact these problems have on the entire watershed and the larger ecological system;

— To teach the students how to monitor fish traps, conduct water quality tests, measure stream flow, operate plant surveys, perform bird density measurements (point counts), sample invertebrate and plankton tows, and perform herpetology surveys along the River;

— To assist local agencies that are concerned with the health of the Carmel River, to maintain existing populations and to assist in restoring those that are in decline through habitat restoration work, community-based education programs and increased local involvement in environmental projects;

— To raise the students' social awareness, stimulate observation, motivate critical thinking, and develop problem-solving skills.

Preparation Phase

All of the 650 students at Monterey Bay High School (grades 9-12) will participate in the preparation phase. At the outset of the preparation phase, we will propose the question: What is the present health of the Carmel River and how does the river's health affect the entire watershed?

During this phase the students and the teachers will do the initial background research on the Carmel River. The students will hear guest lectures delivered by our partner the Carmel River Watershed Council, and other state and local organizations that will be collaborating with us on this project (Carmel River Steelhead Association, Monterey Peninsula Audubon Society, Ventana Wilderness Society and the Monterey County Water Management District, California State Parks). These agencies will also supply us with data previously collected that charts the Carmel River's vegetation, birds, wildlife, and fisheries, invertebrates, and water quality. Certain problems will quickly become apparent to the students such as erosion, vegetation loss, coliform and other water pollutants, loss of wildlife densities and use of habitat, steelhead declines, and invertebrate declines. After discussion and analysis students will make their predictions about other potential problems in the Carmel River Watershed. We will then set up our goals to investigate in greater depth some of the problems associated with the River and the Watershed.

Action Phase
All of the students will participate in this field research phase. After a site visit to four selected Carmel River locations (upper Watershed, middle River site, site adjacent to the High School and Carmel River mouth), we will set up a phenological schedule (timeline) for sampling the sites with the students and the teachers. Each site will be sampled at least once per month over a six month period. Some of the activities and projects conducted at each site will include the monitoring of fish traps, conducting water quality tests, measuring stream flow, operating plant surveys, performing bird density measurements (point counts) and a bird banding program, sampling invertebrate and plankton tows, and performing herpetology surveys. All students will keep extensive journals, and record their observations and experiments at the River sites.

Reflection Phase
Following the six-month research project at the Carmel River Watershed sites, the students and teachers will discuss their findings, refocus on the initial question, analyze the conclusions reached, and evaluate the results. We will also conduct appropriate assessment activities to evaluate the project and the learning achieved. The students will be divided into teams, and each team will write up a different aspect of each finding. The students will present this data to all interested stakeholders, including the Carmel River Steelhead Association, the Monterey County Water Management District, Monterey Regional Parks, California State Parks, the Ventana Wilderness Society and the Monterey Peninsula Audubon Society. Their conclusions will be used by these agencies to help evaluate further watershed studies, to maintain existing populations, and to assist in restoring those that are in decline through habitat restoration work, community-based education programs, and increased local involvement in environmental projects. Local media will report on our results and the projects that will take place as a result of the students' work.

Why

1 - Demonstrate to students that local actions can impact the greater water environment (i.e.. Monterey Bay). An intentional connection needs to be made to water quality, the watershed, and the larger ecological system.

Through this project, the students will learn that the Carmel River is besieged with problems such as erosion, vegetation loss, coliform and other water pollutants, loss of wildlife densities and use of habitat, steelhead declines, and invertebrate declines. Part of

(continued)

the Reflection phase will be used to study the impact these problems in the River have on the entire Watershed and the larger ecological system. As with all of the lessons we teach at our habitat, the students will learn about the connectedness of all living creatures and all actions, both large and small.

2 - Experiences should include activities where questions, problems, and issues are investigated through data collection, observation, and hands-on activities.

Much of the work done throughout this project involves data collection, observation, and hands-on activities. In the preparation phase, the students and teachers will be analyzing data previously collected by our partner organization, the Carmel River Watershed Council, on the vegetation, birds, wildlife and fisheries, invertebrates, and water quality. In the action phase, the students and teachers will perform hands-on activities at four sites along the river, at least once a month at each site, over a six-month period. These activities include the monitoring of fish traps, conducting water quality tests, measuring stream flow, operating plant surveys, performing bird density measurements (point counts) and banding birds, sampling invertebrate and plankton tows, and performing herpetology surveys. In the reflection phase, the students will examine the questions asked in light of the data they collected. They will then share their data with supporting local and state organizations so that some of the problems discovered can begin to be eradicated.

3 - Experiences should stimulate observation, motivate critical thinking, develop problem solving skills, and instill confidence in students.

There is no question but that this project will stimulate observation, motivate critical thinking, and develop problem solving skills. As a result, the students will acquire confidence, not only in their investigative skills, but also to be active participants in the public debate on many environmental issues. It is our experience that projects such as this one not only raise social awareness, but academic skills as well. It is a well known fact that students learn better by doing; by aligning the work done for this project with California state Standards (see number 7) the academic performance of the students will clearly be enhanced.

4 - Activities should encourage students to assist, share, communicate, and connect directly with the outdoors. Experiences can include: (1) Investigative or experimental design activities where students or groups of students use equipment, take measurements and make observations for the purpose of making interpretations and reaching conclusions; (2) Project-oriented experiences, such as restoration, monitoring, and protection projects, that are problem solving in nature and involve many investigative skills. These experiences should involve fieldwork, data collection, and analysis.

As stated above, all of the activities for this project involve investigative or experimental design activities where the students use equipment, take measurements, and make observations for the purpose of making interpretations and reaching conclusions. In addition, many of the activities in the action phase involve monitoring projects that are problem solving in nature and involve many investigative skills. All of the experiences involve fieldwork, data collection, and analysis.

5- The "Meaningful" outdoor experiences need to be part of a sustained activity; the total duration leading up to and following the activity should involve a significant investment of instructional time.

Since many of the activities that will take place during all three phases will be aligned with the California State Standards, the "Meaningful" outdoor experiences will be embedded into the science curriculum (see number 7), and will encompass at least 10% of instructional time over the year period of the grant. This sustained activity will involve a significant outlay of instructional time, not just in the classroom, but in the field as well. As the budget narrative sets forth, all teachers, the principal investigator, and the field coordinators will invest many hours preparing and teaching the lessons as well as helping the students to perform the tasks at the site.

6- An experience should consist of three general parts: a preparation phase—which focuses on a question, problem, or issue and involves students in discussions about it; an action phase—which includes one or more outdoor experiences sufficient to conduct the project, make the observations, or collect the data required; and the reflection phase—which refocuses on the question, problem, or issue, analyzes the conclusion reached, evaluates the results, and assesses the activity and the learning.

This project will consist of a preparation phase, an action phase, and a reflection phase. The details of these are set forth above.

7- "Meaningful" outdoor experiences must be an integral part of the instructional program and clearly part of what is occurring concurrently in the classroom; be aligned with the California academic learning standards; and make appropriate connections among subject areas and reflect an integrated approach to learning. Experiences should occur where and when they fit into the instructional sequence.

All of the sampling and research will be aligned with the California State Standards. For example, water sampling fits into our 9th grade chemistry curriculum; erosion aligns with the earth science requirement for 10th and 11th grade; wildlife monitoring is part of the scientific investigation component that is required for all grade levels. In addition, this Project is a perfect complement to the River of Words program, where the students submit poetry and art to the international contest that is based on their experiences and observations of their Watershed (see description, above).

8- Project should demonstrate partnerships that form a collaborative working relationship, with all partners taking an active role in the project.

The Carmel River Watershed Council (CRWC) will be a collaborative working partner during the course of this grant. The CRWC is a nonprofit, community-based organization

(continued)

founded in 1999 to work with local, state, and federal agencies for improved management of the Carmel River Watershed. The primary mission of the CRWC is the protection of the natural resources that form the Carmel River Watershed.

The CRWC will take an active role in this Project. During the course of this Project, they will provide the students and teachers at Monterey Bay High School with data that has been collected by them and other agencies concerning the vegetation, birds, wildlife and fisheries, invertebrates, and water quality so that the students can assess and analyze the potential problems in the Carmel River. The principal investigator, who is a consultant for the CRWC (and is also a teacher at the High School), will organize and coordinate the project, and provide, on an ongoing basis, expertise, teaching skills, and leadership to guide the staff in working with the students in the sampling of the river, monitoring of fish traps, conducting water quality tests, measuring stream flow, operating plant surveys, performing bird density measurements and bird banding, sampling invertebrate and plankton tows, and performing herpetology surveys. In all of these activities, he and other consultants at CRWC will work with the staff at the Biological Sciences Project and the participating teachers at the Monterey Bay High School to provide the students with a meaningful outdoor experience that will encourage them to have greater pride and ownership of their environment and become more effective and thoughtful community leaders and participants. The students will be doing important research that will benefit not only the Carmel River Watershed Council, but other local and state agencies, such as the Carmel River Steelhead Association, the Monterey County Water Management District, Monterey Regional Parks, California State Parks, the Ventana Wilderness Society, and the Monterey Peninsula Audubon Society. All of the data that is gathered will be shared with these groups through student presentations and be used to evaluate future watershed activities that will help to maintain existing populations and assist in restoring ones that are in decline through habitat restoration work, community-based education programs, and increased local involvement in community projects.

Who

The principal organization is the Monterey Bay High School. The Carmel River Watershed Council will be collaborating as our partner. Both organizations have been described in great detail, above.

The target audience is high school students, grades 9–12. All of the 650 students in the school will participate in all three phases of the project. The students and teachers at Monterey Bay High School, as well as the consultants that work with the Carmel River Watershed Council, have a great deal of experience using outdoor, hands-on activities as an integral part of the educational process. This project is an ideal and natural complement to the activities that take place at our habitat. We are committed to using the environment as a context for learning and improving the understanding of environmental stewardship of students and teachers. Many of the teachers at the school have extensive experience in teaching bird banding, data analysis, and data measurement, and they understand how to teach these types of scientific techniques to this age group of students. For those that do not have expertise in the subject matter that will be emphasized in this Project, the

Principal Investigator and other consultants from local agencies with expertise in these areas will provide **Professional Development** for the teachers.

The Project Manager for the Habitat will be coordinating this project and the Project Intern will assist in the gathering and computation of the data. Four teachers, all from the science department, will participate in all three phases of the Project, and will serve as field coordinators.

A consultant from the Carmel River Watershed Council, who is also a science teacher at the High School, will be directing the study and serve as the Principal Investigator. He will coordinate the three phases, oversee the teachers, and synchronize the visits and data collection during the action phase (see number 8, above).

Where

The Project will take place at Monterey Bay High School and on the Carmel River. The Carmel River is 36 miles long. It drains about 255 square miles while flowing NW out of the valley between the Santa Lucia Mountains on the South and the Sierra del Salinas to the North and East. The river empties into the Pacific Ocean near Carmel, California. The lower river flows from the mouth to the narrows, about 9 miles up stream; the middle river flows from the narrows to Camp Stephani; the upper river flows through rugged canyons.

We will set up four study sites: at the Carmel River mouth; at a site close to the High School; in the mid-Valley area; and at Cachagua Community Park below Los Padres Dam.

Need

This important project could not take place without government financial assistance. The financial health of the state and county budgets preclude funding for projects such as this one; in addition private grant funding is becoming increasingly difficult to secure. Without the financial aid of the federal government, these types of meaningful outdoor experiences for students could not take place.

Benefits or Results Expected

There are a number of benefits and results that will be derived from the proposed activities. First, the students will gather important data that will help to clarify what can be done to help bring the Carmel River and the Carmel Watershed to a state of good health. The students will be doing significant research that will benefit not only the Carmel River Watershed Council, but other local and state agencies, such as the Carmel River Steelhead Association, the Monterey County Water Management District, Monterey Regional Parks, California State Parks, the Ventana Wilderness Society, and the Monterey Peninsula Audubon Society. All of the data that is gathered will be shared with these groups through student presentations and be used to evaluate future watershed

(continued)

activities that will help to maintain existing populations and assist in restoring ones that are in decline through habitat restoration work, community-based education programs, and increased local involvement in community projects.

In addition, the process of progressing through the preparation phase, the action phase, and the reflection phase will provide the students with a meaningful outdoor experience that will encourage them to have greater pride and ownership of their environment and become more effective and thoughtful community leaders and participants. This project will stimulate observation, motivate critical thinking, and develop problem solving skills. As a result, the students will acquire confidence, not only in their investigative skills, but also to be active participants in the public debate on many environmental issues.

Project Evaluation

We will employ various methodologies to ensure that we are meeting the goals and objectives of our project. First, we will give pre and post tests to the students before the project begins and at its conclusion to determine if they have learned the science and skills that we want to teach through this project. In addition, students will be tested periodically throughout the course of the project as part of their coursework requirements.

Second, all students will keep extensive journals in which they will evaluate the data they have collected. At the end of the project, they will be asked to summarize their findings and submit them in a report. In addition, each student will be required to perform an independent science experiment that relates to the work being done at the River (i.e., measure the amount of nitrates in the water).

Third, the results of the data that are gathered will be presented by the students to the local and state agencies that are stakeholders in the Carmel River Watershed. These reports will be evaluated and critiqued by these agencies, along with suggestions for future studies.

Fourth, the students will use their experiences at the river as the basis for the art and poetry they submit to the River of Words competition that takes place each spring.

Finally, interested classes of incoming 9th graders will be invited to take a trip to one of the sites where the High School students can mentor or teach them some of the skills that they have learned.

The Mole

Henry Espinoza knew this proposal was important to his company, Valen Industries. If the proposal was successful, it could be worth millions of dollars in short-term and long-term projects for his employer. For Henry, winning the project meant a likely promotion to vice president and a huge raise in salary. The company's CEO was calling him daily to check up and see if it was going well. That meant she was getting stressed out about the project.

Through the grapevine, Henry knew his company really had only one major competitor for the project. So, the odds were good that Valen Industries would win the contract. Of course, Henry and his team needed to put together an innovative and flawless proposal to win, because their major competitor was going all out to get this project.

Then, one day something interesting happened. One of Henry's team members, Vera Houser, came to him with an e-mail from one of her friends, who worked for their main competitor. In the e-mail, this person offered to give Valen Industries a draft of its competitor's proposal. Moreover, this friend would pass along any future drafts as the proposal evolved. In the e-mail, Vera's friend said he was very frustrated working at his "slimy" company, and he was looking for a way out. He also revealed that his company was getting "inside help" on the proposal, but Henry wasn't sure what that meant.

Essentially, this person was hinting that he would give Henry's team a copy of the competitor's proposal and work as a mole if they considered hiring him at Valen Industries. Henry knew he would gain a considerable advantage over his only competitor if he had a draft of its proposal to look over. The ability to look at future drafts would almost ensure that he could beat his competitor. Certainly, he thought, with this kind of money on the line, a few rules could be bent. And it seemed that his competitor was already cheating by receiving inside help.

Henry went home to think it over. What would be the ethical choice in this situation? How do you think Henry should handle this interesting opportunity?

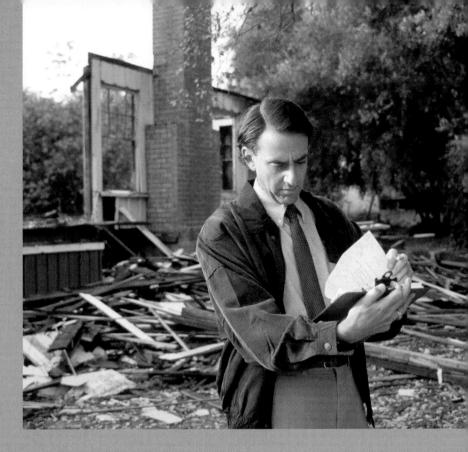

In this chapter, you will learn:

- About the different kinds of activity reports and how they are used in the workplace.

- The basic features of activity reports.

- How to determine the rhetorical situation of an activity report.

- How to organize and draft an activity report.

- Strategies for using an appropriate style.

- How to design and format activity reports.

- The importance of editing and proofreading activity reports.

Today, companies are using computer networks to create management structures that are less hierarchical. As a result, companies require fewer levels of managers than before, because computer networks help top executives better communicate with employees throughout the company.

These "flatter" management structures require more communication, quicker feedback, and better accountability among employees in the company. As a result, activity reports are more common than ever in the technical workplace.

Here are a few examples of kinds of activity reports and how they might be used:

- An electrical engineer would write a progress report that discusses a wire harness her team is developing for a new car.
- A scientist would give a briefing to a government agency on the possibilities of a hydrogen-powered passenger train.
- A chemist would write an incident report to explain a recent accident in the laboratory in which chlorine gas was released.
- A technician might write a lab report that presents his finding of malaria in mosquitoes captured in South America.

Activity reports are used to objectively present ideas or information within a company. As a genre, the activity report has many subgenres, making it adaptable to many situations that you will encounter in the technical workplace.

Basic Features of Activity Reports

An activity report usually includes the following features, which can be modified to suit the needs of the situations in which the report will be used:

- Introduction
- Summary of activities
- Results of activities or research
- Future activities or research
- Incurred or future expenses
- Graphics

Keep in mind that activity reports are used for a variety of purposes. You should adjust this pattern to suit your needs. Figure 22.1 shows the basic pattern for an activity report. Sections can be added and subtracted from this basic pattern, depending on your reasons for writing the report.

Types of Activity Reports

Even though the various types of activity reports are similar in most ways, they are called a variety of names that reflect their different purposes. Activity reports share one goal—to objectively inform readers about (1) what happened, (2) what is happening, and (3) what will happen in the near future.

Basic Pattern for an Activity Report

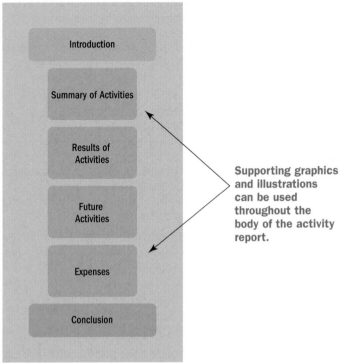

Figure 22.1: Progress reports and briefings emphasize what happened, the results of the activity, and what will happen in the future.

Introduction

Summary of Activities

Results of Activities

Future Activities

Expenses

Conclusion

Supporting graphics and illustrations can be used throughout the body of the activity report.

Progress Reports

A progress report, also called a status report, is written to inform management about the progress or status of a project. These reports are usually written at regular intervals—weekly, biweekly, or monthly—to update management on what has happened since the last progress report was submitted. Your company's management may also periodically request a progress report to stay informed about your or your team's activities.

A typical progress report will provide the following information:

- a summary of completed activities
- a discussion of ongoing activities
- a forecast of future activities

Figure 22.2, for example, shows a progress report that is designed to update management on a project.

Briefings and White Papers

Briefings and white papers are used to inform management or clients about an important issue. Typically, briefings are presented verbally, while white papers are provided in print. Occasionally, briefings will also appear as "briefs" in written form.

Hanson Engineering

March 14, 2008
To: Charlie Peterson, Director
From: Sue Griego, Iota Team Manager
Subject: Progress Report, March 14

Subject, purpose, and main point are identified up front.

This month, we made good progress toward developing a new desalinization method that requires less energy than traditional methods.

Ongoing activities are described objectively.

Our activities have centered around testing the solar desalinization method that we discussed with you earlier this year. With solar panels, we are trying to replicate the sun's natural desalinization of water (Figure A). In our system, electricity from the photovoltaic solar panels evaporates the water to create "clouds" in a chamber, similar to the way the sun makes clouds from ocean water. The salt deposits are then removed with reverse osmosis, and freshwater is removed as steam.

The graphic supports the text.

Figure A: The Desalinator

Results are presented.

We are succeeding on a small scale. Right now, our solar desalinator can produce an average of 2.3 gallons of freshwater an hour. Currently, we are working with the system to improve its efficiency, and we soon hope to be producing 5 gallons of freshwater an hour.

Report ends with a look to the future and a brief conclusion.

We are beginning to sketch out plans for a large-scale solar desalinization plant that would be able to produce thousands of gallons of freshwater per hour. We will discuss our ideas with you at the April 17 meeting.

Our supplies and equipment expenses for this month were $8,921. Looks like things are going well. E-mail me if you have questions. (Suegriego@hansoneng.net)

Briefings and white papers typically present gathered facts in a straightforward and impartial way. They include the following kinds of information:

- a summary of the facts
- a discussion of the importance of these facts
- a forecast about the importance of these facts in the future

An effective briefing presents the facts as concisely as possible, leaving time for questions and answers. When you brief an audience on your subject, try to do so as objectively as possible. Then, interpret the importance of those facts based on evidence, not on speculation.

Briefings and white papers typically do not advocate for any specific side or course of action. Instead, they present the facts in a straightforward way and offer an objective assessment of what those facts mean. It is up to the readers to decide what actions are appropriate.

Figure 22.3 shows the executive summary of a white paper written by senior officers at PayPal. The white paper discusses PayPal's attempts to manage "phishing," a fraudulant form of e-mail spam.

Incident Reports

Incident reports describe an event, usually an accident or irregular occurrence, and they identify what corrective actions have been taken. As with other kinds of activity reports, incident reports present the facts as objectively as possible. They provide the following information:

- a summary of what happened (the facts)
- a discussion of why it happened
- a description of how the situation was handled
- a discussion of how the problem will be avoided in the future

A White Paper

PayPal

A Practical Approach
to Managing Phishing
Michael Barrett, Chief Information Security Officer
Dan Levy, Senior Director of Risk Management – Europe
April 2008

Figure 22.3:
A white paper presents technical information objectively, allowing readers to make decisions based on the facts.

Source: From A Practical Approach to Managing Phishing *by Michael Barrett, Chief Information Security Officer, and Dan Levy, Director of Risk Management–Europe. PayPal, April 2008. These materials have been reproduced with the permission of PayPal, Inc. COPYRIGHT © 2008 PAYPAL, INC. ALL RIGHTS RESERVED.*

GO TO THE NET

To see white papers, go to
www.pearsonhighered.com/johnsonweb3/22.2

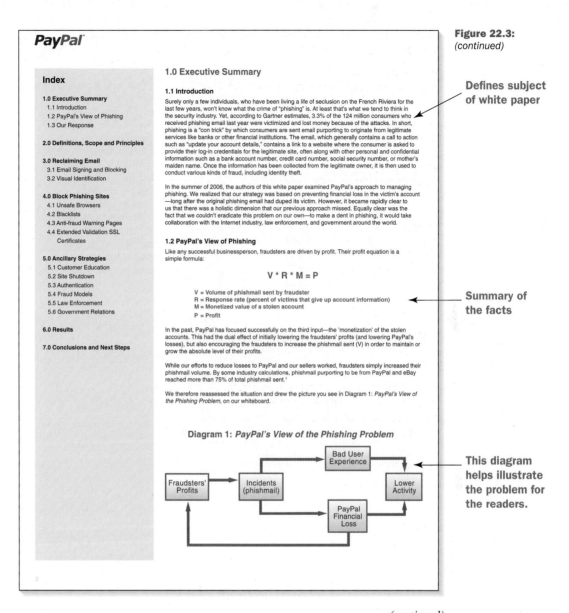

Defines subject of white paper

Summary of the facts

This diagram helps illustrate the problem for the readers.

(continued)

Figure 22.3:
(continued)

We knew that fraudsters' profits drove them to send phishmail. In turn, phishmail drove two responses from our customers: a bad user experience and/or financial loss. Even though the vast majority of users would immediately delete the mail or not click on the phishing links, the negative experience caused residual concern about whether or not PayPal had given up their email addresses or otherwise caused them to be a target.

A much smaller group of users would give up their private financial data (including their passwords), leading to financial loss for PayPal and our customers. And of course, PayPal's financial loss was the fraudsters' profit—thus encouraging even more phishmail and perpetuating the cycle. No matter which path the customer followed, PayPal observed consistently lower activity from these users.

1.3 Our Response

We determined five blocking points (illustrated in Diagram 2: Five Blocking Points) where PayPal could break the business model of fraudsters, and we applied a strategy against each point. There would not be one silver bullet, but perhaps five would do the trick.

Explains the team's solution for addressing the problem →

Diagram 2: *Five Blocking Points*

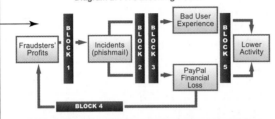

"...phishing is only the first step in a series of crimes that also frequently includes unauthorized account access, identity theft, financial fraud, money laundering, and others. While all of these crimes need to be considered together, this paper focuses primarily on the first step of this crime – phishing, specifically via email."

Block / Strategy	Short Explanation
1. Reclaim email	Prevent phishmail from ever entering customers' inboxes. To be judged by the adoption of email authentication by ISPs who agree not to deliver unsigned email.
2. Block phishing sites	Prevent phishing sites from being displayed to customers.
3. Authenticate users	Prevent stolen login/password combinations from being used on PayPal.com.
4. Prosecute	Create a disincentive by pursuing legal prosecution (in partnership with government and law enforcement).
5. Brand & Customer Recovery	Ensure that targeted customers would still use PayPal.

This paper concentrates on our top two strategies of reclaiming email and blocking phishing sites. For each, we will address both our "passive" and our "active" solutions, corresponding to the different segments of our user base. There are also a few other strategies that we believe are important, and while they may be ancillary to those described above, we'll give some detail about them nonetheless.

3

An Incident Report

Figure 22.4:
An incident report is not the place to make apologies or place blame. You should state the facts as objectively as possible.

Subject and purpose are stated up front. →

What happened is described objectively. →

What was done about it is noted. →

Red Hills Health Sciences Center ⓇH

Testing and Research Division
201 Hospital Drive, Suite A92
Red Hills, CA 92698

March 10, 2008

To: Brian Jenkins, Safety Assurance Officer

From: Hal Chavez, Testing Laboratory Supervisor

Subject: Incident Report: Fire in Laboratory

I am reporting a fire in Testing Laboratory 5, which occurred yesterday, March 9, 2008, at 3:34 p.m.

The fire began when a sample was being warmed with a bunsen burner. A laboratory notebook was left too close to the burner, and it caught fire. One of our laboratory assistants, Vera Cather, grabbed the notebook and threw it into a medical waste container. The contents of the waste container then lit on fire, filling the room with black smoke. At that point, another laboratory assistant, Robert Jackson, grabbed the fire extinguisher and emptied its contents into the waste container, putting out the fire. The overhead sprinklers went off, dousing the entire room.

Even though everyone seemed fine, we decided to send all lab personnel down to the emergency room for an examination. While we were in the waiting room, Vera Cather developed a cough and her eyes became red. She was held for observation and released that evening when her condition was stable. The rest of us were looked over by the emergency room doctors, and they suggested that we stay out of the laboratory until it was thoroughly cleaned.

I asked the hospital's HazMat team to clean up the mess that resulted from the fire. We had been working with samples of *Borrelia burgdorferi* bacteria, which causes Lyme disease. I was not sure if the

(continued)

Want to see other incident reports? Go to
www.pearsonhighered.com/johnsonweb3/22.3

Types of Activity Reports

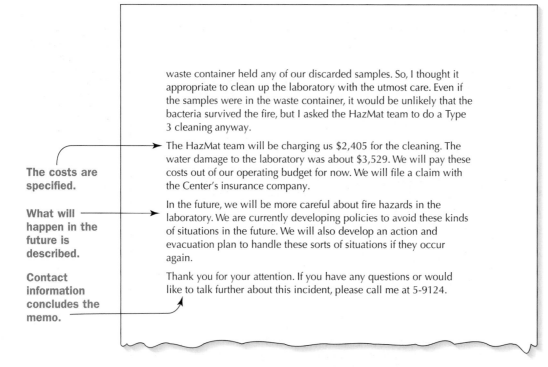

The costs are specified.

What will happen in the future is described.

Contact information concludes the memo.

waste container held any of our discarded samples. So, I thought it appropriate to clean up the laboratory with the utmost care. Even if the samples were in the waste container, it would be unlikely that the bacteria survived the fire, but I asked the HazMat team to do a Type 3 cleaning anyway.

The HazMat team will be charging us $2,405 for the cleaning. The water damage to the laboratory was about $3,529. We will pay these costs out of our operating budget for now. We will file a claim with the Center's insurance company.

In the future, we will be more careful about fire hazards in the laboratory. We are currently developing policies to avoid these kinds of situations in the future. We will also develop an action and evacuation plan to handle these sorts of situations if they occur again.

Thank you for your attention. If you have any questions or would like to talk further about this incident, please call me at 5-9124.

It is tempting, especially when an accident was your fault, to make excuses or offer apologies, but an incident report is not the place. As with other activity reports, you should concentrate on the facts. Describe what happened as honestly and clearly as possible. You can make excuses or apologize later.

Figure 22.4 shows a typical incident report in which management is notified of an accident in a laboratory.

Laboratory Reports

Laboratory reports are written to describe experiments, tests, or inspections. If you have taken a laboratory class, you are no doubt familiar with lab reports. These reports describe the experiment, present the results, and discuss the results. Lab reports typically include the following kinds of information:

- a summary of the experiment (methods)
- a presentation of the results
- a discussion of the results

Lab reports, like other activity reports, emphasize the facts and data. Here is not the place to speculate or develop a new theory. Instead, your lab report should present the results as objectively as possible and use those results to support the reasoned discussion that follows.

Figure 22.5 shows an example of a laboratory report. In this report, the writer describes the results of the testing as objectively as possible.

Figure 22.5:
A lab report walks readers through the methods, results, and discussion. Then, it offers any conclusions, based on the facts.

FEND-LAB, INC.

2314 Universal St., Suite 192
San Francisco, CA 94106
(325) 555-1327
www.fendlabcal.com

Test Address
NewGen Information Technology, LLC
3910 S. Randolph
Slater, CA 93492

Client
Brian Wilson
Phone: 650-555-1182
Fax: 650-555-2319
e-mail: brian_wilson@cssf.edu

Mold Analysis Report
Report Number: 818237-28
Date of Sampling: 091208
Arrival Date: 091408
Analysis Date: 091908
Technician: Alice Valles

Lab Report: Mold Test

The introduction states the subject, purpose, and main point.

In this report, we present the results of our testing for mold at the offices of NewGen Information Technology, at 3910 S. Randolph in Slater, California. Our results show above-normal amounts of allergenic mold, which may lead to allergic reactions among the residents.

Testing Methods

On 12 September 2008, we took samples from the test site with two common methods: Lift Tape Sampling and Bulk Physical Sampling.

Methods are described, explaining how the study was done.

Lift Tape Sampling. We located 10 areas around the building where we suspected mold or spores might exist (e.g., water stains, dusty areas, damp areas). Using 8-cm-wide strips of transparent tape, we lifted samples and pressed them into the nutrient agar in petri dishes. Each sample was sealed and sent to our laboratory, where it was allowed to grow for one week.

(continued)

Bulk Physical Sampling. We located 5 additional areas where we observed significant mold growth in ducts or on walls. Using a sterilized scraper, we removed samples from these areas and preserved them in plastic bags. In one place, we cut a 1-inch-square sample from carpet padding because it was damp and contained mold. This sample was saved in a plastic bag. All the samples were sent to our laboratory.

At the laboratory, the samples were examined through a microscope. We also collected spores in a vacuum chamber. Mold species and spores were identified.

Results of Microscopic Examination

The following chart lists the results of the microscope examination:

Mold Found	Location	Amount
Trichoderma	Break room counter	Normal growth
Geotrichum	Corner, second floor	Normal growth
Cladosporium	Air ducts	Heavy growth
Penicillium spores	Corkboard in bathroom	Normal growth

Descriptions of molds found:

Trichoderma: Trichoderma is typically found in moistened paper and unglazed ceramics. This mold is mildly allergenic in some humans, and it can create antibiotics that are harmful to plants.

Geotrichum: Geotrichum is a natural part of our environment, but it can be mildly allergenic. It is usually found in soil in potted plants and on wet textiles.

Cladosporium: Cladosporium can cause serious asthma and it can lead to edema and bronchiospasms. In chronic cases, this mold can lead to pulmonary emphysema.

Penicillium: Penicillium is not toxic to most humans in normal amounts. It is regularly found in buildings and likely poses no threat.

Results are presented objectively, without interpretation.

2

Discussion of Results

It does not surprise us that the client and her employees are experiencing mild asthma attacks in their office, as well as allergic reactions. The amount of Cladosporium, a common culprit behind mold-caused asthma, is well above average. More than likely, this mold has spread throughout the duct system of the building, meaning there are probably no places where employees can avoid coming into contact with this mold and its spores.

The other molds found in the building could be causing some of the employees' allergic reactions, but it is less likely. Even at normal amounts, Geotrichum can cause irritation to people prone to mold allergies. Likewise, Trichoderma could cause problems, but it would not cause the kinds of allergic reactions the client reports. Penicillium in the amounts found would not be a problem.

The results of our analysis lead us to believe that the Cladosporium is the main problem in the building.

Conclusions

The mold problem in this building will not go away over time. Cladosporium has obviously found a comfortable place in the air ducts of the building. It will continue to live there and send out spores until it is removed.

We suggest further testing to confirm our findings and measure the extent of the mold problem in the building. If our findings are confirmed, the building will not be safely habitable until a professional mold remover is hired to eradicate the mold.

Ignoring the problem would not be wise. At this point, the residents are experiencing mild asthma attacks and occasional allergic reactions. These symptoms will only grow worse over time, leading to potentially life-threatening situations.

Contact us at (325) 555-1327 if you would like us to further explain our methods and/or results.

3

Results are interpreted and discussed.

The conclusion restates the main point and recommends action.

Planning and Researching Activity Reports

One of the nice things about writing activity reports is that you have already developed most of the content. You are probably familiar with your readers. So, minimal planning is required, and the research has been mostly completed. These internal reports, after all, are supposed to describe your activities.

A good workplace practice you might adopt is keeping an *activity journal* or *work log* on your computer or in a notebook. In your journal, start out each day by jotting down the things you need to accomplish. As you complete each of these activities, note the dates and times they were completed and the results.

Computers can also help you keep track of your activities. Software like Llamagraphic's Life Balance and groupware programs like Lotus Notes and Microsoft Outlook can maintain "to do" lists while keeping a calendar of your activities (Figure 22.6). Similarly, personal digital assistants (PDAs) can also maintain a "to do" list and your calendar. When you complete a task, you can cross it off your list. The PDA will keep track of the time and date.

At first, keeping an activity journal will seem like extra work. But, you will soon realize that your journal keeps you on task and saves you time in the long run. Moreover, when you need to report on your activities for the week or month, you will have a record of all the things you accomplished.

Page from an Activity Journal

The "To Do" list keeps track of what needs to be handled each day.

Color-coding allows users to prioritize items on the "To Do" list.

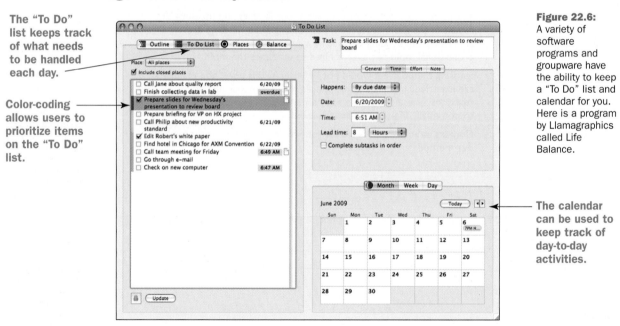

Figure 22.6: A variety of software programs and groupware have the ability to keep a "To Do" list and calendar for you. Here is a program by Llamagraphics called Life Balance.

The calendar can be used to keep track of day-to-day activities.

Source: Llamagraphics, Life Balance.

For links to software packages that help you keep activity journals, go to
www.pearsonhighered.com/johnsonweb3/22.5

Analyzing the Rhetorical Situation

With your notes in front of you, you are ready to plan out your activity report. You should begin by briefly answering the Five-W and How Questions:

> *Who might read or use this activity report?*
>
> *Why do they want the report?*
>
> *What information do they need to know?*
>
> *Where will the report be used?*
>
> *When will the report be used?*
>
> *How might the report be used?*

After considering these questions, you can begin thinking about the rhetorical situation that will shape how you write the activity report or present your briefing.

SUBJECT   The subject of your report includes your recent activities. Include only information your readers need to know. Remove any unneeded want-to-tell information. After all, you did many things that *could* be mentioned in the report, but your readers don't need to hear about all of them.

PURPOSE   The purpose of your report is to describe what happened and what will happen in the future. In your introduction, state your purpose directly:

> In this memo, I will summarize our progress on the Hollings project during the month of August 2008.

> The purpose of this briefing is to update you on our research into railroad safety in north-western Ohio.

You might use some of the following action verbs to describe your purpose:

to explain	*to show*	*to demonstrate*
to illustrate	*to present*	*to exhibit*
to justify	*to account for*	*to display*
to outline	*to summarize*	*to inform*

Link

To learn more about defining a purpose, go to Chapter 2, page 21.

READERS   Think about the people who will need to use your report. The readers of activity reports tend to be your supervisors. Occasionally, though, these kinds of reports are read by clients (lab reports or briefings) or used to support testimony (white papers). An incident report, especially when it concerns an accident, may have a range of readers who plan to use the document in a variety of ways. An insurance adjuster, for example, may use the incident report very differently than the quality control officer at your company would. You need to compose your report to suit both of these readers' needs.

Link

For more ideas about reader analysis, turn to Chapter 3, page 40.

TomWeber

ELECTRICAL ENGINEER, SANDIA NATIONAL LABORATORIES,
ALBUQUERQUE, NEW MEXICO

Tom Weber works as an electrical engineer and previously worked as a technical advisor for the U.S. Department of Defense.

What are white papers and how are they used?

In Washington, D.C., engineers and scientists often act as advisors to the executive and legislative branches and play an important role in briefing policymakers and congressional staff on technical issues. Because time for meetings is limited, these briefings are usually accompanied by a short technical document, or *white paper,* that explains the issue in question.

Although meant to be an objective technical appraisal, white papers will often influence a policy decision or promote a specific program or research agenda. These papers are short, usually four pages or fewer, and cover the technical goals and benefits of a new policy or program. For example, a white paper on power system reliability, with a technical explanation of recent problems and proposed solutions, might be given to administrators at the Department of Energy considering new regulatory policies for the electric power industry. A white paper on the benefits of nanotechnology might be given to congressional staff working on next year's budget priorities for federally funded research and development programs.

In a white paper, the writer must clearly explain, to a mostly nontechnical audience, the relevant technical issues, and come to a logical conclusion as to the best decision and path forward within current political realities. All this must be accomplished in just a few pages. With the incredible information flow in our nation's capital, most government decision makers will have only a few minutes to review any one document.

CONTEXT OF USE The context of use for your activity report will vary. In most cases, your readers will simply scan and file your report. Similarly, oral briefings are not all that exciting. Your listeners will perk up for the information that interests them, but they will mostly be checking to see if you are making progress.

Nevertheless, take a moment to decide whether your activity report discusses any topics that involve troublesome ethical or political issues. Throughout your career, you will find that these kinds of seemingly insignificant reports will surface when things go wrong at your company. When mistakes happen, auditors and lawyers will go through your activity reports, looking for careless statements or admissions of fault. So, your statements need to reflect your actual actions and the results of your work.

Moreover, if you are reporting expenses in your activity report, they need to be accurate. Auditors and accountants will look at these numbers closely. If your numbers don't add up, you may have some explaining to do.

Link

For more help defining the context, go to Chapter 3, page 46.

Worksheets to help you define your purpose, readers, and context of use can be found at
www.pearsonhighered.com/johnsonweb3/22.7

Organizing and Drafting Activity Reports

Remember, organizing and drafting activity reports should not take too much time. If you find yourself taking more than an hour to write an activity report, you are probably spending too much time on this routine task.

To streamline your efforts, remember that all technical documents have an introduction, a body, and a conclusion. Each of these parts of the document makes predictable moves that you can use to guide your drafting of the report.

Writing the Introduction

Readers of your activity report are mostly interested in the facts. So, your introduction should give them only a brief framework for understanding those facts. To provide this framework, you should concisely

- define your subject.
- state your purpose.
- state your main point.

Figures 22.2, 22.4, and 22.5 show examples of concise introductions that include these three common introductory moves.

If your readers are not familiar with your project (e.g., you are giving a demonstration to clients), you might want to expand the introduction by also offering background information, stressing the importance of the subject, and forecasting the body of the report.

Figure 22.7 shows the introduction of a document that would accompany a demonstration. This example includes all the typical moves found in an introduction. The first paragraph offers background information. The second paragraph states the purpose and defines the subject. Then, the author stresses the importance of the subject and states the main point of the demonstration. The last three sentences forecast what will happen during the demonstration.

Link

For more advice on writing introductions, turn to Chapter 8, page 182.

Writing the Body

In the body of the activity report, you should include some or all of the following:

Summary of activities—In chronological order, summarize the project's two to five major events since your previous activity report. Don't dwell too much on the details. Instead, tell your readers briefly what you did and highlight any advances or setbacks in the project.

Results of activities or research—In order of importance, list the two to five most significant results or outcomes of your project. To help a reader scan, you might even use bullets to highlight these results.

Future activities or research—Tell readers what you will be doing during the next work period.

Expenses—If asked, you should state the costs incurred over the previous week or month. Highlight any places where costs are deviating from the project's budget.

The body of the activity report shown in Figure 22.8 includes these four items.

Full Introduction for an Activity Report

Figure 22.7: When readers are less familiar with the subject, you might add background information, stress the importance of the subject, and forecast the rest of the document.

Wilson National Laboratory
Always Moving Forward

Nanotech Micromachines Demonstration for Senators Laura Geertz and Brian Hanson
Presented by Gina Gould, Head Engineer

Background information is offered for readers unfamiliar with the topic.
→ Nanotechnology is the creation and utilization of functional materials, devices, and systems with novel properties and functions that are achieved through the control of matter, atom by atom, molecule by molecule, or at the macromolecular level. A revolution has begun in science, engineering, and technology, based on the ability to organize, characterize, and manipulate matter systematically at the nanoscale.

Purpose and main point are mentioned here.
→ In this demonstration, we will show you how the 5492 Group at Wilson National Laboratory is applying breakthroughs in nanotechnology science toward the development of revolutionary new micromachines. Our work since 2002 has yielded some amazing results that might dramatically expand the capacity of these tiny devices.

Forecasting shows the structure of the briefing.
→ Today, we will first show you a few of the prototype micromachines we have developed with nanotechnology principles. Then, we will present data gathered from testing these prototypes. And finally, we will discuss future uses of nanotechnology in micromachine engineering.

Writing the Conclusion

The conclusion should be as brief as possible. You should

- restate your main point.
- restate your purpose.
- make any recommendations, if appropriate.
- look to the future.

These concluding moves should be made in a maximum of two to four sentences.

> To conclude, in this demonstration our goal was to update you on our progress toward developing nanotechnology micromachines. Overall, it looks like we are making solid progress toward our objectives, and we seem to be on schedule. Over the next couple of months, we will be facing some tough technical challenges. At that point, we will know if micromachines are feasible with current technology.

Link
For more ideas about writing conclusions, go to Chapter 8, page 196.

The conclusion shown in Figure 22.8 is probably more common than this example. It concisely restates the memo's main point, ending the memo.

To see more examples of briefs, go to
www.pearsonhighered.com/johnsonweb3/22.9

Progress Report

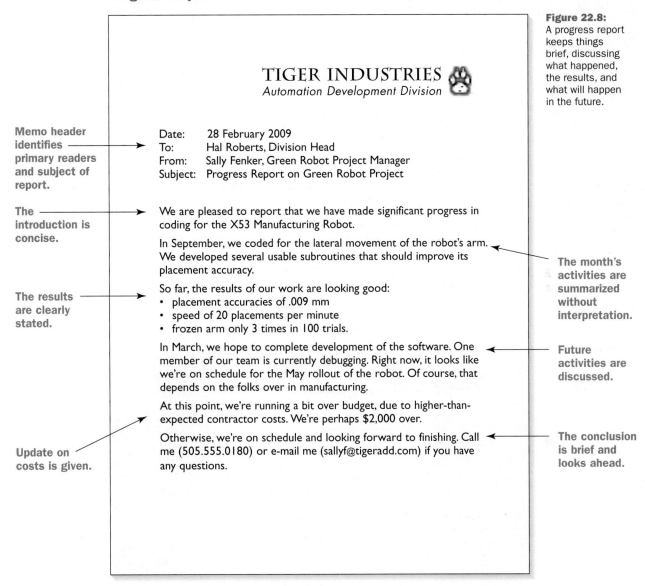

Memo header identifies primary readers and subject of report.

The introduction is concise.

The results are clearly stated.

Update on costs is given.

TIGER INDUSTRIES
Automation Development Division

Date: 28 February 2009
To: Hal Roberts, Division Head
From: Sally Fenker, Green Robot Project Manager
Subject: Progress Report on Green Robot Project

We are pleased to report that we have made significant progress in coding for the X53 Manufacturing Robot.

In September, we coded for the lateral movement of the robot's arm. We developed several usable subroutines that should improve its placement accuracy.

So far, the results of our work are looking good:
• placement accuracies of .009 mm
• speed of 20 placements per minute
• frozen arm only 3 times in 100 trials.

In March, we hope to complete development of the software. One member of our team is currently debugging. Right now, it looks like we're on schedule for the May rollout of the robot. Of course, that depends on the folks over in manufacturing.

At this point, we're running a bit over budget, due to higher-than-expected contractor costs. We're perhaps $2,000 over.

Otherwise, we're on schedule and looking forward to finishing. Call me (505.555.0180) or e-mail me (sallyf@tigeradd.com) if you have any questions.

Figure 22.8:
A progress report keeps things brief, discussing what happened, the results, and what will happen in the future.

The month's activities are summarized without interpretation.

Future activities are discussed.

The conclusion is brief and looks ahead.

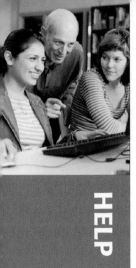

Making and Using PDFs

Increasingly, companies are asking for activity reports to be submitted in portable document format (PDF). Documents in PDF retain the exact look of the original, including the page layout and graphics. However, they do not require the same amount of memory space as other kinds of documents, so they are easily sent through e-mail, stored on a computer, or placed on a website for downloading.

PDFs also have some other helpful features:

- They have an almost identical appearance on all computers, so your file won't look different if someone is using a different kind of computer (PC, Macintosh, PDA).
- They include password protection to keep a file from being altered.
- They can include an electronic signature to signal your ownership or support of the document.

Also, if you are scanning documents, PDF is probably the most efficient way to save them in a usable format (Figure A).

Until recently, you would usually have needed a program called Adobe Acrobat to create and read PDF documents (Adobe Acrobat Reader can be downloaded free of charge from www.adobe.com). Today, word-processing programs like Microsoft Word and Corel WordPerfect will allow you to convert text files into PDF files. With

Scanned Document in Portable Document Format

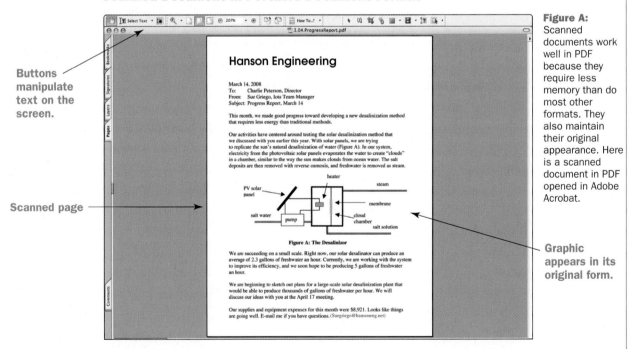

Buttons manipulate text on the screen.

Scanned page

Figure A: Scanned documents work well in PDF because they require less memory than do most other formats. They also maintain their original appearance. Here is a scanned document in PDF opened in Adobe Acrobat.

Graphic appears in its original form.

Want to learn more about using PDFs? Go to
www.pearsonhighered.com/johnsonweb3/22.10

Acrobat, you can also convert other kinds of texts, like PowerPoint or Presentation files, into PDF format.

To save a file as a PDF:

1. Select "Print" in your word processor or presentation program.
2. Select "Print as PDF" in the Print box that pops up.
3. Instead of making a hard copy on your printer, your computer will convert your document or presentation into a PDF file.
4. Save this new PDF document on your hard drive.

Now you can store the PDF file, send it as an attachment, or put it on a website for downloading.

As technical workplaces continue evolving toward the goal of the "paperless office," it is likely that PDFs will become a typical format for sending documents of all kinds, including activity reports.

Using Style and Design in Activity Reports

Generally, activity reports follow a plain style and use simple design. These documents are mostly informative, not overly persuasive, so writers should try to keep them rather straightforward.

Using a Plain Style

As you revise your document with style in mind, pay attention to the following items:

Sentences—Using plain style techniques, make sure that (1) the subject is the "doer" of most sentences and (2) the verb expresses the action in most sentences. Where appropriate, eliminate any prepositional chains.

Paragraphs—Each paragraph should begin with a topic sentence that makes a direct statement or claim that the rest of the paragraph will support. This topic sentence will usually appear as the first or second sentence of each paragraph.

Tone—Since activity reports are often written quickly, you should make sure you are projecting an appropriate tone. It might be tempting to be sarcastic or humorous, but this is not the place. After all, you never know how the activity report might be used in the future. While giving briefings, you want to project a professional tone. If you have negative information to convey, state it candidly with no apologies.

Link

For more information on using plain style, see Chapter 9, page 206.

Link

For more information on using an appropriate tone, go to Chapter 9, page 222.

Using Design and Graphics

The design of your activity report should be straightforward also. Usually, the design of these documents is governed by a standard format, like a memo format or perhaps a standardized form for lab reports. Your company will specify the format for activity reports. Otherwise, you might use the templates available with your word-processing program.

Link

For help using templates, go to Chapter 17, page 472.

If you want to add any visuals, you should center them in the text and place them after the point where you refer to them. Even though activity reports tend to be short, you should still label the graphic and refer to it by number in the text.

If you are presenting a briefing orally, you should look for ways to include graphics to support your presentation. Photographs can help the audience visualize what you are talking about, so use your digital camera to snap some pictures. Graphs are always helpful for showing trends in the data.

Revising, Editing, and Proofreading

Even though activity reports and briefings are usually considered "informal" documents, you should devote some time to revising, editing, and proofreading them. Your reports, after all, might turn up in the future when the stakes are higher. Perhaps they might be reviewed during your performance evaluation. Or, maybe they will be studied by auditors, who are trying to figure out where mistakes were made. An activity report that has not been revised, edited, or proofread will make you look unprofessional in these situations.

Revising and Editing Your Activity Reports

While revising and editing, look back at your original notes about the rhetorical situation in which the activity report might be used:

SUBJECT   Does your report provide all the information that readers need to know? Have you included any additional want-to-tell information that can be removed?

PURPOSE   Does the report achieve your intended purpose and the purpose you stated in the introduction of the document? If you want readers to do something, is it stated up front or clearly in the body?

READERS   Does the report satisfy the needs of the primary readers (action takers) and secondary readers (advisors)? If an auditor, lawyer, or reporter read the document, would he or she be satisfied that it is accurate and truthful?

Link

For more information on editing documents, turn to Chapter 12, page 306.

CONTEXT OF USE   Are the format and design appropriate for the places where the document will be used? Can readers scan the document quickly to locate the most important information? If the report was taken out of context, would it still project a sense of professionalism?

Proofreading Your Activity Report

Proofread your activity report closely before printing it out or hitting the Send button. These documents are typically written quickly, so they often contain small typos, misspellings, and grammar errors. Allow yourself a few minutes to read the text sentence by sentence to locate any errors.

You may even print out a copy of the document and look it over on paper. Often, errors you did not catch on the screen are more noticeable in the paper version.

GO TO THE NET

Need templates for letters and memos? Go to
www.pearsonhighered.com/johnsonweb3/22.12
For worksheets to help you revise, go to
www.pearsonhighered.com/johnsonweb3/22.13

- Activity reports include progress reports, briefings, white papers, reports that accompany demonstrations, incident reports, and lab reports.

- An activity report typically includes the following sections: introduction, summary of activities, results of activities, future activities, expenses, and conclusion.

- While preparing to write an activity report, analyze the rhetorical situation by anticipating the readers and the context in which the report will be used.

- The style and design of activity reports should be plain and straightforward.

- Edit and proofread activity reports carefully. In the future, they may be used in an unexpected context, so you want them to reflect a sense of professionalism.

Individual or Team Projects

1. For a week, keep a journal that tracks your activities related to school or work. Each day, make up a "to do" list. Then, as you complete each task, cross it off and write down the results of the task. In a memo to your instructor, summarize your activities for the week and discuss whether the activity journal was a helpful tool or not.

2. While you are completing a large project in this class or another, write a progress report to your instructor in which you summarize what has been accomplished and what still needs to be completed. The progress report should be submitted in memo format.

3. Think back to an accident that occurred in your life. Write an incident report in which you explain what happened, the actions you took, and the results of those actions. Then, discuss how you made changes to avoid that kind of accident in the future.

4. While completing a larger project for your class (like a proposal or report), give a 2- to 3-minute briefing on that project to your class. Summarize your activities, discuss the results, and discuss what will happen in the future.

Collaborative Project

Your group has been asked to develop a standardized information sheet that will help students report accidents on your campus. Think of all the different kinds of accidents that might happen on your campus. Your information sheet should explain how to report an accident to the proper authorities on campus. Encourage the users of the information sheet to summarize the incident in detail, discuss the results, and make recommendations for avoiding similar accidents in the future.

Of course, numerous potential accidents could occur on campus. Your group may need to categorize them so that readers contact the right authorities.

For additional technical writing resources, including interactive sample documents, document design tutorials and guidelines, and more, go to **www.mytechcommlab.com.**

Revision Challenge

The activity report in Figure A is intended to notify students of the recent changes in computer use policies at their small college. How would you revise this report to help it achieve its purpose?

Smith College

Office of the Provost

Date: August 4, 2008
To: Smith College students
From: Provost George Richards
Subject: File-Sharing

Smith College has worked hard to develop a top-notch computer network on campus to provide access to the Internet for a variety of purposes. As a Smith student, you are welcome to use these computers for legal purposes, You will find the college's computer usage policies explained in the *Computer Policy for Smith College,* version 03.28.03, which is in effect until eclipsed by a revision.

Illegal downloading and sharing of copyrighted materials is a problem, especially music files, over our network. Violating our computer usage policies puts the college at risk for copyright infringement lawsuits. Our information technology experts also tell me that these activities slow down our network because these files require large amounts of bandwidth.

Recently, we have installed network tools and filters that allow us to detect and block illegal file sharing. We are already warning students about illegal use. After October 1, 2008, we will begin disciplining people who share files illegally by suspending their computer privileges. Repeated violations will be referred to the college's Academic Integrity Review Board.

You should know that if a complaint is filed by copyright owners, Smith College must provide your name and address to prosecutors. Our computer usage policy explicitly states that using college computers for illegal activities is forbidden and those who do will be prosecuted. Thank you for giving us this opportunity to stress the seriousness of this situation. We expect your compliance.

Bad Chemistry

Amanda Jones works as a chemical engineer at BrimChem, one of the top plastics companies in the country. Recently, her division had hired a bright new chemical engineer named Paul Gibson. Paul was tall and good-looking, and he was always polite. At lunch during Paul's first week, Amanda and a co-worker teased him about being a "Chippendales guy." Paul laughed a little, but it was apparent that the comment offended him. So, Amanda was careful from then on about her comments regarding his appearance.

A few months after starting at BrimChem, Paul went to a convention and came back somewhat agitated. Amanda asked him what was wrong. After a pause, Paul told her that one of the managers, Linda Juno, had made a pass at him one evening at the convention, suggesting he come up to her room "for a drink." When he declined, she became angry and said, "Paul, you need to decide whether you want to make it in this company." She didn't speak to him for the rest of the convention.

Paul told Amanda he was a bit worried about keeping his job with the company, since he was still on "probationary" status for his first year. Being on probation meant Linda or anyone else could have him fired for the slightest reason.

Later that day, though, Linda came down to Paul's office and seemed to be patching things up. After Linda left his office, Paul flashed Amanda a thumbs-up signal to show that things were all right.

The next week, Amanda was working late and passed by Paul's office. Linda was in his office giving him a backrub. He was obviously not enjoying it. He seemed to be making the best of it, though, and he said, "OK, thank you. I better finish up this report."

Linda was clearly annoyed and said, "Paul, I let you off once. You better not disappoint me again." A minute later, Linda stormed out of Paul's office.

The next day, Paul stopped Amanda in the parking lot. "Amanda, I know you saw what happened last night. I'm going to file a harassment complaint against Linda. If I'm fired, I'll sue the company. I'm tired of being harassed by her and other women in this company."

Amanda nodded. Then Paul asked, "Would you write an incident report about what you saw last night? I want to put some materials on file." Amanda said she would.

A week later, Paul was fired for a minor mistake. Amanda hadn't finished writing up the incident report.

If you were Amanda, what would you do at this point? If you wrote the report, what would you say and how would you say it?

To find links on sexual harassment in the workplace, go to
www.pearsonhighered.com/johnsonweb3/22.16

In this chapter, you will learn:

- About the various kinds of analytical reports used in technical workplaces.

- How to use the IMRaD pattern for organizing reports.

- How to determine the rhetorical situation of your reports.

- How to develop a methodology for collecting and analyzing information.

- Strategies for organizing and drafting an analytical report.

- How to use style and design to highlight important information and make it understandable.

- Strategies for revising, editing, and proofreading analytical reports.

Analytical reports are some of the most common large documents produced in the technical workplace. An analytical report is usually a formal response to a research question. The report typically identifies a research methodology, presents results, discusses those results, and makes recommendations.

Here are just a few ways analytical reports are used in the technical workplace:

- A civil engineer would use a *recommendation report* to suggest ways to strengthen the structural integrity of a bridge after an earthquake.
- A medical researcher would use a *completion report* to present her findings on the effectiveness of a new treatment for diabetes.
- A manager would use a *feasibility report* to explore the use of robots to automate an assembly line.
- An entomologist would write an *empirical research report* to present the results of her research on the migration patterns of monarch butterflies.

Research is the foundation of a good analytical report. To assist you with the research process, you can use computers to access incredible amounts of information. Even a modest search on the Internet quickly unearths a small library of information on any given subject. Meanwhile, the number-crunching capabilities of computers can help you highlight subtle trends in data collected from experiments and observations.

As a result, writing analytical reports involves a combination of (1) managing the large amount of information available on a subject and (2) doing empirical research that clarifies or adds to the existing information. In this chapter, you will learn how to write analytical reports that present information clearly and concisely.

Basic Features of Analytical Reports

Because analytical reports are used in so many different ways, it is hard to pin down a basic pattern that they tend to follow. Nevertheless, you will find that analytical reports typically include the following basic features:

- Introduction
- Methodology or research plan
- Results
- Discussion of the results
- Conclusions or recommendations

Link

For more strategies on organizing information in large documents, see Chapter 8, page 178.

The diagram in Figure 23.1 shows a general pattern for organizing an analytical report. To help you remember this generic report pattern, you might do what many researchers do—memorize the acronym IMRaD (Introduction, Methods, Results, and Discussion). This acronym outlines the main areas that most reports address.

The IMRaD pattern for reports is flexible and should be adapted to the specific situation in which you are writing. For example, in some reports, the methodology section can be moved into an appendix at the end of the report, especially if your readers do not need a step-by-step description of your research approach. In other reports, you may find it helpful to combine the results and discussion sections into one larger section.

Basic Pattern for an Analytical Report

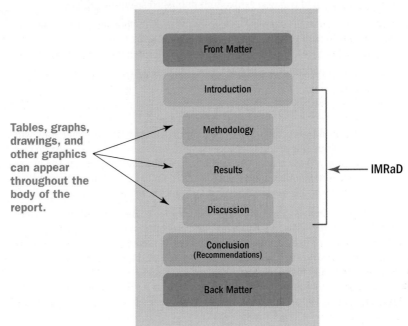

Tables, graphs, drawings, and other graphics can appear throughout the body of the report.

Front Matter

Introduction

Methodology

Results

Discussion

IMRaD

Conclusion
(Recommendations)

Back Matter

Like patterns for other documents, the IMRaD pattern is not a formula to be followed strictly. Rather, it is a guide to help you organize the information you have collected. The IMRaD pattern actually reflects the steps in the scientific method: (1) identify a research question or hypothesis, (2) create a methodology to study the question, (3) generate results, (4) discuss those results, showing how they answer the research question or support the hypothesis.

Link

For more information on the scientific method, turn to Chapter 7, page 158.

Types of Analytical Reports

The main qualities that distinguish analytical reports from activity reports—which you learned about in the previous chapter—are their (1) level of formality and (2) argumentative nature. Analytical reports are formal documents that present findings and make recommendations. Here are a few types:

Research reports—The purpose of a research report is to present the findings of a study. Research reports often stress the causes and effects of problems or trends, showing how events in the past have developed into the current situation.

Empirical research reports—Empirical research reports are written when a scientific project is finished. They first define a research question and hypothesis. Then they describe the methods of study and the results of the research project. And finally, they discuss these results and draw conclusions about what the project discovered. These reports often start out as laboratory reports and can also evolve into published scientific articles.

Completion reports—Most projects in technical workplaces conclude with a completion report. These documents are used to report back to management or the client, assessing the outcomes of a project or initiative.

Recommendation reports—Recommendation reports are often used to make suggestions about the best course of action. These reports are used to study a problem, present possible solutions, and then recommend what actions should be taken.

Feasibility reports—Feasibility reports are written to determine whether developing a product or following a course of action is possible or sensible. Usually, these reports are produced when management or the clients are not sure whether something can be done. The feasibility report helps determine whether the company should move forward with a project.

Figure 23.2 shows a research report on alcohol use among young adults. Like any research report, it defines a research question and offers a methodology for studying that research question. Then, it presents the results of the study and discusses those results. The report does not advocate a particular course of action. If this report were a recommendation report, it would make recommendations for ways to address alcohol use in young adults. (Later in this chapter, Figure 23.7, starting on p. 678, is an example of a feasibility report. Figure 23.10, starting on p. 697, is an example of a recommendation report.)

Planning and Researching Analytical Reports

An analytical report can be a large, complex document, so it is important that you plan properly with a full understanding of the report's rhetorical situation.

Planning

You should start planning the document by first identifying the elements of the rhetorical situation. Begin by answering the Five-W and How Questions:

Who might read this report?

Why was this report requested?

What kinds of information or content do readers need?

Where will this report be read?

When will this report be used?

How will this report be used?

With the answers to these questions fresh in your mind, you can begin defining the rhetorical situation in which your report will be used.

SUBJECT   What exactly will the report cover, and what are the boundaries of its subject? What information and facts do readers need to know to make a decision? What information don't they need?

For examples of analytical reports, go to
www.pearsonhighered.com/johnsonweb3/23.2

A Research Report

A clear title for the report is placed up front.

National Survey on Drug Use and Health

The NSDUH Report

Issue 31   2006

Underage Alcohol Use among Full-Time College Students

In Brief

- The rates of past month, binge, and heavy alcohol use among full-time college students aged 18 to 20 remained steady from 2002 to 2005

- Based on 2002 to 2005 combined data, 57.8 percent of full-time college students aged 18 to 20 used alcohol in the past month, 40.1 percent engaged in binge alcohol use, and 16.6 percent engaged in heavy alcohol use

- Based on 2002 to 2005 combined data, male full-time students in this age group were more likely to have used alcohol in the past month, engaged in binge alcohol use, and engaged in heavy alcohol use than their female counterparts

Main points are placed up front in an easy-to-access box.

During the past decade, increased attention has been directed toward underage alcohol use and binge drinking among college students and the negative consequences related to these behaviors.[1-5] Binge drinking refers to the "consumption of a sufficiently large amount of alcohol to place the drinker at increased risk of experiencing alcohol-related problems and to place others at increased risk of experiencing secondhand effects" (p. 287).[2]

Background information stresses the importance of the subject.

The National Survey on Drug Use and Health (NSDUH) asks respondents aged 12 or older to report their frequency and quantity of alcohol use during the month before the survey. NSDUH defines binge alcohol use as drinking five or more drinks on the same occasion (i.e., at the same time or within a couple of hours of each other) on at least 1 day in the past 30 days. NSDUH defines heavy alcohol use as drinking five or more drinks on the same occasion on each of 5 or more days in the past 30 days. All heavy alcohol users are also binge alcohol users.

The subject is defined.

The NSDUH Report (formerly *The NHSDA Report*) is published periodically by the Office of Applied Studies, Substance Abuse and Mental Health Services Administration (SAMHSA). All material appearing in this report is in the public domain and may be reproduced or copied without permission from SAMHSA. Additional copies of this report or other reports from the Office of Applied Studies are available online: http://www.oas.samhsa.gov. Citation of the source is appreciated. For questions about this report, please e-mail: shortreports@samhsa.hhs.gov.

Source: Office of Applied Studies Substance Abuse and Mental Health Services Administration [SAMHSA], 2006.

Graphs show trends in data while supporting written text.

The methodology is described in detail.

Results are presented and discussed.

The purpose is stated in a direct way.

Headings make the text easy to scan.

NSDUH REPORT: UNDERAGE ALCOHOL USE AMONG FULL-TIME COLLEGE STUDENTS   Issue 31, 2006

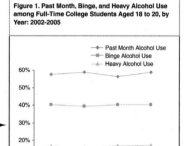

Figure 1. Past Month, Binge, and Heavy Alcohol Use among Full-Time College Students Aged 18 to 20, by Year: 2002-2005

Source: SAMHSA, 2002-2005 NSDUHs.

Figure 2. Past Month, Binge, and Heavy Alcohol Use among Full-Time College Students Aged 18 to 20, by Gender: 2002-2005

Source: SAMHSA, 2002-2005 NSDUHs.

NSDUH also asks young adults aged 18 to 22 about college attendance. For this analysis, respondents were classified as college students if they reported that they were in their first through fourth year (or higher) at a college or university and that they were a full-time student. Respondents who were on break from college were considered enrolled if they intended to return to college or university when the break ended.[6]

Data from the 2005 survey indicate that young adults aged 18 to 22 enrolled full time in college were more likely than their peers not enrolled full time (i.e., part-time college students and persons not currently enrolled in college) to use alcohol in the past month, binge drink, and drink heavily.[7]

This report examines trends and patterns in the rates of alcohol use among full-time college students who have not yet reached the legal drinking age (i.e., college students aged 18 to 20) based on data from the 2002, 2003, 2004, and 2005 NSDUHs.

Demographic Characteristics of Full-Time College Students Aged 18 to 20

From 2002 to 2005, an average of 5.2 million young adults aged 18 to 20 were enrolled full time in college each year. This represents 41.3 percent of young adults in this age range. Full-time college students included an average of 2.8 million women aged 18 to 20 (46.0 percent of women in this age group) and 2.4 million men aged 18 to 20 years (36.9 percent of men in this age group) each year. Over half of full-time college students aged 18 to 20 (58.2 percent) lived in the same household with a parent, grandparent, or parent-in-law, while 41.8 percent lived independently of a parental relative.[8]

Past Month Alcohol Use

From 2002 to 2005, the rates of past month alcohol use among full-time college students aged 18 to 20 remained steady (Figure 1), with an annual average of 57.8 percent (3.0 million

Charts are used to present results visually in ways that are easy to scan.

More results and discussion complete the report.

The authors choose not to include a conclusion because the report is brief and does not make recommendations.

Sources are properly listed.

Figure 3. Past Month, Binge, and Heavy Alcohol Use among Full-Time College Students Aged 18 to 20, by Age: 2002-2005

Source: SAMHSA, 2002-2005 NSDUHs.

students) using alcohol in the past month. Male full-time students in this age group were more likely to have used alcohol in the past month than their female counterparts (60.4 vs. 55.6 percent) (Figure 2). Rates of past month alcohol use among this group increased with increasing age (Figure 3). Among full-time college students aged 18 to 20, those living with a parent, grandparent, or parent-in-law were less likely to have used alcohol in the past month than those who were not living with a parental relative (51.2 vs. 67.0 percent).

Binge Alcohol Use

Rates of past month binge alcohol use among full-time college students aged 18 to 20 also remained steady from 2002 to 2005 (Figure 1), with an annual average of 40.1 percent (2.1 million students) engaging in binge alcohol use. In this group of young adult students, males were more likely to have engaged in binge alcohol use than females (46.9 vs. 34.4 percent) (Figure 2).

Rates of binge alcohol use among this group also increased with increasing age (Figure 3). Full-time college students aged 18 to 20 living with a parent, grandparent, or parent-in-law were less likely to have engaged in binge alcohol use than full-time college students aged 18 to 20 who were not living with a parental relative (34.0 vs. 48.5 percent).

Heavy Alcohol Use

From 2002 to 2005, rates of heavy alcohol use among full-time college students aged 18 to 20 also remained steady (Figure 1), with an annual average of 16.6 percent (866,000 students) engaging in heavy drinking. As is true for past month and binge alcohol use, rates of heavy alcohol use were higher among males than females and increased with increasing age (Figures 2 and 3). Among full-time college students aged 18 to 20, those living with a parent, grandparent, or parent-in-law were less likely to have engaged in heavy alcohol use than those who were not living with a parental relative (12.3 vs. 22.5 percent).

End Notes

[1] Reifman, A., & Watson, W. K. (2003). Binge drinking during the first semester of college: Continuation and desistance from high school patterns. *Journal of American College Health, 52,* 73-81.

[2] Wechsler, H., & Nelson, T. F. (2001). Binge drinking and the American college student: What's five drinks? *Psychology of Addictive Behaviors, 15,* 287-291.

[3] Turrisi, R., Wiersma, K. A., & Hughes, K. K. (2000). Binge-drinking-related consequences in college students: Role of drinking beliefs and mother-teen communications. *Psychology of Addictive Behaviors, 14,* 342-355.

[4] Weingardt, K. R., Baer, J. S., Kivlahan, D. R., Roberts, L. J., Miller, E. T., & Marlatt, G. A. (1998). Episodic heavy drinking among college students: Methodological issues and longitudinal perspectives. *Psychology of Addictive Behaviors, 12,* 155-167.

[5] Wechsler, H., Davenport, A., Dowdall, G., Moeykens, B., & Castillo, S. (1994). Health and behavioral consequences of binge drinking in college: A national survey of students at 140 campuses. *Journal of the American Medical Association, 272,* 1672-1677.

[6] Respondents whose current college enrollment status was unknown were excluded from the analysis.

[7] Office of Applied Studies. (2006). *Results from the 2005 National Survey on Drug Use and Health: National findings* (DHHS Publication No. SMA 06-4194, NSDUH Series H-30). Rockville, MD: Substance Abuse and Mental Health Services Administration.

[8] Living with a parental relative is defined as currently living in the same household with a parent, grandparent, or parent-in-law. Respondents who did not live with a parental relative and had unknown information on one or more household relationships were excluded from this analysis.

PURPOSE What should the report accomplish, and what do the readers expect it to accomplish? What is its main goal or objective?

You should be able to express the purpose of your report in one sentence. A good way to begin forming your purpose statement is to complete the phrase "The purpose of my report is to...." You can then use some of the following action verbs to express what the report will do:

to analyze	*to develop*	*to determine*
to examine	*to formulate*	*to recommend*
to investigate	*to devise*	*to decide*
to study	*to create*	*to conclude*
to inspect	*to generate*	*to offer*
to assess	*to originate*	*to resolve*
to explore	*to produce*	*to select*

READERS Who are the primary readers (action takers), secondary readers (advisors), and tertiary readers (evaluators)? Who are the gatekeeper readers (supervisors) for this report?

Primary readers (action takers) are the people who need the report's information to make some kind of decision. Anticipate the decision they need to make and provide the information they require. If they need recommendations, present your suggestions in a direct, obvious way.

Secondary readers (advisors) are usually experts or other specialists who will advise the primary readers. More than likely, they will be most concerned about the accuracy of your facts and the validity of your reasoning. Give them enough technical detail and data to feel confident about your conclusions and recommendations.

Tertiary readers (evaluators) might be people you didn't expect to read the report, like reporters, lawyers, auditors, and perhaps even historians. Anticipate these kinds of audiences and avoid making unfounded statements that might be used to harm you or your company.

Gatekeeper readers (supervisors) will probably include your immediate supervisor. However, other gatekeepers, like your company's legal or technical experts, might need to look over your report before it is sent to the primary readers.

Link

For more help defining your readers and their characteristics, go to Chapter 3, page 40.

CONTEXT OF USE Where, when, and how will the report be used? What are the economic, political, and ethical factors that will influence the writing of the report and how readers will interpret it?

Physical context—Consider the various places where your report might be used, such as a meeting or at a conference. What adjustments will be needed to make the report more readable/usable in these situations?

Economic context—Anticipate the financial issues that may influence how your readers will interpret the results and recommendations in your report, especially if you are recommending changes.

Political context—Think about the politics involved with your report. On a micropolitical level (i.e., office politics), you should determine who stands to

To find worksheets to help you analyze readers and contexts of use, go to
www.pearsonhighered.com/johnsonweb3/23.3

gain or lose from the information in your report. On a macropolitical level, you should consider the larger political trends that will shape the reception of your report.

Link

For more information on defining the context of use, go to Chapter 3, page 46.

Ethical context—Consider any legal or ethical issues that might affect your report and the methods you will use to collect information. For example, if you are doing a study that involves people or animals, you may need to secure permission release forms. If you need access to sensitive information of any kind, you may encounter legal roadblocks.

If you are writing a report with a team, collaborate on your answers to these questions about the rhetorical situation. Then, print the answers out and give each member a copy. If your team begins the project with a clear understanding of the subject, purpose, readers, and context of use, you will likely avoid unnecessary conflict and wasted time.

Researching

With the rhetorical situation fresh in your mind, you can start collecting information for your report. It is important that you define a research question and develop a plan (a methodology) for conducting research on your subject.

Research in technical fields typically follows a predictable process:

1. Define a research question.

2. State a hypothesis.

3. Develop a research methodology.

4. Collect information by following the research methodology.

5. Analyze gathered information and compare it to the hypothesis.

The most effective methodologies are the ones that collect information from a variety of sources (Figure 23.3).

DEFINE A RESEARCH QUESTION AND HYPOTHESIS   Reports are usually written to answer a specific *research question* or test a *hypothesis*. So, you should begin by defining the question you are trying to answer. Write down the question in one sentence.

Could we convert one of our campus buildings to a renewable heating source, like solar?

Why are the liver cancer rates in Horn, Nevada, higher than the national average?

How much would it cost to automate our factory in Racine, Wisconsin?

Is it feasible to reintroduce wolves into the Gila Wilderness Area?

At this point, you should also write down a one-sentence hypothesis. A hypothesis is essentially an educated guess or tentative explanation that answers your research question. You don't *need* a hypothesis at this point; however, some people like to begin with their best try at answering the research question. With your hypothesis stated,

Link

To learn more about forming a hypothesis, turn to Chapter 7, page 145.

Researching a Subject

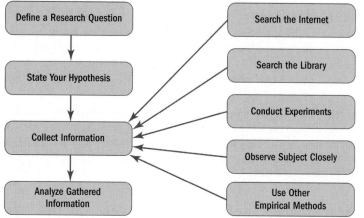

Figure 23.3: A research methodology is a plan for gathering information, preferably from a variety of sources.

you will have a better idea about what direction your research will take and what you are trying to prove or disprove.

> We believe we could convert a building like Engineering Hall to a solar heating source.

> Our hypothesis is that liver cancer rates in Horn, Nevada, are high because of excessive levels of arsenic in the town's drinking water.

> Automating our Racine plant could cost $2 million, but the savings will offset that figure in the long run.

> Reintroducing wolves to the Gila Wilderness Area is feasible, but there are numerous political obstacles and community fears to be overcome.

Remember, though, that a hypothesis is just a possible answer (i.e., your best guess) to your research question. As you move forward with your research, you will likely find yourself modifying this hypothesis to fit the facts. In some cases, you might even need to abandon your original hypothesis completely and come up with a new one.

DEVELOP A METHODOLOGY   Once you have defined your research question and stated a hypothesis, you are ready to develop your research *methodology*. A methodology is the series of steps you will take to answer your research question or test your hypothesis.

A good way to invent your methodology is to use logical mapping (Figure 23.4).

1. Write your research question in the middle of a sheet of paper or your computer screen.

2. Identify the two to five major steps you would need to take to answer that question. For example, if you are researching the use of solar power to generate heat for a building, one major step in your methodology might be to use the Internet to collect information on solar power and its potential applications.

Want help defining a hypothesis? Go to
www.pearsonhighered.com/johnsonweb3/23.5

3. Identify the two to five minor steps required to achieve each major step. In other words, determine which smaller steps are needed to achieve each major step.

4. Keep filling out and revising your map until you have fully described the major and minor steps in your methodology.

Your logical map will help you answer the How question about your methodology. By writing down your major and minor steps, you are describing how you would go about studying your subject.

Allow yourself to be creative at this point. As you keep mapping out farther, your methodology will likely evolve in front of you. You can cross out some steps and combine others. In the end, a reasonable methodology is one that is "replicable," meaning the readers can obtain the same results if they redo your research. As you invent and devise your methodology, ask yourself whether someone in your research area could duplicate your work.

COLLECT INFORMATION   Your methodology is your road map to collecting information. Once you have described your methodology, you can use it to guide your research. The information you collect will become the results section of your report.

Using Logical Mapping to Develop a Methodology

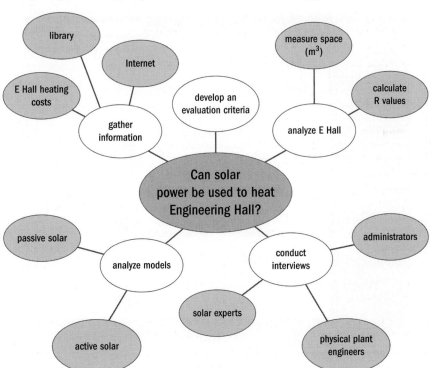

Figure 23.4: When mapping a methodology, ask yourself how you might answer the research question. Then decide on the major and minor steps in your methodology.

For websites that discuss methodologies, go to
www.pearsonhighered.com/johnsonweb3/23.6

Planning and
Researching
Analytical
Reports

669

There are many places to find information:

Link

For more strategies for using search engines, see Chapter 7, page 149.

Internet searches—With some well-chosen keywords, you can use your favorite search engine to start collecting information on the subject. Cut and paste the materials you may need and bookmark helpful websites. Websites like the ones from the U.S. Census Bureau and the Pew Charitable Trust offer a wealth of information (Figures 23.5 and 23.6). Make sure you properly cite any sources when you enter them into your notes.

Library research—Using your library's catalogs, databases, and indexes, start searching for articles, reports, books, and other print documents that discuss your subject. You should make copies of the materials you find or use a scanner to save them on your computer. Again, keep track of where you found these materials so you can cite them in your report.

U.S. Census Bureau Website

Figure 23.5:
The U.S. Census Bureau website is an excellent resource for statistics. The website offers statistics on income, race, trade, and lifestyle issues.

Source: www.census.gov.

For help with Internet research, go to
www.pearsonhighered.com/johnsonweb3/23.7

An Archive on the Internet

Figure 23.6:
The Pew Research Center's Pew Internet & American Life Project's archive offers reports on the Internet and American life. Pew collects data on religion, consumption, lifestyle issues, and many other topics.

Source: http://www.pewinternet.org/reports.asp.

Experiments and observations—Your report might require experiments and measurable observations to test your hypothesis. Each field follows different methods for conducting experiments and observations. Learn about research methodologies used in your field and use them to generate empirical data.

Other empirical methods—You might conduct interviews, pass out surveys, or do case studies to generate empirical information. A report should include some empirical material, preferably generated by you or your company. This kind of data will strengthen the credibility of your report.

To avoid any problems with plagiarism or copyright, you need to carefully cite your sources in your notes. It's fine to take material from the Internet or quote

To find other great sources of information, go to
www.pearsonhighered.com/johnsonweb3/23.8

Link

For more information on plagiarism and copyright, go to Chapter 7, page 170.

passages from print sources, as long as you quote, paraphrase, and cite them properly when you write your report. Chapter 7 offers note-taking strategies that will help you avoid any issues of plagiarism or copyright. Chapter 5 discusses ethical issues involving copyright.

While collecting information, you will likely find that your methodology will evolve. These kinds of changes in the research plan are not unusual. Keep track of any changes, because you will need to note them in the methodology section of your report.

Link

Chapter 7 offers a full discussion of research. You might turn to that chapter for more research strategies; see page 146.

ANALYZE INFORMATION AND COMPARE IT TO YOUR HYPOTHESIS   Here's the hard part. Once you have collected the information you need for your report, you should analyze your materials closely to identify any major issues and themes. From your materials, try to identify two to five major findings about the subject of your research. These findings will be important aspects of your report.

Also, look back at your original research question or hypothesis. Do your findings answer your original research question and/or support your hypothesis? If they do, you are probably ready to start drafting your report. But, if your findings don't sufficiently answer your research question or support your hypothesis, you still need to collect more information and data. In some cases, you may need to abandon your hypothesis and reach a different conclusion about your subject.

Link

For more information about developing and modifying research questions and hypotheses, turn to Chapter 7, page 145.

Don't worry if your original hypothesis ends up being incorrect. It was just a guess anyway. In the end, the facts in your final report are more important than your original guess about what you would discover.

Organizing and Drafting Analytical Reports

Organizing your information and drafting analytical reports will not be difficult if you stay focused on your purpose. Your purpose statement will help you include only need-to-know information in the report. Although you might be tempted to include everything you collected, don't do this. Anything beyond need-to-know information will only muddle your document, making the most important ideas harder for your readers to find.

Writing the Introduction

Let's be honest. Reports are not the most interesting documents to read. Given the slightest excuse, your readers will start thumbing through your report's pages, looking for the main points. That's why your introduction needs to grab their attention and give them good reasons to read your report closely.

Your report should begin with an introduction that sets a framework, or context, for the rest of the document (Figure 23.7 on pp. 678–689). Typically, the introduction will include some or all of the following six moves, though not necessarily in this order:

DEFINE THE SUBJECT   In the introduction, clearly state what your report is about— and perhaps what it is not about. Reports have a tendency to grow far too large when writers are not clear about the scope of the subject.

STATE THE PURPOSE   Reports are usually rather blunt about their purpose. In the introduction, you can make the following kinds of statements:

> The purpose of this report is to show how Engineering Hall can be converted to solar heating with minimal cost.

> In this report, we present the results of our research into high liver cancer rates in Horn, Nevada.

Link

For more information on writing a purpose statement, see Chapter 2, page 21.

In most cases, the more straightforward your purpose statement, the better. After all, you want your readers to know immediately what you are trying to do. If your readers start out reading with a clear idea of your purpose, they will be better able to understand your methods, results, conclusions, and recommendations.

STATE THE MAIN POINT   Your readers will be most attentive at the beginning of the report. So, you should put your main point (i.e., your overall conclusion) here in the introduction. That way, as they read the report, they can retrace your steps as you came to that conclusion.

> We argue that the best way to add solar heating to Engineering Hall is to use a combination of direct gain methods and hydrosolar tubes. We recommend that the university begin working on this project this summer, using faculty and students from the Engineering Department as designers and builders of the system.

> Our conclusion is that Horn's high liver cancer rates are primarily due to excessive levels of arsenic in the town's drinking water.

Of course, you will be giving away the ending by putting your main point in the introduction. But then, reports are not mystery stories. By telling readers your main point up front, you will focus their attention and make their reading of the report more efficient.

STRESS THE IMPORTANCE OF THE SUBJECT   Your readers may not immediately recognize the importance of a report's subject. To avoid this problem, tell your readers in the introduction *why* they should pay attention to your report.

> In the near future, the United States will need to wean itself off nonrenewable energy sources like oil and natural gas. The conversion of Engineering Hall to solar heating will give us a starting point for studying how the conversion to renewable energy could be made on our campus.

> Critics of our findings might point out that Horn is just one small town with high levels of liver cancer—it is not, they might say, the norm. We think they are wrong. Many Western towns have high levels of arsenic in their tap water. If our findings are true, removal of arsenic from tap water could be an important step toward improving health in the region.

OFFER BACKGROUND INFORMATION   Reports often start out with some historical information to familiarize readers with the subject. Background information can be simple, such as reminding readers that they requested the report on a particular date.

Dan Small

SOFTWARE ENGINEER, INTELLIGENT SYSTEMS AND ROBOTICS CENTER, SANDIA
NATIONAL LABORATORIES

*The Sandia National Laboratories in Albuquerque, New Mexico, develop advanced
technology for national defense and progress.*

What is the most efficient way to write a report?

As a software engineering professional working at a major R&D lab like Sandia, a large
part of my duties includes documenting the research performed by my team. I regularly
write analytical reports to our management and our external customers. It's really important that these documents concisely communicate both the technical details and the "big
picture" implications of the results.

When writing an analytical report, the first thing I always do is make an outline,
which usually follows this general form (much like IMRaD):

Introduction—Make sure that you keep the intended audience in mind here. This
needs to be the section where you give the overview and scope of the work you
have done. (Believe me, Dilbert has it right with respect to writing for managers
versus writing for a technical audience.)

Data—Present the data that you have collected, preferably in a graphical form.
Make sure that there is an appropriate legend and markings that support your
analysis.

Analysis—Speak directly to the data that you present. Point out those aspects
that support your conclusions and those that don't. (An analytical report is not
designed to persuade, only to present.)

Conclusions—Draw your conclusions from the analysis presented above. Make
sure that the major points that need to be communicated are handled here and
are supported by your data and analysis. Make sure that the larger implications
(if any) of the results are communicated here as well.

I cannot stress enough the importance of effective writing in the technical workplace. The main difference I see between technical professionals who advance and
those who stagnate is their ability to communicate. Given that you are reading this
book, I can tell you that you are on the right track.

On March 14, the President of Kellen College, Dr. Sharon Holton, asked us to explore ways
to begin converting the campus to renewable energy sources. We decided to use one
building, Engineering Hall, as a test case for studying solar heating options.

More extensive background information might tell the history of the subject, giving
readers some context for understanding the contents of the report.

Horn, Nevada, was settled in 1893 when a rich vein of silver was discovered in the Horn
mountain range. For decades, mining was the primary source of income, though legalized

674   Chapter 23
Analytical Reports

Have writer's block? Go to
www.pearsonhighered.com/johnsonweb3/23.9

gambling later became an alternative source of revenue. Today, Horn is becoming a haven for retirees who live here in the winter.

The people of Horn, especially long-time residents, have always recognized that their cancer rates were especially high. They call it the "Horn Curse." In 1998 medical researchers began to suspect arsenic. They found that levels of arsenic in the drinking water were consistently higher than the recommended limits set by the federal government. Local politicians, however, have resisted recommendations to remove arsenic from the Horn water supply, mostly due to cost and perhaps possible government intrusion.

Usually, background information will tell your readers information they already know or won't find too contentious. The purpose of background information is to give your readers a familiar place from which to begin reading the report.

FORECAST THE REMAINDER OF THE REPORT   Your readers will find forecasting in the introduction helpful as they begin reading the report. Even a quick sentence can help them develop a framework for understanding your document:

> This report includes four sections: (1) our methodology for gathering information, (2) the results of our research, (3) a discussion of those results, and (4) our conclusions and recommendations.

Forecasting is also a nice way to signal the end of the introduction while making the transition into the body of the report.

Describing Your Methodology

Following the introduction, reports typically include a methodology section that describes step-by-step how the study was conducted. The aim of the methodology section is to describe the steps you went through to collect information on the subject. This section should include an opening, body, and closing.

OPENING   In the opening paragraph, start out by describing your overall approach to collecting information (Figure 23.7). If you are following an established methodology, you might mention where it has been used before and who used it.

BODY   In the body of the methodology section, walk your readers step-by-step through the major parts of your study. As you highlight each major step, you should also discuss the minor steps that were part of it.

CLOSING   To close the methodology section, you might discuss some of the limitations of the study. For example, your study may have been conducted with a limited sample (e.g., college students at a small Midwestern university). Perhaps time limitations restricted your ability to collect comprehensive data. All methodologies have their limitations. By identifying your study's limitations, you will show your readers that you are aware that other approaches may yield different results.

Link

For more information on writing introductions, see Chapter 8, page 182.

Link

For help organizing sections in larger documents, see Chapter 8, page 186.

Moves in an Introduction

AT A GLANCE

- Define the subject.
- State the purpose.
- State the main point.
- Stress the importance of the subject.
- Offer background information.
- Forecast the remainder of the report.

Summarizing the Results of the Study

In the results section, you should summarize the major *findings* of your study. A helpful guideline is to discuss only the two to five major results. That way, you can avoid overwhelming your readers with a long list of findings, especially ones that are not significant.

This section should include an opening, a body, and perhaps a closing.

- In your opening paragraph for the results section, briefly summarize your major results (Figure 23.7).
- In the body of this section, devote at least one paragraph to each of these major results, using data to support each finding.
- In the closing (if needed), you can again summarize your major results.

Link

For help making tables, charts, and graphs, see Chapter 11, starting on page 280.

Your aim in this section is to present your findings as objectively and clearly as possible. To achieve this aim, state your results with minimal interpretation. You should wait until the discussion section to offer your interpretation of the results.

If your study generated numerical data, you should use tables, graphs, and charts to present your data in this section. As discussed in Chapter 11, these graphics should support the written text, not replace it. Your written description of your results should be enhanced and clarified by the graphics, but the graphics cannot stand alone.

Discussing Your Results

The discussion section is where you will analyze the results of your research. As you look over your findings, identify the two to five major conclusions you might draw from your information or data. What are your results telling you about the subject?

The discussion section should start out with an opening paragraph that briefly states your overall conclusions about the results of your study (Figure 23.7). Then, in the body of this section, you should devote a paragraph or more to each of your conclusions. Discuss the results of your study, showing what you think your results show.

Stating Your Overall Conclusions and Recommendations

The conclusion of a report should be concise. You should make some or all of the following six moves, which are typical in a larger document:

MAKE AN OBVIOUS TRANSITION A heading like "Our Recommendations" or "Summary" will cue readers that you are concluding. Or, you can use phrases like "In conclusion" or "To sum up" to signal that the report is coming to an end.

RESTATE THE MAIN POINTS Boiled down to its essence, what did your study show or demonstrate? Tell your readers what you proved, disproved, or did not prove.

STATE YOUR RECOMMENDATIONS If you have been asked to make recommendations, your conclusion should identify two to five actions that your readers should

To see samples of well-written methodologies, go to
www.pearsonhighered.com/johnsonweb3/23.11

consider. A good way to handle recommendations is to present them in a bulleted list. Tell your readers exactly what you believe should be done (Figure 23.7).

At a Glance

Moves in the Conclusion

- Make an obvious transition.
- Restate the main points.
- State your recommendations.
- Restress the importance of the study.
- Look to the future.
- Say thank you and offer contact information.

RESTRESS THE IMPORTANCE OF THE STUDY   Tell your readers why you believe your study was important. Tell them why you think the results and recommendations should be taken seriously.

LOOK TO THE FUTURE   You might discuss future research paths that could be pursued. Or, you could describe the future you envision if readers follow your recommendations.

SAY THANK YOU AND OFFER CONTACT INFORMATION   After thanking your readers for their interest, you might also provide contact information, such as a phone number and e-mail address. Readers who have questions or want to comment on the report will then be able to contact you.

Like the introduction, your conclusion doesn't need to make these moves in this order, nor are all the moves necessary. Minimally, you should restate your main point and state any recommendations. You can use the other moves to help end your report on a positive note.

Link

For more ideas on writing conclusions, see Chapter 8, page 196.

Drafting Front Matter and Back Matter

Most reports also include front matter and back matter. Front matter includes the letter of transmittal, title page, table of contents, and other items that are placed before the first page of the main report. Back matter includes appendixes, glossaries, and indexes that are placed after the main report.

Developing Front Matter

Front matter may include some or all of the following items:

LETTER OR MEMO OF TRANSMITTAL   Typically, reports are accompanied by a letter or memo of transmittal. A well-written letter or memo gives you an opportunity to make positive personal contact with your readers before they look through your report.

Link

For more information on writing letters and memos of transmittal, see Chapter 17, page 452.

TITLE PAGE   Title pages are an increasingly common feature in analytical reports. A well-designed title page sets a professional tone while introducing readers to the subject of the report. The title page should include all or some of the following features:

- a specific title for the report
- the names of the primary readers, their titles, and the name of their company or organization
- the names of the writers, their titles, and the name of the writers' company or organization
- the date on which the report was submitted
- company logos, graphics, or rules (lines) to enhance the design

Link

For graphic design strategies, see Chapter 10, page 232.

A Feasibility Report

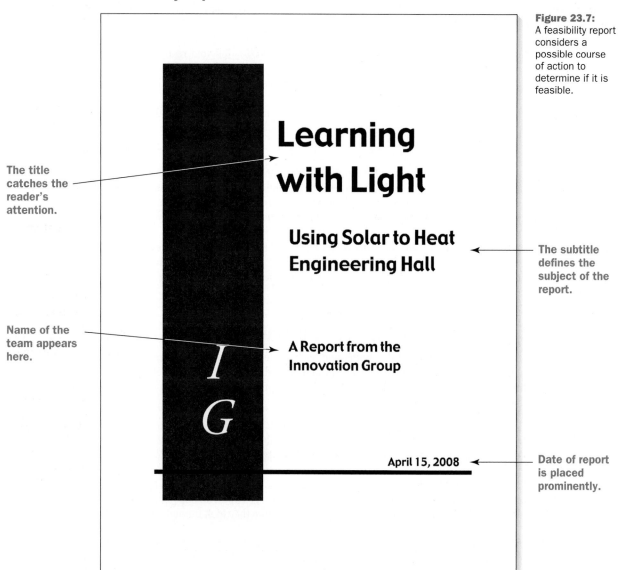

Figure 23.7:
A feasibility report considers a possible course of action to determine if it is feasible.

The title catches the reader's attention.

The subtitle defines the subject of the report.

Name of the team appears here.

Learning with Light

Using Solar to Heat Engineering Hall

A Report from the Innovation Group

I G

April 15, 2008

Date of report is placed prominently.

GO TO THE NET

To see reports with well-written conclusions, go to
www.pearsonhighered.com/johnsonweb3/23.12

Figure 23.7:
(continued)

The table of contents is clearly signaled by a heading.

The headings of the report are listed here as they appear in the report.

The abstract is clearly identified by a header.

The main point is stated up front.

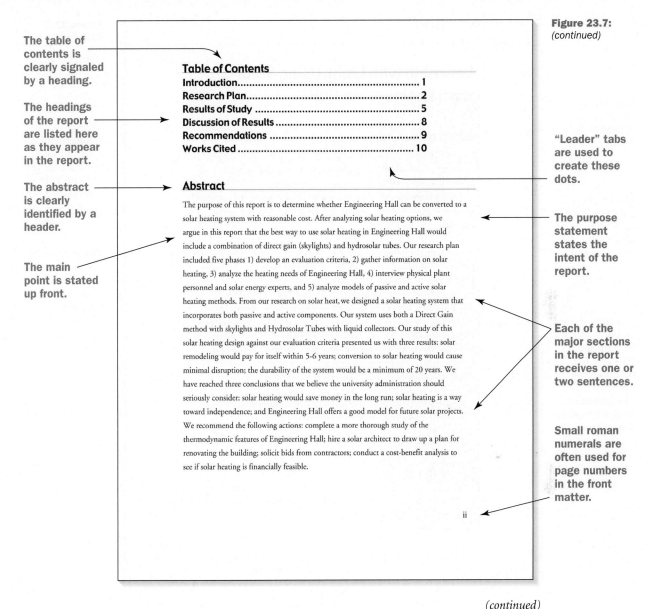

Table of Contents

Abstract
The purpose of this report is to determine whether Engineering Hall can be converted to a solar heating system with reasonable cost. After analyzing solar heating options, we argue in this report that the best way to use solar heating in Engineering Hall would include a combination of direct gain (skylights) and hydrosolar tubes. Our research plan included five phases 1) develop an evaluation criteria, 2) gather information on solar heating, 3) analyze the heating needs of Engineering Hall, 4) interview physical plant personnel and solar energy experts, and 5) analyze models of passive and active solar heating methods. From our research on solar heat, we designed a solar heating system that incorporates both passive and active components. Our system uses both a Direct Gain method with skylights and Hydrosolar Tubes with liquid collectors. Our study of this solar heating design against our evaluation criteria presented us with three results: solar remodeling would pay for itself within 5-6 years; conversion to solar heating would cause minimal disruption; the durability of the system would be a minimum of 20 years. We have reached three conclusions that we believe the university administration should seriously consider: solar heating would save money in the long run; solar heating is a way toward independence; and Engineering Hall offers a good model for future solar projects. We recommend the following actions: complete a more thorough study of the thermodynamic features of Engineering Hall; hire a solar architect to draw up a plan for renovating the building; solicit bids from contractors; conduct a cost-benefit analysis to see if solar heating is financially feasible.

ii

"Leader" tabs are used to create these dots.

The purpose statement states the intent of the report.

Each of the major sections in the report receives one or two sentences.

Small roman numerals are often used for page numbers in the front matter.

(continued)

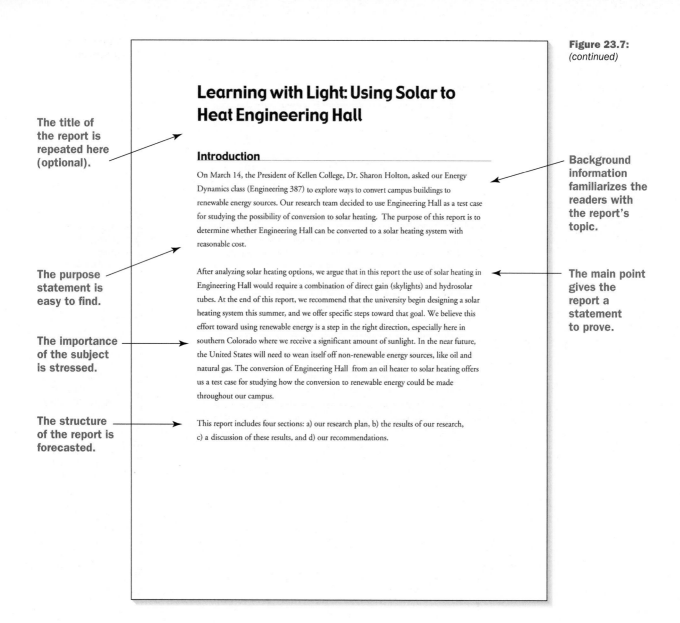

The title of the report is repeated here (optional).

Learning with Light: Using Solar to Heat Engineering Hall

Introduction

On March 14, the President of Kellen College, Dr. Sharon Holton, asked our Energy Dynamics class (Engineering 387) to explore ways to convert campus buildings to renewable energy sources. Our research team decided to use Engineering Hall as a test case for studying the possibility of conversion to solar heating. The purpose of this report is to determine whether Engineering Hall can be converted to a solar heating system with reasonable cost.

After analyzing solar heating options, we argue that in this report the use of solar heating in Engineering Hall would require a combination of direct gain (skylights) and hydrosolar tubes. At the end of this report, we recommend that the university begin designing a solar heating system this summer, and we offer specific steps toward that goal. We believe this effort toward using renewable energy is a step in the right direction, especially here in southern Colorado where we receive a significant amount of sunlight. In the near future, the United States will need to wean itself off non-renewable energy sources, like oil and natural gas. The conversion of Engineering Hall from an oil heater to solar heating offers us a test case for studying how the conversion to renewable energy could be made throughout our campus.

This report includes four sections: a) our research plan, b) the results of our research, c) a discussion of these results, and d) our recommendations.

The purpose statement is easy to find.

The importance of the subject is stressed.

The structure of the report is forecasted.

Background information familiarizes the readers with the report's topic.

The main point gives the report a statement to prove.

The opening of the research section includes a summary of the methodology.

Research Plan

To study whether heating Engineering Hall with solar is feasible, we followed a five-part research plan:

Phase 1: Develop evaluation criteria
Phase 2: Gather information
Phase 3: Study the heating needs of Engineering Hall
Phase 4: Interview physical plant personnel and solar energy experts
Phase 5: Analyze models of passive and active solar heating methods

Phase 1: Develop Evaluation Criteria

The heading states a major step.

In consultation with President Holton and Carlos Riley, Director of Campus Planning, we determined that a successful conversion to solar heating would need to meet three criteria:

The evaluation criteria are listed and defined.

Cost-Effectiveness—the solar heating system would need to be cost-neutral in the long run. In other words, any costs of converting to solar heating would need to be offset by the savings of the system.

Minimal Disruption—conversion needs to cause only minimal disruption to the use of the building. Any construction would need to occur mostly over the summer or in a way that allowed the building to be used.

Durability—the heating system would need to be reliable over time. It would need to last at least 20 years with routine maintenance.

These criteria were the basis of our evaluation of solar heating systems in the Results and Discussion sections later in this report.

2

(continued)

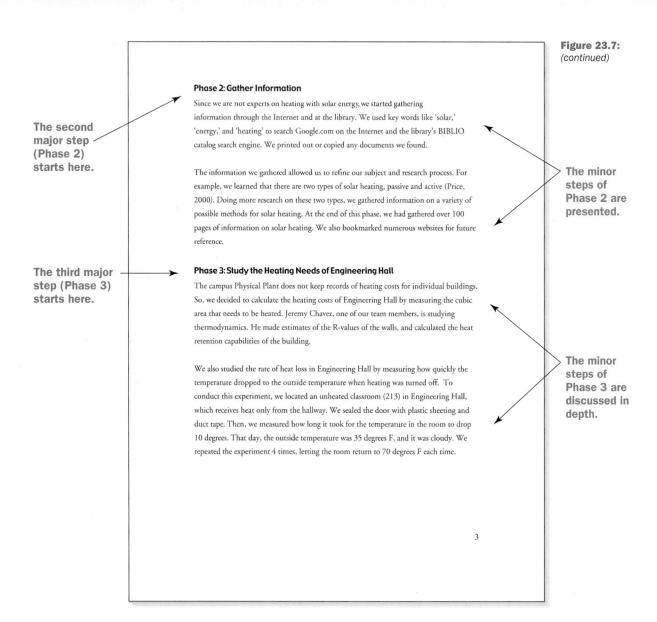

The second major step (Phase 2) starts here.

Phase 2: Gather Information

Since we are not experts on heating with solar energy, we started gathering information through the Internet and at the library. We used key words like 'solar,' 'energy,' and 'heating' to search Google.com on the Internet and the library's BIBLIO catalog search engine. We printed out or copied any documents we found.

The information we gathered allowed us to refine our subject and research process. For example, we learned that there are two types of solar heating, passive and active (Price, 2000). Doing more research on these two types, we gathered information on a variety of possible methods for solar heating. At the end of this phase, we had gathered over 100 pages of information on solar heating. We also bookmarked numerous websites for future reference.

The minor steps of Phase 2 are presented.

The third major step (Phase 3) starts here.

Phase 3: Study the Heating Needs of Engineering Hall

The campus Physical Plant does not keep records of heating costs for individual buildings. So, we decided to calculate the heating costs of Engineering Hall by measuring the cubic area that needs to be heated. Jeremy Chavez, one of our team members, is studying thermodynamics. He made estimates of the R-values of the walls, and calculated the heat retention capabilities of the building.

We also studied the rate of heat loss in Engineering Hall by measuring how quickly the temperature dropped to the outside temperature when heating was turned off. To conduct this experiment, we located an unheated classroom (213) in Engineering Hall, which receives heat only from the hallway. We sealed the door with plastic sheeting and duct tape. Then, we measured how long it took for the temperature in the room to drop 10 degrees. That day, the outside temperature was 35 degrees F, and it was cloudy. We repeated the experiment 4 times, letting the room return to 70 degrees F each time.

The minor steps of Phase 3 are discussed in depth.

3

It is fine to acknowledge the limitations of your study.

Fourth major step begins.

Minor steps are discussed.

Admittedly, our experiment was crude and probably didn't yield us completely reliable results, but it did give us some data that we could use to calculate how much heat was needed to maintain room temperature in the building. We used these results to supplement Jeremy's calculations of heat retention capabilities.

Phase 4: Interview Physical Plant Personnel and Solar Heating Experts

We then set up four interviews with experts in heating issues—two interviews with employees of the campus Physical Plant and two with solar heating experts.

At the physical plant, we interviewed Jerry Madden and Gina Brin on March 21st. Jerry Madden is Director of the Physical Plant, where he has worked for 20 years. Gina Brin is the Physical Plant's energy efficiency expert, and she has worked at the plant for 2 years. We asked them a series of questions about the feasibility of using solar heating in Engineering Hall and their willingness to participate in such a program. During the interviews, both Mr. Madden and Ms. Brin offered us brochures on solar heating. Mr. Madden showed us a plan he had sketched out for using solar heating in Henson Hall, a smaller, more energy-efficient building on campus. Mr. Madden also directed us to some websites about solar heating that we were not aware of. Meanwhile, Ms. Brin looked over our calculations regarding heating needs of Engineering Hall and said we were probably not too far off the correct measurements.

Our interviews with the solar heating experts were conducted on April 21 and 23. We met separately with Jane Karros, who is a solar energy consultant, and Jim Smothers, who is president of the Southern Colorado Solar Energy Institute. They answered our questions about the feasibility of heating an aging building with solar. From these two experts, we learned a couple of new methods for active solar heating that we had not found up to that point. Mr. Smothers gave us photographs and schematic drawings of heating systems used to heat college campus buildings in Japan.

4

(continued)

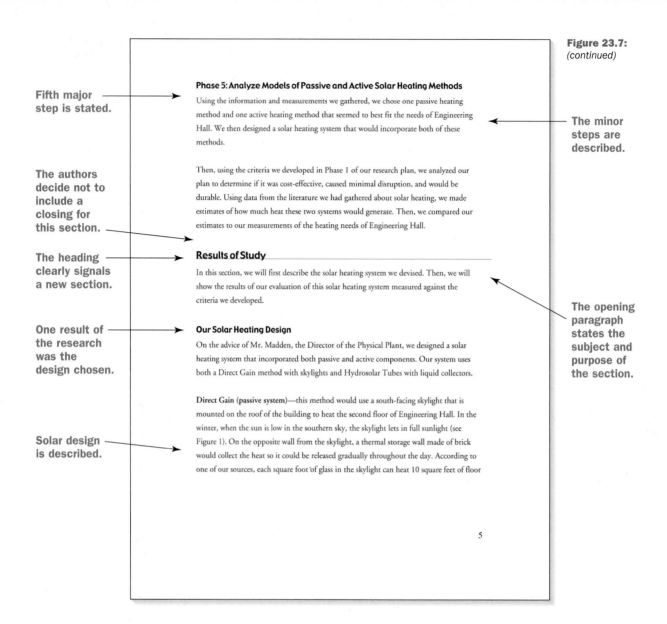

Fifth major step is stated.

The authors decide not to include a closing for this section.

The heading clearly signals a new section.

One result of the research was the design chosen.

Solar design is described.

The minor steps are described.

The opening paragraph states the subject and purpose of the section.

Phase 5: Analyze Models of Passive and Active Solar Heating Methods

Using the information and measurements we gathered, we chose one passive heating method and one active heating method that seemed to best fit the needs of Engineering Hall. We then designed a solar heating system that would incorporate both of these methods.

Then, using the criteria we developed in Phase 1 of our research plan, we analyzed our plan to determine if it was cost-effective, caused minimal disruption, and would be durable. Using data from the literature we had gathered about solar heating, we made estimates of how much heat these two systems would generate. Then, we compared our estimates to our measurements of the heating needs of Engineering Hall.

Results of Study

In this section, we will first describe the solar heating system we devised. Then, we will show the results of our evaluation of this solar heating system measured against the criteria we developed.

Our Solar Heating Design

On the advice of Mr. Madden, the Director of the Physical Plant, we designed a solar heating system that incorporated both passive and active components. Our system uses both a Direct Gain method with skylights and Hydrosolar Tubes with liquid collectors.

Direct Gain (passive system)—this method would use a south-facing skylight that is mounted on the roof of the building to heat the second floor of Engineering Hall. In the winter, when the sun is low in the southern sky, the skylight lets in full sunlight (see Figure 1). On the opposite wall from the skylight, a thermal storage wall made of brick would collect the heat so it could be released gradually throughout the day. According to one of our sources, each square foot of glass in the skylight can heat 10 square feet of floor

5

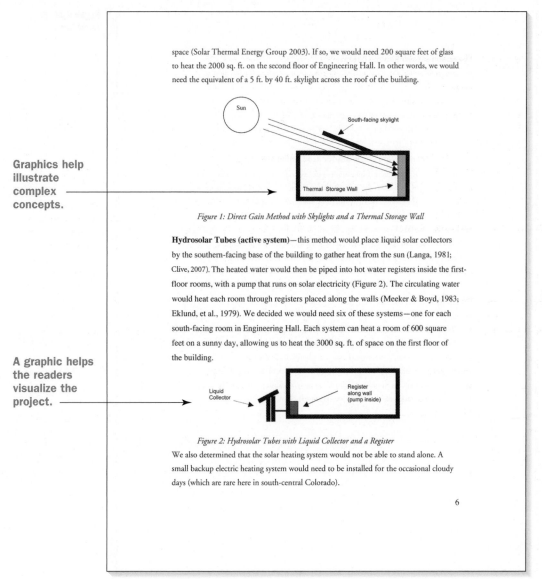

space (Solar Thermal Energy Group 2003). If so, we would need 200 square feet of glass to heat the 2000 sq. ft. on the second floor of Engineering Hall. In other words, we would need the equivalent of a 5 ft. by 40 ft. skylight across the roof of the building.

Figure 1: Direct Gain Method with Skylights and a Thermal Storage Wall

Graphics help illustrate complex concepts.

Hydrosolar Tubes (active system)—this method would place liquid solar collectors by the southern-facing base of the building to gather heat from the sun (Langa, 1981; Clive, 2007). The heated water would then be piped into hot water registers inside the first-floor rooms, with a pump that runs on solar electricity (Figure 2). The circulating water would heat each room through registers placed along the walls (Meeker & Boyd, 1983; Eklund, et al., 1979). We decided we would need six of these systems—one for each south-facing room in Engineering Hall. Each system can heat a room of 600 square feet on a sunny day, allowing us to heat the 3000 sq. ft. of space on the first floor of the building.

A graphic helps the readers visualize the project.

Figure 2: Hydrosolar Tubes with Liquid Collector and a Register

We also determined that the solar heating system would not be able to stand alone. A small backup electric heating system would need to be installed for the occasional cloudy days (which are rare here in south-central Colorado).

6

(continued)

Here begins a section that describes the specific results generated by the researchers.

The first result is described objectively.

The second result is presented with support.

Note how numbers are used to support the presentation of results.

Evaluation of Solar Heating Design

Our evaluation of this solar heating design against our criteria presented us with three results:

- Solar remodeling would pay for itself within 5-6 years.
- Conversion to solar heating would cause minimal disruption.
- The durability of the system would be a minimum of 20 years.

Result: Solar remodeling would pay for itself within 5–6 years

Our measurements show that a combination of passive and active systems like the ones we designed would pay for itself in 5-6 years. With help from Mr. Madden, we estimate that installing the skylights and adding a thermal wall would cost about $24,000. Installing the liquid collectors with tanks, pumps, and registers would cost $23,000. The backup electric heating system would cost about $5,000 and cost about $500 a year to run. The total cost of the system would be $57,000 to heat the building for ten years. This compares favorably with our estimate that the current heating system costs about $4,500 per year for a total of $45,000 over ten years. Renovating the oil heater, Ms. Brin from the Physical Plant estimates, would cost around $20,000. Thus, converting to solar saves us $8000 over ten years.

Result: Conversion to solar heating would cause minimal disruption

Jane Karros, one of the solar consultants we interviewed, estimated that installation of a solar heating system would take approximately 3 months. If the work was begun late in the spring semester and completed by the end of the summer, the disruption to the operations of the building would be minimal. Certainly, the disruption would be no more than a typical renovation of a building's heating system (Ramlow & Nusz, 2006).

7

The third result is presented.

Result: The durability of the system would be a minimum of 10 years

The literature we gathered estimates that the durability of this solar heating system would be a minimum of 20 years (Price 2000). The direct gain system, using the skylight and thermal storage wall, could be used indefinitely with minor upkeep. The hydrosolar tube system, using liquid solar collectors and a solar pump, would likely need to be renovated in 20 years. These systems have been known to go 25 years without major renovation.

Discussion of Results

Based on our research and calculations, we have reached three conclusions, which suggest that a solar heating system would be a viable option for Engineering Hall.

The discussion section starts out with an opening paragraph to redirect the discussion.

The results of the research are discussed.

Solar heating would save money in the long run

Engineering Hall is one of the oldest, least efficient buildings on campus. Yet, our calculations show that putting in skylights and liquid collectors would be sufficient to heat the building on most days. According to our estimates, the solar remodeling would more than pay for itself in 10 years. Moreover, we would eliminate the need for the current oil heating system in the building, which will likely need to be replaced in that time period.

Solar heating is a way toward independence

Our dependence on imported oil and natural gas puts our society at risk a few different ways (US DOE 2004). Engineering Hall's use of oil for heating pollutes our air, and it contributes to our nation's dependence on other countries for fuel. By switching over to solar heating now, we start the process of making ourselves energy independent. Engineering Hall's heating system will need to be replaced soon anyway. Right now would be a good time to think seriously about remodeling to use solar.

8

(continued)

Note how the authors are offering their opinions in this section.

Engineering Hall offers a good model for future projects

Engineering Hall is one of the older buildings on campus. Consequently, it presents some additional renovation challenges that we would not face with the newer buildings. We believe that Engineering Hall provides an excellent model for conversion to solar energy because it is probably one of the more difficult buildings to convert. If we can make the conversion with this building, we can almost certainly make the conversion with other buildings.

It might also be noteworthy that our discussions with Mr. Madden and Ms. Brin at the Physical Plant showed us that there is great enthusiasm for making this kind of renovation to buildings like Engineering Hall.

The recommendations section is clearly signaled here.

Recommendations

In conclusion, we think the benefits of remodeling Engineering Hall to use solar heating clearly outweigh the costs. We recommend the following actions:

- Complete a more thorough study of the heating needs of Engineering Hall.
- Hire a solar architect to draw up a plan for renovating the building.
- Solicit bids from contractors.
- Conduct a formal cost-benefit analysis to see if solar heating is financially feasible for Engineering Hall.

The recommendations are put in bullet form to make them easy to read.

Here is the main point and a look to the future.

If all goes well, remodeling of Engineering Hall could be completed in the summer of 2011. When the remodeling is complete, the college should begin saving money within ten years. Moreover, our campus would house a model building that could be studied to determine whether solar heating is a possibility for other buildings on campus.

9

Offering contact information is a good way to end the report.

The materials cited in the report are listed here in APA format.

Thank you for your time and consideration. After you have looked over this report, we would like to meet with you to discuss our findings and recommendations. Please call Dan Garnish at 555-9294.

Works Cited

Clive, K. (2007). *Build your own solar heating system.* Minneopolis, MN: Lucerno.

Eklund, K. (1979). *The solar water heater workshop manual* (2nd ed.). Seattle, WA: Ecotape Group.

Langa, F. (1981). *Integral passive solar water heating book.* Davis, CA: Passive Solar Institute.

Meeker, J. & Boyd, L. (1983). Domestic hot water installations: The great, the good, and the unacceptable. *Solar Age 6,* 28-36.

Price, G. (2000). *Solar remodeling in southern New Mexico.* Las Cruces, NM: NMSU Energy Institute.

Ramlow, B. & Nusz, B. (2006). *Solar water heating.* Gabriola Island, BC: New Society Publishers.

Solar Thermal Energy Group. (2003). *Solar home and solar collector plans.* Retrieved from http://www.jc-solarhomes.com

U.S. Department of Energy. (2004). Residential solar heating retrofits. Retrieved from http://www.eere.energy.gov/consumerinfo/factsheets/ac6.html

10

Using Google Docs to Collaborate with International Teams

In the universe of Google products, there is a helpful collaboration tool called Google Docs (http://docs.google.com). Google Docs is useful in two ways. First, it includes a free, web-based suite of online software that can be used for writing documents, creating spreadsheets, and making presentations. It's similar to the Microsoft Office package of software.

Second, and more importantly, Google Docs allows you to collaborate with others by storing and "sharing" documents (Figure A). When a file is placed in Google Docs the creator can share it with other members of his or her team. They can work on the document, too.

Figure A shows the Google Docs interface. In the center of the screen, you can see a list of files that are being shared among different collaborative groups. On the left-hand side of the screen, the user names of regular team members are listed, allowing the creator of a document to designate quickly who can read or edit each file.

The Google Docs Interface

Buttons to upload and share files

Types of documents

User names of team members

Files shared among team members

Figure A: Google Docs provides a common space for sharing documents with international teams.

Source: GOOGLE is a trademark of Google Inc.

Want to know more about virtual teaming? Go to
www.pearsonhighered.com/johnsonweb3/23.13

People working on collaborative projects, especially with international teams, find Google Docs to be an amazingly helpful workspace. With Google Docs, everyone can access and edit the most recent versions of any document. This allows teams to avoid the usual questions about whether everyone has the latest version of the file.

Link

For more advice about working in teams, see Chapter 4 page 66.

Another advantage is the free software available to everyone. One of the problems with working internationally is that team members in different parts of the world are often using a variety of software for word processing, spreadsheets, and presentations on a mixture of operating systems. So sharing documents among members of an international team can get bogged down with tricky conversions between software packages and operating systems. But, if the whole team is using Google Docs, everyone has access to the same software.

Even if your team agrees to use a common software package (e.g. MS Word), members may be using different versions or running old versions of Windows, Linux, or Mac OS. Google Docs solves that problem, too, because everyone will be using the up-to-date versions of its software. Since the software is free, you won't hear the common complaints from your team members about how their management won't pay for software or operating system upgrades.

The main drawback to Google Docs is that the software suite is not as advanced as Microsoft or Adobe products (despite Google's claims to the contrary). Also, when documents from other software applications, like MS Word, are placed on Google Docs, they can lose some of their formatting and design. Plus, your should always remember that your files are being stored on a server outside your company. So, any high-security files should not be stored or shared on Google Docs.

Google Docs isn't perfect, but for international teams, its benefits far outweigh the shortcomings. The ability to share documents with Google Docs and use common software makes collaborating internationally much easier.

The title for your report should give your readers a clear idea of what the report is about (Figure 23.8). A title like "Solar Heating" probably is not specific enough. A more descriptive title like "Learning with Light: Using Solar to Heat Engineering Hall" gives readers a solid idea about what the report will discuss.

ABSTRACT OR EXECUTIVE SUMMARY   Your readers usually won't have time to read all the information in your report. So, if your report is longer than ten pages, you should consider including an abstract or executive summary.

An abstract is a summary of the report that uses the phrasing in the report and follows its organizational structure. When writing an abstract, you should draw key sentences directly from the report itself. Start out with the purpose statement of the report, and then state the main point. From there, draw one or two key sentences from each major section. In an abstract for a report, for example, you would probably include these items in the following order:

- Purpose statement (one sentence)
- Main point (one sentence)
- Methodology (one or two sentences)

Designing the Title Page

```
Learning with Light:
Using Solar to Heat
Engineering Hall



A Report from the
Innovation Group



April 15, 2008
```

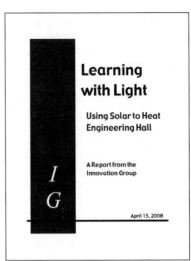

Figure 23.8: Designing an effective title page takes only a few moments. Those few moments of work, though, can make a solid first impression on the readers. Most readers would be attracted to the report on the right.

- Results (one or two sentences)
- Discussion (one or two sentences)
- Recommendations (one or two sentences)

You should modify the sentences in places to make the abstract readable, but try to retain the phrasing of the original report as much as possible.

Abstract

Purpose of report leads the abstract.

The main point comes second.

The remainder of the abstract mirrors the structure of the report (methodology, results, discussion, recommendations).

The purpose of this report is to determine whether the Engineering Hall's heating system can be converted to solar with reasonable cost. After analyzing solar heating options, we argue in this report that the best way to add solar heating to Engineering Hall would include a combination of direct gain (skylights) and hydrosolar tubes. Our research plan included five phases: (1) develop evaluation criteria, (2) gather information on solar heating, (3) analyze the heating needs of Engineering Hall, (4) interview physical plant personnel and solar energy experts, and (5) analyze models of passive and active solar heating methods. From our research, we designed a solar heating system that incorporates both passive and active components. Our system uses both a direct gain method with skylights and hydrosolar tubes with liquid collectors. Our evaluation of this solar heating design against our evaluation criteria presented us with three results: Solar remodeling would pay for itself within 5–6 years; conversion to solar heating would cause minimal disruption; the durability of the system would be a minimum of 10 years. We have reached three conclusions that we believe the university administration should seriously consider: Solar heating would save money in the long run; solar heating is a way toward independence; and Engineering Hall offers a good model for future solar projects. We recommend the following actions: Complete a more thorough study of the thermodynamic features of Engineering Hall; hire a solar architect to draw up a plan for renovating the building; solicit bids from contractors; and conduct a cost-benefit analysis to see if solar heating is financially feasible.

Want to see some sample texts that use good and bad design? Go to
www.pearsonhighered.com/johnsonweb3/23.15

An executive summary is a concise, *paraphrased* version of your report (usually one page) that highlights the key points in the text. The two main differences between an abstract and an executive summary are that (1) the summary does not follow the organization of the report, and (2) the summary does not use the exact phrasing of the report. In other words, a summary paraphrases the report and organizes the information to highlight the key points.

Report Summary

The purpose of the report is placed early in the summary. →

This report was written in response to a challenge to our Energy Dynamics class (Engineering 387) from Dr. Sharon Holton, President of Kellen College. She asked us to develop options for converting campus buildings to renewable energy sources. In this report, we discuss the possibility of converting Engineering Hall's heating system to solar. We conclude that heating Engineering Hall with solar sources would require a combination of direct gain (skylights) and hydrosolar tubes. The combination of these two solar technologies would ensure adequate heating for almost all the building's heating needs. A

The main point is also placed up front. →

backup heater could be retained for sustained cold spells.

The remainder of the summary organizes information in order of importance.

To develop the information for this report, we followed a five-step research plan: (1) develop an evaluation criteria, (2) gather information on solar heating, (3) analyze the heating needs of Engineering Hall, (4) interview physical plant personnel and solar energy experts, and (5) analyze models of passive and active solar heating that would be appropriate for this building.

The results of our research are mostly anecdotal, but they show that solar heating is possible, even for an older building on campus. We believe that our results show that Engineering Hall can be a model for developing solar heating systems around campus, because it is truly one of the more difficult buildings at Kellen to convert to solar heating. Newer buildings on campus would almost certainly be easier to convert. We conclude by pointing out that solar heating would save money in the long run. In the case of Engineering Hall, solar remodeling would pay for itself in 5–6 years.

We appreciate your taking time to read this report. If you have any questions or would like to meet with us, please call Dan Garnish at 555–9294.

The executive summary will often duplicate the contents of the introduction, but it should not replace the introduction. Instead, it should be written so that it can stand alone, apart from the rest of the report.

TABLE OF CONTENTS   If your report runs over ten pages, you should consider adding a table of contents. A table of contents is helpful to readers in two ways. First, it helps them quickly access the information they need in the report. Second, it offers an overall outline of the contents of the report. Since reports tend to be larger documents, your readers will appreciate a quick summary of the report's contents.

In the table of contents, the headings should be the same as the ones used in your report. Then, use tabs or leader tabs to line up the page numbers on the right side. Leader tabs are used to insert a line of dots or dashes from the heading to the page number.

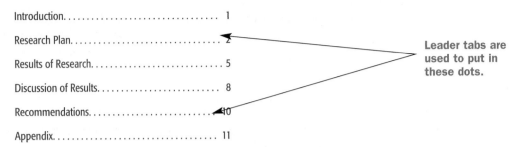

Table of Contents

Leader tabs are used to put in these dots.

Developing Back Matter

Back matter in a report might include appendixes, a glossary of terms, and calculations. Keep in mind, though, that most readers will never look at the back matter. So, if you have something important to say, do not say it here.

APPENDIXES   Appendixes are storage areas for information that may or may not be useful. For example, additional data tables or charts not needed in the body of the report might be displayed in an appendix. You might include clippings from newspapers or magazines.

GLOSSARY OF TERMS   Depending on the complexity of your report and the familiarity of your readers with the subject, you may want to include a short glossary of terms. When creating a glossary, look back over your report and highlight words that may not be familiar to your nonexpert readers. Then, in a glossary at the back of the report, list these terms and write sentence definitions for each.

CALCULATIONS   In highly technical reports, you may want to include your calculations in the back matter. Here is where you can demonstrate how you arrived at your figures in the report.

AT A GLANCE

Front Matter and Back Matter

Front matter—items that appear before the main report:
- Letter or memo of transmittal
- Title page
- Abstract or executive summary
- Table of contents

Back matter—items that appear after the main report:
- Appendixes
- Glossary of terms
- Calculations

Link

For more information on writing sentence definitions, see Chapter 18, page 496.

Using Style in Analytical Reports

Link

For more help with using the plain style, see Chapter 9, page 206.

Your reports don't need to be difficult to read. The most readable reports are written in the plain style, with limited use of persuasion. Reports are persuasive in an unstated way, putting the emphasis on the soundness of the methodology, the integrity of the results, and the reasonableness of the discussion. Plain style will make your reports sound straightforward and clear to the readers.

While revising your report, you might pay close attention to the following plain style techniques:

Make "doers" the subjects of sentences—Reports tend to overuse the passive voice, which makes them harder to read than necessary. If the passive voice is required in your field, use it. But if you want your writing to be more effective, make your sentences active by putting the "doers" in the subjects of the sentences.

> **Passive:** Saplings had been eaten during the winter by the deer we monitored, because they were desperate for food.

> **Active:** The deer we monitored ate saplings to survive the winter because they were desperate for food.

Use breathing-length sentences—Reports are notorious for using sentences that are far too long to be understood. As you write and revise your report, look for places where your sentences are too long to be stated in one breath. These sentences should be shortened to breathing length or divided into two breathing-length sentences.

Eliminate nominalizations—Reports often include nominalizations that cloud the meaning of sentences. You can revise these sentences for clarity:

> **Nominalization:** This report offers <u>a presentation</u> of our findings.

> **Revised:** This report <u>presents</u> our findings.

> **Nominalization:** We made <u>a decision</u> to initiate <u>a replacement</u> of the Collings CAD software.

> **Revised:** We <u>decided to replace</u> the Collings CAD software.

> **Revised further:** We <u>replaced</u> the Collings CAD software.

Link

For more information on eliminating nominalizations, see Chapter 9, page 209.

Nominalizations may make the report sound more formal, but your readers will appreciate their elimination for the sake of clarity.

Define jargon and specialized terms—Jargon should not be completely eliminated from reports. Instead, these words should be defined for nonexperts. When you need to use a specialized term, use a sentence definition or parenthetical definition to clarify what the word means.

> **Sentence definition:** A gyrocompass is a directional finding device that uses a gyroscope to compensate for the earth's rotation and thus points to true north.

> **Parenthetical definition:** Spotting a blue grouse, <u>a plump, medium-sized bird with feathered legs and bluish gray plumage,</u> is especially difficult because they are well camouflaged and live in higher mountain areas.

Link

For more information on writing sentence and parenthetical definitions, see Chapter 18, pages 496–498.

AT A GLANCE

Improving Style in Analytical Reports

- Make "doers" the subjects of sentences.
- Use breathing-length sentences.
- Eliminate nominalizations.
- Define jargon and specialized terms.

Designing Analytical Reports

People rarely read reports word for word. Instead, they scan these documents. Therefore, you should use document design and graphics to highlight your main points and offer readers access points to start reading.

Want to learn about passive voice? Go to
www.pearsonhighered.com/johnsonweb3/23.17
Need a definition for a jargon term? Go to
www.pearsonhighered.com/johnsonweb3/23.18

Designing Analytical Reports   695

Choosing a Document Design

Link

For more help with document design, see Chapter 10, page 232.

Your report's design should reflect the subject of the report and the preferences of your readers. Your report should not look boring if you want people to read it. So, you might experiment with page layout, the design of headings, and uses of graphics. You might look for ways to use multicolumn formats, which allow you to add pull quotes, sidebars, and other page layout enhancements (Figure 23.9).

Page Layouts for Reports

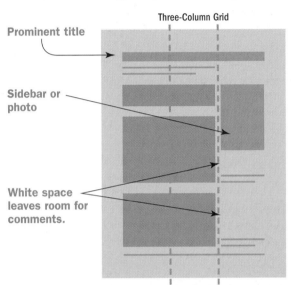

Three-Column Grid

Prominent title

Sidebar or photo

White space leaves room for comments.

Two-Column Grid

Figure 23.9: Reports don't need to look boring. A little attention to design will make your report inviting and easier to read. In these page layouts, grids have been used to balance the texts, leaving room for margin text and other access points.

Pull quote to highlight an important point.

Whatever you do, make sure you choose a document design that fits your subject, readers, and the document's context of use. Often, as a project comes to a finish, writers do not put enough time into the design of their report. As a result, the report makes a bad first impression on readers. With a little effort, you can create a design that ensures a positive first impression, while making need-to-know information easier to find.

Figure 23.10, for example, shows a well-designed report. The writers have designed this document to make it highly scannable. The two-column format and large headings put the information in an accessible format. Meanwhile, the larger font and ample white space helps the reader look quickly through the document.

Link

For more information on making and using graphs, see Chapter 11, page 280.

Using Graphics to Clarify and Enhance Meaning

Increasingly, graphics are essential features of reports. Tables and graphs are especially helpful tools for displaying data. As you write the report, you should actively look for places where tables might be used to better display your data or information. Also, look for places where a graph could show trends in the data.

Want more tips about designing page layouts? Go to
www.pearsonhighered.com/johnsonweb3/23.19

Designing a Report

NIJ | SEXUAL ASSAULT ON CAMPUS

Heather M. Karjane, Bonnie S. Fisher, and Francis T. Cullen

The title is large and easy to locate. →

Sexual Assault on Campus:
What Colleges and Universities Are Doing About It

Campus crime in general and sexual assault in particular have been receiving more attention than in the past, and concern has been expressed at the highest levels of government. On the Federal level, Congress responded by enacting several laws requiring institutions of higher education to notify students about crime on campus, publicize their prevention and response policies, maintain open crime logs, and ensure sexual assault victims their basic rights.[3] The Clery Act, the most notable of these laws, mandates an annual security report from each Federally funded school (see "Recent Federal Laws on Campus Crime").

In 1999, Congress asked the National Institute of Justice to find out what policies and procedures schools use to prevent and respond to reports of sexual assault.[4] The resulting study revealed that schools are making strides

About the Authors

Heather M. Karjane, Ph.D., is coordinator for gender issues at the Commonwealth of Massachusetts Administrative Office of the Trial Court. Bonnie S. Fisher, Ph.D., and Francis T. Cullen, Ph.D., are faculty in the Division of Criminal Justice at the University of Cincinnati. The Police Executive Research Forum conducted some of the field research.

RECENT FEDERAL LAWS ON CAMPUS CRIME

Starting in 1990, Congress acted to ensure that institutions of higher education have strategies to prevent and respond to sexual assault on campus and to provide students and their parents accurate information about campus crime. The major Federal laws pertaining to this study are:

Student Right-to-Know and Campus Security Act of 1990 (the "Clery Act"*) (20 U.S.C. § 1092). This law, Title II of Public Law 101–542, requires that schools annually disclose information about crime, including specific sexual crime categories, in and around campus.

Campus Sexual Assault Victims' Bill of Rights of 1992. This amendment to the 1990 act requires that schools develop prevention policies and provide certain assurances to victims. The law was amended again in 1998 to expand requirements, including the crime categories that must be reported.

*The act was renamed in 1998 the "Jeanne Clery Disclosure of Campus Security Policy and Campus Crime Statistics Act" in honor of a student who was sexually assaulted and murdered on her campus in 1986.

← This "sidebar" offers background information.

1

Source: U.S. Department of Justice, National Institute of Justice, 2005.

(continued)

NIJ RESEARCH FOR PRACTICE / DEC. 05

The two-column format makes the report easy to scan.

in some areas but must continue efforts to increase student safety and accountability. After summarizing what is known about the nature and extent of sexual assault on campus, the researchers highlighted findings regarding response policies and procedures; reporting options; barriers and facilitators; reporter training and prevention programming; victim resources; and investigation, adjudication, and campus sanctions. The study's baseline information can be used to measure progress in how institutions of higher education respond to sexual assault.

The scope of the problem

These large, bold headings help readers find the information they need.

Administrators want their campuses to be safe havens for students as they pursue their education and mature intellectually and socially. But institutions of higher education are by no means crime-free; women students face a high risk for sexual assault.

Just under 3 percent of all college women become victims of rape (either completed or attempted) in a given 9-month academic year. On first glance, the risk seems low, but the percentage

translates into the disturbing figure of 35 such crimes for every 1,000 women students. For a campus with 10,000 women students, the number could reach 350. If the percentage is projected to a full calendar year, the proportion rises to nearly 5 percent of college women. When projected over a now-typical 5-year college career, one in five young women experiences rape during college.[5]

Counter to widespread stranger-rape myths, in the vast majority of these crimes—between 80 and 90 percent—victim and assailant know each other.[6] In fact, the more intimate the relationship, the more likely it is for a rape to be completed rather than attempted.[7] Half of all student victims do not label the incident "rape."[8] This is particularly true when no weapon was used, no sign of physical injury is evident, and alcohol was involved—factors commonly associated with campus acquaintance rape.[9] Given the extent of non-stranger rape on campus, it is no surprise that the majority of victimized women do not define their experience as a rape.

These reasons help explain why campus sexual assault is

2

Figure 23.10:
(continued)

The header at the top of the page makes the text feel consistent.

not well reported. Less than 5 percent of completed and attempted rapes of college students are brought to the attention of campus authorities and/or law enforcement.[10] Failure to recognize and report the crime not only may result in underestimating the extent of the problem, but also may affect whether victims seek medical care and other professional help. Thus, a special concern of the study was what schools are doing to encourage victims to come forward.

Federal law and the schools' response

Institutions of higher education vary widely in how well they comply with Clery Act mandates and respond to sexual victimization. Overall, a large proportion of the schools studied—close to 80 percent—submit the annual security report required by the Act to the U.S. Department of Education; more than two-thirds include their crime statistics in the report. Yet, according to a General Accounting Office study, schools find it difficult to consistently interpret and apply the Federal reporting requirements, such as deciding which incidents to cite in the annual report, classifying crimes, and the like.[11]

For definitions and explanation of terminology such as "acquaintance rape," see Karjane et al., *Campus Sexual Assault: How America's Institutions of Higher Education Respond*, Oct. 2002, NCJ 196676: 2–3; for an analysis of how colleges and universities define sexual assault, see chapter 3.

This screened box offers a source for more information if the reader wants it.

Definitions, even of such terms as "campus" and "student," are often a challenge and contribute to inconsistency in calculating the number of reported sexual assaults. Only 37 percent of the schools studied report their statistics in the required manner; for example, most schools failed to distinguish forcible and nonforcible sex offenses in their reports as required by the Clery Act.

The issues and the findings

Congress specified the issues to be investigated (see "Study Design"). Key areas of concern were whether schools have a written sexual assault response policy; whether and how they define sexual misconduct; who on campus is trained to respond to reports of sexual assault; how students can report sexual victimization; what

Ample white space gives the text a less intimidating feel.

3

Graphics have some other benefits. They tend to break up large blocks of text, making the report more readable. They also offer "access points" that stop readers from scanning by encouraging them to look into the text for an explanation of the graphic.

In most situations, graphics should be titled and numbered so they can be referred to in the text (e.g., "See Table 2"). Place each graphic on the page where it is referenced or very soon afterward.

Revising, Editing, and Proofreading

Always leave at least a couple of days to revise and edit your report. Since analytical reports are usually rather large, you will need to spend extra time editing them into their final form.

Revising the Content, Style, and Design

Try revising the report from three different perspectives—content, style, and design:

Content—When revising for content, look for places where there is too much information or too little. Working paragraph by paragraph, ask yourself whether the included information is need-to-know information. Then, cut out all the information that does not contribute to the overall argument of the report. Also, recheck any calculations to make sure they are accurate.

Style—Since reports are often written in teams, the style can be uneven or inconsistent. Even reports that you write alone will sometimes sound uneven, because your style may have evolved while you were drafting the text. To avoid these uneven spots in your writing, use the "plain style" strategies discussed in Chapter 9 to smooth out your writing. Then, use the "persuasive style" strategies to add color and emphasis to your writing where needed.

Design—Make sure the design is consistent across all the pages. Then, check whether your graphics are properly labeled and referenced in the text. Finally, make sure you have included page numbers on each sheet, except the cover.

Link

For more information on improving style, see Chapter 9, page 206.

Editing and Proofreading the Draft

You and your team should carefully proofread the report before submitting it.

If you're like most people, you are probably up against a deadline at that point. So, it might be tempting to simply skip the editing and proofreading phases of the writing process. Don't do it. Nothing sabotages a report faster than typos, spelling errors, and grammar errors. Your readers will certainly notice your sloppy work, and they will immediately assume that you were equally careless in your research.

Link

For more tips on editing and proofreading, see Chapter 12, page 306.

For worksheets to help you revise and edit your report, go to
www.pearsonhighered.com/johnsonweb3/23.20

- Various types of analytical reports are used in the workplace, including research reports, completion reports, recommendation reports, feasibility reports, and empirical research reports.

- Use the IMRaD acronym to remember the generic report pattern: Introduction, Methods, Results, and Discussion.

- Determine the rhetorical situation for your report by considering the subject, purpose, readers, and context of use of your report.

- Define your research question, and formulate a hypothesis to be tested. Then, develop a methodology and identify the two to five major steps you will need to take to establish or refute your hypothesis.

- Collect information using library, Internet, and empirical sources.

- Organize and draft your report following the separate "moves" for an introduction, body, and conclusion.

- Both the style and the design of analytical reports should aim to clarify the contents of the report.

- It is vital to leave enough time to revise, edit, and proofread an analytical report to ensure professional quality.

Individual or Team Projects

1. On the Internet or elsewhere, find a report on a topic you care about. Analyze the report by paying close attention to its content, organization, style, and design. Does the report include only need-to-know information? Is the report organized according to the IMRaD pattern? Is the style plain and appropriate to the subject of the report? Does the design enhance the reading of the report?

   Write a two-page memo to your instructor in which you critique the report. In your memo, highlight the strengths of the report and offer suggestions for improvements.

2. Write a small report (two or three pages) in which you conduct a preliminary study of a scientific or technical topic that interests you. In your report, create a short methodology that will allow you to collect some overall information about the subject. Then, present the results of your study. In the conclusion, talk about how you might conduct a larger study on the subject and the limitations you might face as you enlarge the project.

3. Devise a research question of interest to you. Then, using logical mapping, sketch out a methodology that would allow you to study that research question in depth. When you have finished outlining the methodology, write a one-page memo to your

instructor in which you describe the methodology and discuss the kinds of results your research would produce.

4. Write an executive summary of an article from a scientific or technical journal in your field. You should be able to find journals in your campus library. Your summary should paraphrase the article, highlighting its main points, methodology, results, discussion, and conclusions.

Collaborative Project: Problems in the Community

With a team of classmates, choose a local problem in your community on which you would like to write an analytical report. The readers of your report should be people who can take action on your findings, like the mayor or the city council.

After defining the rhetorical situation in which your report will be used, develop a step-by-step methodology for studying the problem. Then, collect information on the topic. Here are some ideas for topics that might interest you:

- Driving under the influence of alcohol or drugs
- Violence
- Underage drinking
- Illegal drug use
- Homelessness
- Teen pregnancy
- Water usage
- Pollution
- Graffiti
- Environmental health

In your report, present the results of your study and discuss those results. In the conclusion of your report, make recommendations about what local authorities can do to correct the problem or improve the situation.

For additional technical writing resources, including interactive sample documents, document design tutorials and guidelines, and more, go to **www.mytechcommlab.com.**

Revision Challenge

The introduction to a report shown here is somewhat slow and clumsy. How could you revise this introduction to make it sharper and more interesting?

Introduction

Let us introduce ourselves. We are the Putnam Consulting Firm, LLC, based out of Kansas City, KS. We specialize in the detection and mitigation of lead-related problems. We have been in business since 1995, and we have done many studies much like the one we did for you.

Our CEO is Lisa Vasquez. She has been working with lead-related issues for nearly two decades. In her opinion, lead poisoning is one of the greatest epidemics facing the United States today. The effects of lead on children, especially poor children, is acute. Dr. Vasquez has devoted her life to addressing the lead problem so today and tomorrow's children won't be damaged by continued neglect of the issue.

One of today's silent villains is lead poisoning. It is one of the most prevalent sources of childhood health problems. Each year, several hundred children in Yount County are treated for serious cases of lead poisoning. There are countless others who are suffering silently because their symptoms are either too minor or they go unnoticed by parents, teachers, and other authorities. For example, consider the case of Janice Brown in north-western Yount County. She showed the effects of lead poisoning in her cognitive development, leading to a lower IQ rating. When her house was tested for lead poisoning, it was found that her parents had sandblasted the sides of the home to remove the old paint. Lead had saturated the ground around Janice's home and there was residual lead dust everywhere in the home. The levels of lead in Janice's blood were nearly 10 times the amount that causes problems.

You will find other troubling stories like this one in this report. We were hired by the Yount County Board of Directors to complete this study on the amounts of lead in the county. Unfortunately, we have found that Yount County is in particular trouble. Much of the county's infrastructure and housing was built when lead use was at its peak. Housing in the county relies heavily on lead pipes. Most of the houses were painted inside and out with lead-based paint.

Ironically, lead poisoning is also one of the most preventable problems that face children today. The continual persistence of the problem is a stain on our nation, and it is a particular problem that Yount County must face. In this report, we offer some recommendations.

To see a revision of this introduction, go to
www.pearsonhighered.com/johnsonweb3/23.22

The X-File

George Franklin works as an environmental engineer for Outdoor Compliance Associates. He and his team are responsible for writing analytical reports called environmental impact statement (EISs) on sites that might be developed for housing or businesses. An EIS is required by the government before any work could begin.

George's responsibility on the team is to track down any historical uses of the site. For example, if a gas station had once been on the site, his job is to make sure the old underground holding tanks were removed. If the site once held a factory, his job is to make sure that chemicals had not been dumped in the area. George loved being an environmental detective.

A year ago, George's team had written an EIS for a site where a new apartment complex was planned. The site was about a mile from a major research university. While researching the site, George discovered that it had housed part of the city's waste treatment center until the early 1950s. George's discovery wasn't a problem, though, because any contaminants would have disappeared long ago. So, George's team wrote a favorable EIS, clearing the site for development. The building plan was approved by the city, and construction started soon after.

Then, yesterday, an old, yellowed file mysteriously appeared in George's office mailbox. The file was marked "Confidential" and had no return address. George looked inside.

In the file was a report that nuclear weapons research had once been done at the university during the 1940s. Not recognizing the potential harm of the nuclear waste, the scientists had sent tons of radioactive waste down the drain. The nuclear waste ended up at the city's old waste treatment center. The waste, including the nuclear waste, was then spread around the grounds of the waste treatment plant.

George grabbed his Geiger counter and went out to the building site. The apartment development was now half built. When he pulled out his Geiger counter, it immediately began detecting significant levels of radiation.

The radiation wasn't high enough to violate government standards, but it was close. George knew that the building permit still would have been granted even if knowledge of this site's nuclear past had been known. However, he also knew that as soon as people heard about the radiation, no one would want to live in the apartments. That would be a disaster for the developer building them.

If he reported the radiation, there was a good chance George's company would be sued by the builder for missing this important problem with the property. Moreover, other builders would likely never again hire his company to write an EIS. George would almost certainly lose his job, and the company he worked for might be forced out of business. But then, he could keep it quiet. The radiation, after all, was not above government standards.

If you were George, how would you handle this situation?

More information on nuclear waste disposal is available at
www.pearsonhighered.com/johnsonweb3/23.23

Appendixes

Grammar and Punctuation Guide

This guide is a reference tool to help you handle mechanical issues in writing. The guide has two parts: (1) the Top Ten Grammar Mistakes, and (2) a Punctuation Refresher. The Top Ten Grammar Mistakes will show you how to avoid the ten most common grammar mistakes. The Punctuation Refresher will update you on punctuation rules, showing you how to properly use punctuation marks.

Consistent grammatical correctness and punctuation are essential if you are going to clearly express yourself, especially in technical documents and presentations. Any grammar and punctuation mistakes in your writing will undermine even your best ideas. Moreover, readers will make judgments about you and your work based on your grammar and punctuation. If you send a document littered with grammatical errors to your supervisor or your company's clients, they will question your attention to quality, your commitment to the project, and even your intelligence. That's bad.

Mastering basic grammar and punctuation does not take long. If you haven't worked on improving your grammar since high school, take some time to refresh yourself on these rules.

The Top Ten Grammar Mistakes

Don't be one of those people who is always apologizing with statements like, "I'm not good at grammar." These kinds of statements do not excuse your grammatical problems; they only indicate that you have not taken the time to learn basic grammar rules. If you are one of these grammar apologists, it is time for you to spend the little time necessary to master the rules. An hour or two of study will make those grammatical problems disappear.

Comma Splice

A significant percentage of grammar errors are comma splices. A comma splice occurs when two complete sentences are joined together with a comma.

Incorrect
> The machine kept running, we pulled the plug.
>
> We moved the telescope just a little to the left, the new nova immediately came into view.

In these examples, notice how the parts before and after the commas could stand alone as sentences. The comma is *splicing* these sentences together.

How can you fix these comma splices? There are a few options.

Correct
> The machine kept running, so we pulled the plug.
>
> We moved the telescope just a little to the left, and the new nova immediately came into focus.
>> (Add a conjunction [*so, and, but, yet*] after the comma.)

Correct

The machine kept running; we pulled the plug.

We moved the telescope just a little to the left; then, the new nova immediately came into focus.
>   (Replace the comma with a semicolon.)

The machine kept running. We pulled the plug.

We moved the telescope just a little to the left. The new nova immediately came into focus.
>   (Replace the comma with a period and create two grammatically correct sentences.)

The machine kept running; therefore, we pulled the plug.

We moved the telescope just a little to the left; consequently, the new nova immediately came into view.
>   (Replace the comma with a semicolon and add a conjunctive adverb [*therefore, consequently, however, thus*].)

Since the machine kept running, we pulled the plug.

Because we moved the telescope just a little to the left, the new nova immediately came into view.
>   (Insert a subordinating conjunction [*since, because*] at the beginning of the sentence.)

Run-On Sentence

The run-on sentence error is a close cousin of the comma splice. In a run-on sentence, two or more sentences have been crammed into one.

Incorrect

The computer suddenly crashed it had a virus.

The Orion nebula lies about 1500 light-years from the sun the nebula is a blister on the side of the Orion molecular cloud that is closest to us.

Run-on sentences are corrected the same way comma splices are corrected. You can use conjunctions (*and, but, or, nor, for, because, yet, however, furthermore, hence, moreover, therefore*) to fix them. Or, you can divide the sentences with a semicolon or period.

Correct

The computer suddenly crashed, because it had a virus.

The computer suddenly crashed; we guessed it had a virus.

The computer suddenly crashed. It had a virus.

The Orion nebula lies about 1500 light-years from the sun. The nebula is a blister on the side of the Orion molecular cloud that is closest to us.

The Orion nebula lies about 1500 light-years from the sun; moreover, the nebula is a blister on the side of the Orion molecular cloud that is closest to us.

In most cases, run-on sentences are best fixed by adding a period, thus separating the two sentences completely.

Want more information on comma splices? Go to
www.pearsonhighered.com/johnsonweb3/A.1

**Grammar and
Punctuation Guide**   A-3

Fragment

A fragment, as the name suggests, is an incomplete sentence. A fragment typically occurs when the sentence is missing a subject or a verb, or it lacks a complete thought:

Incorrect   Because the new motherboard was not working.
            (This fragment contains a subject and verb but does not express a complete thought.)

The report missing important data.
            (This fragment contains a subject but no verb.)

The first fragment can be corrected in the following ways:

Correct   The new motherboard was not working.
            (Remove the conjunction [*because*].)

Because the new motherboard was not working, we returned it to the manufacturer.
            (Join the fragment to a complete sentence [an independent clause].)

The second fragment can be corrected in the following ways:

Correct   The report was missing important data.
            (Insert a verb [*was*].)

The report, missing important data, was corrected immediately.
            (Insert a verb [*was corrected*] and an adverb [*immediately*].)

Sometimes writers, especially creative writers, will use fragments for the purpose of jarring their readers. In technical writing, you don't want to jar your readers. Leave the creative uses of fragments to creative writers.

Dangling Modifier

A dangling modifier occurs when a phrase does not properly explain the subject.

Incorrect   While eating lunch, the acid boiled over and destroyed the testing apparatus.
            (The acid is apparently eating lunch while it does damage to the testing apparatus. That's some acid!)

After driving to Cleveland, our faithful cat was a welcome sight.
            (The cat apparently drove the car to Cleveland. Bad kitty!)

These kinds of errors are common—and often funny. To avoid them (and your readers' grins), make sure that the introductory phrase is modifying the subject of the sentence. Here are corrections of these sentences:

Correct   While we were eating lunch, the acid boiled over and destroyed the testing apparatus.

After driving to Cleveland, we were glad to see our faithful cat.

Notice how the information before the comma modifies the subject of the sentence, which immediately follows the comma.

Worksheets on run-ons, fragments, and comma splices are available at
www.pearsonhighered.com/johnsonweb3/A.2

Subject-Verb Disagreement

Subject-verb disagreements occur when the subject of the sentence does not match the verb. Singular subjects should go with singular verbs, while plural subjects should have plural verbs. Here are a few sentences with subject-verb disagreement:

Incorrect

The windows, we discovered after some investigation, was the reason for heat loss in the house.
(The singular verb *was* does not match the plural subject *windows*.)

The robin, unlike sparrows and cardinals, do not like sunflower seeds.
(The subject *robin* is singular, while the verb phrase *do not like* is plural.)

Either my DVD player or my stereo were blowing the fuse.
(The subject *DVD player or my stereo* is singular, while the verb *were* is plural.)

Here are the correct sentences:

Correct

The windows, we discovered after some investigation, were the reason for heat loss in the house.

Robins, unlike sparrows and cardinals, do not like sunflower seeds.

Either my DVD player or my stereo was blowing the fuse.

When *or* is used in the subject, as it is in the third example, the verb needs to agree with the noun that follows *or*. In this sentence, the use of *or* means we are treating the DVD player and the stereo separately. So, a singular verb is needed.

This third example brings up an interesting question: What if one or both of the words flanking the *or* is plural? Again, the answer is that the verb will agree with the noun that comes after the *or*.

The speeding cars or the reckless motorcyclist was responsible for the accident.

Either the falling branches or the high winds were responsible for the damage.

Collective nouns (*crowd, network, group*) that name a group as a whole take a singular verb:

The group uses a detailed questionnaire to obtain useful feedback.

Pronoun-Antecedent Disagreement

Pronoun-antecedent disagreement usually occurs when a writer forgets whether the subject is plural or singular.

Incorrect

Anyone who thinks Wii is better than Sony Playstation should have their head examined.

Like the scientists, we were sure the rocket was going to blast off, but it wasn't long before you knew it was a dud.

In these cases, words later in the sentence do not agree with pronouns earlier in the sentence. In the first sentence, *anyone* is singular while *their* is plural. In the

Some great grammar handbooks are online. Go to
www.pearsonhighered.com/johnsonweb3/A.3

Grammar and Punctuation Guide   A-5

second sentence, *we* is a first-person noun while *you* is a second-person pronoun. Here are a couple of ways to correct these sentences:

Correct   People who think Wii is better than Sony Playstation should have their heads examined.
                 (The subject is made plural [*people*] and an *s* is added to *head* to make it plural.)

                 Anyone who thinks Wii is better than Sony Playstation should have his or her head examined.
                 (The subject [*anyone*] is kept singular, and the plural *their* is changed to the singular *his or her*.)

                 Like the scientists, we were sure the rocket was going to blast off, but it wasn't long before I knew it was a dud.
                 (The *you* was changed to *I,* keeping the whole sentence in first person.)

                 Like the scientists, we were sure the rocket was going to blast off, but it wasn't long before we knew it was a dud.
                 (The *you* was changed to *we,* again keeping the whole sentence in first person.)

In most cases, the secret to avoiding these subtle errors is to check whether the pronouns later in the sentence match the subject earlier in the sentence.

Faulty Parallelism

Lists can be difficult to manage in sentences. A good rule of thumb is to remember that each part of the list needs to be parallel in structure to every other part of the list.

Incorrect   After the interview, we went out for dinner, had a few drinks, and a few jokes were told.
                 (In this sentence, the third part of the list, *jokes were told,* is not parallel to the first two parts, *went out for dinner* and *had a few drinks*.)

                 Our survey shows that people want peace, want to own a home, and that many are worried about their jobs.
                 (In this sentence, the second item in the list, *want to own a home,* is not parallel to the first and third items.)

To correct the sentence, make the items in the list parallel.

Correct   After the interview, we went out for dinner, had a few drinks, and told a few jokes.

                 Our survey shows that people want peace, most want to own a home, and many are worried about their jobs.

To avoid faulty parallelism, pay special attention to any lists you write. Check whether the items are parallel in phrasing.

Pronoun Case Error (*I* and *Me*, *We* and *Us*)

pc

Often, people are confused about when to use *I* or *me* and *we* or *us*. Here is a simple way to make the right decision about which one to use: If you are using the word as the subject of the sentence or phrase, use *I* or *we*. Anywhere else, use *me* or *us*.

Incorrect	Jones's team and me went down to the factory floor to see how things were going. (The word *me* is misused here because it is part of the subject of the sentence. In this case, *I* should have been used.)
	When the roof fell in, the manager asked Fred and I to start developing a plan for cleaning the mess up. (The word *I* is misused in this sentence because the phrase *Fred and I* is not the subject. In this case, the phrase *Fred and me* should have been used.)
	Things were getting pretty ugly in there, so us unimportant people slipped out the back door. (The phrase *us unimportant people* is the subject of the phrase that follows the comma. Therefore, the phrase *we unimportant people* should have been used.)

Remember, if the word is being used as the subject of a phrase or sentence, use *I* or *we*. Anywhere else, use *me* or *us*. Here are the correct versions of these sentences:

Correct	Jones's team and I went down to the factory floor to see how things were going.
	When the roof fell in, the manager asked Fred and me to start developing a plan for cleaning the mess up.
	Things were getting pretty ugly in there, so we unimportant people slipped out the back door.

Shifted Tense

Sentences can be written in past, present, or future tense. In most cases, neighboring sentences should reflect the same tense. Shifting tenses can make readers feel like they are hopping back and forth in time.

Incorrect	Few countries possess nuclear weapons, but many countries tried to build them. (Here, *possess* is in present tense, while *tried* is in past tense.)
	Parts flew everywhere on the factory floor as the robot finally breaks down. (Here, *flew* is in past tense, while *breaks* is in present tense.)
	The advances in microchip technology allowed electronics to become much smaller, which leads to today's tiny electronic devices. (The word *allowed* is in past tense, while *leads* is in present tense.)
	We found ourselves staring in disbelief. The excavation site is vandalized by people who wanted to have a campfire. (Here, *found* is in past tense, while *is vandalized* is in present tense.)

To revise these sentences, make the tenses consistent.

Correct	Few countries possess nuclear weapons, but many countries are trying to build them.
	Parts flew everywhere on the factory floor when the robot finally broke down.
	The advances in microchip technology allowed electronics to become much smaller, which led to today's tiny electronic devices.
	We found ourselves staring in disbelief. The excavation site had been vandalized by people who wanted to have a campfire.

Tense shifts are not always wrong. Sometimes a sentence or phrase needs to be in a different tense than those around it. When checking sentences for unnecessary tense shifts, look for places where tense shifts cause more confusion than clarity.

Vague Pronoun

Occasionally, a writer uses a pronoun, seeming to know exactly who or what the pronoun refers to, while readers are left scratching their heads, trying to figure out what the writer means.

Incorrect

Fred and Javier went to the store, and then he went home.
(Who does *he* refer to, Fred or Javier?)

They realized that the inspection of the building was not going well. It was fundamentally unsound.
(In this sentence, *It* could refer to the inspection or the building.)

We really had a great week. Our program review went well, and we made huge strides toward finishing the project. This is why we are taking all of you to lunch.
(What does *This* refer to? The great week? The program review? The huge strides? All of them?)

The camera captured the explosion as it ripped apart the car. It was an amazing experience.
(In these sentences, the multiple uses of *it* are confusing. In the first sentence, *it* might refer to the camera or the explosion. In the second sentence, *It* might refer to the taking of the picture or the explosion. Or, the final *It* might just be a weak subject for the second sentence and doesn't refer directly to anything in the previous sentence.)

Correcting these sentences mostly involves rewording them to avoid the vague pronoun.

Correct

Fred and Javier went to the store, and then Javier went home.

They realized that the inspection of the building was not going well. The inspection process was fundamentally unsound.

We really had a great week. Our program review went well, and we made huge strides toward finishing the project. For these reasons, we are taking all of you to lunch.

The camera captured the explosion as it ripped apart the car. The explosion was an amazing experience.

A common cause of vague pronoun use is the overuse of "It is..." and "This is..." to begin sentences. These kinds of sentences force readers to look back at the previous sentence to figure out what *It* or *This* refers to. In some cases, two or three possibilities might exist.

To avoid these problems, train yourself to avoid using "It is..." and "This is..." sentences. Occasionally, these sentences are fine, but some writers rely on them too much. You are better off minimizing their use in your writing.

Punctuation Refresher

More than likely, you already have a good sense of the punctuation rules. However, there are quirks and exceptions that you need to learn as your writing skills advance to a new level. So, spend a little time here refreshing your memory on punctuation rules.

Period, Exclamation Point, Question Mark

Let's start with the most basic marks, the period, exclamation point, and question mark. These punctuation marks signal the end of a sentence, or a full stop.

> We need to test the T6 Robot on the assembly line.
>
> The acid leaked and burned through the metal plate beneath it!
>
> Where can we cut costs to bring this project under budget?

The period signals the end of a standard sentence. The exclamation point signals surprise or strong feelings. The question mark signals a query.

Periods can also be used with abbreviations and numbers.

> C.E.
>
> Feb.
>
> Fig. 8
>
> Dr. Valerie Hanks
>
> 56.21 cm

Question marks can also be used in a series.

> How will global warming affect this country? Will it flood coastal cities? Create drought in the southwest? Damage fragile ecosystems?

Using periods, exclamation points, and question marks with quotes

A common mistake with periods, exclamation points, and question marks is their misuse with quotation marks. In almost all cases, these punctuation marks are placed inside the quotation marks.

Incorrect
> He said, "The audit team reported that we are in compliance".
>
> We asked him, "Do you really think we are going to finish the project on time"?

Correct
> He said, "The audit team reported that we are in compliance."
>
> We asked him, "Do you really think we are going to finish the project on time?"

The one exception to this rule is when a quoted statement is placed within a question.

> Did he really say, "The company will be closing the Chicago office"?

Commas

In the English language, commas are the most flexible, useful, and therefore problematic punctuation mark. In most cases, a comma signals a pause in the flow of a sentence.

When he hiked in the mountains east of Ft. Collins, he always took along his compass and map to avoid getting lost.
> (This comma signals that an introductory phrase is finished.)

My PDA is a helpful organizing tool, and it makes a great paperweight too.
> (This comma signals that two independent clauses are joined with a conjunction [*and, yet, but, so, because*].)

Our company's CEO, the engineering genius, is scheduled to meet with us tomorrow.
> (These commas set off information that is not essential to understanding the sentence.)

We are, however, having some luck locating new sources of silicon on the open market.
> (These commas set off a conjunctive adverb [*therefore, however, furthermore, nevertheless, moreover*].)

The archaeological dig yielded pottery shards, scraping tools, and a corn meal grinding stone.
> (These commas separate items in a series.)

Using commas with quotation marks

Using commas with quotation marks can be problematic. Just remember that commas almost always go before, not after, the quotation mark.

Albert Einstein said, "God does not play dice."
> (Here the comma sets off a speaker tag. Notice that the comma is placed before the quotation mark.)

"I'm having trouble hearing you," she said, "so I'm switching over to a new phone."
> (Again, the commas in this sentence offset the speaker tag. Note that both commas come before the quotation marks.)

Using commas in numbers, dates, and place-names

Commas are also used with numbers, dates, and place-names.

Reporters estimated that the rally drew nearly 10,000 people.

We first noticed the problem on January 10, 2009.

For great pizza you need to go to Uncle Pete's in Naperville, Illinois.

Removing excess commas

There is some flexibility in the use of commas. Many editors recommend an "open" punctuation style that eliminates commas where they do not help comprehension. For example, the following sentences are both correct:

In minutes, she whipped up the most amazing apple pie.

In minutes she whipped up the most amazing apple pie.

Here, the comma after *minutes* is not aiding understanding, so it can be removed. Be careful, though. Sometimes the lack of a comma can cause some confusion.

Confusing   Soon after leaving the airplane needed to turn back for mechanical reasons.

Not confusing   Soon after leaving, the airplane needed to turn back for mechanical reasons.

Semicolon and Colon

Semicolons and colons are less common than periods and commas, but they can be helpful in some situations. In a sentence, semicolons and colons signal a partial stop.

The trick to properly using these marks is to remember this simple rule: In most cases, the phrase on either side of a semicolon or colon should be able to stand alone as a separate sentence (an independent clause).

> We were not pleased with the results of the study; however, we did find some interesting results that gave us ideas for new research.
> (The semicolon joins two independent clauses. The second clause [the part starting with *however*] supports the first clause.)

> Commuting to work by car requires nerves of steel: Each mile brings you into contact with people who have no respect for the rules of the road.
> (Here, the colon also divides two independent clauses. In this example, though, the colon signals that the two parts of the sentence are equivalent to each other.)

How do you know when to use a semicolon or a colon? It depends on whether the part of the sentence following the mark is *lesser than* or *equal* to the first part. If lesser, use the semicolon. If equal, use the colon.

If you want to avoid problems, use semicolons and colons only when necessary. When considering these punctuation marks in a sentence, you should ask yourself whether a period or a conjunction (*and, but, so, yet*) would make the sentence easier to understand. After all, joining sentences together with semicolons and colons can often create long, difficult-to-read sentences.

As with most punctuation rules, there are exceptions to the rule that semicolons and colons are used with independent clauses.

Using semicolons in a series

Semicolons can be used to punctuate complicated lists in a sentence.

> We have offices in Boston, Massachusetts; Freetown, New York; and Sedona, Arizona.

Using colons to lead off lists

Colons can be used to signal a list.

> Four steps are required to complete the process: (1) preparing the workspace, (2) assembling the model, (3) painting the model, and (4) checking quality.

Keep in mind the following issues when searching for a job:

- You can't get the job if you don't apply.
- Jobs don't always go to the people with the most experience.
- What makes you different makes you interesting.

Using colons in titles, numbers, and greetings

Colons are commonly used in titles of books and articles, in numbers, and in greetings in letters and memos.

The Awakening: The Irish Renaissance in Nineteenth-Century Boston

Genesis 2:18

11:45 A.M.

They won by a 3:1 ratio.

Dear Mr. Franklin:

Using semicolons and colons with quotation marks

Unlike commas and periods, semicolons and colons should appear after the quotation mark in a sentence.

Land Commissioner George Hampton claimed, "The frogs will survive the draining of the lake"; but he was clearly wrong.

One of my favorite chapters in Leopold's *A Sand County Almanac* is "Thinking Like a Mountain": This essay is his best work.

Whenever possible, though, you should avoid these kinds of situations. In both of these examples, the sentences could be rearranged or repunctuated to avoid these awkward, though correct, uses of the semicolon or colon.

Misusing the colon

A common misuse of the colon is using it with an incomplete sentence.

Incorrect   The reasons for our dissatisfaction are: low quality, late work, and slow response.
(The colon is misused, because the phrase before the colon cannot stand alone as a separate sentence.)

For example:
(Again, the phrase before the colon cannot stand alone as a sentence. In this case, a dash or comma should be used. Or, turn the phrase *For example* into a complete sentence.)

In his report, Bill Trimble claims: "We have a golden opportunity to enter the Japanese market."
(Yet again, the information before the colon cannot exist as a separate sentence. In this case, a comma should have been used instead of the colon.)

As a rule, the information before the colon should *always* be able to stand alone as a complete sentence. Here are the correct versions of these sentences:

Correct

The reasons for our dissatisfaction are the following: low quality, late work, and slow response.

For example, consider these interesting situations:

In his report, Bill Trimble makes this important statement: "We have a golden opportunity to enter the Japanese market."

Notice how all three of these examples have independent clauses (full sentences) before the colon.

Apostrophe

The apostrophe has two important jobs in the English language: (1) to signal contractions and (2) to signal possession.

Using an apostrophe to signal a contraction

An apostrophe that signals a contraction identifies the place where two words have been fused and letters removed.

They're going to the store today.

He really isn't interested in the project.

They shouldn't have taken that road.

Contractions should be used only in informal writing. They signal a familiarity with the readers that could seem too informal in some situations. Some other common contractions include *won't, it's, I'm, you've, wouldn't,* and *couldn't.*

Using an apostrophe to signal possession

An apostrophe is also used to signal possession. With a singular noun, an *'s* is added to signal possession. Joint possession is usually signaled with an *s'*.

We have decided to take Anna's car to the convention.

The players' bats were missing before the game.

When plural nouns do not end in an *s,* you should use an *'s* to create the plural.

We rode the children's bikes.

The men's briefcases were left near the door.

When singular nouns end in an *s,* you should add an *'s* to show possession.

They met in Mary Jones's office.

Charles's computer was shorting out.

Using apostrophes to show possession with two or more nouns

When you are showing possession with multiple nouns, your use of the apostrophe depends on your meaning. If two nouns are acting as one unit, only the last noun needs an apostrophe to signal possession.

> We decided to accept Grim and Nether's proposal.

But if you are signaling possession for several separate nouns, each needs an apostrophe.

> I found it difficult to buy meaningful gifts for Jane's, Valerie's, and Charles's birthdays.

Using apostrophes to signal plurals of numbers, acronyms, and symbols

You can use apostrophes to signal plurals of numbers, acronyms, and symbols, but do so sparingly. Here are a couple of situations where apostrophes would be appropriate:

> The *a*'s just kept appearing when I typed *x*'s.

> Is it necessary to put ©'s on all copyrighted documents?

In most cases, though, do not include apostrophes to show a plural if they do not aid the meaning of the text.

> The police discovered a warehouse full of stolen TVs.

> The 1870s were a tough time for immigrants.

> In the basement, a crate of dead CPUs sat unnoticed.

Quotation Marks

Quotation marks are used to signal when you are using someone else's words. Quotation marks should not be used to highlight words. If you need to highlight words, use italics.

Using quotation marks to signal a quote

Quotation marks are used to frame an exact quotation from another person.

> In *The Panda's Thumb,* Gould states, "The world, unfortunately, rarely matches our hopes and consistently refuses to behave in a reasonable manner."

> "Not true" was her only response to my comment.

> He asked me, "Are you really working on that project?"

Use quotation marks only when you are copying someone else's exact words. If you are only paraphrasing what someone else said, do not use quotation marks.

> In *The Panda's Thumb,* Gould argues that nature often does not meet our expectations, nor does it operate in predictable ways.

She rejected my comment as untrue.

He asked me whether I was working on the project.

Also, when paraphrasing, avoid the temptation to highlight words with quotation marks.

Using quotation marks to signal titles

Titles of works that are part of larger works, like articles, songs, or documents, should be set off with quotation marks.

Time published an article called "The Silicon Valley Reborn."

A classic blues tune covered by the Yardbirds was "I'm a Man."

The report, "Locating Evidence of Ancient Nomads in Egypt," is available online.

Titles of books and other full works should not be set in quotation marks. They should be italicized.

Taking the Quantum Leap, by Fred Wolf, is a very helpful book.

Revolver is one of the Beatles' best albums.

Using single quotation marks to signal a quote or title within another quote

When quoting something within another quote, you should use single quote marks to set it off.

Tim Berra shows the weakness of the creationist argument by quoting one of its strongest advocates: "Morris wrote 'the only way we can determine the true age of the earth is for God to tell us what it is.'"

One of the physicists at the conference remarked, "I cannot believe that Einstein's 1905 paper on special relativity, 'On the Electrodynamics of Moving Bodies,' is already a century old."

Using quotation marks to signal irony

Quotation marks are often used incorrectly to highlight words and slang terms.

Incorrect

One problem with "free-trade policies" is that the laborers who work for Third World countries work almost for free.

I found working with her to be "wonderful," because she is so "attentive" and "understanding."

The sentences above do not need quotation marks to set off these words. If you want to highlight words that are not direct quotes, use italics.

You can, however, use quotation marks to signal irony by quoting another person's misuse of a term or phrase.

Conservation is more than Vice President Cheney's notion of a "personal virtue."

The Matrix is an entertaining film, but it's hard to accept the "biblical significance" that Clarke and others claim for this highly violent movie.

Using quotation marks with in-text citations

One of the exceptions to placing periods inside quotation marks is when quote marks are used with in-text citations.

> In his article on ancient dams, Abbas points out that "water-driven power systems have been around for thousands of years" (p. 67).

Here, note that the period comes after the in-text citation, not within the closing quotation mark.

Dashes and Hyphens

The uses of dashes and hyphens follow some rather specific rules. There are actually two types of dashes, the "em dash" and the "en dash." The em dash is the longer of the two (the width of an *m*), and it is the more widely used dash. The en dash is a bit shorter (the width of an *n*), and it is less widely used. Hyphens are shorter than the two dashes.

Using em dashes to highlight asides from the author or the continuation of a thought

An em dash is typically used to insert comments from the author that are asides to the readers.

> At the meeting, Hammons and Jenkins—this is the ironic part—ended up yelling at each other, even though they both intended to be peacemakers.

> We must recognize the continuing influence of Lamarckism in order to understand much social theory of the recent past—ideas that become incomprehensible if forced into the Darwinian framework we often assume for them.

An em dash can be made with two hyphens (--). Most word processors will automatically change two dashes into an em dash. Otherwise, a series of keystrokes (usually, shift-command-hyphen) will create this longer dash (—).

Using en dashes in numbers and dates

It might seem trivial, but there is a difference between en dashes and em dashes. An en dash is almost always used with numbers and dates.

> Copernicus (1473–1543) was the first European to make a cogent argument that the earth goes around the sun rather than the sun going around the earth.

> Young and Chavez argue conclusively that Valles Bonita is really a dormant sunken volcano, called a *caldera* (pp. 543–567).

As you can see in these examples, the en dash is slightly shorter than the em dash.

Using the hyphen to connect prefixes and make compound words

The hyphen is mainly used to connect prefixes with words or to connect two or more words to form compound words.

neo-Platonists

one-to-one relationship

trisomy-21

four-volume set of books

One thing you should notice is how hyphens are used to create compound adjectives but not compound nouns. You can write "four-volume set of books," where *four-volume* is an adjective. But, you would need to write "the four volumes of books," because the word *volumes* is being used as a noun. Hyphens are usually used to make compound adjectives, but not compound nouns.

Parentheses and Brackets

Parentheses and brackets are handy for setting off additional information, like examples, definitions, references, lists, and asides to the readers.

Using parentheses to include additional information

Parentheses are often used to include additional information or refer readers to a graphic.

> When hiking through the Blanca Mountains, you will be surprised by the wide range of animals you will see (e.g., elk, deer, hawks, eagles, and the occasional coyote).

> The data we collected show a sharp decline in alcohol use when teens become involved in constructive, nontelevision activities (see Figure 3).

> These unicellular organisms show some plant-like features (many are photosynthetic) and others show more animal-like features.

Using parentheses to clarify a list

Parentheses can be used to clarify the elements of a long list.

> When meeting up with a bear in the wild, (1) do not run, (2) raise your arms to make yourself look bigger, (3) make loud noises, and (4) do not approach the animal.

> Only three things could explain the mechanical failure: (1) the piston cracked, (2) one of the pushrods came loose, or (3) the head gasket blew.

Using brackets to include editorial comments or to replace a pronoun

Brackets are less common than parentheses, but they can be helpful for inserting editorial comments or replacing a pronoun in a quote.

> Though pictures of the moon are often spectacular, *any view of the moon from earth is slightly blurred* [emphasis mine].

> Shea points out, "Whether he intended it or not, [Planck] was the originator of the quantum theory."

In this second example, the second *he* was replaced with *Planck* to make the meaning of the quote clearer.

Ellipses

Ellipses are used to show that information in a quote was removed or to indicate the trailing off of a thought. Ellipses are made with three dots, with spaces between each dot (. . .), not (...).

Using ellipses to signal that information in a quote has been removed

Sometimes a passage, especially a longer one, includes more information than you want to quote. In these cases, ellipses can be used to trim out the excess.

> As historian Holton writes, "What Bohr had done in 1927... was to develop a point of view that allowed him to accept the wave-particle duality as an irreducible fact" (117).

Using ellipses to show that a thought is trailing off

At the end of a sentence, you might use ellipses to urge the reader to continue the thought.

> For those who don't want to attend the orientation, we can find much less pleasant ways for you to spend your day....

When ellipses end the sentence, use an additional dot to make four (. . . .). The extra dot, after the last word, is a period that signals the end of the sentence.

English as a Second Language Guide

The English language is a composite of many languages. As a result, English has many maddening exceptions and inconsistencies in spelling, syntax, and usage. If English is not your native language, you should pay special attention to these irregularities so that your writing and speaking will be consistent with the writing and speaking of fluent speakers.

This English as a Second Language (ESL) guide will not help you learn English. Instead, this guide concentrates on three major sources of irregularities in technical English—use of articles, word order of adjectives and adverbs, and verb tenses. As you master the English language, you should first concentrate on understanding these three sources of irregularities. Then, you can turn to other ESL resources to help you refine your use of English.

Using Articles Properly

Perhaps the most significant source of ESL problems is the use of articles (*a, an, the*) in English. For example, the sentences "The computer broke down" and "A computer broke down" have significantly different meanings in English. The use of *the* suggests that a specific computer broke down. The use of *a* suggests that one computer—among many computers—broke down.

Using *the* to refer to specific items

If you are referring to a specific item, use *the* to signal the noun.

The planet Saturn has been bright during the last week.

The car stalled, so we started walking to the nearest service station.

The professor asked us to work harder on the next assignment.

Using *a* to refer to nonspecific items

If you are referring to a nonspecific item, use *a* to signal the noun.

A planet was found circling a star in the Orion system.

A car stalled in the road, so we needed to drive around it.

A professor asked us to attend the party.

Misusing an article with uncountable things

Articles should be used only with things that can be counted, like *the eight cars, the five bikes,* or *an orange.* When items are not counted or cannot be counted, do not use an article.

He surfaced for air.

They decided to have tea with dinner.

Rice grown in Asia tastes better than rice grown in North America.

Putting Adjectives and Adverbs in the Correct Order

Compared to some languages, English is flexible in its syntax. Nevertheless, word order in sentences is important for expressing the meaning you intend. In this section, we will go over the two major sources of word-order problems for ESL writers—adjectives and adverbs.

Using adjectives in the proper order

In English, adjectives should be placed in the proper order.

Improper   The red beautiful sailboat came into the bay.

Proper   The beautiful red sailboat came into the bay.

Fluent speakers of English will still understand the improper sentence, but it will sound odd to them. To properly order adjectives, you can use the following hierarchy of adjectives, which has been modified from *The New Century Handbook*, 4e by Hult and Huckin (© 2008, Longman, p. 903). Hult and Huckin offer a more comprehensive approach.

1. article, determiner, or possessive (*a, an, the, this, that, those, my, our, their, Lisa's*)
2. ordinal (*first, second, third, final, next*)
3. quantity (*one, two, three, more, some, many*)
4. size and shape (*big, tiny, large, circular, square, round*)
5. appearance (*beautiful, filthy, clean, damaged, old, young, ancient*)
6. color (*red, yellow, black, green*)
7. substance (*wool, copper, wood, plastic*).

When properly ordered, adjectives can be strung together indefinitely in a sentence.

A third, large, beautiful, ancient, red, wooden sailboat came into the bay.

Keep in mind, though, that you should not string together too many adjectives. More than three or four adjectives strung together can be difficult to understand.

Using adverbs in proper places

Adverbs usually modify verbs in clauses. Adverbs can be used in a variety of places in the clause. Three guidelines are especially helpful for placing adverbs.

GUIDELINE 1: *Adverbs involving time and place usually go after the verb.*

The cat went *outside* when the children came over.

She arrived *late* to the lecture.

He went *promptly* to his professor's office for help.

Want to practice adjective order? Go to
www.pearsonhighered.com/johnsonweb3/B.2

GUIDELINE 2: *Adverbs that show frequency usually go before the verb.*

The cat *typically* runs outside when the children come over.

She *usually* arrives late to the lecture.

He *always* goes to his professor's office for help.

GUIDELINE 3: *Do not put an adverb between the verb and the object of a clause.*

Improper

Maria drives *recklessly* her car on the interstate.

Kim eats *quickly* his breakfast before he goes to class.

Proper

Maria drives her car *recklessly* on the interstate.

Kim eats his breakfast *quickly* before he goes to class.

Using Verb Tenses Appropriately

Proper use of verbs and verb phrases can be difficult in English, even for fluent speakers. So, learn them as best you can, and be patient while you are mastering the numerous English tenses.

To use tenses appropriately, remember that English, like most languages, has *past, present,* and *future* tenses. Each of these tenses also has four verbal aspects, which are called *simple, progressive, perfect,* and *perfect progressive.*

Simple—indicates whether the event happened, is happening, or will happen.

Progressive—indicates that the event was, is, or will be in progress at a specific time.

Perfect—indicates that the event was, is, or will be completed by a specific point in time.

Perfect progressive—indicates whether the event was, is, or will be progressing until a specific point in time.

Altogether, English has 12 tenses that need to be learned. Let us consider them separately.

Past Tense

Past tense refers to events that have already happened.

Simple Past Tense

Victor walked to the store yesterday.

Past Progressive Tense

Victor was walking to the store yesterday, when he was nearly hit by a car.

Past Perfect Tense

Before going to class, Victor had walked to the store.

Past Perfect Progressive Tense

Victor had been walking to the store before class.

Present Tense

Present tense refers to events that are happening at the moment.

Simple Present Tense

I enjoy talking with my friends.

Present Progressive Tense

I am enjoying talking with my friends this evening.

Present Perfect Tense

I have enjoyed talking with my friends this evening.

Present Perfect Progressive Tense

I have been enjoying talking with my friends this evening.

Future Tense

Future tense refers to events that will happen.

Simple Future Tense

The movie will start at 6:00 this evening.

Future Progressive Tense

The movie will be starting at 6:00 this evening.

Future Perfect Tense

By 6:10, the movie will have started already.

Future Perfect Progressive Tense

By 6:10, the movie will have been playing for nearly 10 minutes.

When learning English, you should begin by mastering the simple past, present, and future forms of verbs. Then, as you grow more comfortable with the language, you can begin using the progressive, perfect, and perfect progressive aspects.

When you are unsure about whether to use progressive, perfect, or perfect progressive, just revert to the simple form. In most cases, these simple sentences will be correct, though perhaps a bit awkward sounding to fluent speakers of English.

Still not sure about verb tenses? Go to
www.pearsonhighered.com/johnsonweb3/B.4

Documentation Guide

Documenting sources is an important part of doing research. As you collect information on your subject, you should keep track of the sources from which you drew quotes and ideas. Then, cite these sources inside your text and use them to create a list of references at the end of your document.

When should you cite and document a source? The answer to this question depends on the kind of document you are writing. A scientific report, for example, requires more citation than a technical description or a set of instructions. The best way to determine the necessary level of documentation is to consider your readers' needs. How much citing and documenting will they need to feel confident in your work?

Some commonly documented materials include the following:

Quotes or ideas taken from someone else's work—If others wrote it or thought it before you did, you should cite them as the owners of their words and ideas. Otherwise, you might be accused of lifting their work. In important cases, you or your company might be sued for using someone else's ideas.

Materials that support your ideas—You can build the credibility of your work by showing that others have discussed the topic before. Readers are going to be highly skeptical of your work if they think you are pulling your ideas out of thin air.

Sources of any data or facts—Any numbers or facts that you did not generate yourself need to be carefully cited and documented. That way, readers can check your sources for accuracy.

Materials that refer to your subject—You want to demonstrate that you are aware of the broader conversation on your subject. By citing sources, including those with which you disagree, you show that you have a complete understanding of the issues involved.

Historical sources on your subject—To build a background for readers to understand your subject, you should include any sources that might help them understand its history.

Graphics taken from online or print sources—On the Internet, it is easy to download graphs, tables, images, and photographs. Sometimes you will need permission to use these items. Minimally, though, you should cite the sources from which you obtained them.

Let us review the three most common documentation styles in technical communication:

- The **APA documentation style** from the American Psychological Association is widely used in engineering and the sciences.
- The **CSE documentation style** from the Council of Science Editors is used primarily in biological and medical sciences, though it is gaining popularity in other scientific fields.
- The **MLA documentation style** from the Modern Language Association is used in the humanities. Although this style is not commonly used in engineering and science, it is sometimes used when scholars approach technical issues from cultural, historical, rhetorical, or philosophical perspectives.

Literally hundreds of other documentation styles are available. So, find out which documentation style is used in the organization or company for which you work.

Citing and documenting a source requires two items, an *in-text citation* and a *full entry* in the References or Works Cited list at the end of the document.

Footnotes and endnotes are not common in APA or CSE style, and they are increasingly rare in MLA style. If you want to use footnotes or endnotes, consult the style guide you are using. Footnotes and endnotes will not be covered here because they are rarely used in technical documents.

We will discuss the most common in-text citations and full-entry patterns. If one of the following models does not fit your needs, you should consult the style guide (APA, CSE, or MLA) that you are following.

APA Documentation Style

APA documentation style is most common in the natural and human sciences, except in fields related to biology and medicine. The official source for this style is the *Publication Manual of the American Psychological Association,* Fifth Edition (2001), and the *APA Style Guide to Electronic References* (2007).

When using APA style, you will need to include in-text citations and a list of alphabetically arranged references at the end of your document.

APA In-Text Citations

APA style follows an author-year system for in-text citations, meaning the author and year are usually cited within the text.

Individual Authors

Individual authors are cited using their last name and the date of the article.

> One study reports a significant rise in HIV cases in South Africa in 1 year (Brindle, 2000).
>
> One study reported a 12.2% rise in HIV cases in only 1 year (Brindle, 2000, p. 843).
>
> Brindle (2000) reports a significant rise in HIV cases in South Africa in 1 year.
>
> Brindle (2000) reports a 12.2% rise in HIV cases in South Africa in 1 year (p. 843).

In most cases, only the author and year need to be noted, as shown in the first example. If you are reporting a specific fact or number, however, you should cite the page from which it was taken.

Multiple Authors

If an article has two or more authors, you should use the ampersand symbol (&) to replace the word *and* in the in-text citation. The word *and* should be used in the sentence itself, however.

To find online sources that discuss APA style, go to
www.pearsonhighered.com/johnsonweb3/C.2

(Thomas & Linter, 2001)

According to Thomas and Linter (2001)...

Technical documents often have more than two authors. If the work has less than six authors, cite all the names the first time the work is referenced. After that, use the last name of the first author followed by "et al."

First Citation of Work

(Wu, Gyno, Young, & Reims, 2003)

As reported by Wu, Gyno, Young, and Reims (2003)...

Subsequent Citations of Work

(Wu et al., 2003)

As reported by Wu et al. (2003)...

If the work has six or more authors, only the first author's last name should be included, followed by "et al." This approach should be used with all citations of the work, including the first in-text citation.

Corporate or Unknown Authors

When the author of the document is a corporation or is unknown, the in-text citation uses the name of the corporation or the first prominent word in the title of the document.

First Citation of Work

(National Science Foundation [NSF], 2004)

("Results," 2002)

(*Silent*, 2002)

Subsequent Citations of Work

(NSF, 2004)

("Results," 2002)

(*Silent*, 2002)

Notice in these examples that the first word of a journal article title should be put in quotation marks, while the first word of a book title should be put in italics.

Paraphrased Materials

When citing paraphrased materials, usually only the year and page number are needed because the authors' names are typically mentioned in the sentence. In many cases, only the year is needed.

Franks and Roberts report that aptitude for visual thinking runs in families (2003, p. 76).

The instinct for survival, according to Ramos (2004), is strong in the Mexican wolf.

Jones (2001) argues that finding a stand of dead trees near an industrial plant is a good indicator that something is seriously wrong (pp. 87–88).

Two or More Works in Same Parentheses

In some cases, several documents will state similar information. If so, you should cite them all and separate the works with semicolons.

> Studies have shown remarkable progress toward reviving the penguin population on Vostov Island (Hinson & Kim, 2004; Johnson & Smith, 2001; Tamili, 2002).

Personal Communication and Correspondence

APA style discourages putting any forms of personal communication in the References list. Personal communication includes conversations, e-mail, letters, and even interviews. So, in-text citations are the only citations for these sources in a document.

> Bathers (personal communication, December 5, 2003) pointed out to me that …

These sources are not listed in the References list because they do not provide information that is retrievable by readers.

The References List for APA Style

When using APA style, your references should be listed in alphabetical order at the back of the document. In your references, you should list only the items that you actually cited in your document. Leave any uncited documents out of the list of references. After all, if a document is important enough to list in your references, it should be important enough to cite in the text itself.

In most cases, the reference list is identified by the centered heading "References." Entries can be single-spaced or double-spaced, although double-spaced is preferred. Also, each reference should use a hanging indent style (i.e., the second line and subsequent lines should be indented). The first line should be flush with the left margin.

The following list includes examples of APA style. This list is not comprehensive. If you do not find a model here for a document you are adding to your references list, you should check the *Publication Manual of the American Psychological Association,* Fifth Edition. Or, check one of many websites that offer examples of APA style.

1. **Website or Webpage, Author Known**

   Jaspers, F. (2001). *Einstein online.* Retrieved from http://www.einsteinonlinetoo.com

2. **Website or Webpage, Corporate Author**

   National Wildlife Service. (2002). *Managing forest on your land.* Retrieved from http://www.nws.gov/manageyourforest.htm

3. **Webpage, Author Unknown**

   Skin cancer treatments debated. (2004, January 1). *CNN.com.* Retrieved from http://www.cnn.com/2004/HEALTH/conditions/01/19/skincancer.treatment.ap/index.html

Having trouble finding an APA model for a source? Go to
www.pearsonhighered.com/johnsonweb3/C.3

4. **Book, One Author**

   Jones, S. (2001). *Darwin's ghost: The origin of species updated.* New York: Balantine Books.

5. **Book, More Than One Author**

   Pauling, L., & Wilson, E. B. (1935). *Introduction to quantum mechanics.* New York: Dover
   Publications.

6. **Book, Corporate or Organization Author**

   American Psychiatric Association. (1994). *Diagnostic and statistical manual of mental
   disorders* (4th ed.). Washington, DC: American Psychiatric Association.

7. **Book, Edited Collection**

   Mueller-Vollmer, K. (Ed.). (1990). *The hermeneutics reader.* New York: Continuum.

8. **Book, Translated**

   Habermas, J. (1979). *Communication and the evolution of society.* (T. McCarthy, Trans.).
   Boston: Beacon Press.

9. **Book, Author Unknown**

   Handbook for the WorkPad c3 PC Companion. (2000). Thornwood, NY: IBM.

10. **Book, Second Edition or Beyond**

    Williams, R., & Tollet, J. (2000). *The non-designer's web book* (2nd ed.). Berkeley,
    CA: Peachpit.

11. **Book, Dissertation or Thesis**

    Simms, L. (2002). *The hampton effect in fringe desert environments: An ecosystem under
    stress.* Unpublished doctoral dissertation, University of New Mexico.

12. **Book, Electronic**

    Darwin, C. (1862). *On the various contrivances by which British and foreign orchids
    are fertilised by insects.* London: John Murray. Retrieved from http://pages
    .britishlibrary.net/charles.darwin3/orchids/orchids_fm.htm

13. **Document, Government Publication**

    Greene, L. W. (1985). *Exile in paradise: The isolation of Hawai'i's leprosy victims
    and development of Kalaupapa settlement, 1865 to present.* Washington,
    DC: U.S. Department of the Interior, National Park Service.

14. **Document, Pamphlet**

    The Colorado Health Network. (2002). *Exploring high altitude areas.* Denver, CO: Author.

15. **Film or Video Recording**

    Jackson, P. (Director), Osborne, B., Walsh, F., & Sanders, T. (Producers). (2002). *The lord
    of the rings: The fellowship of the ring* [Film]. Hollywood, CA: New Line Productions.

16. **Article, Journal with Continuous Pagination**

    Boren, M. T., & Ramey, J. (1996). Thinking aloud: Reconciling theory and practice.
    IEEE Transactions on Professional Communication, 39, 49–57.

 Need more help with online sources? Go to
www.pearsonhighered.com/johnsonweb3/C.4

**Documentation
Guide**   A-27

17. **Article, Journal without Continuous Pagination**

Kadlecek, M. (1991). Global climate change could threaten U.S. wildlife. *Conservationist, 46*(1), 54–55.

18. **Article, Journal with Digital Object Identifier (DOI)**

Tomlin, R. (2008). Online FDA regulations: Implications for medical writers. *Technical Communication Quaterly, 17* (3), 289–310. doi:10.1080/10572250802100410

19. **Article, Edited Book**

Katz, S. B., & Miller, C. R. (1996). The low-level radioactive waste siting controversy in North Carolina: Toward a rhetorical model of risk communication. In G. Herndl & S. C. Brown (Eds.), *Green culture: Environmental rhetoric in contemporary America* (pp. 111–140). Madison: University of Wisconsin Press.

20. **Article, Magazine**

Appenzeller, T. (2004, February). The case of the missing carbon. *National Geographic,* 88–118.

21. **Article, Online Magazine**

Grinspoon, D. (2004, January 7). Is Mars ours? *Slate Magazine.* Retrieved from http://slate.msn.com/id/2093579

22. **Article, Newspaper**

Hall, C. (2002, November 18). Shortage of human capital envisioned, Monster's Taylor sees worker need. *The Chicago Tribune,* p. E7.

23. **Article, Author Unknown**

The big chill leaves bruises. (2004, January 17). *Albuquerque Tribune,* p. A4.

24. **Article, CD-ROM**

Hanford, P. (2001). Locating the right job for you. *The electronic job finder* [CD-ROM]. San Francisco: Career Masters.

25. **Weblog Posting**

Katie. (2007, 17 September). 30 days and tech writing. Message posted to http://techwriterscrum.blogspot.com/

26. **Podcast**

DMN Communications. (Producer). (2008, May 18). Talking wikis with Stewart Mader. *Communications from DMN.* Podcast retrieved from http://dmn.podbean.com/2008/05/

27. **Song or Recording**

Myer, L. (1993). Sometimes alone. *Flatlands* [CD]. Ames, IA: People's Productions.

28. **Television or Radio Program**

Harris, R. (2003, January 6). *Destination: The south pole.* Washington, DC: National Public Radio. Retrieved from http://discover.npr.org/features/feature.jhtml?wfld=904848

29. **Personal Correspondence, E-Mail, or Interview**

In APA style, a personal correspondence is not included in the References list. Instead, the information from the correspondence should be written in the in-text citation:

This result was confirmed by J. Baca (personal communication, March 4, 2004).

Creating the APA References List

In APA style, the References list is placed at the end of the document on a separate page or in an appendix. The sources referenced in the document should be listed alphabetically by author's last name.

References

Assel, R., Cronk, K., & Norton, D. (2003). Recent trends in Laurentian Great Lakes ice cover. *Climatic Change, 57,* 185–204.

Hoffmann, A., & Blows, M. (1993). Evolutionary genetics and climate change: Will animals adapt to global warming? In P. M. Kareiva, J. G. Kingsolver, & R. B. Huey (Eds.), *Biotic Interactions and Global Change* (pp. 13–29). Sunderland, MA: Sinauer.

Houghton, J. (1997). *Global warming: The complete briefing* (2nd ed). Cambridge, MA: Cambridge University Press.

Kadlecek, M. (1991). Global climate change could threaten U.S. wildlife. *Conservationist, 46*(1), 54–55.

Sherwood, K., & Idso, C. (2003). *Is the global warming bubble about to burst?* Retrieved from http://www.co2science.org/edit/v6_edit/v6n37edit.htm

---

CSE Documentation Style (Citation-Sequence)

The CSE documentation style is most commonly used in biological and medical fields, though it is gaining popularity in other scientific fields. The official source for this style is *Scientific Style and Format: The CSE Manual for Authors, Editors, and Publishers,* Seventh Edition (2006).

The *CSE Manual* describes two citation methods. The first method, called the *author-year* system, is very similar to APA style, so it will not be discussed here. The second method, called the *citation-sequence* system, will be discussed here because it offers a good alternative to APA style. If you need to use the CSE author-year system, you can consult the *CSE Manual* or websites that offer examples of this system.

In the citation-sequence system, sources are referred to by number within the text, usually with a superscript number similar to a footnote.

This bacteria has been shown[1] to grow at a significant rate when exposed to black light.

Need to use the CSE author-year system? Go to
www.pearsonhighered.com/johnsonweb3/C.5
For more information on CSE style updates, go to
www.pearsonhighered.com/johnsonweb3/C.6

Documentation Guide   A-29

When referring to multiple sources, a dash is used to signal the range of sources.

Several studies[3–8, 10] have illustrated this relationship.

In some situations, editors will ask for the citations to use numbers in parentheses or brackets instead of superscript numbers:

This relationship between the virus and various illnesses has been demonstrated in numerous studies (3, 12–15).

Franklin and Chou argued this point in their influential research on HIV mutation [3], in which they explained its tendency to seek out new paths for replication.

In the References list at the end of the document, the sources are numbered and listed in the order they were cited in the text. Then, other references to that source in the document will use the same number.

The advantage of the citation-sequence system is that readers feel less disruption than with the author-year system, because the superscript numbers are less intrusive. However, a disadvantage is that readers need to flip back to the list of references to see author names for any sources of information.

The References List for CSE Citation-Sequence Style

The following list includes examples of CSE citation-sequence style. This list is not comprehensive. If you do not find a model here for a document you are adding to your references, you should check *Scientific Style and Format: The CSE Manual for Authors, Editors, and Publishers,* Seventh Edition.

The format of the list of references for CSE citation-sequence style is somewhat different from that of reference lists following APA or MLA style:

- Sources are numbered (1, 2, 3, and so on) to reflect the order in which they were cited.
- The items in the References list are all flush left against the margin (no hanging indent).
- When a citation refers to a specific page or set of pages in a stand-alone document, the full text reference includes the page number(s) after a *p* (e.g., *p 23* or *p 123–36*). If the citation is referring to the whole work, the page numbers are not needed.

Items in the reference list should be single-spaced.

1. Website or Webpage, Author Known

12. Jaspers F. Einstein online [Internet]. Downers Grove (IL): Einstein Inc.; c2001 [cited 2003 Mar 9]. Available from: http://www.einsteinonlinetoo.com

2. Website or Webpage, Corporate Author

34. National Wildlife Service. Managing forest on your land [Internet]. Washington: NWS; c2003 [cited 2004 Sep 8]. Available from: http://www.nws.gov/manageyourforest.htm

3. **Webpage, Author Unknown**

3. Skin cancer treatments debated [Internet]. Atlanta (GA): CNN.com; c2004 [cited 2004 Jan 1]. Available from: http://www.cnn.com/2004/HEALTH/conditions/01/19/skincancer.treatment.ap/index.html

4. **Webpage, Online Periodical**

7. Grinspoon D. Is Mars ours? Slate Magazine [Internet]. 2004 Jan 7 [cited 2004 Jan 19]. Available from: http://slate.msn.com/id/2093579

5. **Book, One Author**

23. Jones S. Darwin's ghost: the origin of species updated. New York: Balantine Books; 2001. p 86–92.

6. **Book, More Than One Author**

2. Pauling L, Wilson EB. Introduction to quantum mechanics. New York: Dover Publications; 1935. p 38.

7. **Book, Corporate or Organization Author**

11. American Psychiatric Association. Diagnostic and statistical manual of mental disorders. 4th ed. Washington: American Psychiatric Association; 1994.

8. **Book, Edited Collection**

22. Mueller-Vollmer K, editor. The hermeneutics reader. New York: Continuum; 1990. p 203–212.

9. **Book, Translated**

14. Habermas J. Communication and the evolution of society. McCarthy T, translator. Boston: Beacon Press; 1979. p 156.

10. **Book, Author Unknown**

13. Handbook for the WorkPad c3 PC Companion. Thornwood (NY): IBM; 2000.

11. **Book, Second Edition or Beyond**

21. Williams R, Tollet J. The non-designer's web book. 2nd ed. Berkeley (CA): Peachpit; 2000. p 123–27.

12. **Book, Dissertation or Thesis**

18. Simms L. The hampton effect in fringe desert environments: an ecosystem under stress [dissertation]. [Albuquerque (NM)]: Univ of New Mexico; 2002.

13. **Book, Electronic**

13. Darwin C. On the various contrivances by which British and foreign orchids are fertilised by insects [Internet]. London: John Murray; c1862 [cited 2002 Sep 5]. Available from: http://pages.britishlibrary.net/charles.darwin3/orchids/orchids_fm.htm

14. **Document, Government Publication**

6. Greene LW. Exile in paradise: the isolation of Hawai'i's leprosy victims and development of Kalaupapa settlement, 1865 to present. Washington: Department of Interior (US); 1985. Available from: U.S. Department of the Interior, National Park Service, Washington, DC.

15. Document, Pamphlet

23. The Colorado Health Network. Exploring high altitude areas. Denver (CO); 2002. Available from: TCHN, Denver, CO.

16. Film or Video Recording

16. The lord of the rings: the fellowship of the ring [DVD]. Jackson P, director. Osborne B, Walsh F, Sanders T, producers. Hollywood (CA): New Line Productions; 2002.

17. CD-ROM

7. Geritch T. Masters of renaissance art [CD-ROM]. Chicago: Revival Productions; 2000. 2 CD-ROMs: sound, color, 4 $^3/_4$ in.

18. Article, Journal with Continuous Pagination

34. Boren MT, Ramey J. Thinking aloud: reconciling theory and practice. IEEE Trans on Prof Comm 1996; 39:49–57.

19. Article, Journal Without Continuous Pagination

32. Lenhoff R, Huber L. Young children make maps! Young Children 2000; 55(5):6–12.

20. Article, Edited Book

1. Katz SB, Miller CR. The low-level radioactive waste siting controversy in North Carolina: toward a rhetorical model of risk communication. In: Herndl G, Brown SC, editors. Green culture: environmental rhetoric in contemporary America. Madison (WI): Univ of Wisconsin Pr; 1996. p 111–40.

21. Article, Magazine

12. Appenzeller T. The case of the missing carbon. National Geographic 2004 Feb: 88–118.

22. Article, Newspaper

6. Hall C. Shortage of human capital envisioned, monster's Taylor sees worker need. Chicago Tribune 2002 Nov 18; Sect E:7(col 2).

23. Article, Author Unknown

3. The big chill leaves bruises. Albuquerque Tribune 2004 Jan 17; Sect A:4(col 1).

24. Article, CD-ROM

21. Hanford P. Locating the right job for you. The electronic job finder [CD-ROM]. San Francisco: Career Masters; 2001. CD-ROM: sound, color, 4 $^3/_4$ in.

25. Song or Recording

12. Myer L. Sometimes alone. Flatlands [CD]. Ames (IA): People's Productions; 1993.

26. Television or Radio Program

4. Harris R. Destination: the south pole [recording]. Washington: National Public Radio; 2003 Jan 6 [cited 2004 Jan 19]. Available from: http://discover.npr.org/features/feature.jhtml?wfId=904848

27. Personal Correspondence, E-Mail, or Interview

These complications seem to have been resolved (2006 e-mail from FH Smith to me) while others seem to have emerged.

(References that refer to personal correspondences or personal interviews should be placed within the text and not in the References list.)

Can't find a CSE model for your source? Go to **www.pearsonhighered.com/johnsonweb3/C.7**

Creating the CSE References List (Citation-Sequence Style)

In CSE style, the References list is placed at the end of the document or in an appendix. The sources are listed by number in the order in which they were referenced in the text.

References

1. Hoffmann A, Blows M. Evolutionary genetics and climate change: will animals adapt to global warming? In: Kareiva P, Kingsolver J, Huey R, editors. Biotic interactions and global change. Sunderland (MA): Sinauerl; 1993. p 13–29.

2. Sherwood K, Idso C. Is the global warming bubble about to burst? [Internet]. Tempe (AZ): Center for the Study of Carbon Dioxide and Global Change; c2003 [cited 2004 Mar 4]. Available from: http://www.co2science.org/edit/v6_edit/v6n37edit.htm

3. Assel R, Cronk K, Norton D. Recent trends in Laurentian Great Lakes ice cover. Climatic Change 2003; 57:185–204.

4. Kadlecek M. Global climate change could threaten U.S. wildlife. Conservationist 1991; 46(1):54–55.

5. Houghton J. Global warming: the complete briefing. 2nd ed. Cambridge (MA): Cambridge Univ Pr; 1997. p 12.

MLA Documentation Style

The MLA documentation style is not commonly used in technical or scientific fields; it is most commonly used in the arts and humanities. Nevertheless, there are occasions where MLA style is requested, because it is a widely used documentation style. The official source for this style is the *MLA Style Manual and Guide to Scholarly Publishing,* Third Edition (2008).

When using MLA style, you will need to use in-text citations and a list of alphabetically arranged references, called "Works Cited," at the end of your document.

MLA In-Text Citations

MLA style follows an *author-page number* system for in-text citations, meaning the author and page number are usually cited within the text.

Individual Authors

Individual authors are cited using their last name and the page number(s) from which the information was drawn. If the year is significant, put it after the author's name in parentheses.

One study reports a significant rise in HIV cases in South Africa in one year (Brindle 834).

One study reported a 12.2 percent rise in HIV cases in only one year (Brindle 843).

Brindle (2000) reports a significant rise in HIV cases in South Africa in one year.

Brindle (2000) reports a 12.2 percent rise in HIV cases in South Africa in one year (834).

In most cases, only the author and page number need to be noted, as shown in the first example above. In MLA style, the year of publication is not usually a large concern, so include the year only if it is necessary.

Multiple Authors

If an article has two or more authors, you should use the word *and* to connect the authors' last names.

(Thomas and Linter 130)

According to Thomas and Linter (2001) the number of mammals in this area was dramatically reduced during the Ice Age (130).

Technical documents often have more than two authors. In these cases, cite all the names the first time the work is referenced. Afterward, you can repeat all the names or use the last name of the first author followed by "et al."

First Citation of Work

(Wu, Gyno, Young, and Reims 924)

Subsequent Citations of Work

(Wu et al. 924)

As reported by Wu et al., the Permian Age…

Corporate or Unknown Authors

When the author of the document is a corporation or unknown, the in-text citation uses the name of the corporation or the first prominent word in the title of the document.

(National Science Foundation 76)

("Results" 91)

(*Silent* 239)

As shown here, if the source is an article, put the first prominent word in quotes. If it is a book, put it in italics.

Paraphrased Materials

Because the authors' names are typically mentioned in the sentence, citing paraphrased materials usually requires only a mention of the page number.

Franks and Roberts report that aptitude for visual thinking runs in families (76).

The instinct for survival, according to Ramos, is strong in the Mexican wolf (198-201).

Jones argues that finding a stand of dead trees near an industrial plant is a good indicator that something is seriously wrong (87–88).

Two or More Works in Same Parentheses

In some cases, several documents will state similar information. In these cases, you should cite them all and separate them with semicolons.

Studies have shown remarkable progress toward reviving the penguin population on Vostov Island (Hinson and Kim 330; Johnson and Smith 87; Tamili 102).

The Works Cited List for MLA Style

When using MLA style, your Works Cited list should be in alphabetical order at the back of the document. In your list, you should include only the items that you actually cited in your document. Leave any uncited documents out of the list of references. After all, if a document is important enough to include in your Works Cited, it should be important enough to cite in the text itself.

In most cases, the list is identified by the centered heading "Works Cited." Entries should be double-spaced. Also, each reference should use a hanging indent style (i.e., the second line and subsequent lines should be indented). The first line should be flush with the left margin.

The following list includes examples of MLA style. This list is not comprehensive. If you do not find a model here for a source you are adding to your Works Cited, you should check the *MLA Style Manual and Guide to Scholarly Publishing,* Third Edition. Or, check the several websites available that offer examples of MLA style.

1. **Website or Webpage, Author Known**

   Einstein Online. Ed. Fred Jaspers. 9 Mar. 2003. Web. 13 Dec. 2008.

2. **Website or Webpage, Corporate Author**

   Managing Forest on Your Land. National Wildlife Service. 8 Sept. 2002. Web. 10 Oct. 2008.

3. **Webpage, Author Unknown**

   "Skin Cancer Treatments Debated." *CNN.com.* 1 Jan. 2004. Web. 9 Aug. 2008.

4. **Webpage, Online Periodical**

   Grinspoon, David. "Is Mars Ours?" *Slate Magazine* 7 Jan. 2004. Web. 19 Jan. 2008.

5. **Book, One Author**

   Jones, Steve. *Darwin's Ghost: The Origin of Species Updated.* New York: Balantine Books, 2001. Print.

6. Book, More Than One Author

Pauling, Linus, and E. Bright Wilson. *Introduction to Quantum Mechanics.* New York: Dover Publications, 1935. Print.

7. Book, Corporate or Organization Author

American Psychiatric Association. *Diagnostic and Statistical Manual of Mental Disorders.* 4th ed. Washington: American Psychiatric Association, 1994. Print.

8. Book, Edited Collection

Mueller-Vollmer, Kurt, ed. *The Hermeneutics Reader.* New York: Continuum, 1990. Print.

9. Book, Translated

Habermas, Jurgen. *Communication and the Evolution of Society.* Trans. Thomas McCarthy. Boston: Beacon Press, 1979. Print.

10. Book, Author Unknown

Handbook for the WorkPad c3 PC Companion. Thornwood: IBM, 2000. Print.

11. Book, Second Edition or Beyond

Williams, Robin, and John Tollet. *The Non-Designer's Web Book.* 2nd ed. Berkeley: Peachpit, 2000. Print.

12. Book, Dissertation or Thesis

Simms, Laura. "The Hampton Effect in Fringe Desert Environments: An Ecosystem Under Stress." Diss. U of New Mexico, 2002. Print.

13. Book, Electronic

Darwin, Charles. *On the Various Contrivances by which British and Foreign Orchids are Fertilised by Insects.* London: John Murray, 1862. Web. 1 Jan. 2008.

14. Document, Government Publication

Greene, Linda W. *Exile in Paradise: The Isolation of Hawai'i's Leprosy Victims and Development of Kalaupapa Settlement, 1865 to Present.* Washington: US Department of the Interior, National Park Service, 1985. Print.

15. Document, Pamphlet

Exploring High Altitude Areas. Denver: TCHN, 2002. Print.

16. Film or Video Recording

The Lord of the Rings: The Fellowship of the Ring. Dir. Peter Jackson. Prod. Barrie Osborne, Peter Jackson, Fran Walsh, and Tim Sanders. New Line Productions, 2002. Film.

17. CD-ROM

Geritch, Thomas. *Masters of Renaissance Art.* CD-ROM. Chicago: Revival Productions, 2000.

18. **Article, Journal with Continuous Pagination**

Boren, M. Ted, and Judith Ramey. "Thinking Aloud: Reconciling Theory and Practice." *IEEE Transactions on Professional Communication* 39 (1996): 49–57. Print.

19. **Article, Journal without Continuous Pagination**

Lenhoff, Rosalyn, and Lynn Huber. "Young Children Make Maps!" *Young Children* 55.5 (2000): 6–12. Print.

20. **Article, Edited Book**

Katz, Steven B., and Carolyn R. Miller. "The Low-Level Radioactive Waste Siting Controversy in North Carolina: Toward a Rhetorical Model of Risk Communication." *Green Culture: Environmental Rhetoric in Contemporary America.* Ed. Carl G. Herndl and Stuart C. Brown. Madison: U of Wisconsin P, 1996. 111–40. Print.

21. **Article, Magazine**

Appenzeller, Tim. "The Case of the Missing Carbon." *National Geographic* Feb. 2004: 88–118. Print.

22. **Article, Newspaper**

Hall, Cheryl. "Shortage of Human Capital Envisioned, Monster's Taylor Sees Worker Need." *Chicago Tribune* 18 Nov. 2002: E7. Print.

23. **Article, Author Unknown**

"The Big Chill Leaves Bruises." *Albuquerque Tribune* 17 Jan. 2004: A4. Print.

24. **Article, CD-ROM**

Hanford, Peter. "Locating the Right Job for You." *The Electronic Job Finder.* CD-ROM. San Francisco: Career Masters, 2001.

25. **Song or Recording**

Myer, Larry. "Sometimes Alone." *Flatlands.* Ames: People's Productions, 1993. CD.

26. **Television or Radio Program**

"Destination: The South Pole." Narr. Richard Harris. *All Things Considered.* National Public Radio. 6 Jan. 2003. Web. 4 Feb. 2004.

27. **Personal Correspondence, E-Mail, or Interview**

Baca, James. Personal interview. 4 Mar. 2004.

Creating the MLA Works Cited List

In MLA style, the Works Cited list is placed at the end of the document on a separate page or in an appendix. The sources referenced in the document should be listed alphabetically, and each entry should be double-spaced.

Can't find an MLA model for your source? Go to
www.pearsonhighered.com/johnsonweb3/C.9

Documentation Guide   A-37

Works Cited

Assel, Robert, Kevin Cronk, and David Norton. "Recent Trends in Laurentian Great Lakes Ice Cover." *Climatic Change* 57 (2003): 185–204. Print.

Hoffmann, Amber, and Marlin Blows. "Evolutionary Genetics and Climate Change: Will Animals Adapt to Global Warming?" *Biotic Interactions and Global Change.* Ed. Paul M. Kareiva, John G. Kingsolver, and Renee B. Huey. Sunderland: Sinauer, 1993. 13–29. Print.

Houghton, James. *Global Warming: The Complete Briefing.* 2nd ed. Cambridge, MA: Cambridge UP, 1997. Print.

Kadlecek, Mary. "Global Climate Change Could Threaten U.S. Wildlife." *Conservationist* 46.1 (1991): 54–55. Print.

Sherwood, Kevin, and Craig Idso. "Is the Global Warming Bubble About to Burst?" 4 Mar. 2004. Web. 8 Nov. 2008.

References

Alcatel-Lucent. (1998). *What is a laser?* Retrieved from http://www.bell-labs.com/history/laser/laser_def.html

Alesis. (2008). *Podcasting microphone.* Retrieved from http://www.alesis.com

Allen, L., & Voss, D. (1997). *Ethics in technical communication: Shades of gray.* New York: Wiley.

American Institute of Aeronautics and Astronautics. (2003). *How design engineers evaluate their education* [Bar chart]. Retrieved from http://www.aiaa.org/tc/de

American Lung Association. (2008). *Top stories.* Retrieved from http://www.lungusa.org

American Orchid Society. (2008). *Orchid information.* Retrieved from http://www.aos.org

American Psychological Association. (2001). *Publication manual of the American Psychological Association* (5th ed.). Washington, DC: American Psychological Association.

American Psychological Association. (2007). *APA style guide to electronic resources.* Washington, DC: American Psychological Association.

American Red Cross. (2008). *Home.* Retrieved from http://www.redcross.org

American Society of Civil Engineers. (2008). *Welcome to the ASCE online library.* Retrieved from http://ascelibrary.org

Archuleta, M. (2007, April). *Control your diabetes for life: What is diabetes?* [Circular 569]. New Mexico State University, College of Agriculture and Home Economics.

Aristotle. (1991). *On rhetoric.* (G. Kennedy, Trans.). New York: Oxford University Press.

Arnheim, R. (1969). *Visual thinking.* Berkeley: University of California Press.

Ask.com. (2008). *Tsunami definition.* Retrieved from http://www.ask.com

Australian Resuscitation Council. (2002). *Basic life support flowchart.* Retrieved from http://www.resus.org.au/public/bls_flow_chart.pdf

Aversa, A. (2003). *Galaxy simulations.* Retrieved from http://www.u.arizona.edu/~aversa/galaxysims.pdf

Barret, M., & Levy, D. (2008). *A practical approach to managing phishing* [White paper]. Retrieved from http://files.shareholder.com/downloads/PAY/

Barrow, L. & Rouse, C.E. (2005). Does college still pay? *The Economist's Voice, 2*(4)1–8.

Belbin, M. (1981). *Management teams.* New York: Wiley.

Benson, P. (1985). Writing visually: Design considerations in technical publications. *Technical Communication, 32,* 35–39.

Berg, A. (2003). *Counseling to prevent tobacco use and tobacco-caused disease* [Brochure]. Washington, DC: U.S. Preventive Services Task Force.

Bernhardt, S. (1986). Seeing the text. *College Composition and Communication, 30,* 66–78 [Survey].

Bledsoe, L., & Sar, B. K. (2001). *Campus survey report: Safety perception and experiences of violence.* Retrieved from http://www.louisville.edu

Blizzard Entertainment. (2008). *World of warcraft anti-harassment policies.* Retrieved from http://us.blizzard com/support

Blum, D., & Knudson, M. (1997). *Field guide for science writers.* Oxford, UK: Oxford University Press.

BMW of North America. (2006). *Cooper Mini unauthorized owner's manual* [Manual]. Retrieved from http://www.motoringfile.com/files/unauth_manual.pdf

Boisjoly, R. (1985). SRM o-ring erosion/potential failure criticality. In *Report of the Presidential Commission on the Space Shuttle* Challenger

Accident. Retrieved from http://science.ksc.nasa.gov/shuttle/missions/51-l/docs/rogers-commission/table-of-contents.html

Boor, S., & Russo, P. (1993). How fluent is your interface? Designing for international users. *Proceedings of INTERCHI '93,* 342–347.

Booth, W., Columb, G. C., & Williams, J. (1995). *The craft of research.* Chicago: University of Chicago Press.

Bosley, D. (2001). *Global contexts.* Boston: Allyn & Bacon.

Brody, M. (1998). *Speaking your way to the top.* Boston: Allyn & Bacon.

Brusaw, C., Alred, G., & Oliu, W. (1993). *Handbook of technical writing* (4th ed.). New York: St. Martin's Press.

Buckinghamshire County Council. (2003). *Flooding—cause and effect.* Retrieved from http://www.southbucks.gov.uk/documents/Flooding%20FACTS3.doc

Burke, K. (1969). *A rhetoric of motives.* Berkeley, CA: University of California Press.

Buzan, T. (1994). *The mind map book: How to use radiant thinking to maximize your brain's untapped potential.* New York: Dutton.

Campbell, G. M. (2002). *Bulletproof presentations.* Franklin Lake, NJ: Career Press.

CareerBuilder. (2008). Home page. Retrieved from http://www.careerbuilder.com

Carey, S. S. (1998). *A beginner's guide to the scientific method* (2nd ed.). Belmont, CA: Wadsworth Publishing.

Carson, R. (2002). *Silent spring.* New York: Mariner.

CBS News (2008). *Are we retreating in the war on cancer?* Retrieved from www.cbsnews.com/stories/2008/05/20/eveningnews

Chaney, L., & Martin, J. (2004). *Intercultural business communication* (3rd ed.). Upper Saddle River, NJ: Pearson Prentice Hall.

Chohan, R., & National Aeronautics and Space Administration. (2005). *Scientists confirm Earth's energy is out of balance.* Washington, DC. Retrieved from www.nasa/vision/earth/environment/earth_energy.html

Cincinnati Children's Hospital Medical Center. (2008). *Hemangiomas and vascular malformations: Conditions and diagnoses.* Retrieved from http://www.cincinnatichildrens.org/health/info/vascular/diagnose/hemangioma.htm

Clark, H., & Haviland, S. (1977). Comprehension and the given-new contract. In R. Freedle (Ed.), *Discourse production and comprehension* (pp. 1–40). Norwood, NJ: Ablex.

Clark, R. (1971). *Einstein: His life and times.* New York: World Publishing.

Clement, D. E. (1987). Human factors, instructions, and warning, and product liability. *IEEE Transactions on Professional Communication, 30,* 149–156.

Consumers Against Supermarket Privacy Invasion and Numbering [CASPIAN]. (2008). *What Is RFID?* Retrieved from http://www.spychips.com/what-is-rfid.html

Council of Biology Editors. (2006). *The CSE manual for authors, editors, and publishers.* Reston, VA: Council of Science Editors.

Curlew Communications. (2003). *Hemyock Castle.* Retrieved from http://www.hemyockcastle.co.uk/bodiam.htm

Davis, D. (2007, Nov. 4). Off target in the war on cancer. *Washington Post.* Retrieved from http://www.washingtonpost.com/wp-dyn/content/article/2007/11/02/AR2007110201648.html

Deming, W. E. (2000). *Out of crisis.* Cambridge: MIT Press.

De Vries, M. A. (1994). *Internationally yours: Writing and communicating successfully in today's global marketplace.* New York: Houghton Mifflin.

Dombrowski, P. (2000). *Ethics in technical communication.* Needham Heights: Allyn & Bacon.

Dragga, S. (1996). A question of ethics: Lessons from technical communicators on the job. *Technical Communication Quarterly, 6,* 161–178.

Dumas, J. S., & Redish, J. C. (1993). *A practical guide to usability testing.* Norwood: Ablex.

Einstein, A. (1939). *August 2, 1939, letter to Franklin Roosevelt* [Letter]. Retrieved from http://www.anl.gov/OPA/frontiers96arch/aetofdr.html

Fagen, W. T., & Coish, D. (1999). *Validating the IN and FOR distinctions of a workplace literacy program.* Retrieved from http://www.mun.ca/educ/faculty/mwatch/fall99/fagancoish.html

Farkas, D., & Farkas, J. (2002). *Principles of web design.* New York: Longman.

Fermi National Laboratory. (2008). *Fermilab.* Retrieved from http://www.fermilab.gov

Field Museum. (2008). Home page. Retrieved from http://www.fieldmuseum.org

Geothermal Energy Association. (2008). Home page. Retrieved from www.geo-energy.org/gea/gea.asp

Gibaldi, J. (2003). *MLA handbook for writers of research papers* (6th ed.). New York: Modern Language Association.

GN Netcom (2005). *User manual for Jabra BT160 Bluetooth headset* [Manual].

Goddard Institute for Space Studies. (2004). *Global temperature trends: 2002 summation.* Retrieved from http://www.giss.nasa.gov/research/observe/surftemp

Google. (2008). *Google.* Retrieved from http://www.google.com

Google Chrome. (2008). *Google chrome internet browser* [Manual]. Retrieved from http://www.google.com/chrome

Google Docs. (2008). *Google docs interface.* Retrieved from http://docs.google.com

Google Translate. (2008). *Translation of purdue university professional writing program home page.* Retrieved from http://translate.google.com

Government of Saskatchewan. (2008). *HIV-AIDS information.* Retrieved from http://www.health.gov.sk.ca/hiv-aids

Guilford, C. (1996). *Paradigm online writing assistant.* Retrieved October 5, 2003, from http://www.powa.org

Gurak, L. (2000). *Oral presentations for technical communication.* Boston: Allyn & Bacon.

Hallberg, B. A. (2003). *Networking: A beginner's guide* (3rd ed.). New York: McGraw-Hill.

Haneda, S., & Shima, H. (1983). Japanese communication behavior as reflected in letter writing. *Journal of Business Communication, 19,* 19–32.

Haviland, S., & Herbert, C. (1974). What's new: Acquiring new information as a process in comprehension. *Journal of Verbal Learning and Verbal Behavior, 13,* 512–521.

Helfman, E. S. (1967). *Signs and symbols around the world.* New York: Lothrop, Lee, and Shepard.

Hemyock Castle. (2008). *Comparison between Hemyock and Bodiam castles.* Retrieved from http://www.hemyockcastle.co.uk/bodiam.htm

Herrington, T. (2003). *A legal primer for the digital age.* New York: Pearson Longman.

Hoft, N. (1995). *International technical communication.* New York: Wiley.

Horton, W. (1993). The almost universal language: Graphics for international documents. *Technical Communication, 40,* 682–683.

Hult, C., & Huckin, T. (2008). *The new century handbook* (4th ed.). Boston: Longman.

Husqvarna. (2002). *Working with a chainsaw* [Manual]. Åsbro, Sweden: Electrolux.

IBM. (2008). *IBM Lotus Notes* [Software]. Retrieved from http://demos.dfw.ibm.com/on_demand/streamed/IBM_Demo_Lotus_Notes_7-1-Nov05.html?S=index

Indiana Department of Transportation. (2007). *Dry flow testing of flowable backfill materials* [Fact sheet]. Retrieved from http://www.in.gov/indot

Indiana Prevention Resource Center at Indiana University. (2001). *Marijuana use among Indiana 6th–12th graders, 2001* [Brochure]. Retrieved from http://www.iprc.indiana.edu/drug_stats/youth2001/s2001-26.gif

InfoSpace. (2003). *Metacrawler.* Retrieved from http://www.metacrawler.com

InfoSpace (2003). *Metacrawler, advanced search.* Retrieved from http://www.metacrawler.com/info.metac/search/advance.htm

Institute of Electrical and Electronics Engineers. (2006). *IEEE code of ethics.* Retrieved from http://www.ieee.org/portal/pages/iportals/aboutus/ethics/code.html

Institute for Social and Economic Research. (2007). *British household panel survey* [Survey]. New Policy Institute. Retrieved from http://www.poverty.org.uk

International Association of Food Protection. (2008). *Sneezing icon* [Graphic]. Retrieved from http://www.foodprotection.org/aboutIAFP/iconmania.asp

International Energy Agency. (2008, May 13). *Oil market report.* Retrieved from http://omrpublic.iea.org/omrarchive

International Institute for Geo-Information Science and Earth Observation (2005). *Characteristics of tsunamis.* Retrieved from http://www.itc.nl/library/Papers_2005/tsunami/Tsunami.pdf

International Olympic Committee. (2004). Symbols for basketball, swimming, and cycling [Graphics].

International Organization for Standardization. (2008). *ISO 9000 and ISO 14000.* Retrieved from http://www.iso.org

Israelson, A. (2007, May 30). Transcript of interview with Bill Gates and Steve Jobs [Interview transcript]. *Ubiqus Reporting.* Retrieved from http://d5.allthingsd.com/d5gates-job-transcript

Jackson, M. (2007). *Memorandum for all DHS employees.* Retrieved from http://www.slate.com

Johnson-Sheehan, R. (2002). *Writing proposals: A rhetoric for managing change.* Boston: Longman.

Johnson-Sheehan, R., & Baehr, C. (2001). Visual-spatial thinking: Thinking differently about hypertexts. *Technical Communication, 48,* 37–57.

Jones, D. (1998). *Technical writing style.* Boston: Allyn & Bacon.

Karjane, H., Fisher, B., & F. Cullen (2005). *Sexual assault on campus: What colleges and universities are doing about it.* Washington, DC: U.S. Department of Justice, National Institute of Justice.

Kiester, E., & Kiester, W. (2003, July). Birdbrain breakthrough. *Smithsonian, 36–37.*

Koffka, K. (1935). *Principles of gestalt psychology.* New York: Harcourt.

Kostelnick, C., & Roberts, D. (1998). *Designing visual language.* Boston: Allyn & Bacon.

Kotler, P. (2002). *Marketing management* (11th ed.). Englewood Cliffs, NJ: Prentice Hall.

Laib, N. (1993). *Rhetoric and style.* Englewood Cliffs, NJ: Prentice Hall.

Lakoff, G. (2004). *Don't think of an elephant.* White River Junction, VT: Chelsea Green.

Lanham, R. (1999). *Revising business prose* (4th ed.). New York: Pearson Longman.

Leopold, A. (1966). *A sand county almanac.* New York: Ballantine.

Lincoln, A. (2001). Lost in translation. *CFO.com.* Retrieved from http://www.cfo.com/article.cfm/3000717

Lipnack, J., & Stamps, J. (2000). *Virtual teams: People working across boundaries with technology.* New York: Wiley.

Llamagraphics. (2008). *Life balance screenshot* [Software]. Llamagraphics, Inc.

Manitoba Conservation Wildlife and Ecosystem Protection Branch. (2004). *Manitoba's species at risk: Ferruginous hawk* [Brochure].

Markel, M. M. (1991). A basic unit on ethics for technical communicators. *Journal of Technical Writing and Communication, 21,* 327–350.

Mathes, J., & Stevenson, D. (1976). *Designing technical reports.* Indianapolis, IN: Bobbs-Merrill.

Medline Plus. (2008). Home page. Retrieved from http://www.medlineplus.gov

Merriam-Webster. (2008). Definition of "gene." *Merriam-Webster Collegiate® Dictionary* (11th ed.) Retrieved from www.merriam-webster.com

Metacrawler. (2008). Home page. Retrieved from http://www.metacrawler.com

Microsoft. (2006). Home page. Retrieved September 6, 2006, from www.microsoft.com

Microsoft. (2006). Middle Eastern web page. Retrieved from www.microsoft.com/middleeast

Microsoft. (2007). *Project Standard 2007 overview* [Software]. Retrieved from http://www.office.microsoft.com/en-us/project/HA101656381033.aspx

MLA. (2008). *Style manual and guide to scholarly publishing.* (3rd ed.). New York: Modern Language Association of America.

Moore, P., & Fitz, C. (1993). Using gestalt theory to teach document design and graphics. *Technical Communication Quarterly, 2,* 389–410.

Mozilla Firefox. (2006). *Help.* Retrieved from www.mozilla.com

National Aeronautics and Space Administration. (2004). *Mars exploration Rover* [Press release]. Retrieved from http://www.jpl.nasa.gov/news/presskits/merlaning.pdf

National Aeronautics and Space Administration. (2005). *Hurricane Katrina approaching the gulf coast* [Graphic]. Washington, D.C. Retrieved from http://rapidfiresco.gsfc.nasa.gov/gallery/?search=Katrina

National Aeronautics and Space Administration. (2008). *Why the moon?* [Poster]. Retrieved from http://www.nasa.gov/pdf/163561main_why_moon2.pdf

National Commission on Writing. (2004). *A ticket to work…or a ticket out: A survey of business leaders.* New York: College Board.

National Foundation for Cancer Research. (2006). *Cancer detection guide.* Bethesda, MD: NFCR.

National Human Genome Research Institute. (2008). Home page. Retrieved from http://www.genome.gov

National Library of Medicine. (2001). *National Library of Medicine recommended formats for bibliographic citation: Supplement: Internet formats.* Bethesda, MD: NLM.

National Library of Medicine. (2003). *Medline Plus.* Retrieved from http://www.nlm.nih.gov/medlineplus/medlineplus.html

National Marine Sanctuaries. (2004). *Project summary.* Retrieved from http://sanctuaries.noaa.gov/news/bwet

National Oceanic and Atmospheric Administration. (2008). Home page. Retrieved from http://www.noaa.gov

National Ocean Service. (2005). *National Ocean Service Accomplishments.* Washington, DC: National Ocean Service.

National Science Foundation. (2002). *Grant proposal guide.* Retrieved from http://www.nsf.gov/pubs/2003/nsf032/032_2.htm

National Science Foundation. (2008). Latest news. Retrieved from http://www.nsf.gov

National Survey on Drug Use and Health. (2005). *Alcohol use and risks among young adults by college enrollment status* [Survey]. Washington, DC: Office of Applied Studies, Substance Abuse, and Mental Health Services Administration.

Netscape Communicator Browser Window. (2008). Netscape browsers.

Nikon. (2002). *The Nikon guide to digital photography with the Coolpix 885 digital camera* [Manual]. Tokyo, Japan: Nikon.

Occupational Safety and Health Administration. (2007). *Fact sheet, flood cleanup* [Fact sheet]. Retrieved from http://www.osha.gov/OshDoc/data_Hurricane_Facts/floodcleanup.pdf

Ocean Alliance. (2008). *Research.* Retrieved from http://www.whale.org/research/index.html

Ocean Alliance. (2008). *Right whale program.* Retrieved from http://www.whale.org/wci/wci_rightwhale.html

Ocean Alliance. (2008). *Welcome to the Ocean Alliance.* Retrieved from http://www.whale.org

Office of Applied Studies Substance Abuse and Mental Health. (2006). *National survey on drug use* [Survey]. SAMHSA.

Ong, W. (1982). *Orality and literacy.* London: Routledge.

Online Ethics Center for Engineering and Science. (2008). Home page. Retrieved from http://www.onlineethics.org

Oxford English Dictionary Online. (2003). *Solstice.* Retrieved from http://www.oed.com

Pacific Northwest Search and Rescue. (2008). *Welcome.* Retrieved from http://www.pnwsar.org

Pakiser, L., & Shedlock, K. (1997). *Earthquakes* [Fact sheet]. Retrieved from http://pubs.usgs.gov/gip/earthq1/earthqkgip.html

Palm. (2005). PalmOne, our most popular handheld is better than ever [Manual]. Tungsten E2. Retrieved from http://www.palmone.com

Pearsall, T. (2001). *Elements of technical writing.* Needham Heights, MA: Allyn & Bacon.

Peckham, G. (2003). Safety symbols. *Compliance engineering* [Graphic]. Retrieved from http://www.ce-mag.com/archive/02/03/peckham.html

Pendergrast, M. (1994). *For god, country, and Coca-Cola.* New York: Collier.

Pew Research Center. (2008). *Reports.* Retrieved from http://www.pewinternet.org/reports.asp

Piven, J., & Borgenicht, D. (1999). How to jump from a moving car. *The worst-case scenario survival handbook.* San Francisco: Chronicle.

Piven, J., & Borgenicht, D. (2003). How to use a defibrillator to restore a heartbeat. *Worst-case scenarios online.* Retrieved from http://www.worstcasescenario.com

Plotnik, A. (1982). *The elements of editing.* New York: Macmillan.

Polkaudio. (2005). *XRt12 reference tuner owner's manual* [Manual]. Baltimore: Polkaudio.

Price, J. (1999). *Outlining goes electronic.* Stamford, CT: Ablex.

Pyle, R. (1981). *National Audubon Society field guide to North American butterflies.* New York: Knopf.

RAND Corporation. (2000). Tranquilizing effect. *RAND review,* Fall 2000. Retrieved from http://www.rand.org/publications/randreview/issues/rr.12.00/tranquil.html

Reynolds, S., & Valentine, D. (2004). *Guide to cross-cultural communication.* Upper Saddle River, NJ: Pearson Prentice Hall.

Rooksby, E. (2002). *E-mail and ethics: Style and ethical relations in computer-mediated communication.* New York: Routledge.

Rubin, J. (1994). *Handbook of usability testing.* New York: Wiley.

Rude, C. (1998). *Technical editing* (2nd ed.). Needham Heights, MA: Allyn & Bacon.

Ryobi. (2000). *510r 4-cycle garden cultivator operator's manual* [Manual]. Chandler, AZ: Ryobi.

Safetyline Institute. (1998). *Gas laws* [Fact sheet]. Retrieved from http://www.safetyline.wa.gov.au/institute/level2/course16/lecture47/l47_02.asp

Schriver, K. (1997). *Dynamics in document design.* New York: Wiley.

Schultz, H. (2000). *The elements of electronic communication.* Needham Heights, MA: Allyn & Bacon.

Segway. (2004). *Discover the Segway HT evolution* [Press release]. Retrieved from http://www.segway.com

Shapiro, G. (1998, March). The ABCs of asthma. *Discover, 35,* 30–33.

Shenk, D. (1998). *Data smog: Surviving the information glut.* San Francisco: HarperCollins.

Silyn-Roberts, H. (1998). Using engineers' characteristics to improve report writing instruction. *Journal of Professional Issues in Engineering Education and Practice, 124,* 12–16.

Silyn-Roberts, H. (2000). *Writing for science and engineering: Papers, presentations, and reports.* Boston: Butterworth-Heinemann.

Smith-Jackson, T. L., Essuman-Johnson, A., and Leonard, S. D. (2003). Symbol primes: Cross-cultural comparison of symbol representation. *Proceedings of the 15th Triennial Congress of the International Ergonomics Association,* Seoul, Korea.

State of Michigan. (2007). *Emerging diseases: 2007 human WNV cases* [Fact sheet]. Retrieved from http://www.michigan.gov/emergingdiseases

State of Utah Office of the Attorney General. (2004). *What is meth?* [Fact sheet]. Retrieved from http://attorneygeneral.utah.gov/Meth/Whatismeth.htm

Stauber, J., & Rampton, S. (1995). *Toxic sludge is good for you! Lies, damn lies and the public relations industry.* Monroe, ME: Common Courage Press.

Stewart-MacDonald. (2003). *Violin kit assembly instructions.* Retrieved from http://www.stewmac.com

Terminello, V., & Reed, M. (2003). *E-mail: Communicate effectively.* Upper Saddle River, NJ: Prentice Hall.

TiVo. (2002). *Start here* [Manual]. Tokyo: TiVo and Pioneer Corporation.

Toastmasters International. (2008). *Become the speaker and leader you want to be.* Retrieved from http://www.toastmasters.org

Tuckman, B. W. (1965). Development sequence in small groups. *Psychological Bulletin, 63,* 384–399.

Tufte, E. (1983). *The visual display of quantitative information.* Cheshire, MA: Graphics Press.

United Nations. (2007, June). *The millennium development goals report.* Department of Economic and Social Affairs.

University of Chicago Press. (2003). *Chicago manual of style* (15th ed.). Chicago: University of Chicago Press.

University of Minnesota Library. (2008). Home page. Retrieved from http://www.lib.umn.edu

U.S. Census Bureau. (2008). Home page. Retrieved from http://www.census.gov

U.S. Centers for Disease Control. (2003). *Fight the bite* [Fact sheet]. Retrieved from http://www.cdc.gov/ncidod/dvbid/westnile/index.htm

U.S. Centers for Disease Control. (2003). *The rabies virus* [Fact sheet]. Retrieved from http://www.cdc.gov/rabies/virus.htm

U.S. Centers for Disease Control. (2003). *West Nile Virus (WNV) infection: Information for clinicians* [Fact sheet]. Retrieved from http://www.cdc.gov/ncidod/dvbid/westnile/index.htm

U.S. Centers for Disease Control. (2007). *Balance scale.* Retrieved from http://www.cdc.gov/diabetes/pubs/images/balance.gif

U.S. Copyright Office. (2008). *About copyright.* Retrieved from http://www.loc.gov/copyright

U.S. Department of Energy. (2003). *Fuel cell technology: How it works* [Fact sheet]. Retrieved from http://www.fe.doe.gov/coal_power/fuelcells/fuelcells_howitworks.shtml

U.S. Department of Health. (2007). *Women's health* [Fact sheet]. Retrieved from http://.www.cdc.gov/lcod.htm

U.S. Environmental Protection Agency. (2000). *National water quality inventory, 2000 report.* Washington, DC: U.S. EPA.

U.S. Environmental Protection Agency. (2002). *Global warming impacts: Forests.* Retrieved from http://yosemite.epa.gov/oar/globalwarming.nsf/content/ImpactsForests.html

U.S. Environmental Protection Agency. (2004). *IPM for schools: A how-to manual* [Manual]. Washington, DC: U.S. Author.

U.S. Federal Bureau of Investigation. (2008). *FBI history.* Retrieved from http://www.fbi.gov/fbihistory.htm

U.S. Geological Survey. (1997). *Predicting earthquakes* [Fact sheet]. Retrieved from http://pubs/usgs/gov/gip/earthq1/predict.html

U.S. Geological Survey. (2002). *Eruption history of Kilauea* [Fact sheet]. Retrieved from http://hvo.wr.usgs.gov/kilauea history/main.html

U.S. Geological Survey. (2003). *A proposal for upgrading the national-scale soil geochemical database for the United States.* Washington, DC: Author.

U.S. Geological Survey. (2004). *Geology.* Retrieved from http://geology.usgs.gov/index.shtml

U.S. National Weather Service. (2004). *Frostbite and hypothermia.* Retrieved from http://www.crh.noaa.gov/unr/edusafe/whys/hypthermia.htm

U.S. Office of Energy Efficiency and Renewable Energy. (2008). *Solar decathlon, 2075.* Retrieved from http://www.solardecathlon.org

U.S. Treasury Department. (2008). *Financial management service.* Retrieved from http://www.fms.treas.gov/mtsindex.html

Velasquez, M. G. (2002). *Business ethics: Concepts and cases* (5th ed.). Upper Saddle River, NJ: Prentice Hall.

Velotta, C. (1987). Safety labels: What to put in them, how to write them, and where to place them. *IEEE Transactions on Professional Communication, 30,* 121–126.

VeriMed Corporation. (2006). *VeriMed* [Brochure]. Retrieved from http://www.verimedinfo.com/files/Patient_VM_003R2(web).pdf

Vucetich, J. (2003). *Population data from the wolves and moose of Isle Royale* [Data file]. Retrieved from http://www.isleroyalewolf.org

Washington State Department of Ecology. (2003). *North Creek cleanup plan* [Fact sheet]. Retrieved from http://www.ecy.wa.gov/programs/wq/tmdl/watershed/north-creek/solution.html

West Virginia Office of Emergency Medical Services. (2007). *EMT-paramedic treatment protocol 4202* [Manual]. Morgantown, WV: Trauma and Emergency Care System, NOROP Center.

Whisper Technology. (2007). Home page [Software]. Retrieved from http://www.whisper.com

White, J. (1988). *Graphic design for the electronic age.* New York: Watson-Guptill.

Williams, J. (1990). *Style.* Chicago: University of Chicago Press.

Williams, R. (2004). *The non-designer's design book* (2nd ed.). Berkeley, CA: Peachpit.

Williams, R., & Tollett, J. (2000). *The non-designer's web book* (2nd ed.). Berkeley, CA: Peachpit.

Writing Center at Rensselaer Polytechnic Institute. (2003). *Revising prose.* Retrieved from http://www.rpi.edu/dept/llc/writecenter/web/revise.html

Yahoo! (2006). *Online calendar.*

Yahoo! Babel Fish. (2008). Home page. Retrieved from http://babelfish.yahoo.com

Yoshida, J. (1996, October 7). A suggestive Woody has Japanese touchy. *Electronic Engineering Times.*

Credits

Text Credits

Page 7 Courtesy: National Human Genome Research Institute. Illustration credit (center of screen shot): Jane Ades, NHGRI. Page 25 © 2003 by TiVo Inc. and Pioneer Corporation. Reprinted by permission of Pioneer Corporation (Tokyo). TiVo and the TiVo logo are registered trademarks of TiVo Inc. TiVo, Season Pass, Wishlist, TiVo Basic, TiVo Plus, and the Series2 logo are trademarks of TiVo Inc. VCR Plus+ is a registered trademark of Gemstar Development Corp. The DVD logos are trademarks of DVD FLLC. Other trademarks are the properties of their respective owners. Page 32 © Copyright 2008 The American National Red Cross. All Rights Reserved. Reprinted with permission. Page 70 Microsoft product screen shot reprinted with permission from Microsoft Corporation. Page 101 World of Warcraft, © 2005–2008 Blizzard Entertainment. All rights reserved. Reprinted with permission. Page 107 Online Ethics Center at the National Academy of Engineering. Reprinted with permission of the National Academy of Sciences, Courtesy of the National Academies Press, Washington, D.C. Copyright © 2003–2007 National Academy of Sciences. Page 130 © 2006 VeriChip Corporation. Reprinted with permission. Page 150 MetaCrawler, © 2008 InfoSpace, Inc. All rights reserved. Page 151 American Society of Civil Engineers Research Library, © 2008 American Society of Civil Engineers. Reprinted with permission. Page 155 Excerpt from transcript of the interview by Kara Swisher and Walt Mossberg conducted with Microsoft Chairman Bill Gates and Apple CEO Steve Jobs at *The Wall Street Journal's* D: All Things Digital conference on May 30, 2007. © 2007 Dow Jones & Company Inc. Reprinted with permission. Pages 156–157 Reprinted with permission of Dr. Linda K. Bledsoe and Dr. Bibhuti K. Sar. This work was supported under award 1999-WA-VX-0012 from the U.S. Department of Justice "Grants to Combat Crimes Against Women on Campuses." The opinions, findings, and conclusions expressed in this document are those of the authors and do not necessarily represent the official position or policies of the U.S. Department of Justice. Page 186 Pacific Northwest Search & Rescue. Reprinted by permission of Pacific Northwest Search & Rescue, Inc. Page 189 Reprinted with permission of the Buckinghamshire County Council. Page 190 Richard Sheppard, Curlew Communications Ltd., UK. Reprinted with permission. Page 191 From Lisa Barrow and Cecilia Elena Rouse, "Does College Still Pay?" *The Economists' Voice,* volume 2, issue 4 (2005), pp. 5–6. Copyright © 2005 by the authors. Reprinted by permission of The Berkeley Electronic Press. Page 193 Reprinted with permission of Alan G. Aversa. Page 211 Reprinted by permission of Geothermal Energy Association. Weather Channel toolbar Courtesy of weather.com. Page 227 Patrick Moore, "Patrick Moore: Going Nuclear Over Global Warming," Special to *The Sacramento Bee,* December 12, 2007. Reprinted by permission of Dr. Patrick Moore, Greenspirit Strategies, Ltd. www.greenspiritstrategies.com. Page 234 Reproduced by permission of The Queen's Printer for Manitoba. The Queen's Printer for Manitoba does not warrant the accuracy or currency of this material. Page 235 Edwin Kiester, Jr., and William Kiester, "Birdbrain Breakthrough," *Smithsonian,* June 2002, p. 36. Text reprinted with permission of Edwin Kiester, Jr. Page 236 Reprinted with permission of Fermilab. Page 246 © 2008 The Field Museum, All Rights Reserved. Reprinted with permission. Photograph of tornado: NOAA Photo Library, NOAA Central Library; OAR/ERL/National Severe Storms Laboratory (NSSL).

"Armoured Pegasus" woodcarving by Joe Leonard; photograph by Andrew Ressetti. Reprinted by permission of Joe Leonard. Page 250 Microsoft product screen shot reprinted with permission from Microsoft Corporation. Page 251 © 2005 Jabra, GN Mobile A/S (GN Netcom A/S). Reprinted by permission of GN Netcom, Inc. Page 256 *Oil Market Report*© OECD/IEA, 2008, pages 24–25. Page 262 Microsoft product screen shot reprinted with permission from Microsoft Corporation. Page 275 2007 Human WNV Cases. Reprinted by permission of Mary Grace Stobierski, DVM/ Michigan Department of Community Health. Page 276 From *The Millennium Development Goals Report 2007* by United Nations Department of Economic and Social Affairs, New York, June 2007, p. 8. © 2007 United Nations. Reprinted with the permission of the United Nations. Page 279 Reprinted by permission of New Policy Institute. Page 281 Data from Figure 1 from John A. Vucetich and Rolf O. Petersen, *Ecological Studies of Wolves on Isle Royale: Annual Report 2007–2008.* Houghton, MI: School of Forest Resources and Environmental Science, Michigan Technological University, March 31, 2008, p. 3. Reprinted by permission of John A. Vucetich/Michigan Technological University. Page 286 Reprinted by permission of the National Foundation for Cancer Research. Page 288 Reprinted with permission of the Australian Resuscitation Council. Page 293 Adobe product screen shot reprinted with permission from Adobe Systems Incorporated. Page 294 Reprinted with permission of the New Mexico Space Grant Consortium. Page 298 Reprinted by permission of the International Olympic Committee. Page 299 Reprinted with permission of The Gallup Organization. Page 309 Microsoft product screen shot reprinted with permission from Microsoft Corporation. Page 310 Microsoft product screen shot reprinted with permission from Microsoft Corporation. Page 312 Microsoft product screen shot reprinted with permission from Microsoft Corporation. Page 315 Microsoft product screen shot reprinted with permission from Microsoft Corporation. Page 332 Microsoft product screen shot reprinted with permission from Microsoft Corporation. Page 333 Microsoft product screen shot reprinted with permission from Microsoft Corporation. Page 337 Microsoft product screen shot reprinted with permission from Microsoft Corporation. Page 350 Reprinted with permission © 2008 American Lung Association. For more information about the American Lung Association or to support the work it does, call 1-800-LUNG-USA (1-800-586-4872) or log on to www.lungusa.org. Page 356 Original English web page © 2008 Purdue University. Reprinted with permission. Translated with Google Translate. Page 359 © 2002 Ocean Alliance. Reprinted with permission. Page 361 © 2002 Ocean Alliance. Reprinted with permission. Page 362 © 2002 Ocean Alliance. Reprinted with permission. Page 363 Adobe product screen shot reprinted with permission from Adobe Systems Incorporated. Page 479 Microsoft product screen shot reprinted with permission from Microsoft Corporation. Page 495 © 2008 Ask.com. Photo by David Rydevik. Page 499 © 1999–2008 Cincinnati Children's Hospital Medical Center. Reprinted with permission. Page 504 Reprinted by permission of the International Institute for Geo-Information Science and Earth Observation. Figure 3 reprinted by permission of the illustrator, Bretwood Higman, graduate student, University of Washington, in 2006. Pages 505–506 Karen Halderson and Martha Archuleta, *Control Your Diabetes for Life: What Is Diabetes?* New Mexico State University, College of Agriculture and Home Economics, Circular

569 A, Revised December 2007. Reprinted by permission of Martha Archuleta/Department of Family & Consumer Sciences, New Mexico State University. **Page 524** Reproduced from ISO web site, http://www.iso.org/iso/iso_catalogue/management_standards/iso_9000_iso_14000.htm with the permission of the International Organization for Standardization, ISO. www.iso.org. **Page 536** From www.spychips.com/what-is-rfid.html. © 2003–2006 Katherine Albrecht and Liz McIntyre. Reprinted with permission. **Pages 537–538** © 2005 palmOne, Inc. Reprinted by permission of Palm, Inc. **Pages 554–556** Reprinted by permission of the West Virginia Trauma & Emergency Care System, NOROP Center, Morgantown, WV. **Page 568** Reprinted with permission of Stewart-MacDonald. **Page 570** Reprinted with permission from Nikon. **Pages 572–573** From **Working with a Chainsaw**. Husqvarna, 2002, pp. 6–7. Reprinted by permission of Husqvarna. **Page 576** From **Ryobi 510r 4-Cycle Garden Cultivator: Operator's Manual.** Reprinted by permission of Ryobi. **Page 578** Reprinted by permission of Polk Audio. **Page 596** Screen shot courtesy National Science Foundation. Banner photo: Edward Brook/Oregon State University. Thumbnails: top, Roger Winstead/North Carolina State University; middle, Porter Hoagland/WHOI; bottom, Zina Deretsky/National Science Foundation. **Page 607** Microsoft product screen shot reprinted with permission from Microsoft Corporation. **Page 608** Microsoft product screen shot reprinted with permission from Microsoft Corporation. **Page 646** Life Balance. Reprinted by permission of Llamagraphics, Inc. **Page 652** Adobe toolbar reprinted with permission from Adobe Systems Incorporated. **Page 671** © 2000–2008 Pew Internet & American Life Project. Reprinted with permission. Windows Internet Explorer browsers reprinted with permission from Microsoft Corporation. Netscape browsers © 1998–2008 Netscape Communications Corporation. Used with permission. Netscape Communications has not authorized, sponsored, endorsed, or approved this publication and is not responsible for its content.

Photo Credits

Page iii Warren Morgan/Corbis. **Page v** Javier Larrea/PowerStock. **Page vii** Kate Mitchell/Sefa/Corbis. **Page xiii** Royalty Free/Corbis. **Page xvi** Holger Winkler/Zefa/Corbis. **Page 2** Javier Larrea/PowerStock. **Page 9** Javier Larrea/PowerStock. **Page 12** Paul Conklin/PhotoEdit, Inc. **Page 16** Ashley Cooper/Corbis. **Page 17** Color Blind Images/Blend Images/Corbis. **Page 29** Kristjan Maack/Nordic Photos/Getty Images. **Page 37** image 100/Corbis. **Page 38** Michael Newman/PhotoEdit, Inc. **Page 39** Jon Feingersh/Blend Images/Corbis. **Page 53** Dorling Kindersley Media Library. For further information see www.dk.images.com. **Page 55** Lucas Lenci Photo/The Image Bank/Getty Images. **Page 59** Advertising Archves. **Page 61** Michael Newman/PhotoEdit, Inc. **Page 65** Dynamic Graphics Group/Creatas/Alamy. **Page 82** Dynamic Graphics Group/Creatas/Alamy. **Page 90** Pete Leonard/Zefa/Corbis. **Page 91** Herbert Kehrer/Zefa/Corbis. **Page 93** Alfred Eisenstaedt/Time Life Pictures/Getty Images. **Page 100** Photograph Robert McCabe. Courtesy of the Aldo Leopold Foundation Archives and the Robert McCabe Family Collection. **Page 110** ThinkStock/agefotostock. **Page 116** Jose Luis Pelaex Inc./drr.net. **Page 118** David Buffington/Blend Images/Corbis. **Page 119** Helen King/Corbis. **Page 135** David Buffington/Blend Images/Corbis. **Page 139** Royalty Free/Corbis. **Page 140** Powerstock/Superstock. **Page 141** Powerstock/agefotostock. **Page 154** Powerstock/Superstock. **Page 173** E. Dygas/Digital Vision/Getty Images. **Page 174** Kate Mitchell/Sefa/Corbis. **Page 204** Patrik Giardino/Corbis. **Page 205** KIMBERLY WHITE/Landov. **Page 230** Maximillan Stock LTD/Phototake. **Page 231** David Young-Wolff/PhotoEdit, Inc. **Page 235T** Robert Maier/Animals Animals-Earth Scenes. **Page 235B** Fernando Nottebohm. **Page 266** ALL: Dorling Kindersley Media Library. For further information see www.dk.images.com. **Page 270** Will Sanders/Stone/Getty Images. **Page 274** LWA-Dann Tardif/Corbis. **Page 277** Corel. **Page 291** Corel. **Page 292** The Newark Museum/Art Resource, NY. **Page 297L** Pittsburgh Poison Center at the University of Pittsburgh Medical Center. **Page 300** Courtesy of Alesis. **Page 303** Mike Iannace/Corbis. **Page 305** Michael Newman/PhotoEdit, Inc. **Page 325** moodboard/Corbis. **Page 328** Helen King/Corbis. **Page 330** Alamy. **Page 343** Pinnacle Pictures/Digital Vision/Getty Images. **Page 346** Royalty Free/Corbis. **Page 346** Royalty Free/Corbis. **Page 348** Royalty Free/Corbis. **Page 368** NASA. **Page 373** Royalty Free/Corbis. **Page 374** Bloom Productions/Getty Images. **Page 399** Dorling Kindersley Media Library. For further information see www.dk.images.com. **Page 402** Dorling Kindersley Media Library. For further information see www.dk.images.com. **Page 408** Veer. **Page 409** Warren Morgan/Corbis. **Page 410** Royalty Free/Corbis. **Page 417** Doug Martin/Photo Researchers, Inc. **Page 418** Dorling Kindersley Media Library. For further information see www.dk.images.com. **Page 419** Frank Herholdt/Stone/Getty Images. **Page 426** Flying Colors/Digital Vision/Getty Images. **Page 431** Ghislain & Mare-David deLossy/The Image Bank/Getty Images. **Page 437** Walter Hodges/Stone/Getty Images. **Page 444** Stone/Airedale Brothers/Getty Images. **Page 448** Ralf Schultheiss/zefa/Corbis. **Page 450** David Young-Wolff/PhotoEdit, Inc. **Page 451** ColorBlend Images/Blend Images/Getty Images. **Page 485** Royalty Free/Corbis. **Page 487** Marc Steinmetz/VISUM/The Image Works. **Page 511** SOHO/ESA/NASA. **Page 513** Veer. **Page 515** Marc Steinmetz/Aurora. **Page 528** NASA. **Page 540** From Google Chrome: Behind the Open Source Browser Project, words by the Google Chrome team, comics adaptation by Scott McCloud <http://www.scottmccloud.com/. **Page 548** Helen King/Corbis. **Page 550** Toshifumi Kitamura/AFP/Getty Images. **Page 551** Bob Daemmrich/PhotoEdit, Inc. **Page 589** Richard Megna/Fundamental Photographs. **Page 590** Matt Cardy/Getty Images. **Page 633** iStockPhoto. **Page 634** Elena Rooraid/PhotoEdit, Inc. **Page 658** iStockPhoto. **Page 659** Holger Winkler/Zefa/Corbis. **Page 704** Veer.

Index

Metaphors, 225–226; in cross-cultural documents, 319; in technical descriptions, 517 (fig.), 535

Methodology, research, 142, 142 (fig.), 146–148, 146 (fig.), 668–669, 668 (fig.), 669 (fig.)

Methodology section, in analytical report, 675

Methods, results, discussion pattern, in presentation, 427

Microform/microfiche, as research sources, 153

Microsoft: Outlook, 84, 646; Word, 255, 290

Middle East, direction of reading in, 262, 262 (fig.)

Mission statement, purpose as, 68

MLA (citation-sequence): documentation of sources, A23, A33–A38; for in-text citations, A33–A35; for works cited list, A35–A38

Modifier, dangling, A4

Monitor evaluator, in team, 83

Mossberg, Walt, 155

Motion: instruction and, 564; technical description and, 525

Motivation, in instructions, 567

MP3 players, 150, 300

Murnan, Blake, 335

Music, 300

Names, in introductions for cross-cultural markets, 565

Narrative pattern, in presentation, 427

NASA Godard Institute, reasoning in document by, 124, 125 (fig.)

National Science Foundation (NSF) website, 596 (fig.)

Nauman, Mary, 131

Navigation, of website, 350 (fig.)

Navigation bar: for balance, 239, 242 (fig.); on home page, 361

Navigation pages, 350, 361–362

Needs: goals and, 129; of readers, 42–44

Needs and goals compared, 130

Need-to-know information, 159, 159 (fig.), 453, 458, 460

Negation, defining through, 501

Netiquette, for e-mail, 338–339

Networking, personal and professional, 376–378

Newspapers, as research sources, 152–153

Node pages, 350, 350 (fig.); drafting, 360, 361 (fig.)

Nominalizations, eliminating, 209, 695

Nonlinear organization, of website, 352

Nonsequential lists, designing, 253–254, 254 (fig.)

Norming stage of teaming, 66–67, 67 (fig.), 80

Notebook, for note taking, 30

Notes: keeping on computer, 160 (fig.); in steps in instructions, 574; using during presentation, 438–439

Notes View in presentation software, 439, 440 (fig.)

Note taking, 159; paraphrasing as, 161–162; for quotations, 161

Nouns, apostrophes for possession with, A14

Numbered list, in instructions, 571–574

Numbers: colons and, A12; commas and, A10; dashes in, A16

Objectives: career, 380, 381 (fig.), 388 (fig.); of meeting, 80; in planning process, 120; of project mission, 69 (fig.); for usability testing, 326

Objects, photographing, 291–292

Observations: for instruction research, 563, 564; persuasion with, 128; in research, 671; as technical description, 516

Online calendars, 122–123, 123 (fig.)

Online copyediting, 320–321, 320 (fig.)

Online dictionaries, 493–494

Online documentation, instructions and, 579, 579 (fig.)

Online Ethics Center for Engineering and Science (Case Western), 108, 109 (fig.)

Online Help features, 579, 579 (fig.)

Online sources, for technical definitions, 493–494, 493 (fig.)

Opening, 187, 187 (fig.); of current situation section (proposals), 7; of project plan (proposals), 604

Oppenheimer, Robert, 93 (fig.)

Organization: of activity reports, 649–653; of analytical reports, 672–677; of application letter, 392–393, 394 (fig.), 395 (fig.), 397; basic, 175–176, 176 (fig.); of conclusion, 196–198; culture and, 54; and drafting of body, 186–196; and drafting of introduction, 182–185; editing for, 309–311; by genre, 176, 178; of instructions, 552 (fig.), 566–577; international audience and, 199; of letters and memos, 457–461; outline and, 178–180, 180 (fig.); of presentation, 419–431, 428 (fig.); presentation software for, 180–181, 181 (fig.); of proposals, 600–618; of technical definitions, 496–503; of technical descriptions, 530–535

Organizing stage, 23–24, 24 (fig.)

Outcomes, identifying, 68–69, 69 (fig.)

Outline View, in word processor, 179, 180 (fig.), 321

Outlining, 27–28, 28 (fig.); of document, 178–180, 180 (fig.)

Outlook (Microsoft), 84

Overhead projector, with transparencies, 417

Oxford English Dictionary (OED), as online source for technical definitions, 493–494, 493 (fig.)

Pacing, of reader, 226

Page layout, 620, 620 (fig.), 621; for analytical report, 696 (fig.); balancing with grids, 237–239 (fig.); grids for, 237–239 (fig.); for instructions, 581–584; for technical description, 536–539

Pamphlets: in APA style, A27; in CSE style, A32; in MLA style, A36

Panasonic, web browser at, 317

Paone, Karen, 425

Paper, for document, 267

Paragraphs: in active and passive voice, 220–222; aligning sentence subjects in, 218–219; copyediting, 312; elements of, 215–217; given/new method of writing, 219–220; in plain style, 215–220, 653; using sentence types in, 216; on websites, 364

Paragraph style instructions, 572–574, 572 (fig.)

Parallelism, faulty, A6

Paraphrased materials, 160 (fig.); APA style for, A25; MLA style for, A34–A35; recording, 161–162

Paraphrasing, 162

Parentheses, A17

Parenthetical definitions, 489, 496

Partitioning of subject: with logical mapping, 527, 527 (fig.); in technical description, 517 (fig.), 525–529

Parts list, in instructions, 567–568, 568 (fig.)

Parts of a sentence, 207

Passive voice, 220–222; in application letter, 396

Past perfect progressive tense, A22

Past progressive tense, A21

Past tense, A21–A22

"Patchwriting," 170

Patent and Trademark Office, 109

Patents, ethics and, 109

Patterns of arrangement: for application letter, 393 (fig.); for documents, 188–196

PDF format: activity reports in, 652–653; for online documentation, 580; scanned document in, 652 (fig.)

People, photographing, 291

Perfect binding, 265–266, 266 (fig.)

Perfect progressive tenses, A21

Perfect tenses, A21

Performance tests, 325–326, 584–585

Performing stage of teaming, 66–67, 67 (fig.), 86–87

Period, A9

Periodical index, 152–153

Perkins, Jane, 622

Permission: copyright and, 112–113; for using source material, 170

Persona, and presentation style, 432

Personal digital assistants (PDAs), 3, 30; and activity reports, 646; keeping notes on, 160

Personal ethics, 96–97, 96 (fig.), 103

Persuasion, 119; in ethical conflict with company, 105; in high-context cultures, 134–136; project planning used for, 119–123; with reason, 124–129; reasoning-based, 124–129; types of, 124, 124 (fig.); values-based, 124, 129–133

Persuasive style, 31; balancing with plain style, 226–228; for proposals, 618; techniques for, 124, 226 (fig.); types of, 206; uses of, 222–226

Peterson, Chris, 526

Pew Research Center, archive of, 671 (fig.)

Photographs, 291–292 (fig.), 293 (fig.); in activity reports, 654; adding to websites, 368 (fig.); inserting into document or presentation, 292–293, 293 (fig.)

Physical context of use, 47; for analytical reports, 666; for presentation, 414; for proposals, 598

Physicians, graphics used by, 283 (fig.)

Pictures, 289–296

Pie charts, 287, 287 (fig.)

Place names, commas in, A10

Plagiarism, 108, 113, 162–163; avoiding, 170, 671–672; on websites, 362

Plain style, 31, 206, 212 (fig.); for activity reports, 653; for analytical reports,

References, on résumé, 386–387, 387 (fig.)
References list: in APA style, 165–166, 166 (fig.), A26–A29; in CSE style, A30–A33; MLA works cited list and, A35–A37; sources recorded in, 165
Reframing, 133, 133 (fig.)
Refusal letters and memos, 466, 469, 469 (fig.)
Rehearsal: to overcome fear of public speaking, 425; of presentation, 439–441
Reliability, of sources, 142, 142 (fig.), 149, 151
Religious symbols, 298
Report: activity, 634–657; analytical, 659–704; ethics in preparing, 204; genre, 24; writing process for, 32–34
Request for proposals (RFP), 592, 592 (fig.)
Research, 22; for activity reports, 646–648; for analytical reports, 666–672; avoiding junk science on Internet, 166–167; beginning, 143–144; bias in, 168–169; into culture, 58; defining subject, 143; documenting sources in, 165–166; electronic sources for, 149–151, 150 (fig.); empirical (primary), 141, 153–154, 154 (fig.); ethics in, 173; information appraisal in, 168–169; information management in, 141, 159–167; for instructions, 565–566; interviews for, 153; for letters and memos, 453–457; methodology for, 146–148, 146 (fig.); narrowing subject of, 144–145; need-to-know information in, 159, 159 (fig.); note taking in, 160–165; plagiarism and, 162–163, 170; for presentations, 411–415; print sources for, 151–153; process of, 142 (fig.); for proposals, 599–600; quantitative and qualitative, 153; questionnaires, 157; question or hypothesis for, 145; source materials in, 142, 142 (fig.), 148; for technical definitions, 494–495, 495 (fig.); for technical descriptions, 524–525; triangulating materials in, 142, 142 (fig.), 148–154; want-to-tell information in, 159, 159 (fig.); for website, 352–356
Research angle, 144
Research databases, 150, 151 (fig.)
Researchers: logical mapping used by, 144; responsibilities in team, 70
Research methodology, 668–669, 669 (fig.)
Research question, in analytical reports, 667–668
Research reports, 661, 663–665 (fig.)
Research triangle, 148 (fig.)
Resource investigator, of team, 83
Response letters and memos, 463, 464 (fig.)
Results: in activity report, 649, 651 (fig.); section in analytical report, 676–677; in usability testing, 326
Résumé, 378–379; chronological, 379–387, 387 (fig.); design of, 390–391; scannable/searchable, 390, 391 (fig.); types of, 387
Revising, 306–309, 307 (fig.); activity reports, 654; analytical reports, 700; application letter, 397–398; of design, 265; of instructions, 584–585; letters and memos, 480; proposals, 621–622; technical definitions, 507–508; technical descriptions, 539–543; websites, 368–369
Rhetorical situation: in activity reports, 647; in analytical reports, 666; defining, 20–21, 21

(fig.), 412–415; in letters and memos, 453–457; proposal writing and, 599
Rights, 97
Roberts, D., 279
Roles, in teams, 83–84
Roosevelt, Franklin D., Einstein letter to, 94–95
Rules, in design, 249–250, 249 (fig.), 250 (fig.)
Run-on sentences, A3

Safety information, in instructions, 574–575
Safety symbols, in instructions, 577 (fig.)
Safety tests, 326, 585
Sans serif typeface, 252, 252 (fig.), 255 (fig.)
Sarcasm, 338
Sayings, in cross-cultural documents, 319
Scannable résumé, 390, 391 (fig.)
Scanned documents, in PDF format, 652 (fig.)
Scanning, by culture, 56–57, 57 (fig.)
Scheduling, in groupware, 86
Science, avoiding junk science on Internet, 166–167
Scientific method, 158, 158 (fig.)
Screen-based documents: design balance in, 239–243, 242 (fig.); types of images in, 366–368, 368 (fig.)
Screen shots, 295–296, 296 (fig.)
Search engines, 45–46, 45 (fig.), 149, 152, 422; advanced search of, 150 (fig.); for jobs, 376, 376 (fig.), 377; online calendars and, 122; for profiling readers, 45; for research, 150 (fig.); for websites, 361–362
Searches, Internet, 670
Secondary readers (advisors), 42; of analytical reports, 666; of instructions, 561; of letters and memos, 456; for presentation, 413; of proposals, 598; of technical definitions, 494; of technical descriptions, 523
Secondary sources, 142
Second-level headings, 247
Security clearance, 110
Semicolons, A11, A12; with quotation marks, A12; in series, A11
Senses: instructions and, 564; technical description and, 54, 533
Sentence. See also Subject of sentence in active and passive voice, 220–222; aligning subjects in paragraphs, 218–219; basic parts of, 207; breathing-length, 695; computer for, 213–214; controlling reader pacing with long and short, 226; copyediting, 311; fragment, A4; guidelines for writing, 206–215; lengths of, 209–210; in paragraph, 215–220, 216 (fig.); in plain style, 206–215, 653; on websites, 364
Sentence definitions, 489, 490 (fig.), 496–498; diagram for, 498 (fig.); extending, 500 (fig.)
Sequential lists, 253–254, 254 (fig.)
Sequentially ordered steps, in instructions, 552–560, 555 (fig.), 568–574
Series, semicolons in, A11
Serif typefaces, 252, 252 (fig.), 255 (fig.)
Setting a tone technique, 619–620, 620 (fig.)
Sewn binding, 266–267, 266 (fig.)
Shading, in design, 257–258, 258 (fig.)
Shaper, in team, 83
Shared values, 129, 130–131

Shifted tense, A7–A8
Sidebars, for design balance, 242, 242 (fig.)
Signature file, for e-mail, 335, 337, 340
Signatures: for letters, 454 (fig.), 475 (fig.), 476; memos and, 454, 455 (fig.), 478
Silent Spring (Carson), 166–167
Silyn-Roberts, H., 5, 12
Similes, 223–225; technical descriptions and, 517 (fig.), 533
Simple future tense, A22
Simple past tense, A21
Simple tense, A21
Site maps, 361
Skills: in application letter, 394, 394 (fig.); in portfolio, 400; on résumé, 381 (fig.), 386
Slander, 110–111
Slang, in cross-cultural documents, 319
Slides, for presentation, 421 (fig.), 430 (fig.), 434 (fig.)
Slogans: in cross-cultural documents, 317; in instructions for cross-cultural markets, 565
Small, Dan, 674
"Smoking Gun" Challenger memo (Boisjoly), 175, 177 (fig.)
Social ethics, 96 (fig.), 97–99, 98 (fig.); ethical dilemma and, 103–104
Software: for activity reports, 646; for authoring websites, 363–364, 363 (fig.); for boxing, 28, 28 (fig.); groupware as, 84, 84 (fig.); for layouts, 249; for mapping, 26; for note organizing, 160; for online calendars, 122; Outline View in, 179, 180 (fig.); outlining with, 27, 28 (fig.); presentation, 410, 433, 440 (fig.); project planning, 70 (fig.), 289; spreadsheet programs for tables, 285; styles and templates in, 255–256, 255 (fig.); translation, 317
Solicited proposals, 591–592; internal, 593–595 (fig.)
Songs: in APA style, A28; in CSE style, A32; in MLA style, A37
Spam, in e-mail, 338
Spatial organization: in presentation, 427; of website, 352
Specialist, in team, 83
Specialized terms, 695
Specifications, 551, 557–560 (fig.). See also Instructions
Spell checker, 34 (fig.), 315–316
Spelling: proofreading and, 315–316, 315 (fig.); spell checker, 315 (fig.)
Spiral binding, 266, 266 (fig.)
Splash page, on website, 350
Spreadsheet program: making visuals with, 290, 290 (fig.); for tables, 285
Stage fright, 425
Statement of importance, 359
Status report. See Progress reports
Stauber, John, 167
Stevenson, D., 41
Stitched binding, 266–267, 266 (fig.)
Storming stage of teaming, 66–67, 67 (fig.), 75–80
Storytelling, with graphics, 276–277, 277 (fig.)
Style, 31; of activity reports, 653–654; of analytical reports, 694–695, 700; of application letters, 396–397; culture and, 54–56;

SAMPLE DOCUMENTS